Fritz Deumlich
Surveying Instruments

Title of the Original Edition

Instrumentenkunde der Vermessungstechnik
7., überarbeitete Auflage
VEB Verlag für Bauwesen, Berlin 1980.

Author

Fritz Deumlich, Dr. sc. techn., Dr. h. c.
Professor of Surveying
Technische Universität Dresden

translated by

Wolfgang Faig, Dr.-Ing., P. Eng.
Professor of Surveying Engineering,
Department of Surveying Engineering
University of New Brunswick
Fredericton, N. B., Canada

With 783 figures and 38 tables

Library of Congress Cataloging in Publication Data

Deumlich, Fritz.
Surveying instruments.

Translation of: Instrumentenkunde der Vermessungstechnik.
Bibliography: p.
Includes index.
1. Surveying-Instruments. I.Title.
TA 562. D 413 526.9'028 81-5536
ISBN 3-11-007765-5 AACR2

CIP-Kurztitelaufnahme der Deutschen Bibliothek

Deumlich, Fritz:
Surveying instruments/Fritz Deumlich. [Transl. by Wolfgang Faig]. – Berlin; New York: de Gruyter, 1982.
Einheitssacht.: Instrumentenkunde der Vermessungstechnik 〈engl.〉
ISBN 3-11-007765-5

Printed in the German Democratic Republic.

Surveying Instruments

Fritz Deumlich

W DE G

Walter de Gruyter
Berlin · New York · 1982

Preface

Surveying instruments are the "tools" of surveyors, engineers and geodesists. Solid knowledge of their design, function and operation enable them to select and to effectively utilize and handle the proper instrument for their measurements. It is therefore indispensible for them to study surveying instruments.

Like many of my colleagues involved in teaching surveying. I have been painfully aware of the need for a comprehensive treatment in the English language. Although certain common instrument types are adequately covered in a number of books, these are generally written for different purposes, and thus do not offer the deep understanding of the instruments, which is needed if one is to fully comprehend the factors behind certain field procedures and measuring accuracies.

Having appreciated Prof. *Deumlich's* book as information source and reference, and having occasionally translated the odd section to a graduate student. I whole-heartedly agreed to translate the new edition when approached by the publisher.

Prof. *Deumlich's* book was first published in 1957 and has been thoroughly revised and updated in 1963, 1967, 1972 and 1974. This new 7th edition provides up-to-date and complete information on surveying instrumentation on a wordwide scale. After short remarks on purpose and classification of surveying instruments, and a sketch of their historical development, emphasis is placed on the design of surveying instruments, standardization as well as further developments and the treatment of instruments. As a basis for subsequent sections, optical elements and spirit levels are discussed.

Presently produced instruments are covered in detail, starting with instruments for horizontal angles. For instruments like compasses or rectangle prisms, the coverage reflects their practical importance, while the theodolite as most important instrument of this category receives prime treatment. The most modern developments in automation for circle readings are high-lighted. This section is rounded out by presentations of gyroinstruments, optical precision plummets, and alignments, especially the ones using laser.

Although simple instruments are mentioned, levelling instruments are central to the section on instruments for measuring elevations and elevation differences. The different types of automatic levels and their design receive special attention. When dealing with instruments to measure vertical angles, the respective device of the theodolite is prosented, especially the development of the automatic height index. An overview of instruments for barometric levelling as used in remote areas is also given.

In view of modern developments regarding instruments to measure distances, the treatment of mechanical and optical instruments has been shortened as to provide better coverage to electronic distance meters, which are the most modern and efficient instruments for that purpose.

The closing section deals with the various types of stadia- and tacheometer instruments. Here again, the treatment of optical instruments has been shortened in favour of electronic tacheometers which presently represent the most "complete" surveying instruments.

Although many instruments are listed, the emphasis is placed on instrument types and design concepts rather than particular units. Similarly, methods of instrument testing are not intended to be rigidly adhered to like book recipes, but rather as guides to inspire the user. Some instruments, which conceivably could have been treated in the historical section, are included because they are in practical use.

Numerous instructive illustrations simplify studying of the book. Tables with technical data of instruments produced in countries all over the world provide a complete overview for the respective instrument type.
In recognition of Prof. *Deumlich's* accomplishments I have tried to retain his style and form of presentation as much as possible. Without distracting from the main topics, this provides the reader with the additional advantage of discovering—quasi between the lines—European approaches to surveying problems thus offering alternatives to his own way of thinking. Lie may—in places—find an unfamiliar term, e.g. angular units in "gons" as well as in the familiar "degrees", or "metres" instead of "feet", but then again, he has to cope with this in practice as well.
With its comprehensive treatment of all surveying instruments, this book should be welcomed by the surveying community and serve as a reference for professionals and para professionals, as well as a text for students. It is my privilege to have been able to make a small contribution to the English speaking surveying profession in helping to make this book available.
On behalf of Prof. *Deumlich*, I would like to thank all the companies, agencies and colleagues who have contributed to this book by generously supporting him with materials, illustrations and advise.

Fredericton

Dr. *W. Faig*, P. Eng.
Professor of Surveying Engineering
University of New Brunswick

Contents

Introduction

I. Purpose and Classification of Surveying Instruments

Good knowledge of surveying instruments enables the professional surveyor to select the most suitable instrument for his measurements, to operate it correctly and to utilize it efficiently. This requires not only sufficient theoretical knowledge of structure and function of the instruments, but also practical experience in handling and use.

The surveying instruments discussed here are used for surveying of terrain or objects with the aid of selected points for numerical or cartographic representation. According to their purpose, different types of instruments are being distinguished.

For the determination of the horizontal projection of points, instruments are used which measure directions or horizontal distances. Furthermore, instruments to determine the position of a point along the plumbline are available.

For the determination of heights, instruments to measure elevations and elevation differences are utilized.

In many cases position and height of points are needed. Tacheometric instruments are then used for simultaneous determination of both.

The characteristics mentioned indicate the main purpose for the instruments. Points which are referred to a horizontal plane can be determined via orthogonal or polar coordinates. The necessary distance measurement can be done directly or indirectly (e.g. optically or electronically).

The main sections of this book are arranged according to these groups. This arrangement based on purpose is further subdivided according to the capabilities of the instruments. Low accuracy instruments are simple and are used for simple technical measurements, e.g. on construction sites.

Instruments of medium and high precision are used for triangulation, traversing and levelling, to densify control nets, and for engineering surveys.

Precision instruments are required for astro-geodetic measurements like azimuth, latitude and time determinations, for triangulation and 1st and 2nd order levelling, as well as occasionally for engineering surveys.

Every instrument designed to measure some quantity has errors which affect the results. In order to obtain results free of the influences of instrument errors, one can deal with it as follows:

a) the error is adjusted at the instrument such that its influence can be considered negligible for a specific purpose of the measurement.

b) one can select a surveying method which yields results free of the influence of the instrument error.

c) the magnitude of the instrument error is determined and its influence is compensated by some corrections.

Depending on the required accuracy one or more of these possibilities can be used. The modern instruments are usually produced with such high precision that adjustments are seldom needed.

II. History of Surveying Instruments

The oldest surveying instruments served to determine distance when erecting shelter, diverting water, staking fields and determining their area. The construction of roads, canals, and buildings of art as well as developing war geometry indicate the use of more complete instruments, while later surveys for mapping and the origins of geodetic measurements point to further improvements. While the oldest instruments (for distances, plumb bob, set

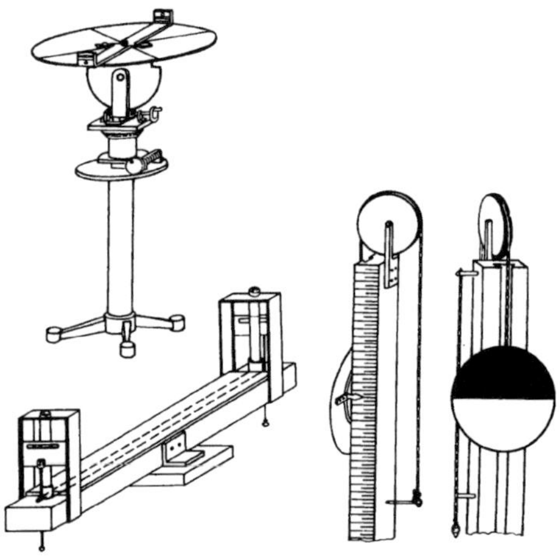

Figure 1 Dioptra (dioptra on tripod; sighting ruler for levelling; 4 yard long hydrostatic level; shifting rod for dioptra)

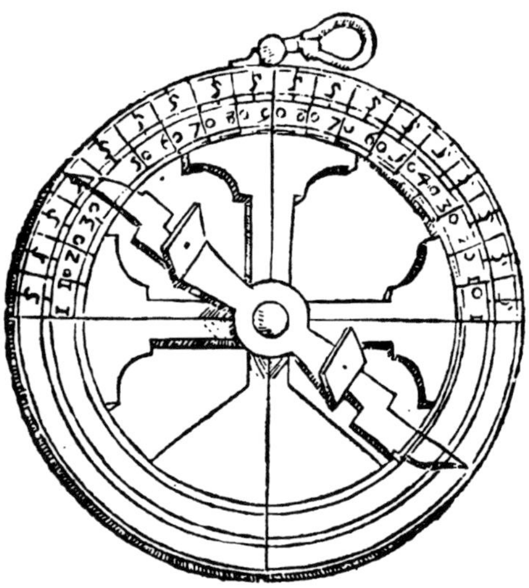

Figure 2 Astrolabe

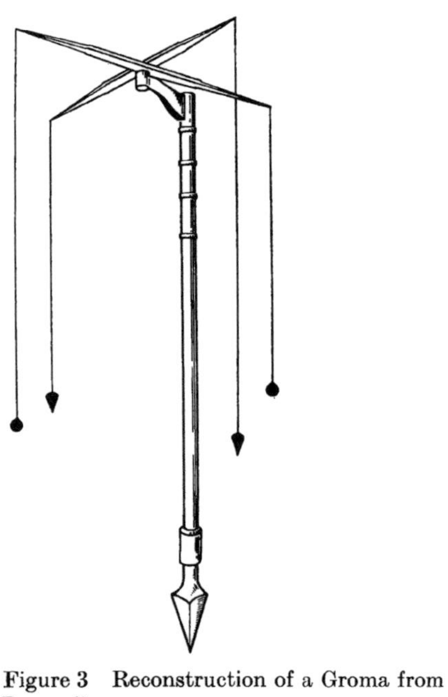

Figure 3 Reconstruction of a Groma from Pompeji

square) are of surveying origin, others, especially instruments to measure angles, were taken from the field of astronomy.

Early surveying instruments have been established in Sumaria, Babylonia, Chaldea, Egypt, China and India. Around 3000 B.C. rulers, measuring ropes and rods were known in Babylonia and Egypt. Similar instruments were used in China around 1100 B.C. Plumb bob and spirit level were utilized by the Sumarians, Egyptians and Chinese. The set square was known to the Egyptians, the magnetic compass to the Chinese. The accuracy of surveys at that time was quite high, for instance for levelling in Egypt ± 8 cm/200 m. The knowledge of the Asiatic and Egyptian people was enhanced by the Greeks. The scientists of these times were often not only distinguished mathematicians or surveyors, but also excellent mechanics and designers. Around 560 B.C. *Anaximander* introduced in Greece, the "Gnomon" (shadow square) which probably was already known to Babylonians and Egyptians. It was used by *Meton* around 440 B.C. to determine the north direction and around 200 B.C. by *Erathostenes* to determine the circumference of the earth. Around 100 B.C. *Heron* from Alexandria wrote his volume "about the Dioptra", collecting material from his predecessors as well as expanding it himself. For nearly 2000 years, this book remained the best text on practical surveying. The dioptra served as cross staff with diopter ruler (figure 1). Its main part was a simple water levelling device consisting of a U-shaped pipe which could be rotated around a vertical axis. The appropriate levelling rod had a movable target. *Heron* also mentioned the design of an automatic distance meter, which determines the distances from revolutions of a wheel.

Ptolemäus was the first to describe the quadrant as used for astronomical observations. Its name was derived from a quarter circle (radius up to 3 m) plotted on a plate. For vertical angles the Ptolemäus rulers, (approx. 150 B.C.) made from tangent – and chord rulers was used until the Middle Ages. The Greek scientist *Hipparch* is considered to be the inventor of the astrolabe (approx. 150 B.C.), a circular disc of 10 to 20 cm diameter with degree graduation hanging on a small ring (figure 2). This forerunner of the theodolite was originally an astronomical device.

Although the Romans added little to the Hellenistic instruments, they spread the Greco-Roman knowledge to Central Europe. Besides measuring rods for distances, the surveyor's cross "Groma" was the most important instrument of the Roman land surveyors, the Agrimensores. The groma is a cross fastened eccentrically on a wooden staff with plumb bobs (later diopters) for aiming (figure 3). *Vitruvius* (approx. 15 B.C.) mentiones besides the Dioptra the Chorobates (figure 4), which is an approximately 2 meter long spirit level and was probably known to the Greeks. He also mentions carts with counters for distance measurements. Large distances were also determined by pacing. Pace counters (Bematists), whose paces were "calibrated" could be found in most army units.

In Europe, scientific development was hindered for thousand years by the church. The Arabs, however, who penetrated into France, had a high level of astronomic-geodetic knowledge. They took over the geometric knowledge from the Greeks and had astrolabes divided to 5 minutes of arc. Usbeke *Biruni* (973–1048) designed the prototype of a circle graduation machine. After the Arab Empire collapsed, the knowledge continued in the Schools of Baghdad, France and Spain and influenced the Europe of the Middle Ages, which previously only had contact with the remains of Roman land surveying.

Geographic discoveries, the expansion of shipping and world wide trading connected with it, increased the demand for maps and geodetic data. Growing productivity led to an uplift for arts and science. The new physical findings also influenced the development of surveying instruments. Even towards the end of the Middle Ages, instruments known for ages, perhaps slightly improved, were still used in Europe, e.g. measuring strings and rods for distances, cross staffs with diopters for staking right angles, horizontal straight edges and open water levelling devices for elevations as well as astrolabes and quadrants for angles. Around 1300, the Jacob staff (figure 5), a device for indirect distance – and angular measurement appeared and was first described by the Arab *Levi ben Gerson* (1288–1344). Here the parallactic angle was obtained by moving a lateral rod along the main rod until its ends appeared to coincide with the end points of the line to be measured. Then the angle was read on the graduation of the main rod.

Leonardo da Vinci (1452–1529) designed carts with distance meters, even a pace counter. The French doctor *Fernel* (1497–1558) measured distances with a measuring wheel attached to his coach for geodetic degree measure-

Figure 4 Chorobates (Reconstruction)

ments in 1525. One geographic degree represented 17,024 revolutions. He used a quadrant to determine the latitude. Mapping of the countries created a boom for the builders of cart-distance meters. Directions were usually measured with a compass.

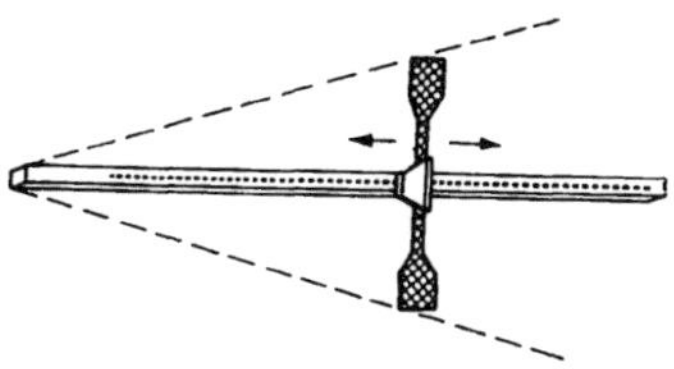

Figure 5 Jacob staff

Such compasses were specially designed for surveying purposes, and most likely came from China to the Arabs. It is a fact, that Chinese ships were equipped with magnetic compasses 100 years prior to their use in the Western World. In 1187 the Scotch monk *Alexander Neckham* mentions a compass. *Petrus Peregrimus*, probably of Norman origin, who in 1269 was suited up with the Duke of Anjou, first described the wet and dry compasses. In these times the housing was made from boxtree wood (figure 6) from which its name in several languages is derived (e.g. German "Bussole" or French "Boussole"). In Italy the versatile *Leonardo da Vinci* sketched around 1500 a compass in a circular housing. In 1650 the compass theodolite was developed, and in 1812 the mechanic *Schmalcalder* invented the prism compass which carries his name.

Around 1530 the surveyor's chain, the predecessor of the tape, was first utilized in the Netherlands. Around 1550 tripods appeared, and around 1600 devices to measure horizontal angles were developed from astrolabes (figure 7). The Englishman *Digges* described such a device in 1552 and used for the first time the term theodolite. The universal instrument by *Josua Habermel*, a device based on the theodolite principle with compass was built in 1576 in Germany.

It is believed that the Dutch astronomer *Gemma Frisius* (1508–1555) invented the plane table. It became wider known through the German professor *Johann Praetorius* (1537–1616). Instead of an alidade, a diopter ruler (figure 8) was used in that period. Important inventions lead to significant improvements of surveying instruments at the beginning of the 17th century. The construction of the first telescope in 1608 is attributed to the Dutch eye glass maker *Hans Lipperhey* (1560–1619). Apparently, the Italian physicist and mathematician *Galileo Galilei* (1564–1642) heard of it and build in 1609 an improved version of the same telescope, which is called Dutch or Galilei telescope. It did not gain much significance in surveying, since it cannot be fitted with a cross hair. In 1611 *Johannes Kepler* (1571 to 1630) presented the lens arrangement for the astronomic or Kepler telescope, which was first built by the Suabian Jesuit Father *Christoph Scheiner* (1575–1650). The terrestrial telescope with inverted lens also comes from *Kepler* in 1611.

Generini introduced around 1630, the aiming telescope with ocular diopter. The English astronomer *William Gascoigne* (1620–1644), who in 1640 invented the screw micrometer, fitted cross hairs into the focal plane of the telescope of his height quadrant. After these improvements, the aiming telescope slowly started to displace the up to then dominant diopter devices. In 1670 for instance, the French astronomer *Picard* (1620–1682) used a quadrant for his degree measurements, whose telescope had crosshairs for aiming (figure 9).

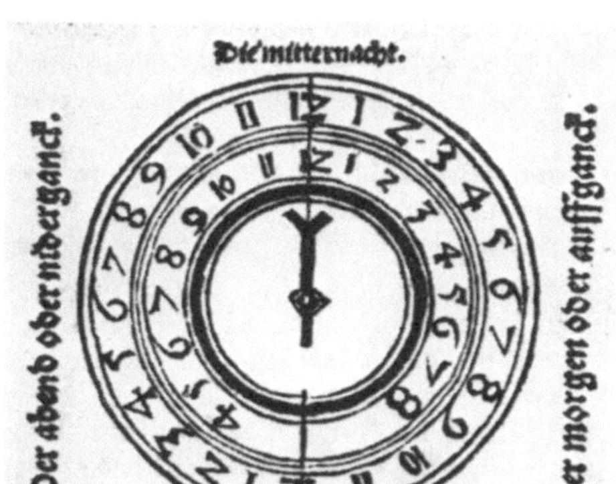

Figure 6 Compass of *Rülein* (1505)

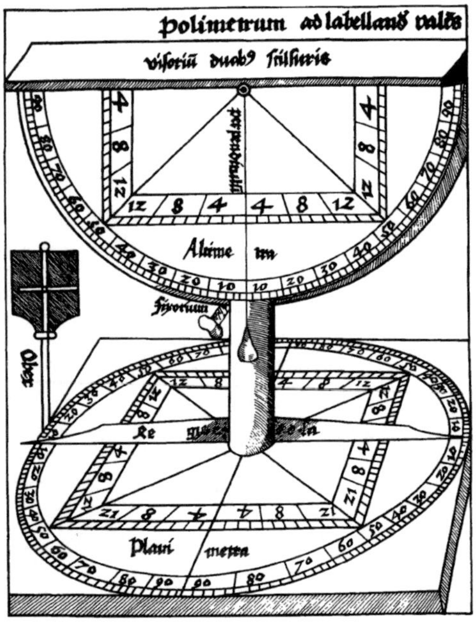

Figure 7 Polimetrum—the first European fore-runner of the theodolite (1512)

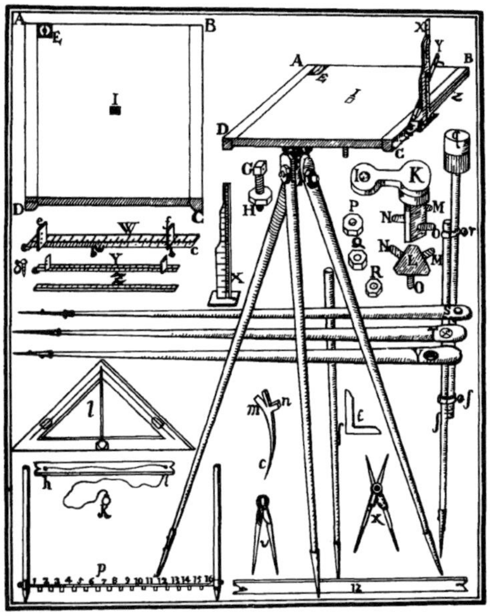

Figure 8 Plane table equipment designed by *Praetorius*

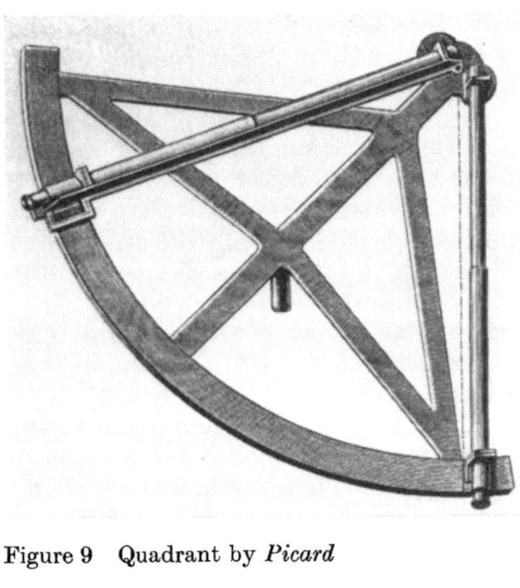

Figure 9 Quadrant by *Picard*

For astronomical purposes *Tobias Mayer* (1723–1762), Professor and director of the observatory in Göttingen, did not use real threads in micrometers but instead drew lines on glass in 1748. The mechanic and instrument builder *Georg Friedrich Brander* (1713–1783) from Augsburg scribed with diamonds fine lines into the glass.

Magnification of the telescopes built in the 17th century was small (9 to 30 times). With the invention of the achromatic objective lens in 1729 by the English lawyer *Moor Hall* (1704–1771) and its introduction by *John Dollond* in 1758 as well as improvements in glass techniques by *Louis Guinand* and *Fraunhofer*, further improvements were possible. *Josef Fraunhofer* (1787–1826) put the production of optical instruments onto a scientific base. Instead of the experimentations of the tradesmen, he utilized optical calculations. He also was a pioneer in the precise mechanical production of instruments. The Italian Major *Ignazio Porro* (1801–1875) became known for his inverted prism systems (1850), the Porro telescope (1823), his self reducing tacheometer (1858) and others.

The oldest aid to refine readings is probably the principle of transverse graduation, used around 1300 by *Levi ben Gerson* and by the Danish astronomer *Tycho Brahe* (1546–1601) for his quadrant. In 1542 the Portuguese *Pedro Nuñez* (1492–1577) suggested a reading device for a quadrant. Each of 46 concentric circles was divided into $(n - 1)$ parts of the previous one so that the radius as index in any position nearly coincided with a gradual mark of one of the circles. The "vernier" principle, which is still used was invented by the mathematician *Clavius* in 1593. However, he used the device for setting out and not for measuring angles and distances. Later the Dutch *Peter Werner* (*Pierre Vernier*, 1580–1637) used in 1631 the "vernier" in its presently used form.

In 1629 *G. Branca* developed in Rome the hydrostatic level. However, it took until 1849 when *Geiger* from Stuttgart utilized rubber hoses, that it gained practical importance. This hydrostatic level superseeded the earlier levelling device with U-pipe and water, even though the latter was developed as a hand held instrument by *Bolz* and by *Kahle* in the middle of the 19th century.

In order to level the line of sight of his telescope used for trigonometric height measurements, *Picard* used in 1674 a pendulum with 1.30 m length. Already the Romans had used a pendulum for levelling purposes, and so did *Huygens*. The development of levels as well as all other devices for levelling of geodetic instruments was significantly influenced by the invention of the tubular bubble in 1662 by the Parisian mechanic *Thévenot* (1620–1692). *Mallet's* reports (Paris, 1702) indicate, that the levels of that time already were equipped with eyepiece slide, tilting screw and adjustment devices. In Germany, the tubular bubble was first mentioned by *L. Christoph Sturm* in 1715.

Only towards the end of the 18th century it reached a practically suitable shape. *Amsler-Laffon* built in 1857 the first level with a reversible bubble. The tilting screw was named after the Karlsruhe fine mechanic *Sickler* (figure 10). In 1770, *Johann Mayer* of Göttingen invented the bull's eye level used for approximate levelling of instruments. In 1904 *Mollenkopf* produced the whole body of the level bubble of glass to avoid evaporation of the liquid.

The English mechanic *John Sisson* built in 1730 the first theodolite. Up to the end of the 18th century it was improved in England by *James Short*, *Adams*, and especially by *Jesse Ramsden* (1735–1800), who invented a microscope with screw micrometers for circle readings as well as in 1783 an eye piece, named after him.

The Dutch physicist *Christian Huygens* (1629–1695) had designed an eyepiece in 1684. From the Ramsden eyepiece the mechanic *Kellner* from Wetzlar developed in 1849 a three-lens eyepiece. In the 20th century, eyepieces with large apparent field of view appeared.

In 1763 *Ramsden* built the first circle graduation machine in England. A predecessor of this machine was produced by *Hidley* in 1760 in York. In 1684 *Hooke* had introduced a semi automatic method of circle graduation. In 1803 *Reichenbach* built a circle graduation machine based on the copy method whose principle is still in use today. Towards the end of the 19th century *Heyde* introduced the globoid spiral.

In 1785 the French astronomer *Borda* (1733–1799) introduced a new axial system. *Reichenbach* (1804), with his repetition theodolite, utilized another one (figure 11), and the mechanic *Repsold* from Hamburg used yet another one around 1830. In order to achieve forced centering, *Breithaupt* utilized

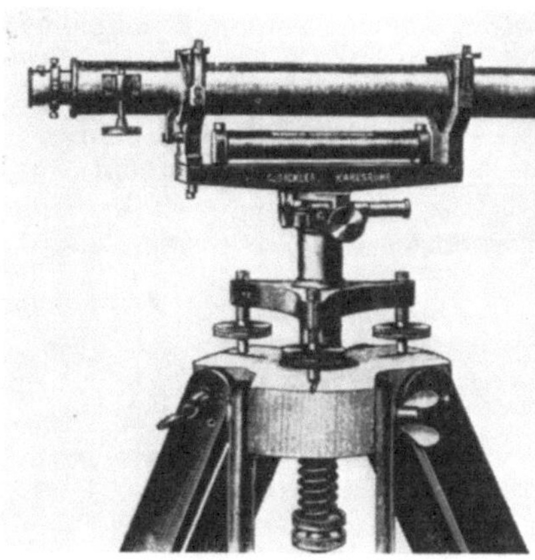

Figure 10 Level (19th century)

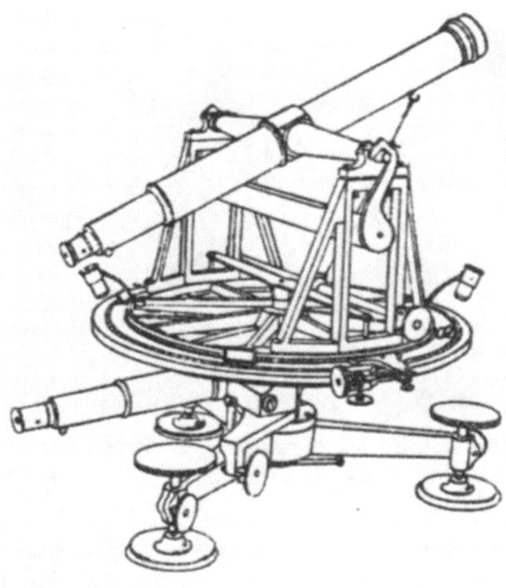

Figure 11 Reichenbach's precision theodolite (with circle diameter of 12 Parisian inches)

Figure 12 Optical Square

in 1840 a socket arrangement and *Hildebrand* in 1876 the Freiberg sphere. Only after World War I, forced centering started to be used in surveying practice, especially for precise traversing in urban areas.

Hildebrand published in 1878 – based on *Nagel's* notes from 1873 – the first optical plumbing instrument with a central telescope. Since approximately 1960, optical precision plummets are produced, and used primarily in construction surveys.

Around 1740, the London mechanic *Adams* (1720–1773) built an optical square (figure 12), while the double optical square was described in 1844 by *Berlin*. The three sided right angle prism, discovered in 1851 by the Munich Professor *Bauernfeind* eventually superseded the mirror instruments. *Prandtl* introduced in 1890 the pentagon prism, mentioned by *Goulier* in 1864, into the surveying field. Since 1924, the totally reflecting prism, mentioned in 1812 by the English physicist *Wollaston* (1766–1828) is used in two prism squares.

The mercury barometer, used for barometric levelling was invented in 1638 by *Galilei* and *Torricelli*, and practically used by *Pascal* ten years later. Through improvements, especially by the French physicist *Gay-Lussac*, the siphon barometer was created around 1800. In 1847, the French *Lucien Vidi* (1805–1866) invented the aneroid barometer, which now is mainly used in the field. Probably it was first built by *Naudet* and later perfected by *Bourdon*.

Geminiano Montanari carried out the first optical distance measurements in 1674 in Italy using 12 to 15 parallel and equidistant crosshairs in his telescope. The steam engine producer *James Watt* built in 1771 an optical distance meter with two horizontal and one vertical crosshair. The optician *William Green* described in 1778 a similar one. From 1812 on however, optical distance measurements gained widespread use, starting with cadastral surveys in Bavaria by *Georg von Reichenbach* (1771–1826) who in 1810 added stadia hairs to his alidade.

In 1823 *Porro* moved the vertex of the parallactic angle into the vertical axis with the aid of placing a positive lens into the optical path. From 1839 on he calls his instruments tacheometer and thus introduces the term tacheometry. The increasing demand for topographic maps and plans for design and construction of extended structures, such as railroads and canals, contributed to the distribution of optical distance meters, especially in Italy, France, Germany and Austria.

In 1800 the engineer-Colonel *Hogrewe* invented in Hannover the tangent screw which he used as clinometer when levelling. *Stampfer* introduced in 1839 in Austria the chord screw for optical distance measurement, which is based on the tangent principle. This tangent principle which is used for screw distance meters led to tangent – and contact tacheometers. With these instruments horizontal distances and elevation differences could automatically be obtained with the aid of reduction devices rather than by subsequent calculations. The French *Sanguet* constructed in 1866 the first contact tacheometer with vertical scale (figure 13) which sold to several thousand in France, Italy and Switzerland. In *Charnot's* tacheometer (1889) the contact device was horizontal. In 1898 the Max Hildebrand Company in Freiberg produced an instrument with horizontal scale according to *Vogler's* design. Further known are the instruments by *Doergens* (1900) and the one designed by the French engineer *Balu* in 1912 which was produced by Kern, Aarau. In France, contact tacheometer are still used and produced. *Eckhold's* omnimeter (1868), a tangent tacheometer with horizontal scale was widely spread in England and her colonies. In Hungary and England the tangent tacheometer designed by *Szepessy*, produced by Süss, Budapest was used in practice. A further development of this graduation in the field of view (by *Bors*) was used by MOM, Budapest in 1956 for their alidade MF, while in 1954 Filotecnica Salmoiraghi in Milano produced the tangent tacheometer Tari.

At about the same time as the contact tacheometers, shift – or projection tacheometers appeared, which are especially suitable for plane table work. Its predecessor of this type was already built by *G. F. Brander* in 1780. In 1865, the surveyor *Kiefer* from Cologne suggested the construction of a shift tacheometer, and Breithaupt built it in 1873. The best known one is the Tachygraphometer by *Wagner*, using a tilted rod built in 1867 by Fennel in Kassel and known as Wagner-Fennel shift tacheometer (figure 14). In 1873 Ertel in Munich built *Kreuter's* design. *Peaucellier* and *Wagner* designed in France a shift tacheometer as alidade with a circular plotting table, called Homolograph. In 1878 Kraft & Son, Vienna produced *Stern's* tacheo-

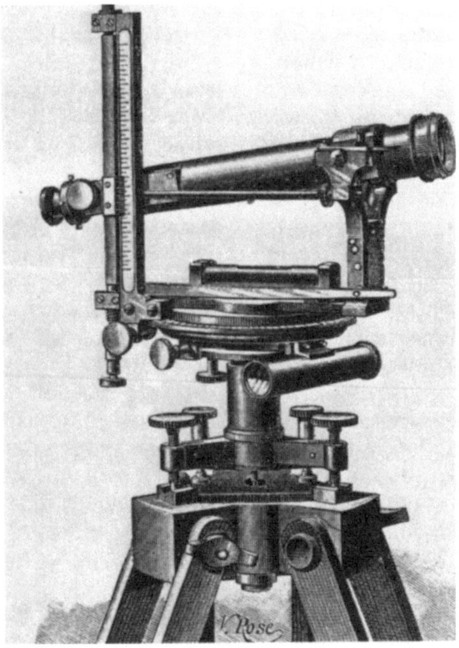

Figure 13 Contact tacheometer by *Sanguet*

meter with a vertical rod. *Puller* introduced a new projection principle with vertical rod, which was utilized in an instrument built by Breithaupt, Kassel in 1901 and used until about 1930.
A diagram tacheometer as alidade with computation sector according to *Soldati* was produced in 1900 in Italy by Salmoiraghi. An instrument of this type is the tacheometer with Ewing-Esdaile-curve drum built by Hilger & Watts, London in 1954, based on an invention by *J. A. Ewing*, Australia (figure 15).

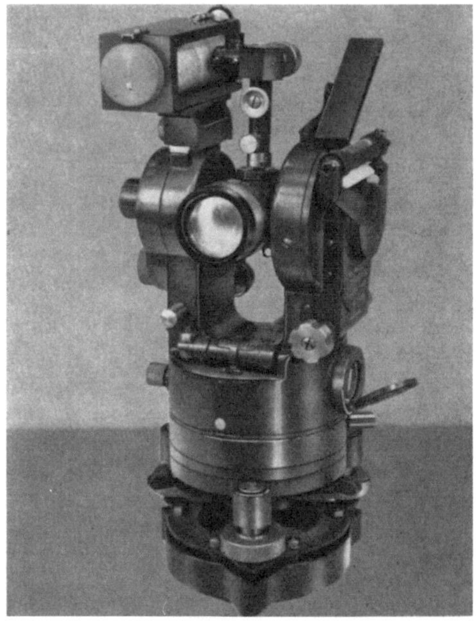

Figure 15 Ewing distance- and height meter attached to Watts theodolite

In 1878 the Austrian forester *A. Tichy* developed logarithmic tacheometry. His first instrument with an eye piece micrometer was built by *Starke*, Vienna in 1884, another with an optical part in 1890 by *Ott*. This method became interesting again in 1955 with the Lotakeil (logarithmic tacheometer wedge) by Carl Zeiss, Jena, did however, not make an impact (figure 16). Several ideas to obtain automatic reduction by changing the focal length (*Porro* 1858) or the distance between the crosshairs (*Jeffcott* 1912 with Cooke, Troughton & Simms, London; *G. Heyde* 1934) have not gained practical acceptance. Only recently the monitoring device could be produced with the required accuracy of 1 µm (DK-RV by *Conzett* and *Hinden* 1955, and Kl-RA 1963, both produced by Kern, Aarau).
The optical monitoring of the crosshair separation gained great importance. The basis for the development of diagram tacheometers was the plan of the Italian engineers *Roncagli* and *Urbani* in 1890 to use two pairs of distance lines instead of fixed crosshairs in the tacheometer. These line pairs were to be edged into a glassplate located horizontally in the image plane of the telescope and which could be shifted normal to the line of sight (figure 17).
Professor *von Hammer* (1858–1925) from Stuttgart developed with this a diagram with distance curve and added a two branch height curve. Instead of the straight zeroline, *Fennel* used a circular arc as base line for the diagram. He produced in 1900 the first diagram tacheometers, utilizing the principle of the "anallactic telescope" according to *Porro* (figure 18). Subsequently, newer improved models of the Hammer-Fennel-Tacheometer appeared.
This Hammer-Fennel principle was further developed by Carl Zeiss, Jena with the Dahlta, based on a patent of the Norwegian *Dahl* (prototype 1932, in series 1942). A principle reported by the Swiss Canton surveyor *Leemann* in 1930 was realized with an alidade in 1936, the tacheometer DKR in 1939, the DKRM in 1946, and the alidade RK all produced by Kern, Aarau in 1951.
In 1953, the Fennel Company presented the Fenta with undivided field of view. Other further developments are the RDS from Wild, Heerbrugg (1951), the alidade RKl (1962), the TA-Dl from MOM, Budapest with *Bességh's* diagram (1959), the TA-2 from the Soviet Union (1959) and the RTa4 from

Figure 14 Shift tacheometer by *Wagner-Fennel*

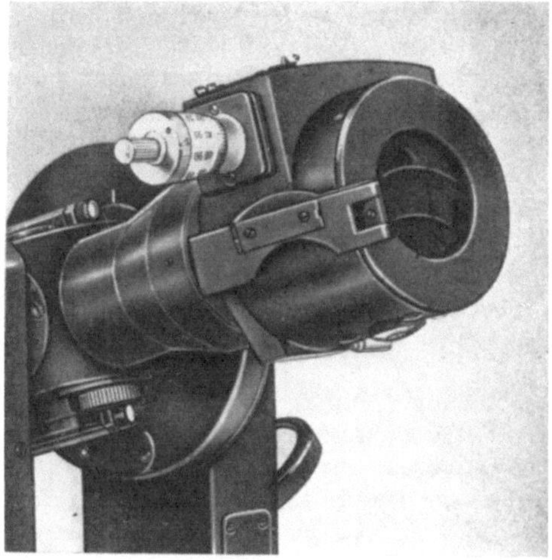

Figure 16 Lotakeil

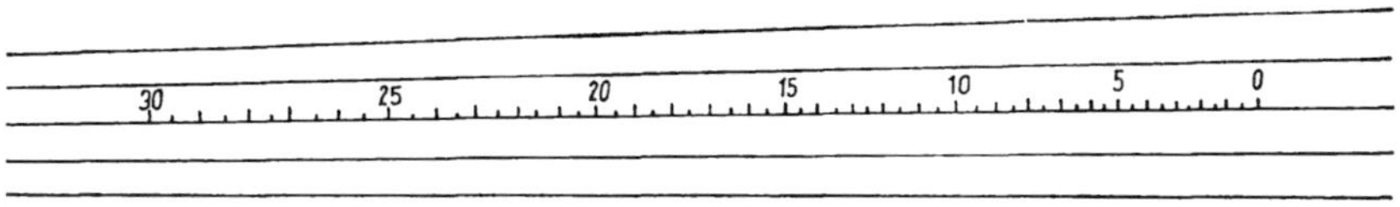

Figure 17 Stadia hairs by *Roncagli* and *Urbani*

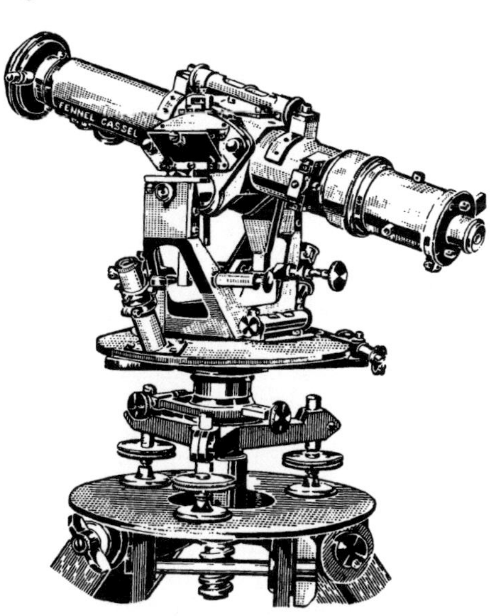

Figure 18 The first Hammer-Fennel tacheometer

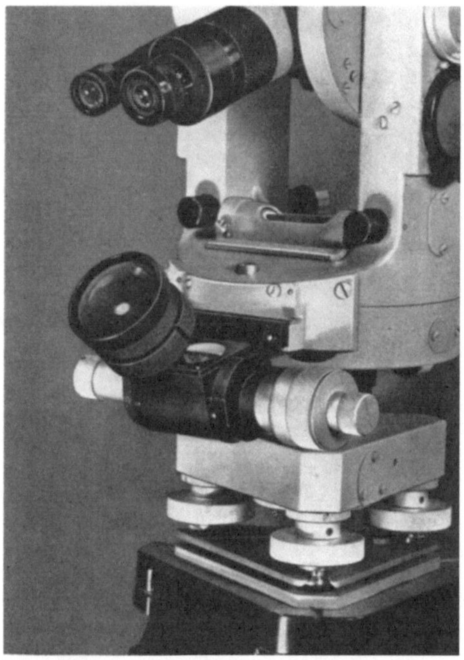

Figure 19 Tangent Screw at the Zeiss IV theodolite (about 1930)

Carl Zeiss, Oberkochen (1967). Since 1904 the diagrams are used accordingly for self reducing alidades.

Towards the end of the 19th century numerous range finders appeared. Their base was at the station, thus requiring no rod at the target. *Brander* built in 1781 a one station range finder with mirrors at the ends of a cross pipe. The images created by two objectives are made to coincide in the field of view of the common eyepiece, and onto a vertical line (coincidence telemeter). In 1790, *Ramsden* presented a half-image range finder. Since 1888 the English company Barr & Stroud builds half image range finders. Later Carl Zeiss, Jena, and C. P. Goerz, Berlin-Friedenau built them too.

The first stereo range finder (stereo telemeter) originated from *Pulfrich* (1858–1927) at Carl Zeiss, Jena in 1899 according to a patent by *Grousilliers* (1893). These range finders gained only little importance in surveying. Even the stereotachygraphs, especially designed for surveying purposes by *Hugershoff* in 1931/32 and produced by Heyde in Dresden could not gain practical acceptance. Only the teletop, built since 1937 by Carl Zeiss, Jena according to *Eppenstein's* design, is still being produced. Other range finders for surveying purposes are produced by Breithaupt & Son, Kassel (Todis, 1952), Wild, Heerbrugg (TM 10, 1959), Hilger & Watts, London, and Works of the Soviet Union (DWT, 1965).

Since about 1880, wooden subtense bars were used. In 1906, *Pulfrich* at Carl Zeiss, Jena used for the first time a subtense bar made of a steel pipe, in connection with a phototheodolite with tangent screw (figure 19). Optical distance measurement with subtense bar gained importance after the production of Pulfrich's distance measuring theodolite (1921) and the invar subtense bar (1923) at Carl Zeiss, Jena, and a 2 m invar subtense bar together with a one second theodolite which is the present standard equipment from Wild, Heerbrugg built since 1969.

Development has been at a rapid pace. In former times the various instrument types were large and heavy and therefore awkward for use and transportation, the telescope was very long, the rough circle reading had to be refined by special methods (figure 20).

In the last decades it was the aim of the instrument manufacturers to make the instruments more handy, smaller and lighter. Furthermore, the reading accuracy had to be increased to fully utilize the optics. Circle readings should be simple and fast, possibly right near the eye piece of the telescope, and both horizontal and vertical circles should be visible at the same time. Sensitive instrument parts should be protected against damage, dust, and humidity.

The Swiss *Heinrich Wild* (1877–1951), who in 1908 as collaborator of Carl Zeiss, Jena introduced interior focussing, earned great merit in modernizing surveying instruments. While the first levels with interior focussing had a positive focussing lens, *Wild* switched later to the negative lens as presently used, and thus could reduce the length of the telescope. He also designed cylindrical axes, split level bubbles, plane parallel micrometer and invar rod – all integral parts of modern surveying instruments. The parallel plate micrometer can be traced to *Clausen* (1841), and *Porro* used it already in 1854 for the micrometer microscopes of his theodolites. In 1918 *Wild* discovered the optical coincidence micrometer with parallel plates, and in 1922 Carl Zeiss, Jena produced the first optical theodolite (Th I) with the above mentioned modernizations (figure 21). Also of importance are the graduated glass circles, which for the first time were used in theodolites produced in series. In 1884 glass circles were built into a mining theodolite by *Josef* and *Jan Frič*, Prague. In 1936, *Smakula* at Carl Zeiss, Jena was successful in improving optical glass by vaporization of a film in vacuum, which significantly reduces the light loss in optical systems, caused by reflection. Similar methods of improving optical glass were invented at other places at around the same time.

The following inventions contributed to simplifying circle readings: in 1879 *Hensoldt* reported on the scale microscope; in 1912 Fennel introduced the vernier microscope, Breithaupt utilized in 1925 *Heckmann's* combination

Figure 20 Theodolite with screw microscope (about 1900)

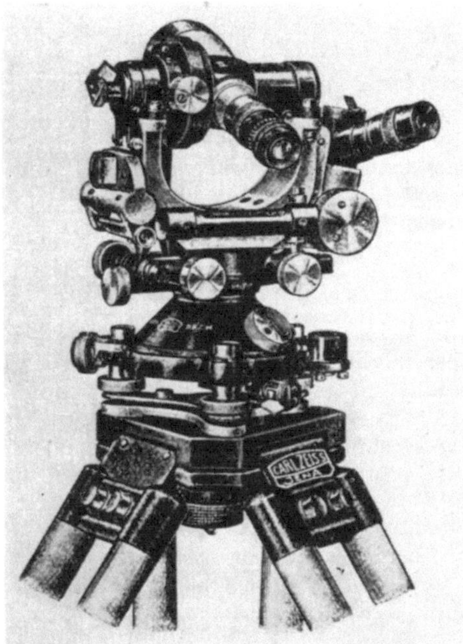

Figure 21 Th I—the first optical theodolite

microscope which is similar to it, and in 1920 Fennel introduced to plane glass microscope. Until 1964 Askania, Berlin equipped a microtheodolite with an inclined scale micrometer based on transverse graduation.
Further novelties from *H. Wild* reached practical importance with the DK-theodolites of Kern, Aarau, with levelling knobs instead of foot screws and ball bearing axes, disktype guidance surfaces (1937), and the mirror telescope which was used for the first time in surveying instruments in the DKM-3 (1939). Only for astronomical purposes, mirror telescopes had been used prior to that. Other types are mentioned by *Maxutow* in 1941 (utilized partly in the Russian TT2/6), and the mirror telescope in *Cassegrain's* (1682) arrangement as computed by *H. Köhler* and utilized in the Theo 010 (figure 22) of Carl Zeiss, Jena.
Based on a suggestion by *Gigas*, the Askania Company added in 1942 to its 27 cm triangulation theodolite a device for photographic registration of the readings, in order to better take advantage of the most suitable measuring conditions. This later became the precision theodolite Tpr. Since 1950, Wild Heerbrugg produces the theodolite T3, and Carl Zeiss JENA the Theo 002 with photographic registration. Theodolite attachments for electro-optical rather than visual registration of illuminated targets ("electric eye", Askania, 1958) have not yet gained practical acceptance (figure 23).
In 1886 *Sanguet* invented a distance meter with an optical wedge in front of the whole objective lens. Two crosshair readings, with and without the wedge had to be taken. The Englishmen *A. Barr* and *W. Stroud* used in 1889 a wedge, which can be shifted inside the telescope (mentioned in 1777 by *Maskelyne*), and in 1890 a prism that can be snapped in front of the objective lens of the telescope. The American *Richards* used a measuring wedge in 1890 in connection with a vertical rod. From 1910 on *Oltay* produced in Budapest distance meters with a wedge in front of the whole objective lens, whose models were built in 1915 and 1922 by Süss, Budapest. Wild's design from 1921, features an attachment which can be fixed onto the objective lens of the theodolite, containing two achromatic glass wedges, operating in opposite directions and covering half of the lens each, as well as a parallel plate micrometer. A suggestion by the Swiss *Engi* in 1923 to place tacheometer and rod at the station and a pair of wedges with constant deflection at the target, was not picked up. On order to eliminate personal errors, Kern, Aarau placed the wedge attachment in the central strip of the objective lens, as suggested by *Aregger*. This arrangement is still common. Hildebrand, Freiberg produced in 1928 a theodolite attachment with Aregger-Wedge. At Fennel's distance meter, model 1929, the wedge covered the bottom half. Since 1931, Zeiss, Jena has been producing the wedge attachment "Dimess" with the *Aregger* arrangement (figure 24). Since 1942, Kern, Aarau produces the DM-M with parallel plate, and Wild, Heerbrugg a similar type wedge since 1949.
Uhink's design for Breithaupt & Son, Kassel (1929), using two mirror prisms in front of the telescope objective was no practical success, neither were the plumbing staff distance meter Lodis (1930) and Kipplodis (1932) by *Gröne*, both produced by Carl Zeiss, Jena (figure 25).
According to the literature, the French *Lugeol* was the first one in 1859 to utilize the heliometer principle, as described by *Bouguer* in 1752, for measuring terrestrial distances. *Belizyn*, in Russia used it in 1954 for the theodolite attachment DNB-2 to measure the varying parallactic angle to a fixed base. Also in Russia, *Greim* and *Tschurilowski* utilized wedges that can be shifted. In the Russian differential distance meters (1969), the parallactic angle is created with the aid of a lens compensator.
In the third decade of this century, the practical use of double image tacheometers started. These were predominantly equipped with rotational wedges (1777) according to *Bošković* (1711–1787).
In 1924 a prototype of these instruments was the product of the cooperation of the Swiss land surveyor *R. Bosshardt* (1884–1967) with Carl Zeiss, Jena (figure 26). Heyde, Dresden built in 1930 a three image tacheometer designed by *Hugershoff* with vertical rod and simultaneous reduction to horizontal distance. This instrument was no success, neither was one designed by *Barot* and produced by Wild, Heerbrugg in 1935. In this case, the interval is measured with a parallel plate, which is rotated by the amount of the angle of telescope inclination, via cog-wheels, and thus reduces the micrometer range. In 1947, Kern, Aarau introduced the DK-RT, and Wild, Heerbrugg in 1950 the RDH, which in addition provides elevation differences. In 1954 the reducing wedge attachment DR was built in series at Kern, Aarau, and in 1960 Carl Zeiss, Jena introduced the BRT 006 which has the base at the sta-

Figure 22 Theo 010 with mirror telescope

Figure 24 Wedge attachment Dimess

Figure 25 Plumbing staff distance meter "Kipplodis"

Figure 23 Electric eye on top of Askania theodolite

tion. In 1961, the DAR-100 for vertical rod with a wedge pendulum in front of the objective lens appeared in Russia.
Among the suggestions, which did not gain practical success are attempts by *Nestler* (1912) and the Swiss *Werffeli* to use the vernier principle for horizontal rods with wedge type graduation, as built in 1919 by Kern, Aarau, and the idea of the Dutch *Dieperink* (1920) to improve the measuring accuracy for distance measurements with stadia hairs, as well as in levelling by using a rod with transverse graduation. *Heckmann's* precision distance meter built by Breithaupt (1933) with a vertical, and a 1 : 10 inclined stadia hair for reading a horizontal rod with transverse graduations (figure 27) did also not become popular. Carl Zeiss, Jena accepted *Candidos'* suggestion in 1966 to refine the rod reading with small plate attachments.
In recent years, mechanization and automation of observation and evaluation processes for surveying instruments involves electronics as well as precision mechanics and optics.
For a long time, attempts have been made to automatically level the line of sight. The pendulum telescope of the Mining Academy in Clausthal built in 1790 is supposed to provide a standard deviation of 10 mm per kilometre. In 1878 *Couturier* designed a reflecting level with a cardanically suspended vertical telescope, which serves also as pendulum. The levelling errors of these pendulum instruments were however too large as to compete with spirit levels. Around the turn of the century the French *Claude* and *Drien-*

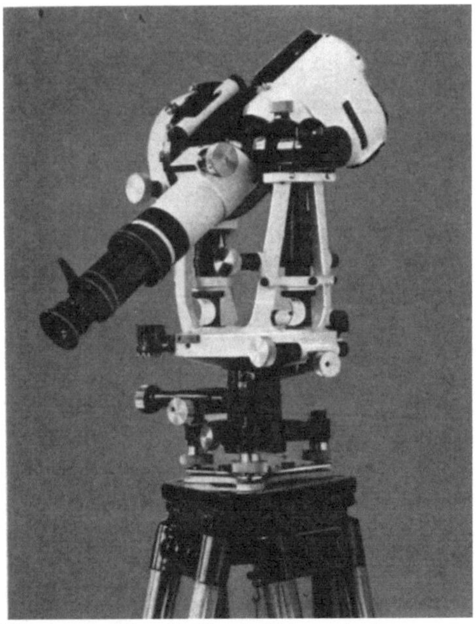

Figure 26 First Boßhardt-Zeiss reduction tacheometer

Figure 27 Rod graduations by *Heckmann*

Figure 28 Compensator attachment

court designed an instrument which is levelled by autocollimation with a mercury horizon and a pentagon prism. *H. Wild* at Zeiss, Jena also attempted to include a mercury horizon into the optical train of a level in 1923. *Heckmann* suggested in 1932 an instrument where the level bubble, reflected into the field of view, serves to determine the correction for an inclined line of sight. In U.S.S.R., a compensation level was built in series in 1938. *Drodofsky* utilized in 1940 at Carl Zeiss, Jena the bubble of a 30″ bull's eye level as lens element in the optical train. The Italian *Bonechi* patented in 1940 a liquid compensation for levelling. At Carl Zeiss, Oberkochen a level was designed in 1946 where the image of half the bubble arc of a focus-bubble was used instead of a crosshair to determine the line of sight (when using a focus-bubble, its radius corresponds to the focal length of the objective lens).

The self levelling level with bubble compensator, developed by *Stodolkjewich* in U.S.S.R. in 1946 was the first to be used in practice. In 1950, the Ni 2 from Carl Zeiss, Oberkochen started a new epoch, because this instrument contains a mechanical compensator instead of a tubular bubble. Since then, 70,000 levels of this design have been produced, and numerous other automatic levels appeared on the market. They simplify and speed up the work, and have partly replaced the spirit levels.

The types of compensators used vary significantly. Besides pendulum arrangements with mirrors, even the Abat wedge (1777), a glass wedge with a variable refraction angle, is used. Compensator attachments which make it possible to use spirit levels like automatic ones (1960 Feinmess, Dresden (figure 28) and Askania, Berlin) were hardly used.

Occasionally, theodolites are also built without tubular bubble; the principle of automatic levelling of the line of sight is then applied to the reading of the vertical circle. Askania, Berlin produced in 1956 the first theodolites with automatic vertical index, although the Th 3 from Carl Zeiss, Oberkochen had utilized the end of the level bubble as automatic vertical index in 1953.

In 1852 the French physicist *Foucault* discovered that a gyro with two degrees of freedom will point towards North (figure 29). In 1908, the gyrocompass which is based on this principle, was introduced into marine navigation. *M. Schuler* build in 1921 the first surveying gyro, which however, could not cope with rough transportation conditions because of its sensitivity. In 1949, a surveying gyro, named meridian pointer, was applied by the Mining Academy Clausthal for the first time underground. Similar use is reported from U.S.S.R. in 1950. There gyrotheodolites were equipped with autocollimation telescopes around 1952–54. In 1960, Fennel & Sons, Kassel produced the first gyro theodolite KT 1 in series. Since 1963 gyros are also constructed as attachments for theodolites.

In 1905 *Bykow* presented the first suggestion for fully automated height measuring instruments in Russia. *Leontowski* designed in 1915 a cart with four wheels whose instrumentation plotted the profile of the path of the cart. Since 1947 fully automatic height meters installed in automobiles or trailers are used in the United States and in Russia.

After invar was discovered in 1897 by the French *Benoit* and *Guillaume*, invar wires were produced in 1898, and the Jäderin method became popular. In the following years, many baselines for triangulation were measured in this manner. More recently, measurements with long invar—or steel tapes have become quite common.

A new step in the development started with the use of electromagnetic waves for distance measurements. The Finn *Väisälä* presented in 1923 a method for highest precision distance measurement with the aid of light interference. Since then it is used to calibrate invar wires and to measure test bases. In 1945 the Shoran method was introduced in the United States for surveying purposes by *Aslakson*. After it was tested, the Hiran method followed in 1950, which developed into the Shiran approach by 1965. In 1936 the first electro-optical distance meter was built in U.S.S.R. at the Governmental Optical Institute (GOI). This instrument type reached practical importance with *Bergstrand's* Geodimeter, built by AGA in Stockholm in 1948 (figure 30). In 1957 the South African *Wadley* presented the Tellurometer (figure 31), a distance meter operating with micro waves, which quickly became widely used in surveying and geodesy. In 1968 electro-optical distance meters using lasers were presented with a range of 30 km in daylight—"Quartz" (U.S.S.R.) and AGA Geodimeter 8. In 1968, the first instrument with a GaAs-diode for close range (up to 3 km) was built in series. It was the Wild DI 10, a light instrument of small dimensions. Attached to a theodolite, this becomes an EDM tacheometer. The Electronic

Distance Measurements (EDM) have laed to significant economic benefits and high accuracy for planimetric nets as well as topographic and engineering surveys.
Progress in mechanizing and automation led to the construction of code theodolites and – tacheometers (1963 Fennel, Kassel; 1965 Kern, Aarau). The observations were directly registered on film in a code which, after processing, was evaluated with photoelectric sensors which transferred the results onto punched tape. A further step in the development is the digital theodolite Digigon, designed in 1965 by Breithaupt & Son, Kassel. Graduated circle and optical reading device are replaced by incremental encoders. The angles are presently digitally. Only a prototype was produced.
With the development of electro optical distance meters for close range and the possibility to detect angles sufficiently accurate by code or incremental methods and store them electronically, self registering electronic tacheometers the–at present–most complete instruments could be designed. In 1968, Carl Zeiss, Oberkochen introduced the Reg Elta 14, the first such instrument built in series. In 1977, Hewlett-Packard, United States presented with the HP 3820 A a new generation of EDM with laserdiode and microcomputer.

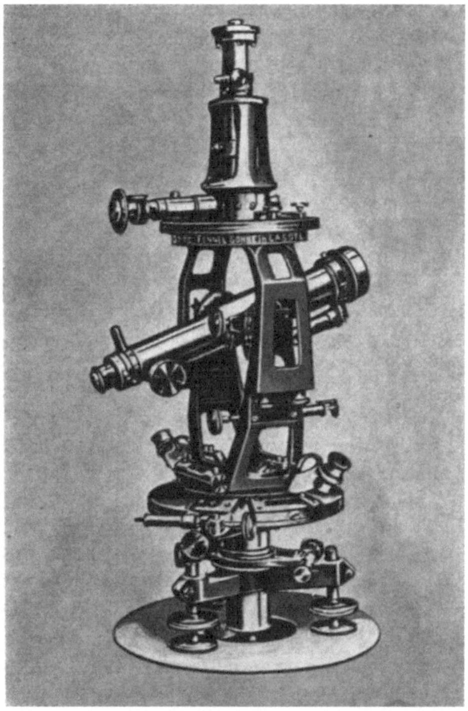

Figure 29 Declinatorium-fore-runner of the gyrotheodolite (2nd of the 19th century)

Figure 30 First geodimeter (1948)

Figure 31 Tellurometer MRA 1 (1957)

III. Manufacturers of Surveying Instruments

In the course of time companies were formed in various countries which are primarily involved in the production of surveying instruments.

In 1846, *Carl Zeiss* (1816–1888) founded in Jena a company, whose successor in Jena, the VEB Carl Zeiss Jena is the leader in the precision mechanical-optical-electronic industry of the German Democratic Republic. Since 1909 it has a separate division for surveying instruments. *Gotthelf Studer* founded in 1791 a workshop in Freiberg, which was continued by *Friedrich Lingke*. In 1873 *Max Hildebrand* joined the company, which obtained his name, but was changed to VEB Freiberger Präzisionsmechanik in 1950. In 1872 the Gustav Heyde Company originated in Dresden and is now named VEB Feinmess Dresden. In 1762 the company F. W. Breithaupt and Son was founded in Kassel. In 1802 *Georg von Reichenbach* opened a mechanical-optical institute in Munich, which was continued by *Ertel* and to which, among others *Soldner*, *Utzschneider* and *Fraunhofer* were associated (since 1935 "Ertel-Werk für Feinmechanik"). In 1848 Dennert and Pape was founded in Hamburg-Altona, and in 1851 Otto Fennel, Kassel. In 1871 the company of Carl Bamberg, later Askania Werke, was founded, and in 1946 Carl Zeiss, Oberkochen/Württemberg started production. Theis Gmbh & Co., Breidenbach/FRG (wich comtinues the production of Fennel-instruments), Pentax, Hamburg, and G. Nestle KG, Dornstetten are also mentioned.

In Switzerland surveying instruments are produced by Kern & Co. AG in Aarau (founded 1819), and by Wild Heerbrugg AG (founded 1921). Altimeters are made by Revue Thommen AG, Waldenburg (founded 1936), and levels also by Visomat AG, Rümlang.

In Austria, Gebr. Müller GmbH, Innsbruck produced surveying instruments a few years ago.

Up until 1917, surveying instruments were produced in Russia primarily by Schwabe in Moscow. This company became Geofizika in the U.S.S.R. In 1929 Aerogeopribor (later Aerogeoinstrument, since 1960 EOMS) was formed, and in 1938 the Charkow Works for Mining Instruments. At present the factories of the Main Administration Geodesy and Cartography of the Council of Ministers of U.S.S.R. produce approximately 25,000 theodolites and 35,000 levels per year.

In Hungary, *Ferdinand Süss* founded in 1876 a company for optical instruments, today called MOM = Magyar Optikai Murek or Hungarian Optical Works.

In the Č.S.S.R., Meopta, Prague – since 1945 the successor of the Srb and Stys Company, founded 1923 – produced surveying instruments until 1965.

In Poland and Bulgaria, surveying instruments are produced by the Polish Optical Works (PZO = Polskie Zaklady optyczne), respectively RPGP, Sofia (founded 1962.)

With Filotechnica Salmoiraghi, Milano (founded 1865 by *Porro*) and Officine Galileo, Florence (founded 1866) Italy has well known producers of surveying instruments.

The two largest French companies for optics and precision mechanics, ESSEL and SOPELEM, combined their production in 1955 as SLOM (Société d'Optique, Précision, Electronique et Mécanique), a Division of the Essilor group. Furthermore, the companies Chasselon, Cachau (Seine) and Coppin, Paris should be mentioned.

In Great Britain, surveying instruments are produced by Vickers Ltd., Vickers Instruments, York (formerly Cooke, Troughton and Simms, founded 1922, dating back to 1686), Rank Precisions Industries Ltd., England, of which Watts (founded 1856) is a part and W. F. Stanley and Co., London (founded 1853).

Svenska Aktiebolaget Gasaccumulator (AGA), Stockholm Lidingö (founded 1904) produces EDM equipment in Sweden.

The production of surveying equipment in the U.S.A. takes place at Berger and Sons Inc., Boston, Brunson Instrument Comp., Kansas City, Dietzgen Corp., Chicago, Ill., W. & L. E. Gurley and Co., Troy, N.Y. (founded 1845), Keuffel & Esser Comp., Morristown, N.J. (founded 1867), Lietz Comp., South San Francisco (founded 1882) and Path Instr. Intern., New York.

In Japan there are Fuji Surveying Instr. Co., Ltd., Tokyo (founded 1929), Sokkisha Ltd., Tokyo (founded 1920), Tokyo Optical Co. (Topcon, founded 1932) and Nippon Kogaku (NIKON, founded 1917).

India has the National Instruments and Ophthalmic Glass in Calcutta (founded 1830).

The People's Republic of China has government factories for surveying equipment.

Electronic Distance Meters and tacheometers are produced by Tellurometer Ltd., belonging to the Plessey International Group in South Africa, U.K., U.S.A., Canada and Australia, by Cubic Int. Ltd., Hewlett-Packard L.S.E. and Precision Int. Inc., Tullahoma (U.S.A.).
Among the producers or industrial groups, connections and sales arrangements to compliment their own offerings are common. In foreign countries sales companies are formed. More recently, instruments are also often produced under licence from or for other companies. The influence of electronics has also led to connection with companies of that type.

IV. Standardization

The technical development has led to a steadily increasing number of technical terms, dimensions, materials and products. This rather uneconomical multitude caused efforts towards standarization in Germany as early as the middle of the 19th century.
Under standards in this context should be understood: unification regarding the quality of tools and products (Quality standards, markings with information on dimensions, materials and surface characteristics), production-, test- and other methods (method standards). Unification related to the amount of work are referred to as effort norms.
In the G.D.R., for instance TGL regulations have been delcared as state standards (TGL are technical norm, quality regulations and delivery conditions). A state standard is a legal standard for the area of the G.D.R. and has to be recognized by the Bureau of Standardization, Mensuration and Merchandize Control (ASMW) as well as entered into the Central Register of the G.D.R.
Therefore the DIN as presented by the German Institute for Norms (Deutsches Institut für Normung e.V. (DIN) in the Federal Republic are no standards according to the above definition, since their use is only recommended.
National standards of other states are for instance the State Union Norm (GOST) in the U.S.S.R., ČSN in the Č.S.S.R., PN in the People's Republic of Poland, BS in England, NP in Portugal, NS in Norway, SIS in Sweden, IS in India, NF in France, UNI in Italy, NGN in the Netherlands, ASTM and ASA in the U.S.A. The efforts to unify national standards and to arrive at new international standards led in 1928 to the constitution of the International Organization for Standardization (IS), so named since 1946.

V. Further Developments

There are two main possibilities for further improvements of surveying instruments, namely: Improvements of present instruments by purposefully utilizing known building elements, or the development of completely new instruments and methods.
Ideas and suggestions which are worth consideration will not all be suitable. The practical use is the main criterion. Here the novelty has to prove itself and to obtain confirmation of its suitability.
H. Wild wrote in 1939 about the aim of novelties in the design of surveying instruments:
"The new instrument designs shall not lead to a reduction of the given tolerances, e.g. the permissible error of the measuring results, because claims in this regard are in part already exaggerated. Rather they should make it possible to obtain these results in a simpler manner, in less time, and with less effort. It should also not be necessary anymore, that the user has to adjust the instrument prior to measuring, because we have known methods for a long time, which permit simple elimination of possible instrument errors."
The new instruments therefore, should simplify the work and enable better performances. Furthermore, accuracy and economy have to be considered. Accuracy should not be exaggerated. Finally, the stability of the instruments should be improved, their optics refined and more robust types produced. This would make the instruments less sensitive for external influences. In recent years, more efforts are directed to mechanize and automate the working procedures, and to make them objective.
The industry strives to deliver the best instruments. The professional surveyor has to strive to master these highly developed instruments and to utilize them purposefully.

VI. Operation and Care of Surveying Instruments

Surveying instruments are only then completely effective, if they are carefully and conscientiously treated as well as professionally operated. The directions which are included should not only be read but also be followed. More than ever, this holds true for EDM instruments and other surveying equipment with electronic parts.
If the methods match the characteristics of the instruments, then they are effectively utilized.

Instrument Storage

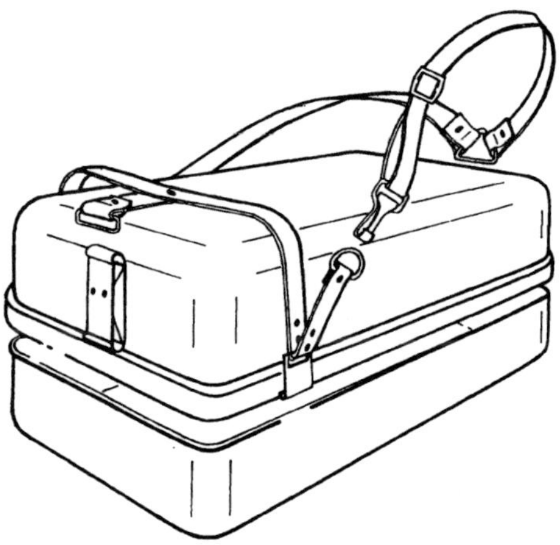

Figure 32 Instrument container

Surveying instruments should be stored inside their respective container (figure 32) in dust free rooms without large temperature changes. In humid climate, they have to be removed from the tightly sealed containers so that the air can circulate freely around them.
In large collections, instruments and equipment are usually registered in a card file. It is expedient to prepare an instrument passport. It contains producer, type, description and technical data as well as a record on the calibration results.
In extremely cold areas the instrument should not be taken into a heated shelter as long as it is needed for measurements, but rather remain exposed to the outside temperature at some protected location. This prevents vapour formation on the optical and interior elements of the instrument when work is being resumed.
When storing a compass for a longer period of time, the position of the unlocked needle should be checked in order to maintain its magnetic characteristics.

Instrument Inspection and Checks

Instruments should be carefully inspected and checked for their suitability for a specific task before they are issued from the storage area. Furthermore, it should be checked whether the auxiliary equipment in the container is complete and operational.
At the beginning of each field season, the instruments should be tested and, when needed, adjusted according to the directions provided. It is recommended to repeat this test after completion of the field work, or after extended work interruptions or long distance transportation. In this manner, work stoppage due to faulty equipment can be prevented.
Instruments should only be adjusted when it is really necessary and only according to directions. When tightening adjustment screws, care has to be taken as to not create stresses.

Transport of Instruments

Before transporting instruments, one should first check whether the clamps are evenly tightened and then close the container and perhaps lock it. The keys have to be kept at a safe place.
When lifting or moving in vehicles, jerky movements should be avoided and shocks dampened.
It is best to hold the container with instrument upright on ones lap, possibly wrapped in a soft blanket. For longer transportation on land—sea—or air the container with the instrument is placed inside a padded crate (figure 33). During transport the crate has to stand upright. When using pack animals, the instrument is fastened upright, usually hanging. Generally the instruments are to be protected against fall, shock and heavy vibrations.
Rods are to be transported in their crates. In any case they have to be packed like instruments so they are not exposed to sudden blows.
When walking, the pointed ends of tripods, range poles etc. have to be kept in view. Rods should not be touched at their graduation, and protected from heavy blows. This is especially important for invar rods.

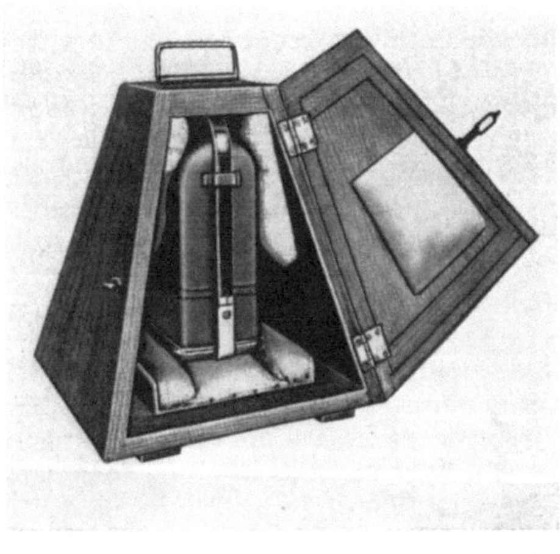

Figure 33 Padded transport crate

Setting-up of Instruments

At the survey location, warning signs and -flags should be erected, and industrial safety rules have to be observed. It is of advantage to have the back of the rods and the tripods painted in bright colour, such as red and white.
The tripod is to be set up solidly and in such a way that its legs are not in the way when observing certain directions. The tripod legs should be ex-

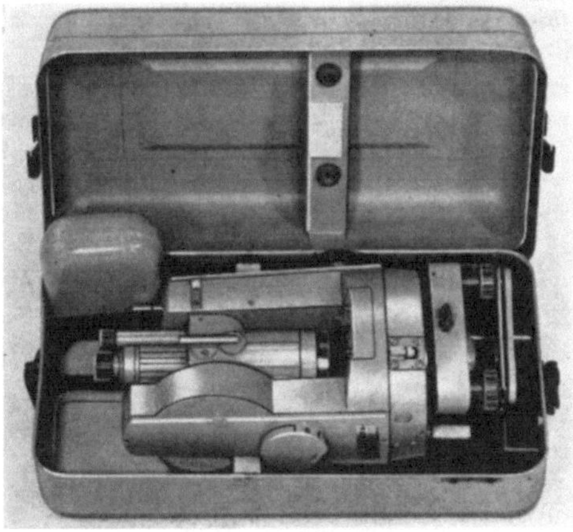

Figure 34 Position of instrument inside the container

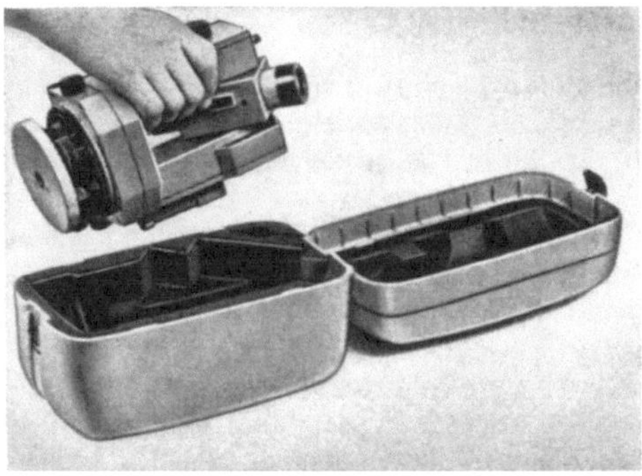

Figure 35 Packing and unpacking of an instrument

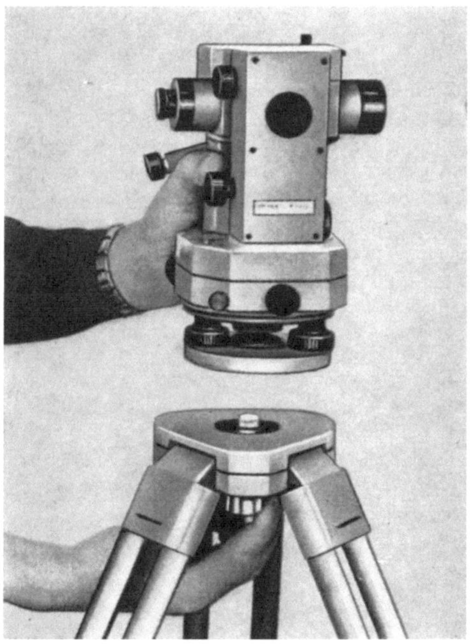

Figure 36 Placing the instrument onto the tripod

tended such that the observations can be made with ease. Their pointed ends have to be firmly pressed into the ground. One should also watch that the top of the tripod is approximately level and, for angular and distance measurements, is centered above the station.
When unpacking an instrument one should note its arrangement within the container (figure 34). A figure showing the arrangement of the various parts when properly packed should be inside the container. In any case, the given directions should be followed. Prior to unpacking, all clamps should be loosened. Then theodolites and tacheometers are lifted at the right standard – never on the side which houses the index bubble! – levels at the tribach (figure 35).
The instrument is then placed on the tripod and fastened to the tripod head while still being held with one hand. An instrument may never stand loosely on the tripod. Only after it is fastened, the hand can be taken away (figure 36). For centering and levelling, the fastening screw is to be loosened somewhat to reduce the pressure on the thread of the foot screws. It will be tightened again afterwards.
If the instrument temperature differs significantly from the field temperature, the instrument has to be left on the tripod until its temperature conforms. For 10 K temperature difference this requires about 5 minutes.

Care of the Instrument during Measurements

The instrument and the whole tripod are to be protected from direct sunlight and rain with an umbrella (figure 37). If work is interrupted due to rain, the instrument has to be protected with a cover. Drops of water should be blotted with a soft clean rag. Optical parts may not be touched with fingers. Dust should be carefully removed with a soft hair brush to the edge and then with a dust- and spot free soft cloth or soft chamois. Other dirt should be removed with hygroscopical cotton, never with liquid.
Whenever work stops, the instruments are to be protected against rain or dust with a hood or other cover. Prior to measuring, instruments with graduated circles should be rotated several times around both vertical and horizontal axes, so that the lubrication in the bearings is distributed. When measuring, touch only the solid parts of the instruments, never the eyepiece.
Stress on the instrument should be avoided. Clamps should be tightened slowly and evenly. When measuring horizontal directions, the vertical clamp does not need to be tightened. Finite fine motions should only be operated clockwise, so that the part is moved by the screws and not the spring, thus avoiding back lash.
When connecting upper and lower theodolite part with a repetition clamp of the *Mahler* type, one should press vertically on the clamp and counteract this movement with a fingertip. If the clamp is not needed, it should remain open.
Bends in tapes, caused for instance by vehicles driving have to be avoided. When rolling up the tape extra loops cannot be tolerated. Invar wires have to be protected from shock and reeled carefully. Their metal parts have to be cleaned daily with a soft rag and then rubbed with acid free grease. Prior to measuring, the grease has to be removed again.
Even though the human eye is not directly endangered when working with laser instruments, because of their low power (construction laser up to 5 mW, laser diodes), one should never look directly into a laser beam. If the laser beam is directed through a telescope, one should not look into the eyepiece as long as the laser is operating. If necessary, protective eye glasses should

Figure 37 Field umbrella

be worn. The laser should be screened off as much as possible, and never be without supervision when running.
When using instruments with mercury, one has to be extremely careful. Especially inside closed rooms, mercury should not be spilled. If this should happen, mercury drops can be lifted with copper sticks and then treated with sulfur for chemical bonding.

Instrument Transport between Stations

The observer should transport the instrument from one station to the next. If the distance is only a few hundred metres, the instrument can remain on the tripod. One has to check however, whether the fastening screw and the instrument clamps are tightened. For the reduction tacheometer Redta, however the side clamp has to be opened.
Then each hand holds one tripod leg, while the 3rd one hangs over the shoulder (figure 38), so that the vertical axis remains vertical.
Horizontal rods for optical distance meters are turned to vertical, but remain on their support. If needed the braces are loosened.

Figure 38 Transporting the instrument between stations

Care of Instruments after the Measurements

Before packing of the instrument, its clamps should be loosened. Once the instrument is correctly in the container, they are tightened evenly.
The pointed ends of tripods and rangepoles are to be cleaned with rag, brush or bushel of grass.
Tapes which have gotten wet have to be dried with a cloth, and then greased lightly with acid free oil or fat. If there are rust spots, do not use sand or sand paper but wood ash soaked with kerosene.
Instruments that got wet have to be unpacked at home and left out until they are completely dry. Occasionally the producer supplies with the instrument a little bag of Silicon-gel (highly hydroscopic grains of amorphous quartz). The grains are blue when dry and pink when saturated. Since they absorb water from the air, the instrument container has to be closed except when packing or unpacking. Pink grains can be regenerated when placing them directly on a heatable plate and heating it above the boiling temperature (check with a water drop: hiss-test). If the temperature is too high, the grains crack. The now blue grains are placed again into their bag after they are cooled down.
The tripod head, and the threads of the foot screws and the fastening screw should be kept clean and lightly oiled.
The bottom parts of rods have to remain free of dirt and dust and should be greased lightly.
After completion of the field season, the instruments should be thoroughly inspected. Damages are to be fixed by the mechanic.

1. Optical Equipment and Level Bubbles

1.1. Optical Equipment

1.1.1. Fundamentals of Optics

Geometric Optics and Wave Optics

Most surveying instruments are optical instruments whose operation is based on different characteristics of light. More and more, electronic parts are used.

Natural light is composed of electro magnetic waves starting from the light source; it penetrates straight into all directions with the same velocity depending on the matter. The direction of the energy flow of light is called light ray, the total amount of light rays is known as bundle of rays.

In vacuum the light velocity is 3×10^{10} cm/s. In air it is only slightly less, but in other matter much slower, e.g. 2×10^{10} cm/s in glass. In geodesy the velocity of electro-magnetic waves in vacuum is used as 299,792.458 km/s.

Optics is the science of light. Wave optics considers the wave characteristics of light. It is used to describe phenomena occuring with light energy, such as diffraction, interference, polarization and others. Geometric optics or ray optics is concerned with the path of rays in optical systems. With the aid of light rays the operation principle of optical systems can be presented much easier than with wave optics.

Laws of Geometric Optics

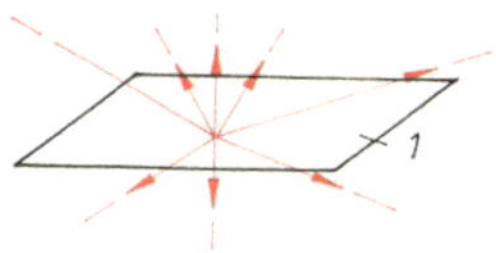

Figure 39 Diffuse reflection
1 rough surface

Reflection. If a bundle of light rays hits the boundary surface between two media, part of the light is thrown back (reflected), while the rest is refracted into the other medium.

Diffuse reflection (reflected rays into all directions) can be observed on rough surfaces (figures 39), while directional reflection (rays are reflected into one direction) occurs on polished bodies (figure 40). The intensity of the reflected light is always less than that of the incoming light, partly because of partial penetration into the other medium, partly because of some diffuse reflection. The intensity loss depends on the incident angle, on the intensity and on the characteristics of the reflecting surface.

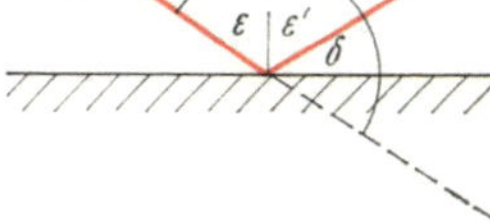

Figure 40 Directional reflection

The law of reflection (figure 41) states: Incoming ray, normal to the reflecting surface and reflected ray lie in one plane. The incident angle ε equals the reflection angle ε'.

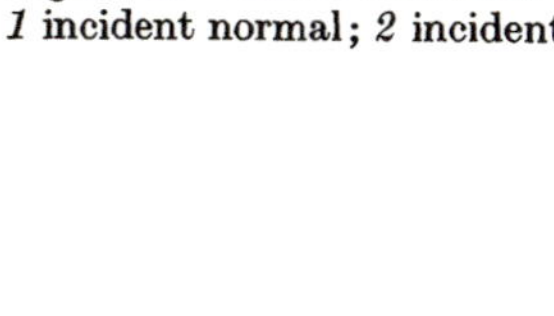

Figure 41 Mirror reflection
1 incident normal; *2* incident ray; reflected ray

Refraction. If a light ray passes from one medium to another, e.g. air to glass (figure 42a) or glass to air (figure 42b), its direction of the ray is reversible.

When passing from an optically thinner medium (air) to a denser one (glass), the incident ray is refracted towards the normal. In reverse direction it is refracted away from the normal (figure 42). In the first case $\varepsilon > \varepsilon'$ in the second $\varepsilon < \varepsilon'$. Again, incident ray, normal and refracted ray span one plane.

Light rays passing from air into another transparent body are refracted the more, the larger the inclination to the boundary surface. At the point of entry into the denser medium, the velocity of propagation of the light is instantaneously reduced. The ratio between light velocities c_L in air and

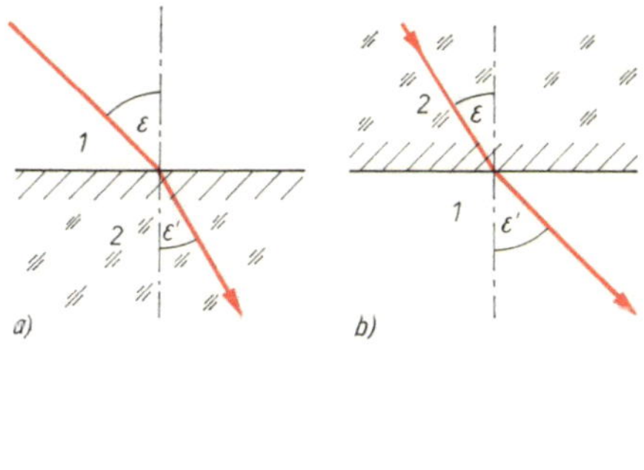

Figure 42 Refraction of a ray when passing (a) from air to glass, resp. (b) from glass to air
1 air; *2* glass

c_M in the medium represents the refractive ratio $n_{L,M}$. It corresponds to the ratio of the sines of the incident and refraction angles. This ratio remains constant for two particular media.

$$n_{L,M} = \frac{c_L}{c_M} = \frac{\sin \varepsilon}{\sin \varepsilon'} = \text{const.} \tag{1}$$

The refractive ratio referred to vacuum is called the refractive index n:

$$\frac{\sin \varepsilon_0}{\sin \varepsilon} = n \tag{2}$$

The refractive index of air at 20°C is 1.00028. Optical glass types have refractive indexes ranging between 1.47 and 1.92, e.g. light crown glass has 1.5153, while heavy flint glass has 1.7515.
With the same angle of incidence ε_0 the refraction angle ε' results for a medium with n' as refractive index, because $\sin \varepsilon_0 / \sin \varepsilon' = n'$.
Dividing this with equation (2) one obtains after rearranging

$$n \sin \varepsilon = n' \sin \varepsilon' = \text{const.} \tag{3}$$

This is the law of refraction, discovered around 1618 by the Dutch *Snellius* (1581–1626). It states that the product between refractive index and the sine of the refraction angle remains constant. If refractive index and incident – or refraction angle are known, the unknown angle can be computed.
For ray tracing in optical instruments the refractive indices of air – practically equal to 1 – and glass – approximately 1.5 – are mainly needed. When passing from air to glass (figure 42a), $\sin \varepsilon / \sin \varepsilon' = 3/2$, and reversed (figure 42b) $\sin \varepsilon / \sin \varepsilon' = 2/3$. According to equation (3) the refraction angle of light ray passing from air to glass at 50 gon from the normal is:

$$\sin \varepsilon' = \sin 50 \text{ gon} \times 2/3 = 0.707 \times 2/3; \ \varepsilon' \approx 31 \text{ gon}$$

For a ray passing from glass at 25 gon to the normal, the refraction angle is:

$$\sin \varepsilon' = \sin 25 \text{ gon} \times 3/2 = 0.383 \times 3/2; \ \varepsilon' \approx 39 \text{ gon}$$

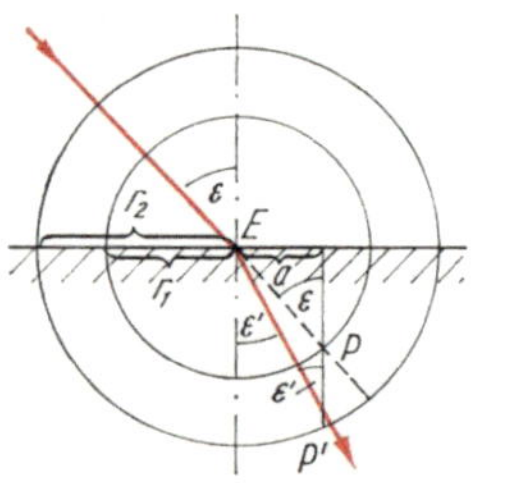

Figure 43 Graphical determination of refraction

The ray tracing can be done by simple graphical means (figure 43). Here, a light ray falls at E onto the plane surface of the refracting body. Two circles are drawn around E, whose radii correspond to the ratio of refractive indices, e.g. air–glass as 3 : 2. The incident ray intersects the arc with r_1, at P. The intersection of a parallel to the surface normal at P with the arc of r_2 is point P'. The straight line connecting P' and E represents the ray after refraction. The following proves the validity of this graphical approach:

$$\sin \varepsilon : \sin \varepsilon' = \frac{a}{r_1} : \frac{a}{r_2} = r_2 : r_1 = \frac{3}{2}$$

The ray tracing from the optically denser medium to the thinner one (glass to air) can be performed graphically in a similar manner.
If a light ray falls perpendicularly onto the boundary between two media, its direction will not change (figure 44).

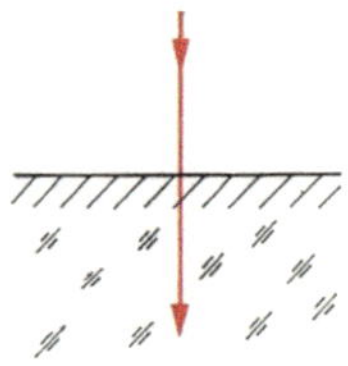

Figure 44 A ray hitting a boundary surface orthogonally

Total Reflection. According to the law of reflection the values $\sin \varepsilon \cdot n/n'$ and $\sin \varepsilon' \cdot n'/n$ cannot become larger than 1 because $\sin \varepsilon$ and $\sin \varepsilon'$ cannot exceed 1.
When light passes from the optically denser glass into the optically thinner air it is split at the boundary into a refracted part ($E_1 S'$ in figure 45) and a reflected part ($E_1 S'_1$). At an incident angle of ε_2 in glass, the refracted ray ($E_2 S_2$) passes along the boundary surface. The refraction angle is 100 gon, therefore $\sin \varepsilon_2' = 1$. According to equation (3) the limiting angle ε_2 can be computed from $\sin \varepsilon_2 = 1/n$. With $n = 1.47$ ε_2 becomes about 48 gon while with $n = 1.92$ it becomes approximately 35 gon.
If the incident angle is larger than ε_2, e.g. ε_3, then no light can pass into the optically thinner medium. The ray is totally reflected. Total reflection therefore occurs, when a light ray in an optically denser medium hits the boundary surface to a thinner medium at an indicence angle larger than the limiting angle.

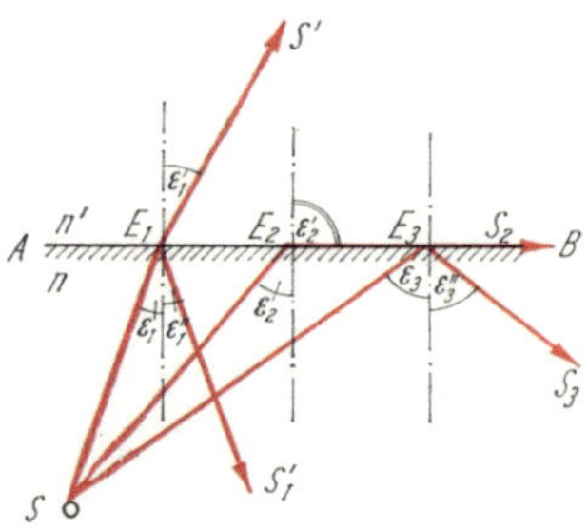

Figure 45 Refraction and total reflection

Application of the Laws of Geometric Optics to Mirrors and Prisms

Plane Mirrors. Mirrors are bodies with an optically reflecting surface. They may be made of material that can or cannot be penetrated by light. Light rays falling from a point L onto a plane mirror under various incident angles are reflected according to the law of reflection (figure 46).

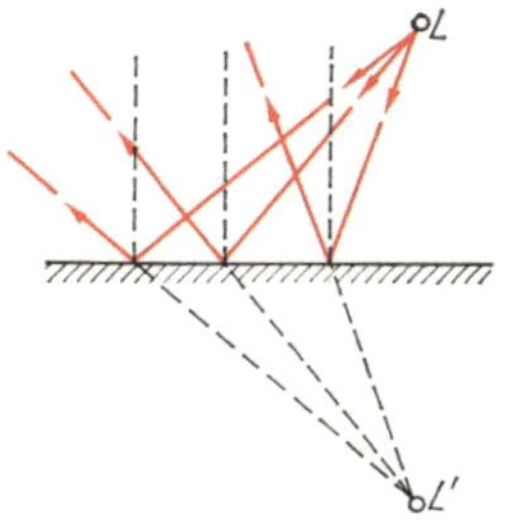

Figure 46 Generation of an image with a plane mirror

Since the reflecting angle always equals the corresponding incident angle, the rays diverge after reflection. The rays which hit the eye of an observer appear to originate at the point L' behind the mirror. The image of the light point L appears to be there. The image L' is a virtual image of L. It does not exist in reality and therefore cannot be captured on a screen. The line LL' is normal to the mirror surface. The virtual image is exactly the same distance behind the mirror as the object is in front of it.
The mirror image of a surface or body consists of the mirror images of its points and is mirror reversed.
The direction of this mirror reversal depends on the position of both mirror and observer. With a vertical mirror, a horizontal reveral (left–right) is obtained, while a horizontal mirror creates a vertical reversal (top–bottom).
A light ray hitting a mirror under an angle ε (figure 47) is reflected such, that angular deflection from its original direction is

$$\delta = 200 \text{ gon} - 2\varepsilon \tag{4}$$

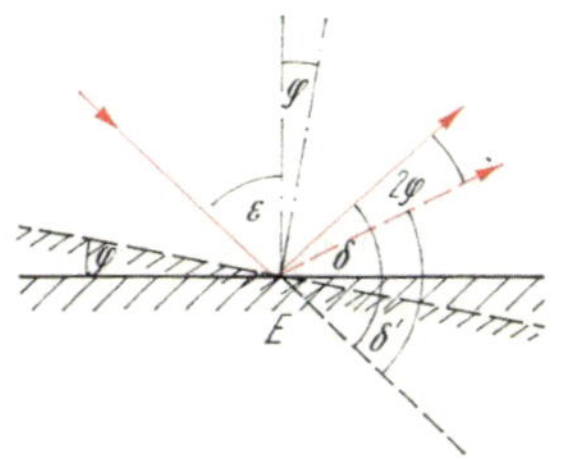

Figure 47 Propagation of rays when rotating a mirror

If one rotates the mirror around point E by a random angle φ, then the reflected ray is diverted by

$$\delta' = 200 \text{ gon} - 2(\varepsilon + \varphi)$$

$(\varepsilon + \varphi)$ is the incident angle for the rotated mirror. With this one gets (figure 47)

$$\delta - \delta' = 200 \text{ gon} - 2\varepsilon - [200 \text{ gon} - 2(\varepsilon + \varphi)] = 2\varphi \tag{5}$$

Therefore when turning the mirror by an angle φ the reflected ray is deflected by 2φ. This characteristic of the plane mirror is utilized in some magnetic instruments (declinators and compasses) as well as compensator levels.
A light ray who is reflected on two parallel mirrors retains its direction but is shifted (figure 48). The direction is even retained if both mirrors are rotated by the same amount in the same sense. Only the amount of the parallel shift changes (from E_2S' to E_3S''). If however, only one of the mirrors is rotated, then the reflected ray is deflected by twice the amount of the angle between the mirrors. The ray passing in direction E_2S' (figure 49) is rotated into the position $A''B''$. According to equation (4) it is deflected by the angle $S''E_2S' = 2\varphi$. The rotation angle φ equals the angle between the two mirrors.
Therefore

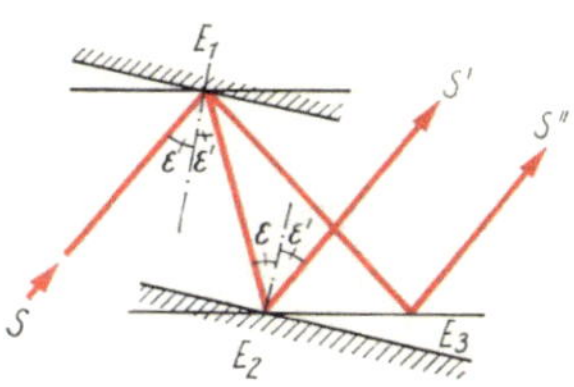

Figure 48 Propagation of rays when rotating two parallel mirrors

$$\delta = 2\varphi \tag{6}$$

The angle of deflection δ is thus independent of the angle of incidence at the first mirror. This is utilized in the mirror sextant.
An angle-mirror consists of two mirrors at an angle φ (figure 50). In surveying, angle mirrors with $\varphi = 50$ gon are used, which according to equation (6) results in $\delta = 100$ gon.

With this arrangement the following is achieved:

1. The incoming ray is deflected by 100 gon after two reflections, and
2. the image remains stationary when turning the instrument.

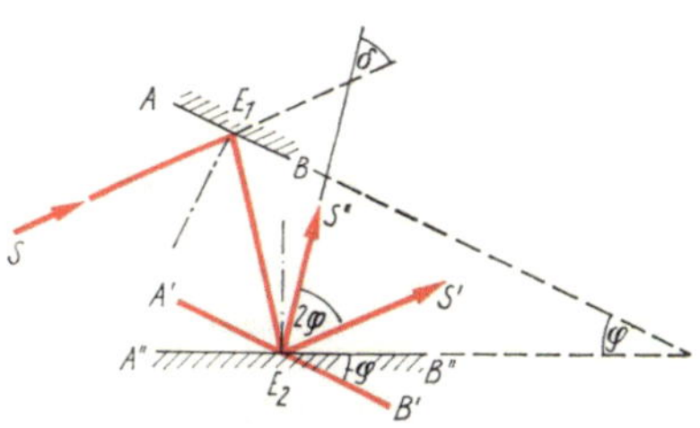

Figure 49 Propagation rays when rotating one of two parallel mirrors

This holds true for all mirror combinations which have an even number of reflecting surfaces.
Figure 50 can also be used for another proof:

$$\begin{aligned}(100 \text{ gon} - \alpha) + (100 \text{ gon} - \beta) + \varphi &= 200 \text{ gon}\\ \alpha + \beta &= \varphi\\ \delta &= 2\alpha + 2\beta\\ \delta &= 2\varphi\end{aligned}$$

A light ray between two parallel mirrors and two reflections on each mirror is again shifted while retaining its original direction (figure 51). If the first mirror is rotated by an angle φ, the twice reflected ray on each mirror is deflected by 4φ from its original direction. This characteristic is utilized in compensation levels.

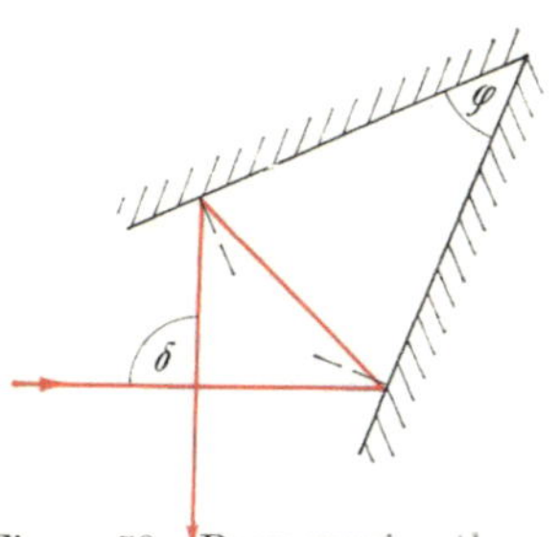

Figure 50 Rays passing through an angle mirror

Curved Mirrors. The laws of plane mirrors also hold for small elements of curved surfaces, e.g. concave and convex mirrors. In this case, the normal to the curved surface provides the reference.
The optical axis or primary axis is defined as the straight line connecting the centre of curvature and the centre of the mirror, as given by the vertex

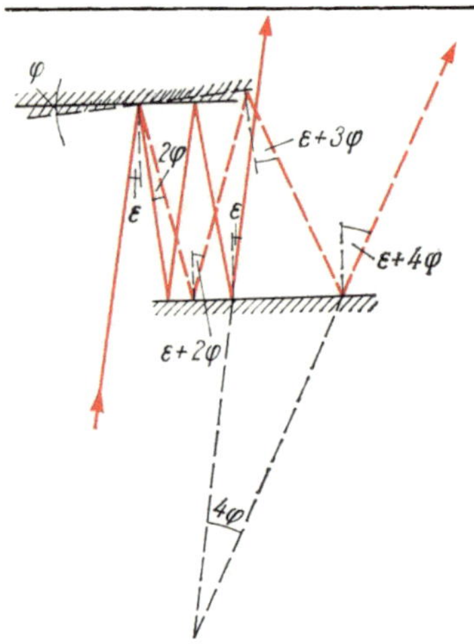

Figure 51 Double reflection

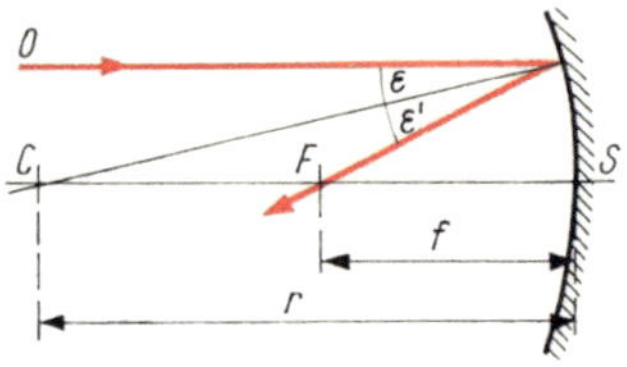

Figure 52 Spherical concave mirror

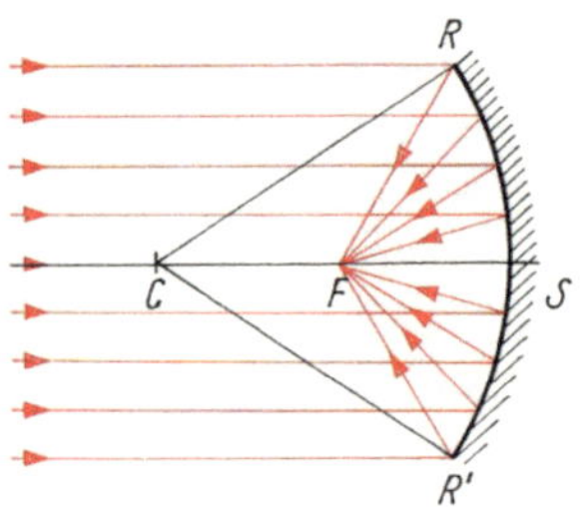

Figure 53 Effect of a spherical concave mirror

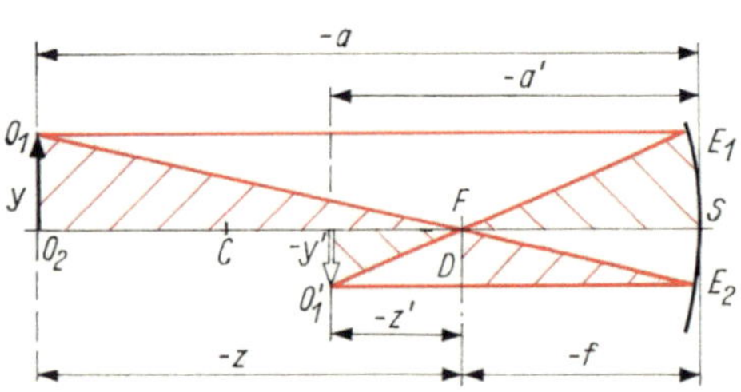

Figure 54 Graphical tracing of an image using a spherical concave mirror

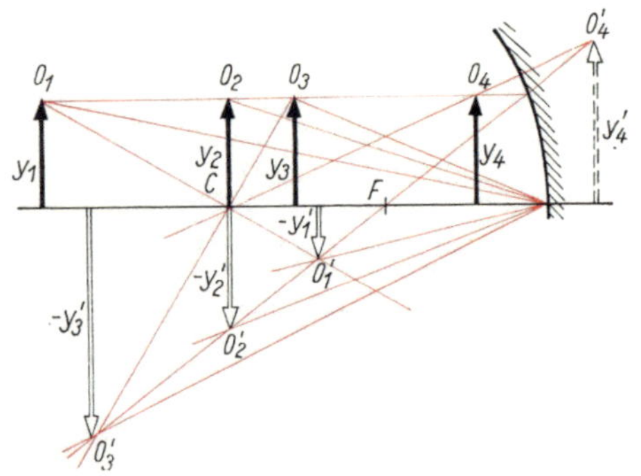

Figure 55 Image locations for different object locations, when using a spherical concave mirror

S (figure 52). The angle RCR' from the centre of curvature to the mirror edges is referred to as field angle (figure 53).

If a ray OE parallel to the primary axis hits a parabolic mirror, it is reflected to the point F ($\varepsilon = \varepsilon'$). This also holds as a first approximation for spherical mirrors with small field angle (figure 52).

The rays parallel to the axis converge into the focal point F. All rays passing through the focal point will be parallel to the optical axis after reflection at a concave mirror. For a spherical concave mirror the focal point is located half way between C and S. The distance f between focal point and vertex is called focal length. If the radius of curvature of the mirror is r, then

$$f = \frac{r}{2} \tag{7}$$

For larger field angles the rays parallel to the axis will not converge to a point any more. The points of intersection of neighbouring rays form a focal area which is pointed at the focal point. When intersected with a plane containing the mirror axis, this focal surface provides a rather oddly shaped line called catakaustic.

In order to graphically display the image of a point O_1, who is not located on the primary axis of the spherical mirror (figure 54), one first has to determine the focal point halfway between C and S. Then the image point O_1' is obtained at the intersection of the axis-parallel ray $O_1\, E_1$ after reflection through the focal point and the focal ray O_1FE_2 or with a ray passing through C (figure 55), which is reflected into itself. $SO_2 = -a = -z - f$ is the object distance while $SO_2' = -a' = -z' - f$ denotes the image distance. z and z' are the distances to the focal points of the object—resp. image point.

From $FD = -y'$ and the similarity of the triangles O_1O_2F and FDE_2 the relationship $-y'/y = -f/-z$ follows. For small y values the triangles FE_1S and $FO_1'O_2'$ are approximately similar, then $-y'/y = -z'/-f$.

Combining the two equations results in Newton's Imaging Equation for concave mirrors.

$$z \times z' = f^2 \tag{8}$$

With $z = a - f$ and $z' = a' - f$ equation (8) which is referred to the focal point, can be transformed to the lens equation which in this case is referred to the vertex.

$$\frac{1}{a} + \frac{1}{a'} = \frac{1}{f} \tag{9}$$

Depending on the distance of the object y from the mirror, the image y' can be real and reversed or virtual and upright.

From figure 55 the following can be derived:

1. Real upright objects outside twice the focal length of a concave mirror result in real, reversed and reduced images located between single and double the focal length.
2. Real upright objects at twice the focal length of a concave mirror result in real reversed images of equal size at double the focal length.
3. Real upright objects between single and double focal length of a concave mirror result in real, reversed and enlarged images beyond the double focal length.
4. Real upright objects within one focal length of a concave mirror result in virtual enlarged images.

For spherical convex mirrors, where the outside of the sphere is the reflecting surface, the same imaging equations (8) and (9) hold true. Focal point and focal length are virtual. Real upright objects when reflected by a convex mirror result in virtual, upright and reduced images within one virtual focal length (figure 56).

Although curved mirrors have been utilized in astronomic telescopes for more than 300 years, they only appeared within the last decades in surveying as elements of mirror telescopes.

Parallel Plate. A transparent body bounded by two plane parallel surfaces is called a (plane)parallel plate (figure 57). A light ray entering from air the upper surface under an angle ε_1 is refracted towards the normal at an angle ε_1'. It continues on a straight path and hits the glass-air surface at an angle ε_2. Due to the parallelity of the two surfaces, ε_1' equals ε_2. When entering the air, the ray is refracted away from the normal. Since ε_1' equals ε_2, ε_1

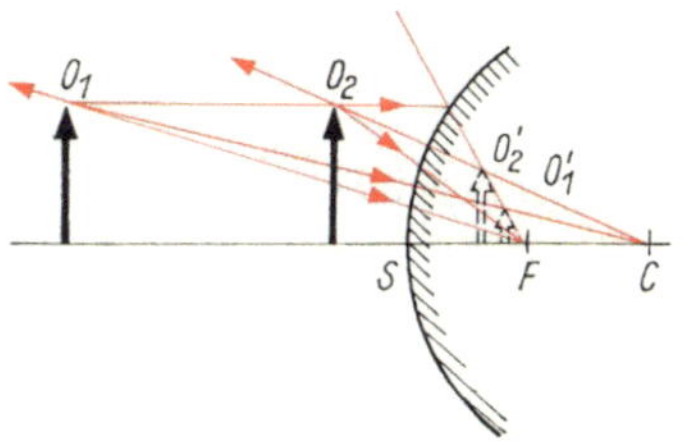

Figure 56 Image positions for a spherical convex mirror

equals ε_2'. Therefore, when penetrating a parallel plate a lightray is shifted parallel. The amount of the parallel shift q is $q = E_1E_2 \times \sin(\varepsilon_1 - \varepsilon_2)$. With $E_1E_2 = d/\cos\varepsilon_1$ and $\varepsilon_1' = \varepsilon_2$ one gets

$$q = \frac{d\sin(\varepsilon_1 - \varepsilon_2)}{\cos\varepsilon_2} \tag{10}$$

where $\sin\varepsilon_2 = \sin\varepsilon_1/n$, according to equation (1). The approximate formula

$$q \approx \frac{n-1}{n} d\tan\varepsilon_1 \tag{11}$$

derived from equation (10) and used by *H. Wild* is sufficiently accurate for practical purposes.

The parallel shift q depends on the thickness d of the glass plate, the incident angle ε_1 and the refractive index n of the glass. In order to shift a horizontal ray by 5 mm, parallel plates of 10, 20 and 30 mm thickness have to be rotated by the following angles from their vertical starting position according to equation (11) (figure 58):

q	d	n	ε_1
5 mm	10 mm	1.52	61.8 gon
5 mm	20 mm	1.52	40.2 gon
5 mm	30 mm	1.52	28.9 gon

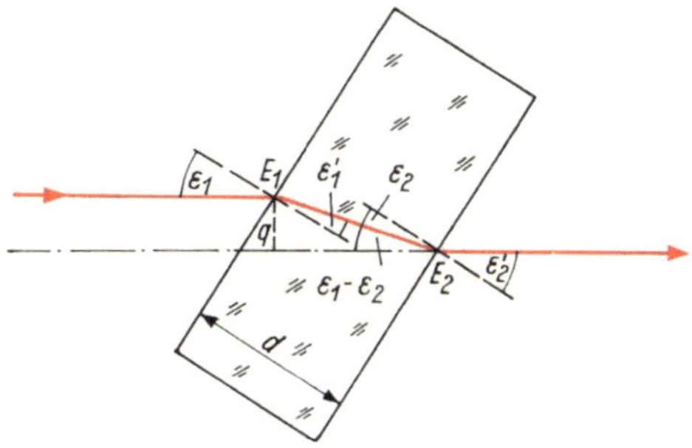

Figure 57 Rays passing through a plane parallel plate

For surveying instruments, parallel plates are used for reading of graduations by measuring small intervals, e.g. parallel plate micrometers at levels and optical micrometers at theodolites.

Prism (*Glass Wedge*). An optical prism is a body of transparent material with at least two boundary surfaces. The plane perpendicular to the reflecting edge K (figure 59) is called primary section, and the angle within this primary section is called prism angle. A glass wedge is a prism with a small prism angle.

A light ray falling onto the prism in the plane of the primary section is deflected after two reflections by the angle δ away from the reflecting edge. The amount of the deflection depends on the refractive index n, the material of the prism as well as the prism angle and the incident angle ε_1. According to figure 59

$$\delta = \varepsilon_1 - \varepsilon_1' + \varepsilon_2' - \varepsilon_2 \text{ and } \alpha = \varepsilon_1' + \varepsilon_2 \tag{12}$$

Thus the deflection becomes

$$\delta = \varepsilon_1 + \varepsilon_2' - \alpha \tag{13}$$

According to the law of refraction the following holds with n being the refractive index of glass

$$\sin\varepsilon_1 = n\sin\varepsilon_1' \tag{14a}$$

$$n\sin\varepsilon_2 = \sin\varepsilon_2' \tag{14b}$$

With $\varepsilon_2 = \alpha - \varepsilon_1'$ from equation (12) and $\varepsilon_2' = \delta + \alpha - \varepsilon_1$ from equation (13), equation (14b) becomes

$$n \cdot \sin(\alpha - \varepsilon_1') = \sin(\delta + \alpha - \varepsilon_1) = \sin[(\delta + \alpha) - \varepsilon_1].$$

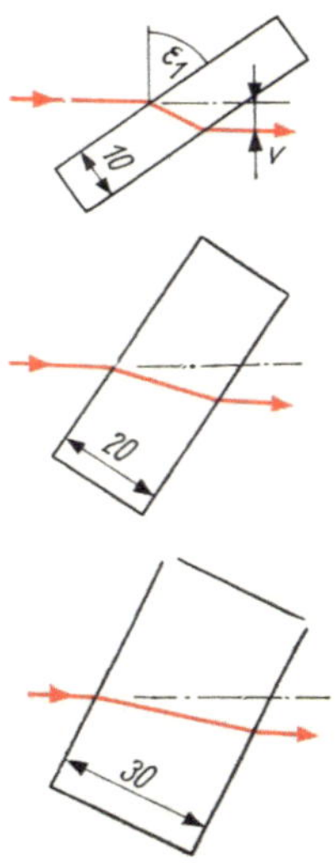

Figure 58 Shift of rays by 5 mm caused by plane parallel plates of various thicknesses

Since α and δ are small angles for a wedge, the sine can be replaced by the arc and the cosine by one. With this as well as equation (3) and $n\alpha\cos\varepsilon_1' = (\delta + \alpha)\cos\varepsilon_1$ one gets

$$\delta = \left(\frac{n\cos\varepsilon_1'}{\cos\varepsilon_1} - 1\right)\alpha \tag{15}$$

Therefore if the prism angle α, the refractive index n of the wedge material, and the incident angle ε_1 are known, the deflection δ can be computed according to equations (14a) and (15). If in addition the incident angle ε_1 is small, one obtains the approximate formula

$$\delta \approx (n-1)\alpha \tag{16}$$

with $\cos\varepsilon_1'/\cos\varepsilon \approx 1$, which can be used for incident angles up to ± 1 gon.

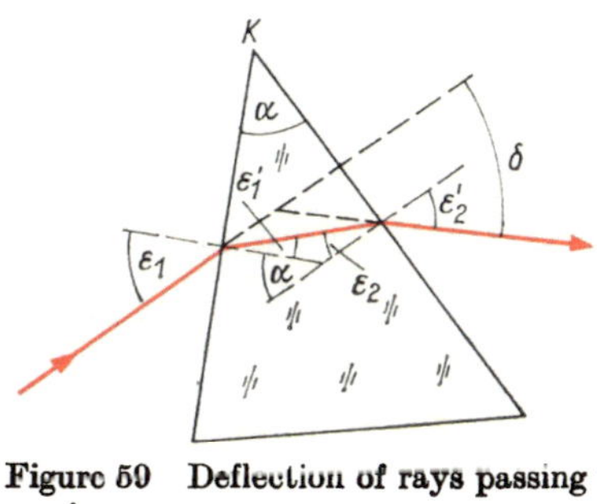

Figure 59 Deflection of rays passing through a prism

The biprism (figure 60) is a combination of two glass wedges with equal prism angles. At the left side two lightly inclined refracting surfaces meet at

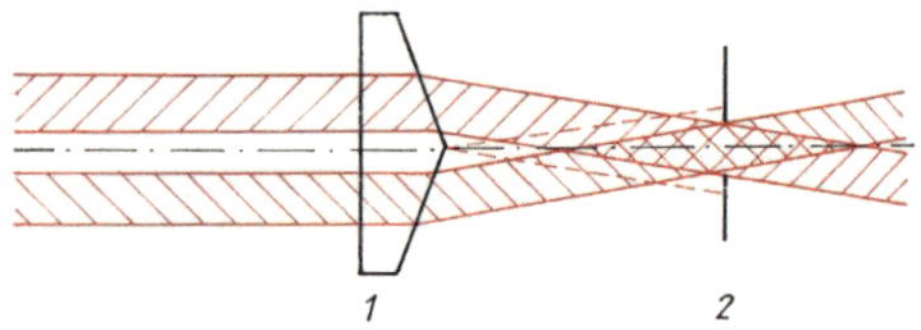

Figure 60 Biprism
1 biprism; *2* diaphragm

a carefully machined edge, while the right reflecting surface is plane. Combined with a converging lens, a crossed path of rays can be generated which results in two images shifted against each other.
Among others, the glass wedge is used in optical micrometers and double image distance meters, while the biprism is used in the latter.

Reflection Prism. Reflection prisms (total reflecting prisms) are prisms with several polished plane surfaces where total reflection is used for the deflection of the light ray. Prisms have the following advantages over mirrors:

1. The angle between reflecting surfaces is fixed.
2. The reflecting surfaces are not subjected to wear.
3. The images are clear and bright.
4. With a prism, several reflections of a ray can be obtained.
5. Prisms can easily be fastened and built into instruments.

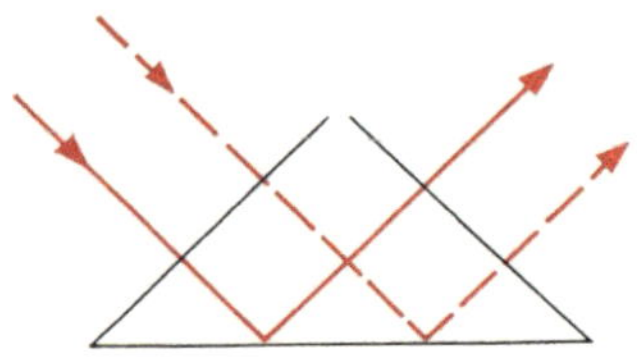

Figure 61 Deflection of rays by 100 gon when passing through a rectangle prism with equal sides

A prism with an even number of reflecting surfaces gives a straight image, one with an odd number, a reversed image like with a plane mirror.
If the incident angle is smaller than the boundary angle than the reflecting surface has to be silver coated which has the disadvantage of light loss.
A rectangle prism with equal sides is used to deflect a light ray by 100 gon (reflection at the hypothenusis, figure 61), or by 200 gon (reflection at the sides, figure 62) and to reverse the image (reverse prism, also called Dove- or Amici prism, figure 63). Prisms as in figures 61 and 62 are used for inclined sights and for circle readings.
The rectangular prism with equal sides and silver coded hypothenuses is used as angular prism (figure 64). If a light ray passing in a plane normal to the side edges, hits the edge AB, it is refracted at E_1, reflected at E_2 and again refracted at E_3. The deflection of the ray from its original direction is as follows:
The sum of angles in the quadrilateral E_1AE_3D is

$$(100\text{ gon} - \varepsilon_1) + 100\text{ gon} + (100\text{ gon} - \varepsilon_3') + (200\text{ gon} - \delta) = 400\text{ gon}$$

Therefore $\delta = 100\text{ gon} - (\varepsilon_1 + \varepsilon_3')$

Since $\varepsilon_1' = \varepsilon_3$ and $\varepsilon_1 = \varepsilon_3'$, δ becomes $100\text{ gon} - 2\varepsilon_1$.

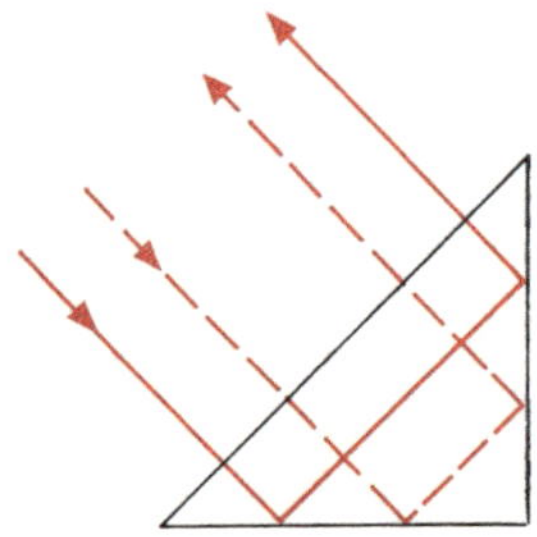

Figure 62 Deflection of rays by 200 gon when passing through a rectangle prism with equal sides

Thus the amount of δ depends on the incident angle. If the prism is rotated around an axis normal to its cross-sectional plane, the image moves. It should be noted, that only one reflection occurs.
If the ray hits the side close to the 100 gon edge (figure 65), it is refracted at E_1, totally reflected at E_2, reflected at E_3 and refracted at E_4. The deflection δ from its original direction is as follows:
The sum of angles in triangle CE_3E_4 amounts to

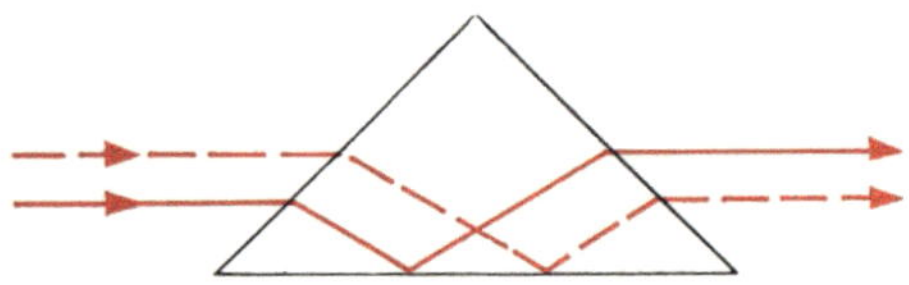

Figure 63 Reverse prism

$$(100\text{ gon} - \varepsilon_3) + (100\text{ gon} - \varepsilon_4) + 50\text{ gon} = 200\text{ gon}$$
therefore $\varepsilon_3 = 50\text{ gon} - \varepsilon_4$.

From triangle CE_2E_3 it is apparent that $\varepsilon_3 + \varepsilon_1' = 50$ gon, therefore $\varepsilon_4 = \varepsilon_1'$ and $\varepsilon_4' = \varepsilon_1$.
Triangle DE_4F supplies $\delta + (100\text{ gon} - \varepsilon_4') + \varepsilon_1 = 200$ gon. With $\varepsilon_1 = \varepsilon_4'$, δ becomes 100 gon.
This proof can also be found by simple considerations: Since AC and BC enclose an angle of 50 gon, and the ray is twice reflected on each surface like in an angle-mirror, the direction E_3E_4 is deflected by 100 gon from E_1E_2. Since the sides of ε_4 and ε_1 are normal to each other, $\varepsilon_1' = \varepsilon_4$. This equality of the angles in glass, commands equality of the incident angles $\varepsilon_1 = \varepsilon_4'$. This means that the ray entering the prism at E_1 and exiting at E_4 forms a 100 gon angle.
The deflection angle δ is independent of the incidence angle ε_1. If the prism is rotated around an axis normal to its cross sectional plane, the image remains static.

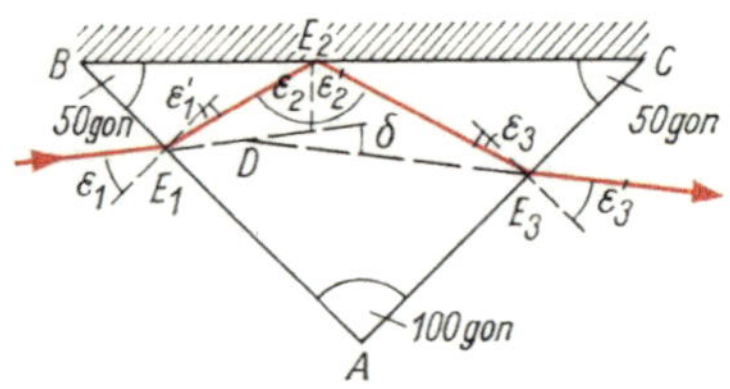

Figure 64 Movable image from a triangle prism

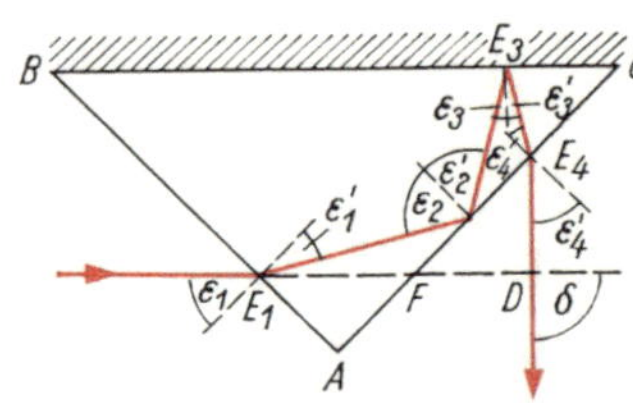

Figure 65 Fixed image from a triangle prism

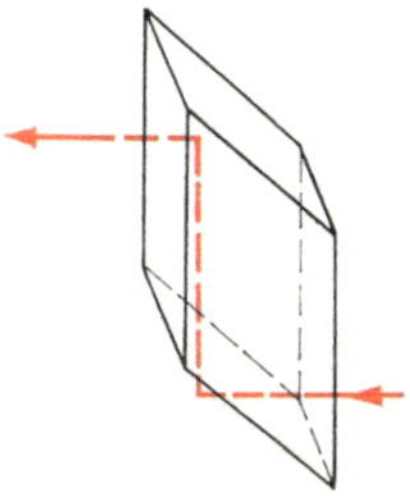

Figure 66 Rhomb-prism

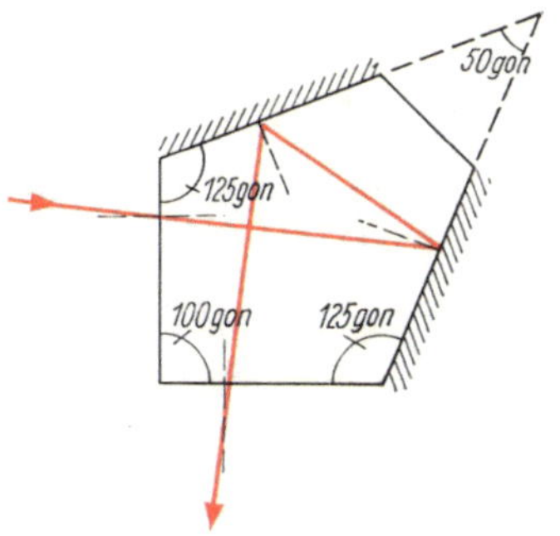

Figure 67 Ray passing through pentagon prism

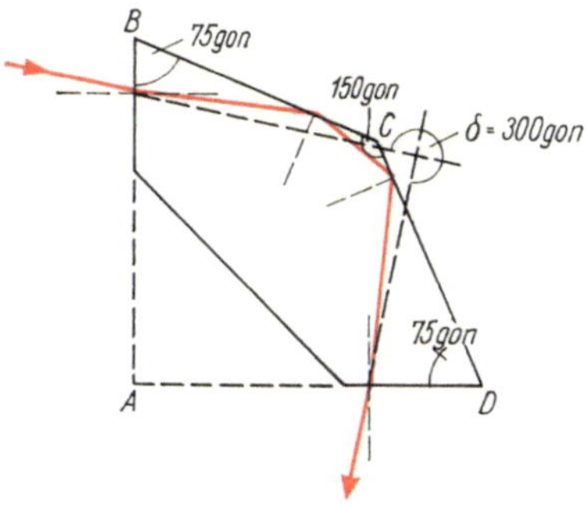

Figure 68 Ray passing through Wollaston prism

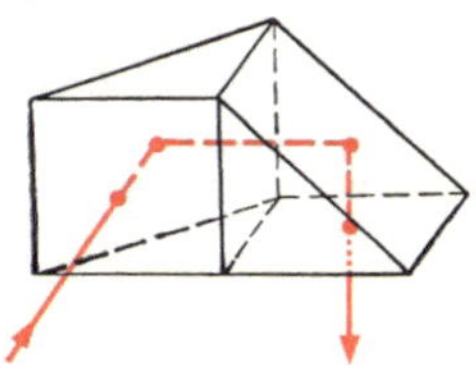

Figure 69 Sphenoid

It should be noted that two reflections take place in this case.
Therefore one has to distinguish two types of images with a triangle prism: Moving images, which wander by rotation of the prism around its vertical plane, and images which remain stationary when rotating the prism. Only the latter are of interest for practical purposes.
With a rhomb prism the light ray can be shifted by the length of its side without directional change (figure 66). It is used for graduated circle readings and others.
The pentagon prism (figure 67) has a quadrilateral cross-section symmetrical to one of its diagonals. It acts like an angle mirror whose mirror surfaces enclose an angle of 50 gon. It thus deflects the ray by 100 gon from its original direction, independent from the incident angle on the prism surface. The pointed prism edge is cut off, which results in its handy form as pentagon prism. Since total reflection does not occur, the reflecting sides are mirror cowated.
The Wollaston prism (figure 68) acts like an angle mirror whose reflecting sides *BC* and *CD* enclose an angle of 150 gon and therefore deflect the ray by 300 gon, according to equation (5). The complementing angle amounts again to 100 gon. For this prism the reflecting surfaces need not to be coated. The rectangular edge is usually cut off, so that it becomes another five sided prism. The pentagon- and Wollaston prisms are utilized as rectangle prisms.
A combination of two rectangular prisms with equal sides, called sphenoid (figure 69) causes a light ray to be not only shifted sideways but also deflected by 100 gon, because of one total reflection in each prism. Sphenoids are used for instance with a combination of other prisms for split bubble readings of a spirit bubble.
Porro's reversal systems serve to reverse images. The first Porro system (figure 70a) which is mainly used in prism field glasses and surveying instruments, consists of two rectangular prisms with equal sides. It can also be used to create an upright and correct image with an astronomical telescope. The second Porro system (figure 70b) is a combination of three rectangle prisms and is used in compensator levels.
Roof prisms serve to simultaneously obtain ray deflection by 100 gon and image reversal. If one of the reflecting surfaces is replaced by a roof edge, triangle-, pentagon- or Wollaston prisms can be made into roof prisms. In surveying instruments, primarily the *Amici* roof prism (figure 71) is utilized, which is a triangle prism with roof edge at its hypothenusis. A ray passing through the prism is twice refracted (at entrance and exit, with usually 0 gon incidence- and refraction angles) and twice totally reflected (at the roof surfaces). The roof prism is used for graduated circle readings and compensator levels.
A triple mirror prism has three prism surfaces normal to each other, thus forming a corner is space (figure 72). Any ray coming from the outside is totally reflected at all three surfaces and returns parallel to the incident direction and symmetrically to the central ray. When turning the triple mirror around any axis the direction of the returned ray will not change.
Triple mirror prisms are used with EDM instruments to return the light from the reflector station.

Lenses

A lens is a body of transparent material, mainly glass, which is bounded by at least one spherical surface. The other surface may be plane or spherical.
A lens can be considered to consist of numerous prisms with the outside ones having a larger reflective angle than the centre ones. By steadily increasing the number of the prisms which become progressively smaller the

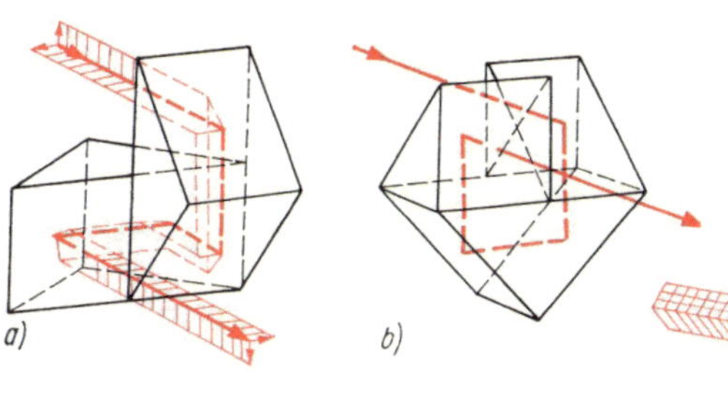

Figure 70 Porro's first (a) and second (b) reversal systems

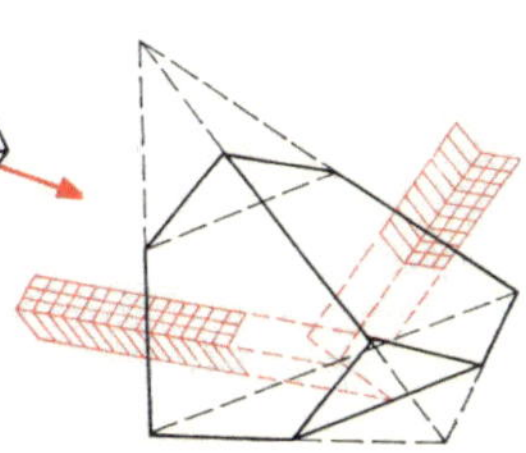

Figure 71 Roof prism by *Amici*

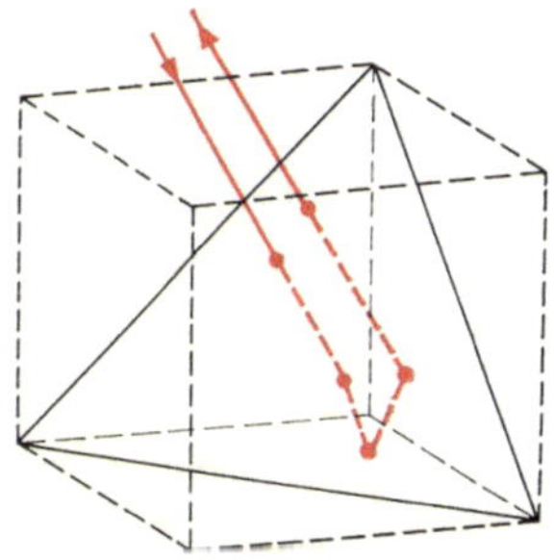

Figure 72 Triple mirror prism

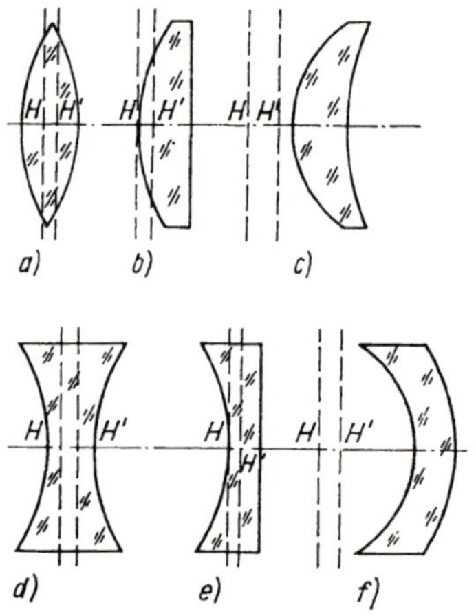

Figure 73 Different lens types:
a) bi-convex; b) plane-convex; c) concave-convex (collecting meniscus); d) bi-concave; e) plane-concave; f) convex-concave (dispersing meniscus)

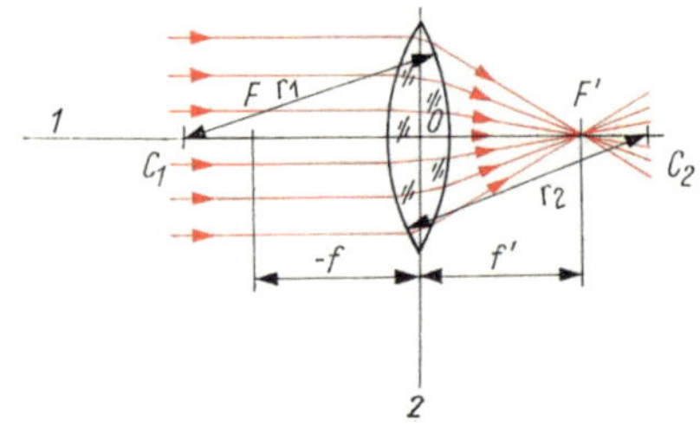

Figure 74 Collecting lens
1 optical axis; *2* central plane

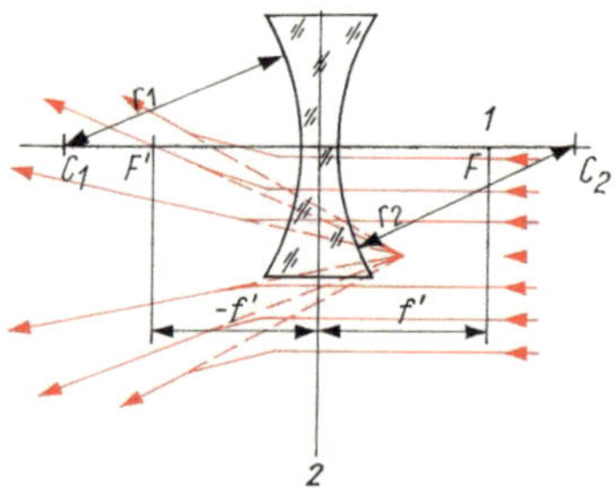

Figure 75 Dispersing lens
1 optical axis; *2* central plane

spherical lens surface is finally obtained. Depending whether the lenses are thicker at the centre or edge, one distinguishes between collecting or dispersing lenses (figure 73). The line connecting the two centres of curvature C_1 and C_2 represents the optical axis of the lens, with r_1 and r_2 being the radii of curvature (figure 74).

Collecting lenses cause convergency of parallel rays (figure 74), dispersing lenses divergency (figure 75). Rays entering the lens parallel and near the optical axis intersect after exiting at the focal point. For the collecting lens, this occurs directly (figure 74), for the dispersing lens virturally at the backwards extension (figure 75). The back focal point F' (image focal point) represents the image of an infinitely distant object point located on the axis (in figure 74 a real image, in figure 75 a virtual one).

Similarly the front focal point F (object focal point) is the axis point who is imaged at infinity.

The distances of the focal points from the central plane of the lens (the thickness of thin lenses can be neglected for most practical purposes) are the image focal length f' and the object focal length f. For thin lenses f and f' can be considered as equal.

A parallel bundle of rays entering the lens inclined to the optical axis converges to a point in the respective focal plane (figure 76). The focal planes are perpendicular to the optical axis and contain the respective focal point.

A ray passing through the centre of a thin lens will not change direction, except for a slight parallel shift.

Mirror lenses are a special form of lenses with one surface mirror coated. Thus refracting and reflecting effects occur simultaneously and supplement each other.

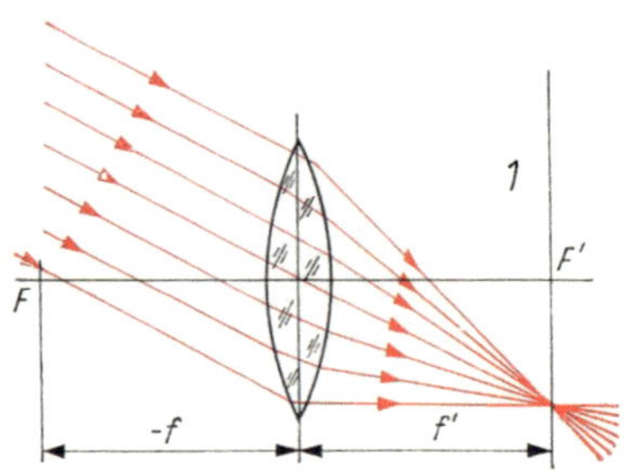

Figure 76 Parallel bundle of rays, inclined to the optical axis
1 focal plane

Imaging with Lenses and Lens Systems

Imaging with Thin Single Lenses. A light ray parallel to the lens axis originating at the object point O_1 passes after refraction through the image focal point of the lens (figure 77). A ray passing through the object focal point is parallel to the lens axis after refraction. The point of intersection is also common for all other rays. In the figure, this point is the image of the arrow head, while y' is the image of the object y.

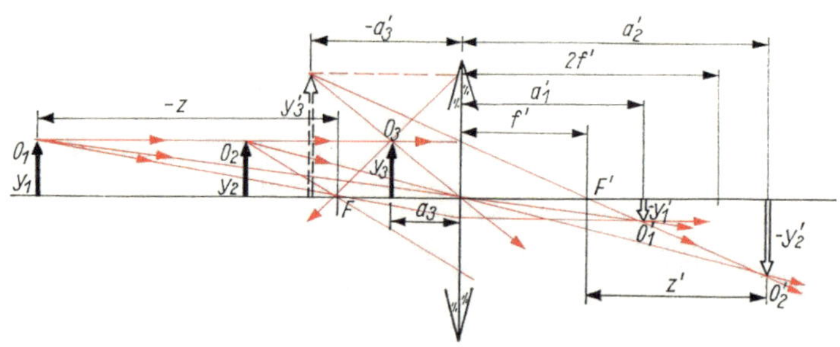

Figure 77 Positions of object and image when imaged with a collecting lens

For the graphical construction of this image point one can also use the ray which passes unchanged through the centre of the lens together with one of these fundamental rays.

Like for the curved mirror, three cases of object location have to be distinguished for imaging which result in different image, locations and sizes.

1. If the object y_1 is located beyond twice the focal length in object space ($a > 2f$), a real, reversed and reduced image located between one and two focal lengths in the image space is obtained.

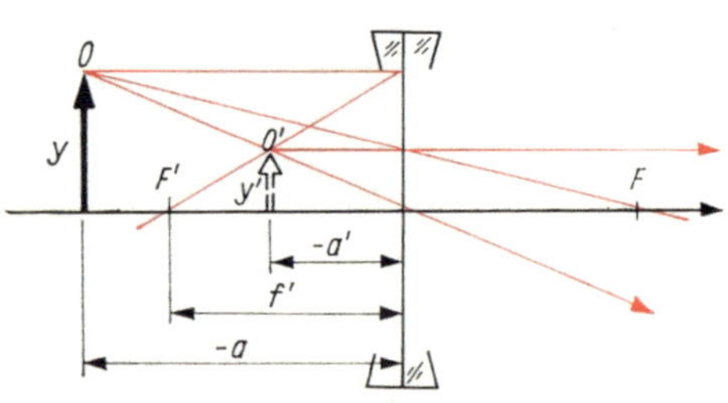

Figure 78 Graphical tracing of an image for a dispersing lens

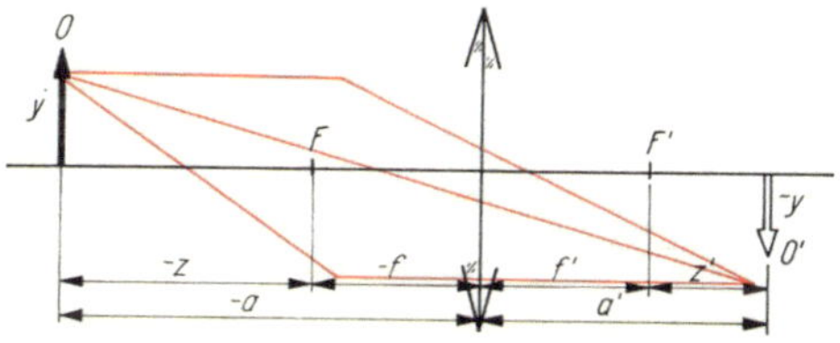

Figure 79 Graphical tracing of an image for a collecting lens

2. If the object y_2 is located between one and two focal lengths in the object space ($2f > a > f$), a real, reversed and enlarged image, located beyond two focal lengths in the image space is obtained.
3. If the object y_3 is located between lens and focal length in the object space ($a < f$), a virtual, upright and enlarged image is obtained in the object space at a distance from the lens larger than that of the object.

The image created by a concave lens is constructed similarly (figure 78). It always is virtual, upright and reduced.

The locations of object and image are marked in reference to the refracting surfaces, i.e. all distances extending from the refracting surface in direction of the light are computed as positive, opposite ones as negative.

Distances perpendicular to the optical axis are considered positive for upwards and negative for downwards.

According to figure 79: $y/-y' = -a/a' = f'/z' = -z/-f$. This leads to Newton's lens equation

$$zz' = ff' = f'^2. \tag{17}$$

This equation can be used to calculate rather than draw the image position, provided the object position and the focal length are known and the signs are obeyed. Similarly, the focal length can be computed if object- and image positions are known.

When manipulating equation (17) with $-z-f = -a$ and $z' + f' = a'$ as well as $z = a + f'$ and $z' = a' - f'$ and dividing it by $aa'f$ the well known lens equation is obtained:

$$\frac{1}{a'} - \frac{1}{a} = \frac{1}{f'} = -\frac{1}{f}. \tag{18}$$

The refractive power of a lens is

$$\frac{1}{f'} = D \tag{19}$$

with f' expressed in metres. D is expressed in dioptrias. A lens with $f' = 20$ cm has $D = 1/0.2 = 5$ dpt.

The imaging scale is the ratio between image size and object size.

$$\beta' = \frac{y'}{y} \tag{20}$$

For $+\beta'$ image and object are similar, for $-\beta'$ the image is reversed. According to figure 79:

$$\beta = \frac{y'}{y} = \frac{f}{z} = \frac{z'}{f'}. \tag{20a}$$

For $z = -f$, β' becomes -1, e.g. image and object are of same size but reversed and are positioned at twice the focal length from the lens. According to equation (20a), the image appears enlarged for $|z| < |f|$ and reduced for $|z| > |f|$.

Imaging with Compound Lenses Consisting of Thin Lenses

Instead of single lenses, compound lenses consisting of single lenses combined to one unit are used. The image of an object with a compound lens is graphically constructed according to the method for single lenses. This may be explained with a system consisting of a collecting lens L_1 and a dispersing lens L_2 separated by e, and having the focal points F_1, F_1', F_2, F_2' (figure 80). All incident rays parallel to the axis converge to the image focal point F' (figure 80a).

A ray parallel to the axis and coming from O is refracted in L_1 towards F_1' and is again refracted in L_2. The ray of the bundle which passes through the centre of the lens intersects the focal plane of F_2' in C. The rays of the bundle, including BF' intersect in C. The point of intersection F' of this ray and the optical axis is therefore the image focal point of the compound lens.

The plane containing the intersection D between incident ray and the extension of BF' is called principal plane, its intersection with the axis is referred to as principal point. The principal plane through H' and the distance $H'F'$ on the image space of a compound lens correspond to central plane and focal length of a thin lens.

In order to find the object focal length, a ray parallel to the axis on the right hand side of the system is traced (figure 80b). It is refracted at E in L_2

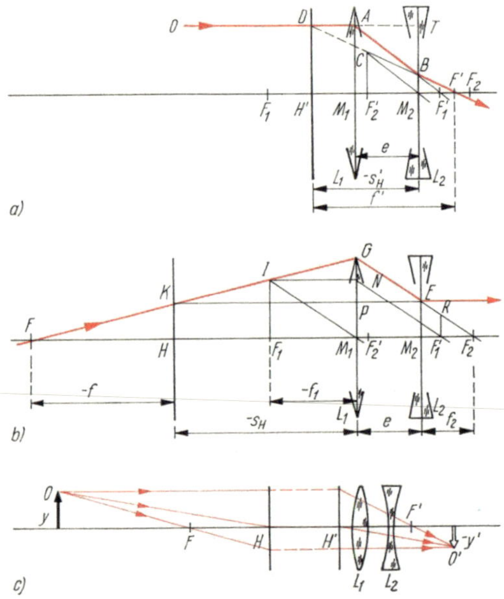

Figure 80 Graphical tracing of cardinal points and image for a lens system
a) of the image focal point and the principal point H'
b) of object focal point and principal point H
c) of image

towards G (extension through the focal point F_2), and in L_1 towards I, and intersects the optical axis in the object focal point F.
Its intersection K with the extension of the axis-parallel ray contains another principal plane which is normal to the optical axis.
The intersection H of this plane with the axis is another principal point of the system. The principal plane through H and the distance HF on the object space of a compound lens correspond to the central plane and the object focal length of a thin lens.
The effect of a compound lens is therefore like that of an equivalent single lens. $HF = -f$ and $H'F' = f'$ are therefore called equivalent focal lengths of the system.
The image of an object through a compound lens can be obtained with the aid of the two focal points F and F' and the two principal points H and H' (figure 80c).
The imaging equations of the single lens also hold for compound lenses. Some auxiliary lines in figure 80 aid in the computation of the distances s_H and s'_H between the principal points and the lenses, as well as for the determination of the equivalent focal lengths f' and $-f$ of a compound lens with focal lengths f'_1, $-f_1$, $-f'_2$, f_2 and separation e. From similar triangles KPG and ING as well as PGE and F'_1RF_2 follows (figure 80b)

$$-s_H/-f_1 = PG/NG = PG/F'_1R \text{ and } PG/F'_1R = e/F'_1F_2.$$

Since $F'_1F_2 = e + f_2 - f'_1 = e - f'_2 - f'_1$,

$PG/F'_1R = e/(e - f'_2 - f'_1)$ and $s_H/f_1 = e/(e - f'_2 - f'_1)$.

With $f_1 = -f'_1$

$$s_H = \frac{ef'_1}{f'_1 + f'_2 - e}. \tag{21}$$

From similar triangles GPK and GM_1F as well as GPE and GM_1F_2 follows $-s_H/(-f - s_H) = e/(f_2 + e)$. With $f = -f'$ and $f_2 = -f'_2$ one obtains $f = s_Hf_2/e$ and $f' = s_Hf'_2/e$. Using s_H from equation (21) the formula for equivalent focal length becomes

$$f' = \frac{f'_1f'_2}{f'_1 + f'_2 - e} \tag{22}$$

or in reciprocal form

$$\frac{1}{f'} = \frac{1}{f'_1} + \frac{1}{f'_2} - \frac{e}{f'_1f'_2}. \tag{23}$$

From similar triangles DBT and $F'M_2B$ as well as ATB and $F'M_2B$ (figure 80a) one obtains $-s'_H/(f' + s'_H) = e/(f'_1 + e)$ and $s'_H = -ef'/f'_1$. Substituting the value of f' from equation (22):

$$s'_H = \frac{-ef'_2}{f'_1 + f'_2 - e}. \tag{24}$$

Equations (21) to (24) derived here for a compound lens consisting of one collecting- and one dispersing lens are generally valid for a compound lens consisting of two single lenses (duplet).

Example: The compound lens shown in figure 80 may have the following data:

$f_1' = 185.5$ mm, $f_2' = -60$ mm, $e = 150$ mm.

The equivalent focal length is according to equation (22):

$$f' = \frac{f_1' f_2'}{f_1' + f_2' - e} = \frac{185.5(-60)\text{mm} \cdot \text{mm}}{185.5\text{ mm} - 60\text{ mm} - 150\text{ mm}} = 454.3\text{ mm}.$$

The principal point separations s_H and s_H' are computed using equations (21) and (24):

$$s_H = \frac{e f_1'}{f_1' + f_2' - e} = \frac{150 \cdot 185.5\text{ mm} \cdot \text{mm}}{185.5\text{ mm} - 60\text{ mm} - 150\text{ mm}} = -1136.0\text{ mm}$$

$$s_H' = \frac{-e f_2'}{f_1' + f_2' - e} = \frac{(-150)(-60)\text{ mm} \cdot \text{mm}}{185.5\text{ mm} - 60\text{ mm} - 150\text{ mm}} = -367.4\text{ mm}.$$

The separation d of the two principal planes amounts to

$d = -s_H + s_H' + e = 918.6$ mm.

For practical purposes, the distance of the focal point F' from the front lens is important, because it determines the length of the telescope which has such a compound lens, called telelens. In the case at hand this distance is $f' + s_H' + e = 236.9$ mm

If a single convex length with equal focal length ($f' = 454$ mm) would have been used instead of the compound lens, the telescope would be longer by nearly 220 mm.

Telelenses are used in modern theodolites and levels, because they shorten the telescopes significantly.

If the focal points F_1' and F_2 coincide, objects at infinity are imaged at infinity. Such a system is called telescopic or afocal.

Thick Lenses. The derivations so far are only valid for very (more accurately: infinitely) thin lenses. In thick lenses, the central plane is replaced by two principal planes like for compound lenses. Principal planes are normal to the optical axis and contain the intersection of the extension of axis-parallel rays with rays emerging from the lens and passing through the focal point (figure 81). In figure 73 the locations of the principal planes for the most important types of lenses are shown. Figure 81 illustrates the graphical construction of the cardinal elements which are needed for the graphical construction of the image (figure 80c). Equations (21) to (24), developed for compound lenses also hold for thick lenses, provided the optical path length of the distances within glass is considered, namely the focal length f_1' has to be replaced by f_1'/n, and the separation e of the principal planes by e/n.

The focal length f' and the separations of the principal planes s_H and s_H' from the lens vertices S_1 and S_2 can often be easier computed with the radii of curvature r_1 and r_2 for the thick lens and the refractive index n of the particular glass. Without deriving it, the final formulae may be given as

$$f' = \frac{n r_1 r_2}{(n-1)[n(r_2 - r_1) + (n-1)e]} \qquad (25)$$

$$s_H = \frac{e r_1}{n(r_1 - r_2) - (n-1)e} \qquad (26)$$

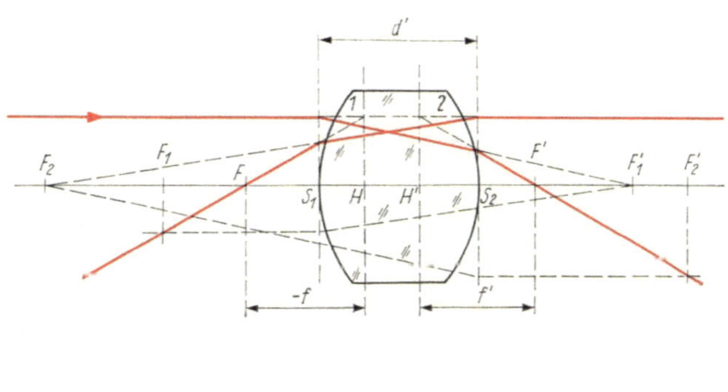

Figure 81 Graphical tracing of the cardinal points of a thick lens
1 surface *1*; *2* surface *2*

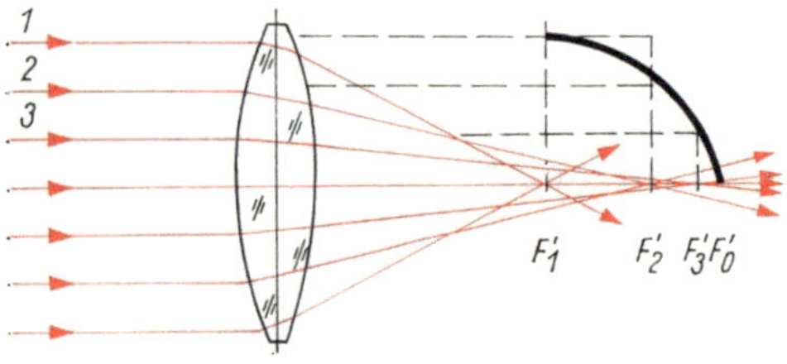

Figure 82 Spherical aberration

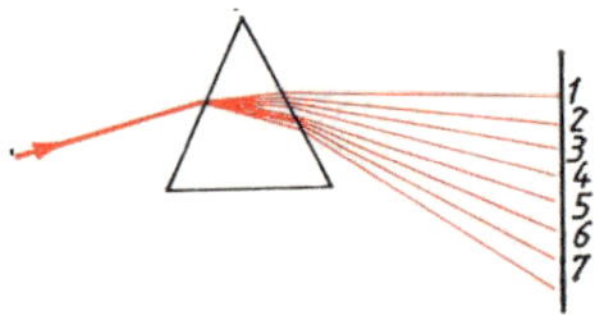

Figure 83 Dispersion
1 red; *2* orange; *3* yellow; *4* green; *5* blue; *6* ultra marine; *7* violet

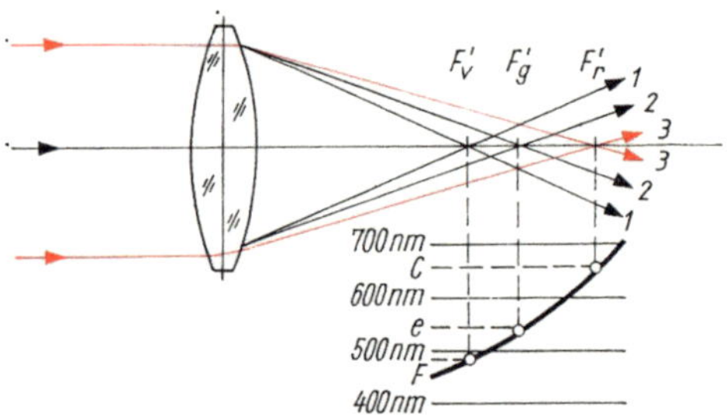

Figure 84 Chromatic aberration
1 violet; *2* green; *3* red

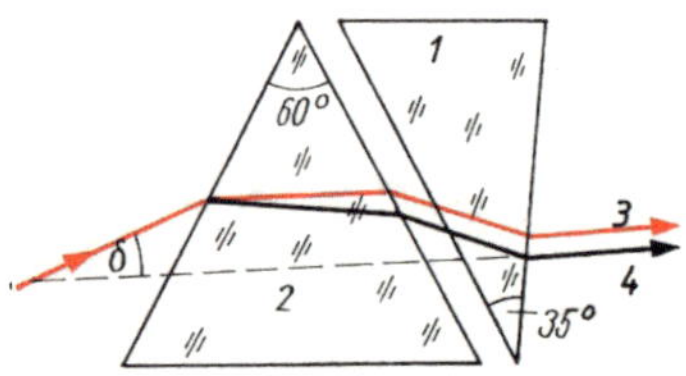

Figure 85 Achromatic prism
1 flint glass; *2* crown glass; *3* red; *4* violet

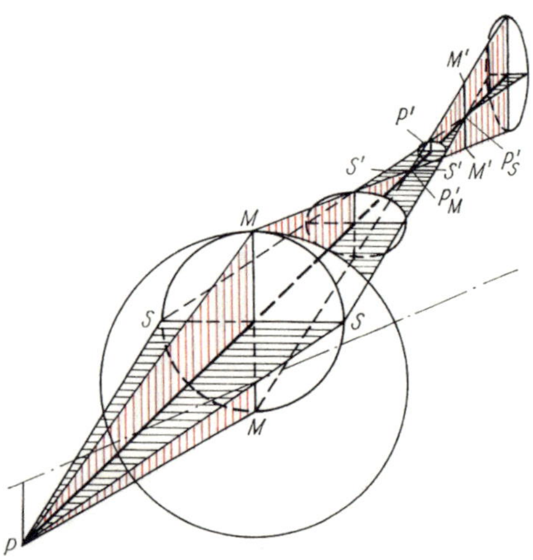

Figure 86 Astigmatism

$$s'_H = \frac{e r_2}{n(r_1 - r_2) - (n - 1)\,e}. \qquad (27)$$

When combining two equal-axis thick lenses in air, equations (21) to (24) are applicable. Then e represents the separation of the image principal point of the first lens from the object principal point of the second one, s_H represents the separation of the object principal point of the system from the one for the first lens, and s'_H the separation of the image principal point of the system from the one of the second lens.

Imaging Errors of the Lenses and Corrective Measures

When considering lenses and lens systems, the imaging bundle of rays was assumed to pass close to the axis (in the paraxial region). This assumption does not hold for practical situations, which leads to imaging errors (aberrations). The rays do not converge to a point but to circles of confusion, which creates an unsharp image. The spherical and chromatic aberrations are the most important ones.

Spherical Aberration. It can be proven experimentally, that the rays hitting the edge of a lens intersect at a point which differs from the point of intersection for the central rays. The principal point of the outside lens zones is closer towards the lens than for the central part of the lens (figure 82). Thus an object point is imaged as a circle of confusion rather than as a point. The spherical aberration can be reduced by using several positive lenses instead of one, or by achromatic objective lenses (see figure 102) by combining a convex with a concave lens.

Chromatic Aberration. Light of different wave lengths is refracted differently when passing from one medium to another with different refractive index. With increased wave length the refractive index of glass becomes smaller. It is therefore less for red light with long wave length than for violet light with short wave length.
Since white sunlight is a mixture of electromagnetic waves with different wave lengths, a ray falling onto a prism is spread like a fan and forms a coloured band, the spectrum (figure 83).
This dispersion characteristic results for optical systems in spherical aberration: All objects imaged with single lenses have coloured edges, since the violet rays of a parallel bundle intersect in a point closer to the lens than the red rays, which creates circles of confusion for the differt colours (figure 84). The distance between the intersection points of the rays depend on the wavelength, or the colour of light. If it is plotted in respect to the wavelength of the parallel bundle of rays, one obtains a curve for chromatic (longitudinal) aberration. Due to chromatic aberration, the focal length of single thin lenses can be reduced by up to $^1/_{20}$th.
To obtain an achromatic system (free of colour errors) two prisms of flint glass and crown glass with a prism angle ratio of approximately 1 : 2 are connected together with opposite refracting edges. In flint glass the dispersion and also the length of the spectrum is twice as large as for crown glass, which gives an achromatic prism, that deflects light rays practically without colour dispersion (figure 85).
An achromatic lens system (achromat is a compound lens of one lens each of flint- and crown glass, figure 102) is nearly free of chromatic aberration. It can be completely free of it only for two spectral colours, e.g. for red and blue. For the other colours, remaining chromatic errors exist, called the secondary spectrum. One can see it for example when one observes with a telescope a small dark object against the light sky. Out of focus it appears either purple or green depending on whether the focussing distance is too long or too short. It is the task of the numerical optician to reduce the chromatic errors as much as possible for a specific purpose by properly combining lenses. Systems with reduced secondary spectrum are referred to as apochromatic.

Astigmatism and Curvature of Field. Light rays originating far away from the axis are refracted less in the meridian plane of the object point and more in the sagittal—(or equator) plane, normal to it. Therefore they do not converge in one point but in a horizontal and vertical line (figure 86). Because of astigmatism a plane object surface is imaged in respect to the axis point in two curved image shells, the meridional and the sagittal one—curvature of field.

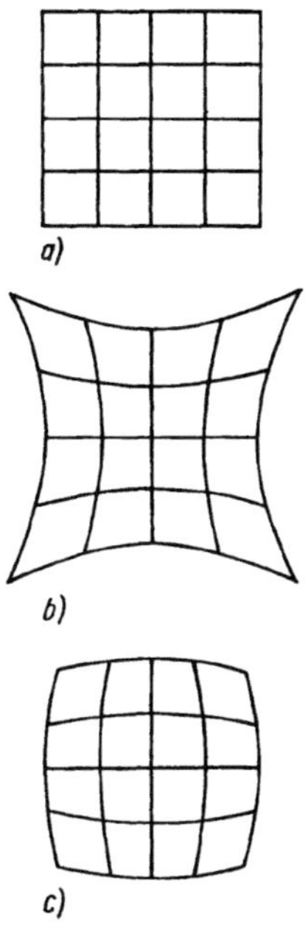

Figure 87 Distortion
a) distortion free image
b) pillow shaped distortion
c) barrel shaped distortion

Astigmatism and curvature of field are corrected by proper selection of radii of curvature, of thickness and refractive index of the lenses as well as of their separation. Such systems are referred to as anastigmates.

Coma and Distortion. When imaging a point far away from the axis with the aid of a bundle of rays with finite opening, coma is added to astigmatism. This is a figure similar to the tail of a comet and appears instead of an image point as a result of spherical aberration with inclined bundles of rays.
To get rid of coma, the optical system is computed such that the sine condition

$$\frac{y'}{y} = \beta' = \frac{n \sin \sigma}{n' \sin \sigma'} = \text{const.} \tag{28}$$

is fulfilled. This condition was found by *Ernst Abbe* (1840–1905) for the image scale [equation (20)], which is the ratio between image size y' and object size y.
With surveying telescopes the observed object is practically at infinity. This simplifies equation (28) to

$$f' = \frac{h}{\sin \sigma'} \tag{29}$$

where h represents the height of entry of an axis parallel ray, and σ' represents the angle between ray and axis.
An optical system, which is corrected for spherical aberration and which fulfils the sine condition is referred to as aplanatic.
Distortion occurs when the imaging process does not maintain a uniform scale, e.g. when image scale β', telescope magnification Γ_F or effective focal length change with imaging angle or image height. A square grid is distorted pillow shaped by positive distortion and barrel shaped by negative distortion (figure 87).
For surveying telescopes, distortion is not critical. It is corrected by combining lenses with different radii of curvature, refractive indices and thicknesses. Distortion free systems are referred to as orthoscopic.

Lens Manufacturing

Lenses are produced from optical raw glass. Such glass is composed differently than normal everyday glass and contains additives besides the usual materials (quarz sand, soda or potash and lime), which have to be very pure. With these additives the refractive index and the dispersion of the glass can be influenced.
Optical glasses are categorized with Abbe's number ν, which relates the refractive indices of the glass to certain wavelengths of light. This number is the reciprocal value of the relative dispersion:

$$\nu = \frac{n_d - 1}{n_F - n_C} \tag{30}$$

n_C, n_d, n_F are the refractive indices corresponding to the Fraunhofer lines C, d and F of the solar spectrum. The somewhat longer wave length of the d-line signifies the "mean refraction" n_d, while the difference of the refractive indices for the F-line (blue line of the hydrogen spectrum) and for the C-line (red spectral line of hydrogen), between which the optically lightest part of the spectrum is located, signifies the mean colour dispersion. For most glasses, n_d is between 1.5 and 1.65.
Abbe's number ranges from 10 to 100, for technical glasses between 20 and 70. Large ν-values point to a small colour dispersion, and vice versa.
Optical glass is distinguished by complete homogenity in large volumes. To accomplish that, the melting mixture in the glasspot is stirred and purified mechanically, then cast on casting tables and cooled in cooling ovens such that internal tensions completely disappear. The main types of optical glasses are the lead free crown glasses with low mean refractive index and small colour dispersion ($\nu > 55$), and the leaded flint glasses with high refractive index and large colour dispersion ($\nu < 50$).
A glass block is first cut into square plates with circular saws equipped with diamond splinters or -dust and cooled with an oil kerosene mixture. The plates are then cemented together and pressed to form a column. Afterwards they are cut with diamond tools into round disks similar in size to the desired lenses. The lens edge (facette) is ground centrically. For fine grinding the disks are cemented onto a mushroom for convex lenses and

a metal bowl for concave ones. The mushroom rotates around its axis on a grinding machine. The disk rests on it lightly, weighted with lead and undergoes swinging or rotating movements stimulated by eccenters. Sanding material is inserted between disk and mushroom. In the first process, a few tenths of a millimetre are ground away in about 30 minutes with rough grain while in the third and last process, a few hundredths of a millimetre are sanded away in the same time with extra fine grain (sanding compound # 11). This gives a silky matt surface. Afterwards the lenses are polished for several hours using polishing disks lined with felt or pitch and a specially prepared iron oxyde as polishing chemical. Then they are cemented onto the end of the spindle of centering machines, or centering automats for centering and sanding of the cylindrical lens edges.
When used in compound lenses, the lenses are either combined free with air separation or else cemented together. This is done either with polymere plastic cements or else with Canadian balsam. The latter is transparent terpentine, with excellent refractive behaviour, obtained primarily in Canada from blazed balsam firs. It remains clear after hardening. Glass surfaces with equal radii can be connected by blasting them together without cement (optical contact) and thus remove all air between them. Another process is the coating with reflection reducing films. All these processes are subject to strict quality control.
Lenses or compound lenses are placed in cylindrical metal casings and held at the opposite end with either a screw ring or a tension ring or the bent-in edge of the cylinder. The fit between lens and cylindrical casing has an accuracy of about 0.1 mm for objective lenses, and 0.01 mm for more sensitive optical elements.

Diaphragms

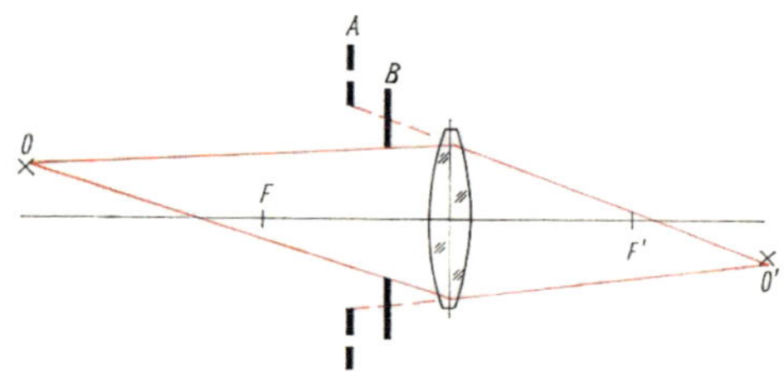

Figure 88 Effect of a diaphragm

To reduce the influence of imaging errors, the outside part of the bundle of rays is screened off.
A diaphragm B in front of a lens (figure 88) limits the aperture angle of a bundle of rays originating at point O and creating point O'. If the diaphragm is inside the focal length of the lens, a virtual image A of it is obtained, which determines the conical bundle of rays originating at O'. A diaphragm behind the lens limits the bundle of rays originating at O', while its virtual image has the same effect on the one from O. In both cases the same effect has been accomplished. It is therefore immaterial whether the bundle of rays originating at the object is bounded by the diaphragm or its virtual image. The diaphragm determining the opening cone is called aperture stop, its image in object space is the entrance pupil (*EP*), its image in image space the exit pupil (*AP*). *EP* and *AP* react to the system like object and image. For a telescope the location of the diaphragm is important.
If there are several diaphragms, the diaphragm or its image who appears under the smallest angle from O forms the entrance pupil. The exit pupil is the diaphragm (or its image) appearing from O' under the smallest angle. Even lens casings can serve as aperture stops and represent then the entrance- and exit pupils.
While the aperture stop determines the illumination of the image, a diaphragm in the image plane of an optical system limits the field of view in object and image space. It is therefore called field stop and screens off the rays which are too much inclined to the axis.
Sun shields have no particular influence on the imaging process. They can be attached to the telescope to screen off the light of intense sources outside the field of view.

Eye

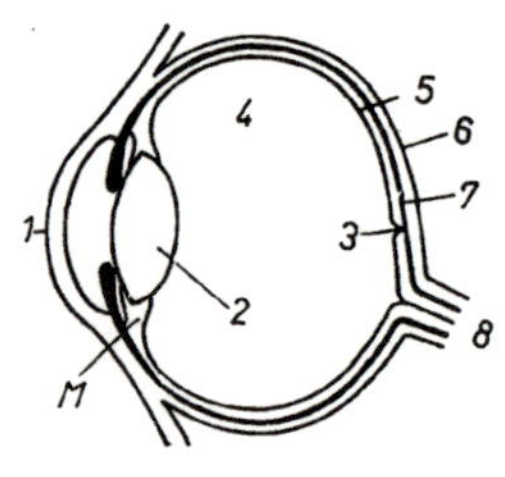

Figure 89 Longitudinal section through the human eye
1 cornea; *2* crystal lens; *3* retina depression; *4* glass body; *5* retina; *6* horny skin; *7* veiny skin; *8* visual nerve

The eye (figure 89) is a spherical transparent body, covered in tough white leather skin as protection against injuries. Its front part is also transparent and referred to as cornea. Below it is the veiny skin which contains the blood vessels to nourish the eye. It is dark to protect the interior of the eye from diffuse light. The front part of the veiny skin, the iris has a hole, the pupil. The eye lens, consisting of a horntype transparent material with a refractive index of $n \approx 1.41$ throws real reduced and reversed images of the viewed object onto the retina. These images enter consciousness as upright. Focussing on objects at different distances (accomodation) is accomplished by changes in curvature and partly in position of the lens.

The light intensity entering the eye is monitored by the retina by changing the pupil size, thus protecting itself from over-exposure. The pupil has a diameter of about 2 mm in bright daylight, and of about 8 mm for poor illumination.
A normal seeing eye images infinitely far objects on the retina without accommodation. The far point is therefore at infinity. The closest point, which can be accommodated, is the near point. A young normal seeing person can see objects sharply at 10 cm distance. When accommodating real short distances, the muscle M who deforms the lens is so strained, that it aches. The object distances which can be accommodated without strain is about 25 cm for normal eyes and is called clear sight distance.
The causes of abnormal sight are either that the axis of the eye is too long or too short or that the eye lens has a different refractive index than normal. When the rays intersect behind the retina, far sightedness occurs, while for short sightedness they intersect in front of it.
By placing a collecting or dispersing lens in front of the eye, one can accomplish that the rays intersect at the retina and give a sharp image.
The retina consists of the branches and ends of the visual nerve. As light sensitive part, it has a mosaic type layer consisting of a large number of light- and colour sensitive little pegs and of light sensitive but colour insensitive little rods. The little pegs transmit the sharp sight when the object is illuminated normally. The largest amount of little pegs is located opposite the pupil in the retina depression. Here the retina is most sensitive and the sighting sharpest. It decreases rapidly towards the edge of the retina. The diameter of the little pegs and rods is about 1.5 µm in the centre of the retina, and 5 µm at its edge.
If only one little peg is excited when seeing, the object is recorded by the eye as a point. Two objects can only then be separated, if separation of their images on the retina is larger than the diameter of one little peg (figure 90). Thus the limiting angle for separation and resolution of the eye is, depending on its quality, 12 to 18 mgon. For computations one utilizes as unit of eye resolution the round value of 18.5 mgon (= 1 minute of arc). It corresponds to a separation of about 5 µm of the little pegs at the retina. The best resolution is obtained for a pupil diameter of 3 to 4 mm.
A shift between two straight lines however, can be resolved to about 3 mgon, even 1 mgon for some persons. Thus the "vernier sharpness" is six times better than a point resolution, because the images of the lines excite several little pegs and are such easily registered by the eye (figure 90). For graduation readings intervals of 1.0 to 1.5 mm with line thickness of 0.1 to 0.15 mm are most suitable, provided the line lengths are about 1.6 times the intervals.

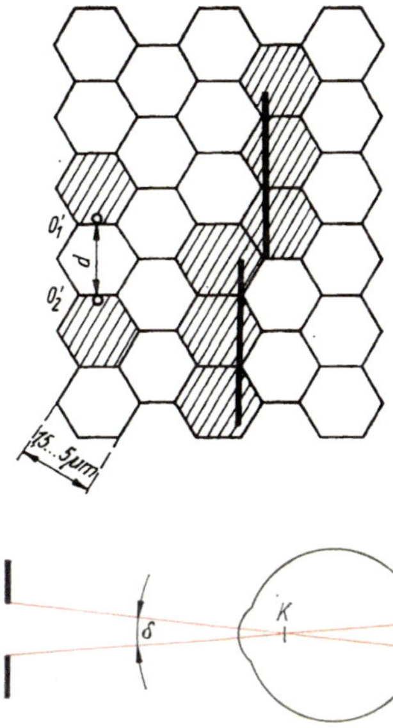

Figure 90 Resolution of the eye

1.1.2. Optical Instruments

Due to the clarity of the imaging process absolute values without signs for directions are given for subsequent formulae for magnifying glass, microscope etc.

Magnifying Glass

The magnifying glass consists of either a single collecting lens with small focal length or else of a compound lens.
If the magnifying glass is placed such that the object to be viewed is within the focal distance near the focal point, an upright enlarged and virtual image is obtained.
The eye needs to accommodate. To avoid tiring of the eye because of continuous accommodation when using a magnifying glass, it is placed such between object and eye, that all rays passing from the object through it emerge as parallel rays. This is the case if the object is located at the object focal length (figure 93). The image created by the magnifying glass is then at infinity. Often unexperienced observers accommodate unnecessarily when working with a magnifying glass (also with the telescope and the microscope). For best effects, all optical instruments should be used without accommodation of the eye.
The magnification Γ_L of the magnifying glass (with relaxed, non accommodated eye) is the ratio of the angle σ' under which the object appears with the magnifying glass, and the angle σ, as it is seen with the naked eye at the clear sigth distance s (figure 91): $\Gamma_L = \tan\sigma'/\tan\sigma$. Since $\tan\sigma = y/s$ and $\tan\sigma' = y/f'$,

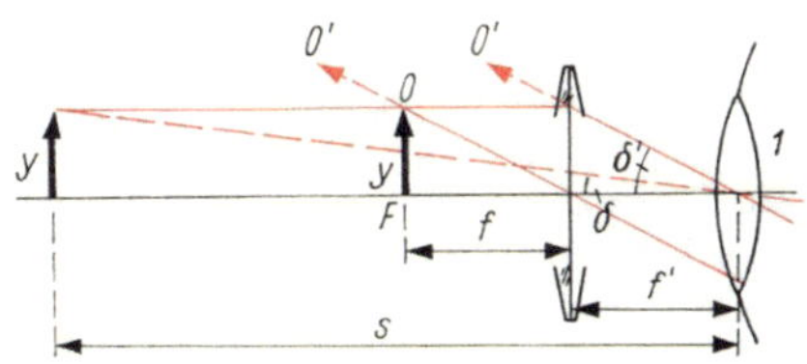

Figure 91 Magnification with magnifying glass
1 eye

$$\Gamma_L = \frac{s}{f'}. \tag{31}$$

With $s = 250$ mm the magnification of the magnifying glass becomes

$$\Gamma_L = \frac{250}{f'}. \tag{32a}$$

In practical use, the eye is not normally accommodated to infinity, but to short distances. The user magnification Γ'_L is then with $s = 250$ mm

$$\Gamma'_L = \frac{250}{f'} + 1. \tag{32b}$$

Satisfying images can be obtained with a bi-convex magnifying glass for up to 5 times magnification, and with a plan-convex magnifying glass for up to 14 times magnification. For surveying graduation readings, magnifying glasses are selected such, that the image of a graduation segment appears to be 1.5 to 2 mm. The field of view of a magnifying glass is 30 gon.

Microscope

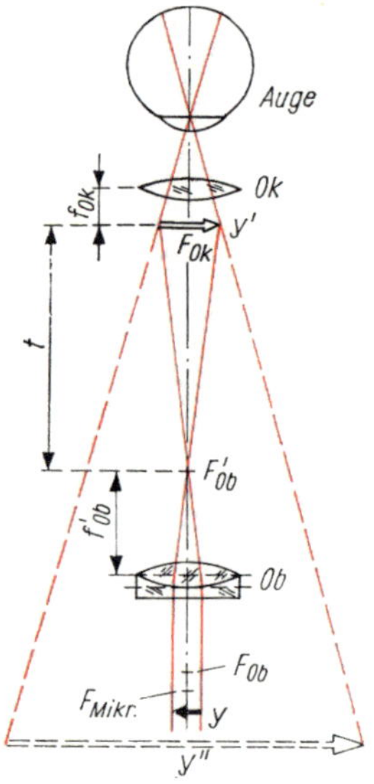

Figure 92 Optical train in microscope

Reading devices of modern surveying instruments contain microscopes with a much higher magnification than possible with a magnifying glass. A microscope consists of an objective (*Ob*) lens and an eye piece (*Ok*), both collecting lenses (figure 92). The objective lens with short focal length generates an enlarged, reversed, and real image y' of the object y beyond twice its focal length, provided the object is located between one and two focal lengths. This image y' is in the object focal plane of the eye piece, which acts as magnifying glass. Thus the microscope gives a reversed, real and extremely enlarged image y''.
The separation t of the matched focal points F'_{Ob} and F_{Ok} is called tubal length. The combined magnification results from the magnification of the objective lens (imaging scale) plus the magnification of the eye piece according to equation (20).

With $\beta'_{Ob} = y'/y = t/f'_{Ob}$ and $\beta'_{Ok} = y''/y' = 250/f_{Ok}$ one obtains

$$\Gamma_M = \frac{y''}{y} = \frac{y''}{y'}\,\frac{y'}{y} = \beta'_{Ob}\,\beta'_{Ok} = \frac{250t}{f_{Ok}\,f'_{Ob}}. \tag{33}$$

Reading microscopes in surveying instruments do not have more than 50 times magnification and not more than three times objective lens magnification. This relatively small magnification is sufficient to clearly view intervals of a few hundredths of a millimetre and thus permits a simple construction of the microscope.
The combined focal length of the microscope is according to equation (22)

$$f'_M = \frac{f'_{Ob}\,f'_{Ok}}{f'_{Ob} + f'_{Ok} - e} = \frac{f'_{Ob}\,f'_{Ok}}{t}. \tag{34}$$

Equation (34) indicates, that for large tubal length the combined focal length can be obtained much smaller than the focal lengths of the single components.
The distance between object y and focal point of the objective lens can be determined with Newton's equation (17) using $z' = t$:

$$z = \frac{f'^2}{t}. \tag{35}$$

In microscopes for surveying instruments, a measuring mark, graduation or a similar device is placed at the location of the real image.

Simple Astronomic Telescope

There are several basic types of lens telescopes, namely

- the astronomic telescope
- the Dutch or Galilei telescope, consisting of a collecting lens as objective and a dispersing lens as eye piece
- the terrestrial telescope with collecting lenses for both objective and eye piece and a reversal system with lenses in between
- the telescope with prism reversal system.

With a few exceptions, only the astronomic telescope (figure 93) is used in surveying instruments. It consists of a collecting objective lens with large focal length, and an eye piece with small focal length.

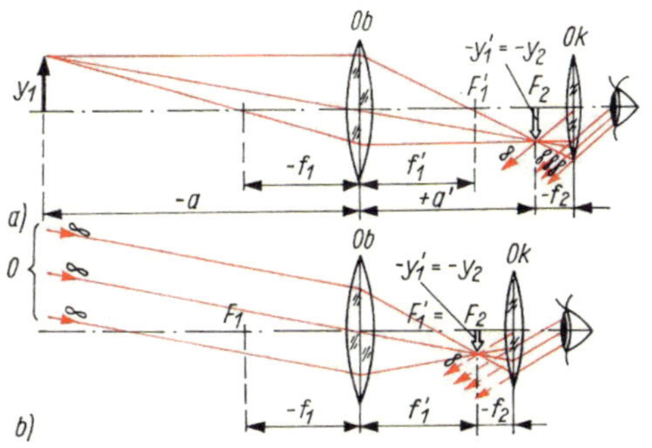

Figure 93 Rays passing through simple astronomic telescope

The objective lens (*Ob*) generates a real reduced and reversed image $-y_1'$ of the object y_1. Since y_1 is always located beyond twice the focal length of objective lens, its image $-y_1'$ is near the back focal point F_1'. This image is viewed with the eye piece (*Ok*) like with a magnifying glass. Thus of this image ($-y_1' = -y_2$) an enlarged, virtual but reversed image $-y_2'$ is generated at infinity.
For short distances, F_1' and F_2 are close together (figure 93a). If a normal seeing person accommodates to a large distance (practically infinity) the image of a distant object falls in the focal plane common to objective lens and eye piece (figure 93b). The image focal point F_1' of the objective lens coincides with the object focal point F_2 of the eye piece (telescopic arrangement).

Telescopes of Surveying Instruments

Telescopes for surveying instruments, also called aiming- or measuring telescopes, contain attachments for aiming and for reduction of imaging errors. The designer tries to fulfill the following demands: short telescope, large magnification, i.e. long focal length, large entrance pupil, sharp image, few lenses. For rough sighting, notch and bead or a small collimator on the telescope casing is used.

Cross Hairs, Diaphragms

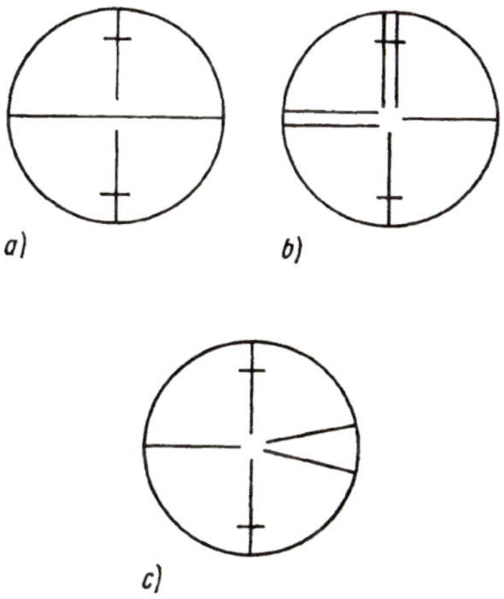

Figure 94 Different types of cross hairs (c: for precision level)

For aiming to the target, the telescope contains a collimating axis defined by two fixed points. These are the optical centre of the objective lens and a crosshair of fine lines (1 to 2 μm). The latter is either engraved into a parallel glass plate or photographically edged and subsequently blackened. Years ago spider threads or hair was used, which led to the name "cross hairs". The cross hair plate is protected against dust by a glass coverplate cemented against it. Dust or water drops would appear much enlarged in the eye piece and thus disturb the telescope image. In figure 94 some of the common cross hair plates are illustrated.
The cross hair has to appear sharp together with the sighted object. It therefore has to be located in the plane of the real image, generated by the objective lens. The cross hair plate is therefore positioned in the front focal plane of the eye piece. It is held by a ring (figure 95), which restricts the incoming cone of rays and thus determines the field of view—it acts as a diaphragm. This ring can be somewhat shifted with the aid of adjustment screws. Since the centre of the cross hair partly defines the collimation axis, its direction can be changed by such a shift.
The casing of the front element of the objective lens represents the entrance diaphragm, it is therefore the entrance pupil. The exit pupil is the real image of the objective lens casing as generated by the eye piece. It appears in the area of the rear focal point of the eyepiece as a light circle ("Ramsden Circle") and can be intercepted and measured with a scale.

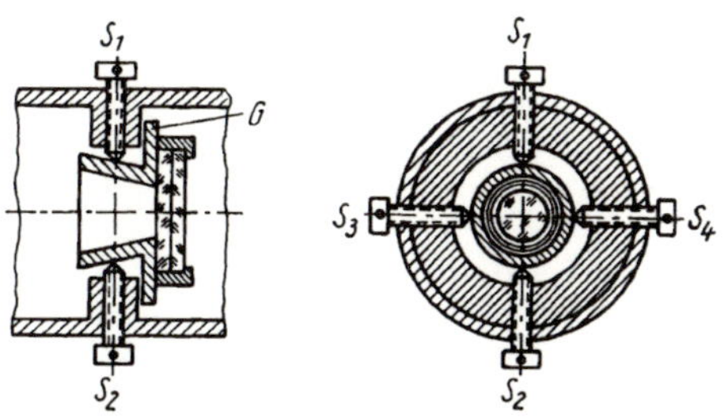

Figure 95 Adjustment screws for reticule

Eye Piece—and Focussing Motions

Since the image distance changes with the object distance, the separation between the rear focal point F_{Ob}' of the objective lens and the front focal point F_{Ok} of the eye piece (location of the cross hair) has to change as well. Thus the distance between objective lens and eye piece has to increase for closer objects. This change of the telescope length can be accomplished with an eye piece motion (objective motions are hardly ever used). With knob and cog wheel arrangement the eye piece can be moved along the longitudinal axis of the main telescope (figure 96). For triangulation theodolites, this may be a free hand motion because of all targets being at infinity. The cross hairs get thus to the location where the image appears. Both main telescope and eye piece tube are blackened inside to avoid disturbing reflections from the metal. The length of the eye piece tube determines the shortest object distance.
Up to a few decades ago, eye piece motion was exclusively used in surveying telescopes. Now it has been superseeded by the interior focussing lens, positioned inside the telescope (figure 97). A telescope with interior focussing has constant length, is dust free, and can be shortened significantly by proper selection of focal lengths. This is a big improvement over the eye piece shift, which also became loose with use.

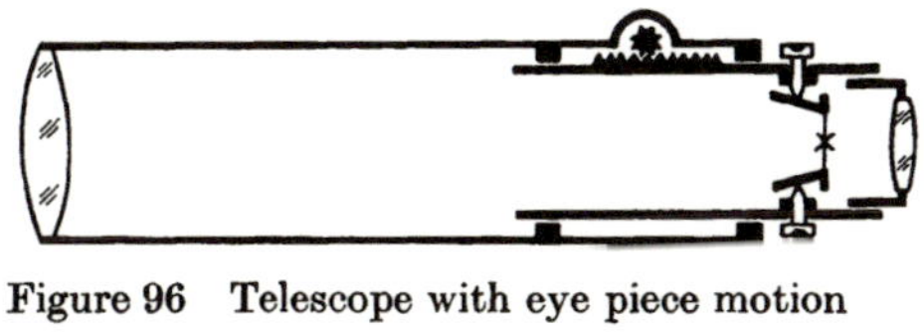
Figure 96 Telescope with eye piece motion

The usually slightly negative focussing lens is counted as rear elements of the objective lens. Thus, it is built like a tele lens and has a much shorter focal length than a regular telescope (see figure 80). When computing such systems, one has to consider that each of the two lens elements has two principal planes separated by 2 to 3 mm and that the focal length is to count from them.

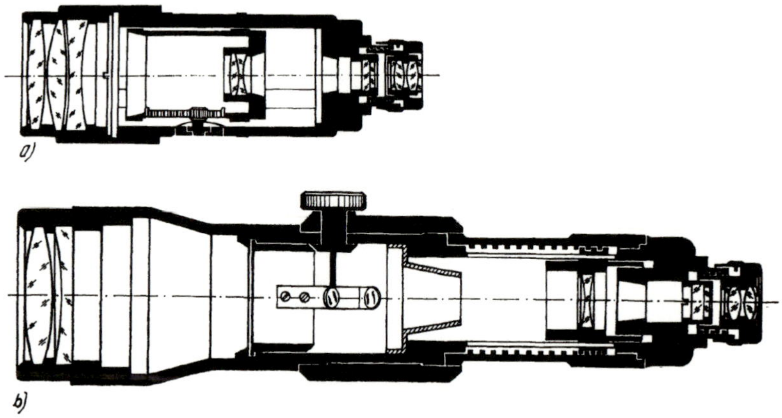

Figure 97 Section through telescope with focussing motion. Operation for focussing motion by turning a) a knob at the telescope standard
b) part of the telescope housing the knob shown is used to set a small mirror within the telescope for artificial illumination of the cross hair

The focussing lens is moved along the longitudinal axis of the telescope by turning part of the telescope tube or with a knob (figure 97). When the lens is properly positioned, the light rays are refracted such that the real image of the object falls into the front focal plane of the eye piece. For closer objects, it has to be shifted more towards the eye piece, while for distant objects, it is near the objective lens (figure 98). Its separation e from the principal plane of the front objective lens element can be determined to

$$e = \frac{z+d}{2} - \sqrt{z\left(f_z - \frac{d}{2} + \frac{z}{4}\right) + d\left(\frac{d}{4} - f_z\right)}, \qquad (36)$$

where s is the object distance, d the separation of the cross hair plate from the principal plane of the front objective lens element, and f_{Ob} and f_z the focal lengths of both objective lens elements with $z = f_{Ob}/(1 - f_{Ob}/s)$.

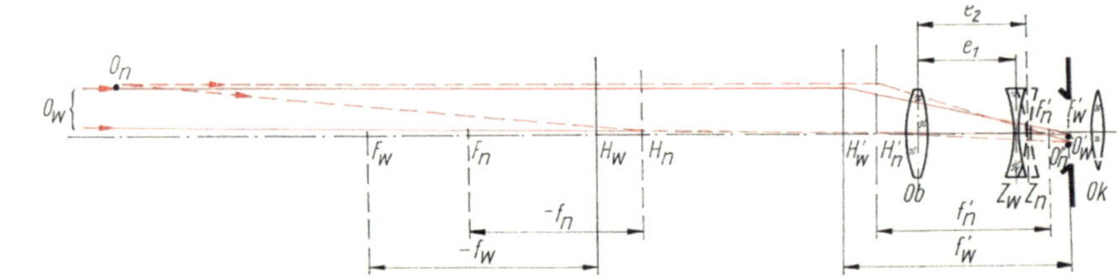

Figure 98 Imaging of a distant object (O_w) and a close object (O_n) in the cross hair plane by shifting of the focussing lens
Z (image was plotted using the principal plane)

Very rarely a convex lens is used instead of a concave one, which results in a focal length shorter than the one for the front element. For close range, the positive focussing lens would have to be moved close to the front objective element.
Some compensation levels (see figures 434, 457, 464, 470) have a fixed negative and a movable positive lens between objective lens and cross hair plate. With this arrangement, the length of the telescope is only slightly shorter than for an old fashioned telescope, which provides space in front of the cross hair for inserting the compensator. This is accomplished without giving up the advantages of a telescope with negative focussing lens. In some cases, compensation levels are focussed by shifting the cross hair or a prism in the refracted path of the rays (figure 461); the anallactic point remains unchanged.
The collimation axis of a telescope is the straight line connecting an infinitely distant point centered on the cross hairs, and the object focal point of the objective lens. Generally it does not coincide with the optical axis of the telescope. Practically only tele lenses are used at present, where the separation between cross hair and front objective lens element remains constant. When focussing on various distances, the focussing lens has to be shifted along its axis, which changes the equivalent focal length of the system. When moving the focussing lens, the trace of the image of the fixed cross hair in object space is considered to be the collimation axis.

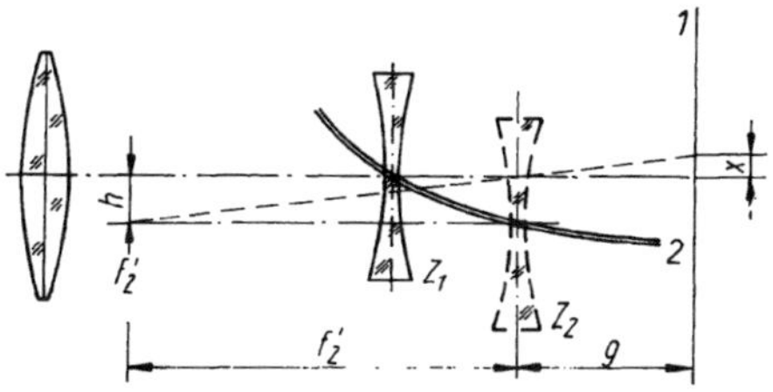

Figure 99 Influence of a loose focussing lens for a tele lens
1 crosshair plane; *2* path of the focussing lens

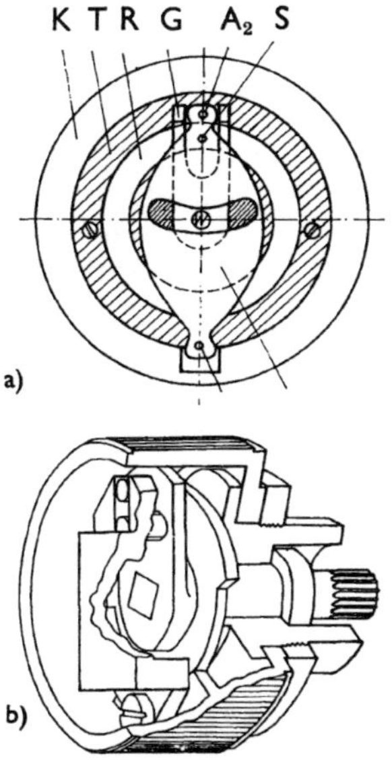

a)

b)

Figure 100 Rough and fine focussing
a) in the Ni 2 (Zeiss, Oberkochen) b) in the S 77 (Vickers), ADCE = rough drive, ABCE = fine drive

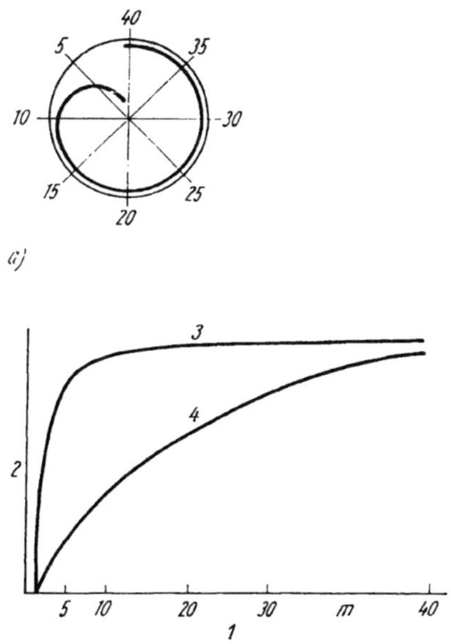

a)

b)

Figure 101 Functional drive
a) curve disk with sighting distance in m
b) comparison to usual focussing system
1 sighting distance [m]; *2* rotation of focussing knob = shift of focussing lens; *3* focussing system with linear transmission; *4* functional drive with curve disk

Even with careful adjustment, it is nearly impossible to have the optical axes of objective lens and focussing lens coincide. Due to the changes of the equivalent focal length, the collimation axis does not coincide with the optical axis. It is not even a straight line but a hyperbola, whose slight curvature becomes noticeable for extremely short distances only. When observing a point in both telescope positions, the mean lies on the optical axis. When levelling, this error influence remains, because only one telescope position is utilized.
A looseness of the lens guide is by far not as critical as with eye piece motion, where the cross shift of the cross hair centre causes a respective shift of the image against the cross hair.
When moving the focussing lens from position Z_1 for $s = \infty$ to position Z_2 for a close object, its focal point changes from F_1' to F_2'. With its separation h from the axis as defined by cross hair and front element of the objective lens, and with the distance g from the cross hair, the image shift becomes (figure 99):

$$x = h\frac{g}{f_2'}. \tag{37}$$

This has to be considered especially for short sights, because then the focussing lens has to be moved by large amounts. If possible, precise measurements are made with infinite focus setting. When observing in both telescope positions, this error is eliminated.
For fast focussing and parallaxe free sighting, several newer telescopes have rough and fine motions with a transmission ratio of 1 : 5. Instead of the generally used convenient five revolutions of the focussing screw, only two revolutions of the rough screw are necessary to cover the range. When focussing, the rough setting is turned somewhat beyond the sharpest image, which is afterwards obtained by turning the fine setting, which is automatically engaged. The range of the fine setting is usually $^1/_3$ to $^1/_4$ of one revolution, then the rough gear sets in again.
Such an arrangement is based for example on the following principle (figure 100a): The external focussing knob K can be rotated against the part T which is fixed to the telescope. Within a circular groove in T, the ring R can be turned with relatively large friction. On the stick S of the ring R a lever H is fixed so it can swivel. This lever grabs into a groove of the knob K at A_1 and onto the fork G at A_2 which is fixed onto a turable shaft W. A point on the shaft connects with a rack at the casing of the focussing lens, so that it is shifted proportionally to the revolution of W, resp. G. The swivelling range of the lever is limited by a slot within, into which the other end of the shaft fits. When rotating the knob K by φ, the axis A moves into the same direction because of the friction of the ring R and takes fork G along, while the axis A_2 moves in opposite direction.
The fork and the shaft rotate by $\varphi' = \varphi b/a$, which is slower than the knob (fine gear). If the shaft hits the slot end of the lever, the ring R is taken along with the stick S when the rotation is continued. The shaft then rotates with the same angular velocity as the lever (rough gear).
Figure 100b illustrates another design of rough- and fine settings.
At levels of the Ertel Co, Munich, the linear focussing gear is replaced with a functional gear. With the focussing knob, a curved disk is rotated which is connected to the shaft of the pulling pipe. The changing curvature of the curved disk nearly linearized the hyperbolic function between rotation of the focussing knob and the sighting distance (figure 101).

Objective Lenses

To eliminate imaging errors as much as possible, objective lenses and eye pieces of telescopes are compound lenses. The objective lenses of geodetic instruments have focal lengths ranging from 100 to 700 mm. The simplest objective lens consists of a positive crown glass lens and a negative flint glass lens cemented together (figure 102a). It can be corrected for spherical and chromatic aberrations. The same arrangement, but with air separation as indicated by *Fraunhofer*, had been used in most surveying instruments (figure 102b). It can be better corrected for aberrations than the cemented lens system.
Modern instruments have telelenses, consisting of a two- (figures 102a and b) or three lens (figure 102c) front element, and a shiftable one or two lens rear element serving as focussing lens. This permits a rather short tele-

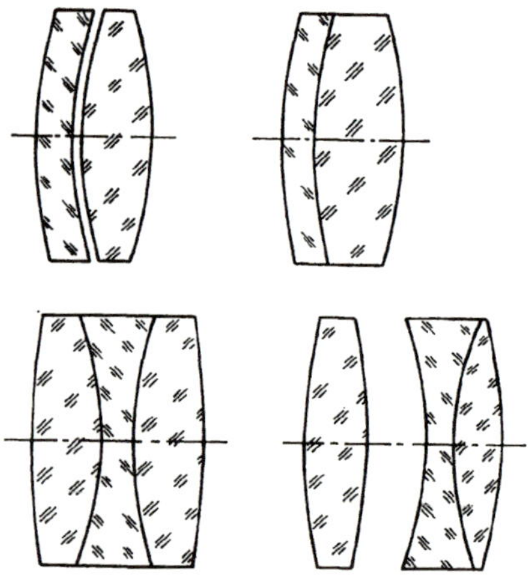

Figure 102 Objective lenses
a) common achromat
b) two lens objective with small air space ("Fraunhofer-Objective")
c) heavy flint achromats (Heavy flint glass = glass with high refractivity); nearly complete correction of the secular spectrum

scope length. The relative opening of the objective lens is the ratio of free diameter d_{EP} of the objective lens and the focal length f:

$$q = \frac{d_{EP}}{f'}. \tag{38}$$

The larger the relative opening, the higher the light strength. The latter increases with the square of q. Telelens objectives of surveying instruments have relative openings up to about 1 : 5, e.g. for theodolites 1 : 5.8 for the Wild T3, 1 : 7.9 for the T05, 1 : 5.8 for the T5 (U.S.S.R.); for tachometers 1 : 6.3 for the Dahlta, 1 : 4.8 for the Redta (Zeiss, Jena), 1 : 5.3 for the Wild RDS and 1 : 5.5 for the Wild RDH.
The ratio is smaller for instruments which are used together with illuminated targets, than for instruments for sighting extended objects such as range poles or levelling rods. An increase in q leads to more complicated lens systems.
Telescopes with bent path of rays are even shorter than the ones with straight arrangements and apochromats (figure 329, 330).
Mirror telescopes also are very short, but have high light strength and are nearly free of the secondary spectrum as well as providing upright images. Figure 103a illustrates a straight telescope in the *Cassegrain* arrangement. Rays passing through the focal compound lens L (for aplanatic correction) are reflected from the drilled main mirror S_1 to the convex capturing mirror S_2, and from there, through the focussing lens Z to the cross hair plate G. The length of the system is $0.4f$. Another telescope with bent path of rays (figure 103b) consists of a collecting lens system with positive lens L_1 and negative lens L_2 (which facilitates spherical correction and fulfils the sine condition), and two mirror lenses S_1 and S_2 (concave mirror). L_1, L_2, and S_1 generate a real image inside the prism P_1, of which S_2 and the focussing lens Z create a second image via the prism P_2, which is real and enlarged at the plane of the cross hairs G. Similar to these mirror telescopes, are the optical arrangements for the telescopes of some compensator levels (figure 425, 428).

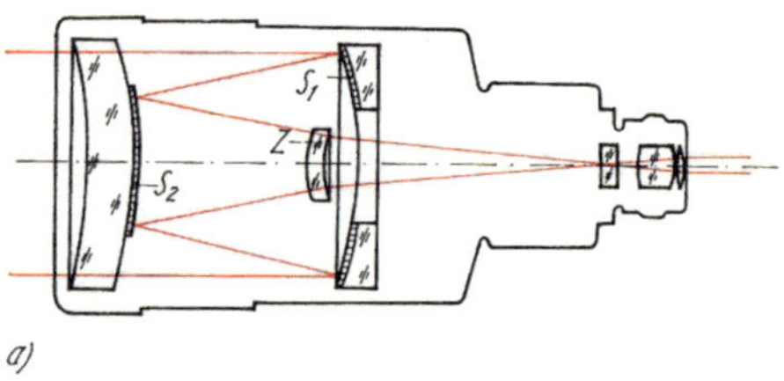

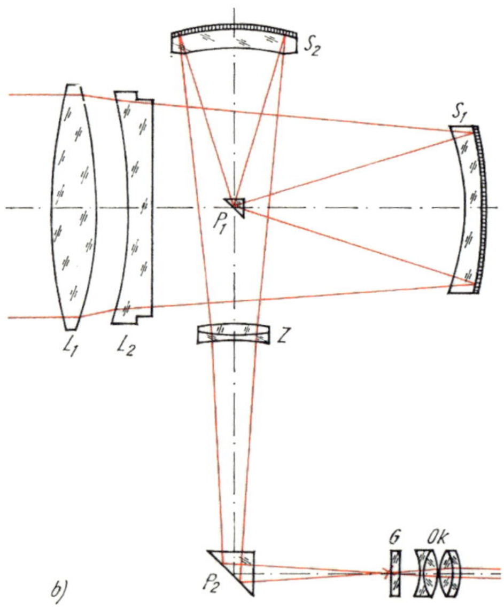

Figure 103 Mirror lens telescopes
a) according to *Köhler* (Theo 010 by Zeiss Jena; l = 135 mm)
b) according to *Wild* (DKM 3 by Kern, Aarau; l = 150 mm)

Eye Pieces

The eye piece is the part of optical instruments which is closest to the eye of the observer. It consists of a collecting—or field lens, facing the objective lens, and of the eye lens, facing the eye. Eye pieces of surveying instruments have focal lengths of about 8 to 10 mm. The exit pupil is separated from the last glass surface by up to 4 mm. This separation is of interest because the pupil of the eye should be placed at this position. If that is not possible, the field of view will be reduced.
The simplest eye piece, mentioned by *Ramsden*, consists of two plan-convex lenses with the plane surfaces at the outsides (figure 104a). The field lens L_1

is located near the focal plane of the eye lens L_2, which means that the equivalent focal length of the eye piece depends primarily on the focal length of the eye lens. The object focal plane, located outside the eye piece, contains the cross hair plate. Also, the image y' is generated in this plane by the objective lens. The field lens images y' virtually as y'' in the focal plane of the eye lens. This virtual image is then projected by the eye lens to the distant point, which for a normal sighted eye falls to infinity. This type of Ramsden eye piece can be found in numerous geodetic instruments of previous times, but presently is not anymore of practical importance. The occurance of chromatic aberration is the main disadvantage.

This chromatic error is avoided for the Kellner eye piece (figure 104b), where the eye lens is a compound lens, cemented together to become nearly achromatic. The field of view is 45 gon. The exit pupil is located at a distance of $0.5f'$ from the last glass surface. Since it is nearly free of distortion, *Kellner* designated it as orthoscopic (correct appearance). It has often been utilized in earlier microscope theodolites.

Newer instruments mainly have orthoscopic eye pieces with small air spaces between the elements (figures 104c and d). The large distance between exit pupil and last glass surface is remarkable, as it is not less than the focal length of the eye piece. The field of view amounts up to 45 gon. The three lens eye piece, designed by *Koenig* is also often used (figure 104e).

The images from these astronomical eye pieces are mirror reversed and upside down, while in newer telescopes an upright image is preferred.

Figure 104f illustrates Fraunhofer's terrestrial eye piece, where the lenses L_1 and L_2 operate as reversal system, while the lenses L_3 and L_4 function as eye piece. Since such reversal systems increase the length of the telescope, prism systems are now used to reverse the images, because they reduce the telescope length (figure 105).

Since human eyes are not equal, but often suffer from sighting errors, a normal sighting person will see a sharp image through the eye piece when accommodated to infinity, while it may be blurred for a wrong sighted person.

In order to enable short- and far sighted people to obtain a sharp image of the cross hairs, the eye piece can be moved by about 1 mm towards the cross hair plate by rotating it with its screw casing. Its position can be read against a fixed index with a scale graduated in dioptrias. This scale is usually

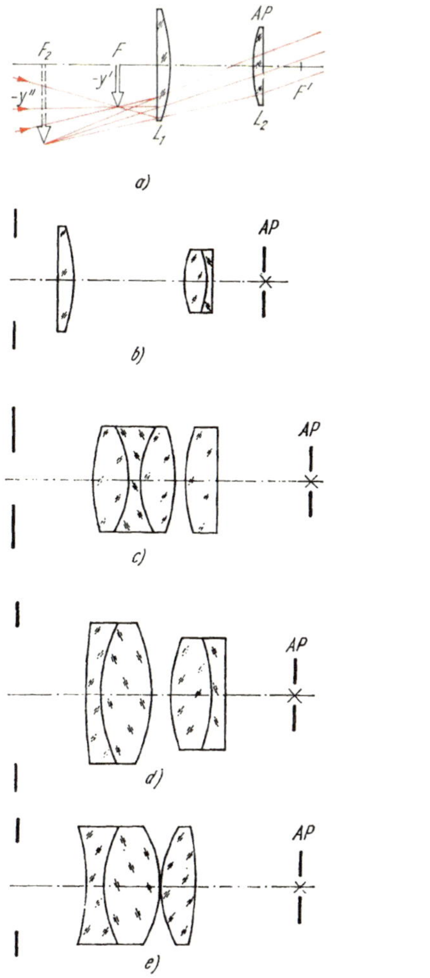

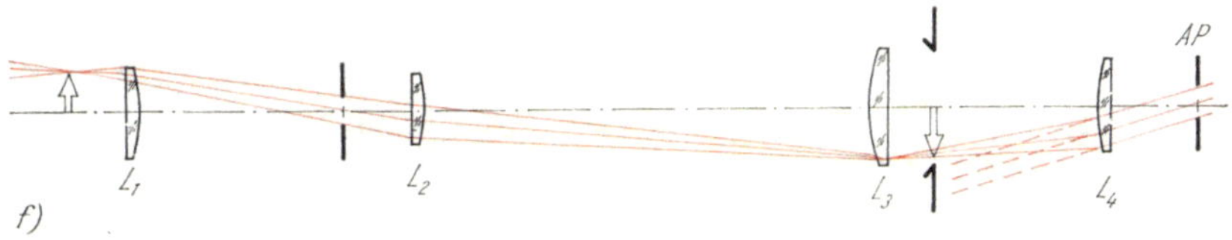

Figure 104 Eye Pieces
a) Ramsden's eyepiece
b) Kellner's eyepiece
c) orthoscopic eyepiece according to *Abbe*
d) orthoscopic eyepiece according to *Ploessl*
e) eyepiece according to *Koenig*
f) terrestrial eye piece

fastened to the movable eye piece with a ring. It is properly mounted when it gives a zero reading for a normal sighted person, when accommodated to infinity and sharply seeing the cross hair. Nowadays it is assumed that the released accommodation is between near and distant points. Therefore inexperienced observers often set negative dioptria values.

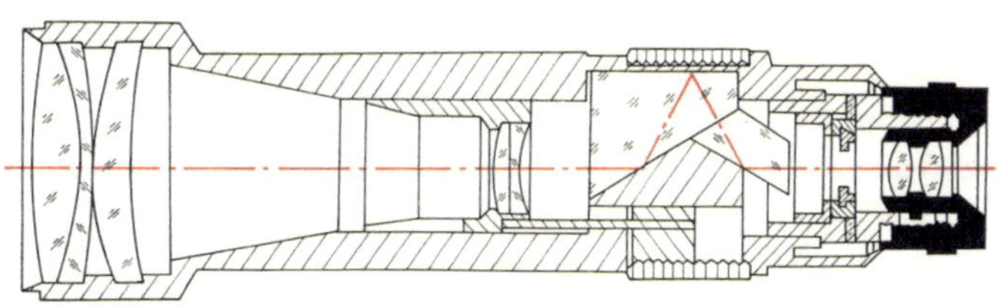

Figure 105 Creation of an erect image using prisms (telescope of the STN 27 by SLOM, Paris)

In this way, short- or far sighted people can observe without their eye glasses. If however, the eye also suffers from astigmatism, eye glasses have to be used for observing and reading through the eye piece.

If the eye piece is rotated outwards, the cross hair is beyond the focal plane and appears blurred to a normal sighted person. By rotating inwardly, the cross hair becomes sharp. It is advisable to do this when pointing the telescope against a light back ground. The eye is set for the distant point, and the cross hair is located in the focal plane of the objective lens.

If one continues to rotate the eye piece, the cross hair gets closer than the focal length. The eye would have to accommodate for a progressively shorter distance, which is not desirable.
Only after the cross hair has been set sharply, the focussing lens is shifted until a sharp appearance of the object is achieved. Then image and cross hair are located in the same plane and do not shift against each other. If the cross hair is not located in the image plane of the objective, a change in position between image and cross hair can be noted when moving the eye sideways within the exit pupil.
This parallax can lead to sighting errors. When the eye is moved to the edge of the exit pupil, the centre of the image plane appears at the direction of the optical axis (solid line in figure 106), as before. The centre of the cross hair S, however, is seen at a different location, because of the incorrect position of the cross hair (dotted line). The separation of the points of intersection between these rays and the image plane B of the objective lens corresponds to the sighting error δ at the eye, which is the amount of the apparent shift between cross hair and image.
When dividing this error by the telescope magnification, the object sighting error $\Delta\alpha$, a directional error results:

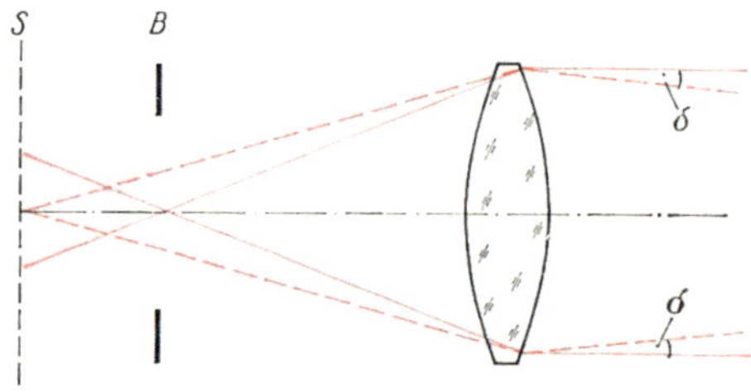

Figure 106 Sighting errors due to parallax caused by insufficient focussing of the telescope

$$\Delta\alpha = \frac{d_{AP}\,x}{2000\text{ mm }\Gamma_F}\varrho \qquad (39)$$

where x is entered in dioptrias.
With $d_{AP} = 1.4$ mm, $x = {}^1/_2$ dpt. and $\Gamma_F = 25$, $\Delta\alpha$ becomes 0.88 mgon at the object. Generally one considers one third of this amount, because an experienced eye is well centred in front of the exit pupil.
The cross hair has to be set sharply only once for the same observer, namely at the outset of a job. Due to varying object distances, however, one has to focus for each sight.

Effect of the Telescope

The effect of the telescope is the ratio of the sharp sights of the eye with and without the telescope; ideally it equals the magnification. It results from resolution and contrast depending on the magnification. There are other geometric characteristics, which are of interest to the surveyor.

Telescope Magnification

The telescope magnification Γ_F indicates how much larger an infinite object appears to be when viewed through the telescope rather than seen with the naked eye. It is the ratio of the angle σ by which the object is seen from the centre of the entrance pupil (EP), and the angle σ' by which the observer sees its image in the telescope.

$$\Gamma_F = \frac{\tan\sigma'}{\tan\sigma} \approx \frac{\sigma'}{\sigma} \qquad (40)$$

Since the object is far away, the image is formed at the common focal plane of objective lens and eye piece. According to figure 107 one obtains with $\tan\sigma = -y'/f'_{Ob}$ and $\tan\sigma' = -y'/f_{Ok}$

$$\Gamma_F = \frac{\tan\sigma'}{\tan\sigma} = \frac{f'_{Ob}}{f_{Ok}}. \qquad (41)$$

Therefore the magnification equals the ratio between the focal lengths of objective lens and eye piece for the astronomic telescope.
For telescopes with focussing lens, the equivalent focal length for the tele-system replaces the one for the objective lens.
If the object is close to the observer, one has to consider the length of the telescope- or more precisely the separation of entrance- and exit pupils. Because in this case, the apparent size of the object is larger for $EP(\sigma)$ than for $AP(\bar{\sigma})$. The ratio

$$\Gamma_A = \frac{\tan\sigma'}{\tan\bar{\sigma}} \qquad (42)$$

is called reading magnification. This reading magnification is the ratio of apparent size of the object in the telescope to amount of the angle, under which it appears to the naked eye remaining at the same position. Based on equations (41) and (42) as well as $\tan\sigma = y/s'$ and $\tan\bar{\sigma} = y/(s' + l)$, one obtains.

$$\Gamma_A = \Gamma_F\left(1 + \frac{l}{s'}\right) \tag{43}$$

where y represents the object size, s' the distance from object to EP, and l the separation of AP and EP.
The reading magnification is a bit larger than the telescope magnification. The difference becomes smaller, when the object distance s' is large in comparison to the telescope length l.
Since the diameters d_{EP} and d_{AP} of entrance- resp. exit pupils conform to the ratio of their respective focal lengths (figure 107) one obtains

$$\Gamma_F = \frac{f'_{Ob}}{f_{Ok}} = \frac{d_{EP}}{d_{AP}}. \tag{44}$$

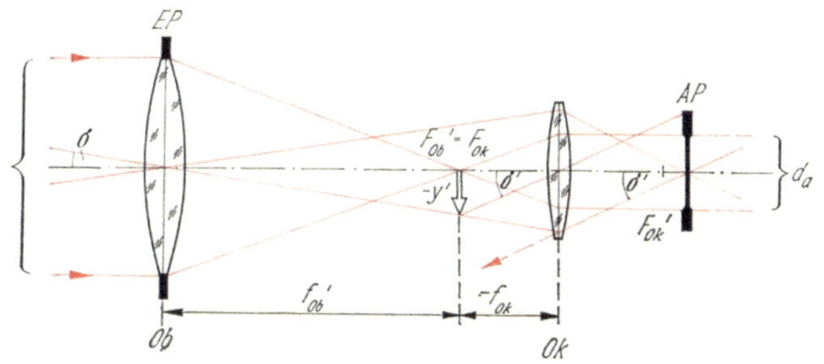

Figure 107 Telescope magnification

This equation (44) is especially well suited for the determination of the telescope magnification.
Surveying instruments of lower accuracy have generally a 12 to 20 times magnification, while for medium, high and highest accuracy up to 60 times magnification is used. This depends on the purpose of the instruments.

Determination of Telescope Magnification with *EP* and *AP*

In order to determine the telescope magnification using equation (44), d_{EP} and d_{AP} have to be measured when the telescope is focussed to ∞.
Most of the time the free opening of the objective lens is also the entrance pupil. Its diameter d_{EP} can be measured with sufficient accuracy using a millimetre ruler or compass and ruler.
In order to obtain d_{AP}, the telescope is directed against a light background (e.g. the sky), so that the objective lens is well illuminated. The image of the entrance pupil appears behind the eye piece as a light sharply defined circular disk (Ramsden circle). Its diameter can be fixed and measured on a piece of transparent paper placed normal to the telescope axis at this location. Since measuring errors are rather large with this method, the diameters are measured several times and the arithmetic mean is taken.
Example: Determine the telescope magnification Γ_F if $d_{EP} = 45$ mm and $d_{AP} = 1.5$ mm. According to equation (44):

$$\Gamma_F = \frac{45\ \text{mm}}{1.5\ \text{mm}} = 30.$$

The telescope has a 30 times magnification.
The measurements are more accurate, if a glass scale with millimetre graduation is placed directly in front of the objective lens focussed to ∞, and the AP is set with a microscope. Within the field of view of the microscope, the edge of the objective lens housing as well as the millimetre graduation can be seen sharply. Thus d_{EP} can be read at the graduated ruler, while d_{AP} can be measured simultaneously. This determines Γ_F.
The accuracy for measuring *EP* can be improved by placing a diaphragm with predetermined opening d_{EP} in front of the objective lens of the telescope. The exit pupil as image of the entrance pupil will be proportionally smaller; while d_{AP} can be obtained as mentioned. Another excellent way of measuring the diameter of AP utilizes Ramsden's dynameter, a magnifying glass with a fine graduation fastened onto the eye piece (figure 108).
The telescope magnification can also be determined from the ratio of apparent field of view to true field of view (see equation (41)).

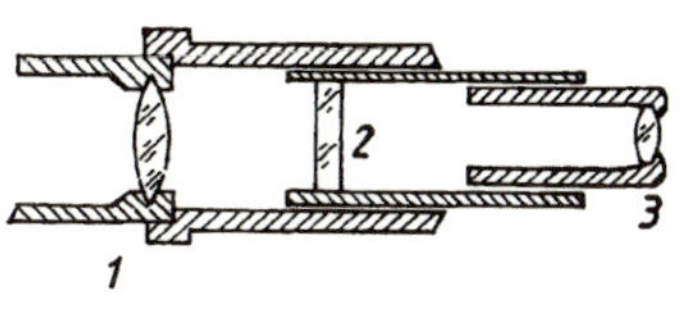

Figure 108 Dynameter
1 eyepiece; *2* scale; *3* magnifying glass

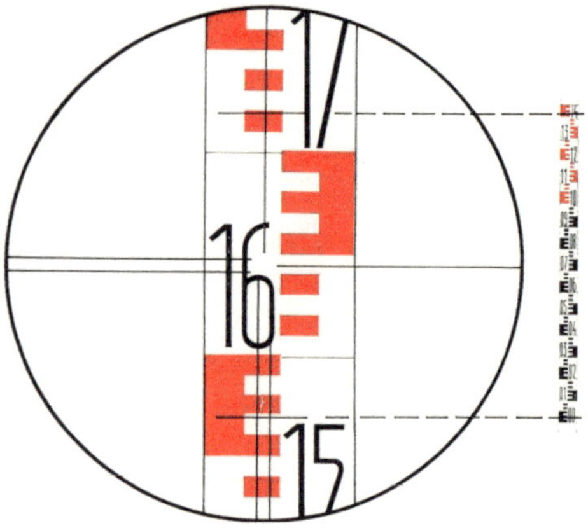

Figure 109 Determination of reading magnification by comparison

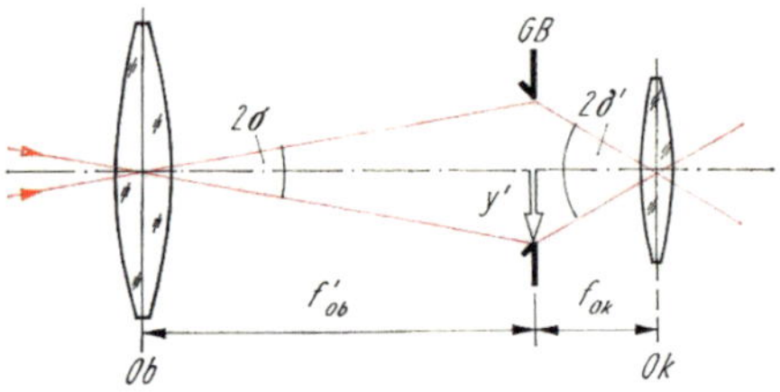

Figure 110 Field of view of a telescope

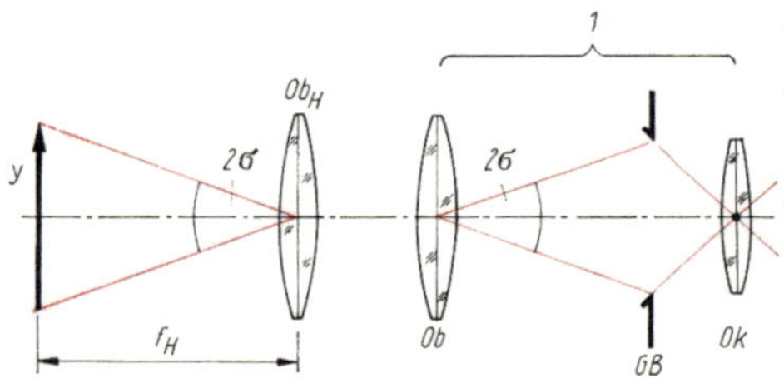

Figure 111 Determination of the true field of view of a telescope using an auxiliary objective lens
1 telescope

Determination of the Reading Magnification

The reading magnification of a telescope can be determined simply by comparing an object viewed with the naked eye with its image viewed through the telescope.
With one eye, one looks through the telescope to an object with regular graduation (e.g. brick wall, levelling rod) while with the other the object is viewed directly. The number of directly seen sections which are covered by the field of view of the telescope indicates the reading magnification Γ_A (figure 109).
This value can also be considered an approximation to the telescope magnification Γ_F. By simple rearrangement of equation (43), Γ_F can be obtained from the reading magnification.

Field of View

The field of view of a telescope is the conical space which is seen through the telescope when focussed to infinity.
The bundle of rays, which passes through the centre of the entrance pupil and is bounded by the edge of the field stop GB, describes in its back extension the true (or objective) field of view (figure 110). It is designated by the field angle 2σ.
If the observer's pupil of the eye is located in the exit pupil, he sees the field stop under the angle $2\sigma'$. This angle designates the apparent (or subjective) field of view. For telescopic arrangement, one obtains with equation (41):

$$\frac{\tan \sigma'}{\tan \sigma} = \frac{f'_{Ob}}{f'_{Ok}} = \Gamma_F .$$

The ratio of apparent to true field of view again gives the telescope magnification. The apparent field of view depends on the quality of the eye piece. For surveying instruments it is 25 to 30 gon, while the true field of view, in space covers 1 to 2 gon.

Determination of the True Field of View

In order to determine the true field of view, a distant point is aimed at with one edge of the field stop, and then with the diametrally opposite one. Depending on whether the telescope is rotated around the horizontal or the vertical axis, one reads the vertical or horizontal angle. The field angle is the difference between the two readings.
The field of view can also be determined with a scale (e.g. levelling rod), positioned normal to the collimation axis at about 50 m. One has to read the graduation at the edges of the field stop. If the rod reading is denoted with l, and the distance between rod and instrument with s, the field angle becomes

$$2\sigma = \frac{l}{s}\varrho . \tag{45}$$

Example: To simplify the computation, the rod was placed at 63.7 m from the instrument. At the edges of the field stop the values a_u = 0.600 m and a_o = 2.044 m were observed. According to equation (45)

$$2\sigma = \frac{2.044 \text{ m} - 0.600 \text{ m}}{63.7 \text{ m}} \cdot 63.7 \text{ gon} \approx 1.4 \text{ gon}.$$

The field angle of the tested instrument is 1.4 gon.
Another possibility to obtain the true field of view utilizes an objective lens Ob_H (for photo or telescope) with known focal length f_H and a mm ruler (figure 111). The ruler is placed in the focal plane of Ob_H by shifting it along the optical axis until its image becomes sharp.
The true field angle 2σ can be determined using the graduated lines at the edge and the following equation:

$$\tan \sigma = \frac{y}{2 f_H} . \tag{45a}$$

If the ruler is subdivided with radians, the field angle is directly obtained without computations. Ruler and objective lens can be combined to a separate instrument, the field of view collimator.

Example: The focal length of the photographic objective lens was $f_{II} = 250$ mm, the readable section of the ruler was 6 mm. With equation (45a) one obtains:

$$\tan \sigma = \frac{6 \text{ mm}}{2 \cdot 250 \text{ mm}} = 0.012; \quad \text{and} \quad 2\sigma \approx 1.43 \text{ gon}.$$

The field angle amounts to 1.4 gon.

Determination of the Apparent Field of View

Like the true field of view, the apparent one can be measured with the aid of a ruler. To do so, the telescope is inserted into the system with reversed direction.

The apparent field angle can be derived from the true one, using equation (41) and the magnification, as

$$2\sigma' = \Gamma_F \cdot 2\sigma. \tag{46}$$

The apparent field angle can be directly determined with the use of an additional telescope (e.g. a theodolite). Both telescopes are arranged such, that their objective lenses face each other and their optical axes coincide if possible (figure 112). A scale is placed at a distance s from the exit pupil of the telescope to be tested. The section of the scale covered by the field stop GB_1 is read with the second telescope. If the field stop GB_2 of the second telescope is smaller than the one of the first one (GB_1), the edges of GB_1 are sighted in succession and read on the scale.

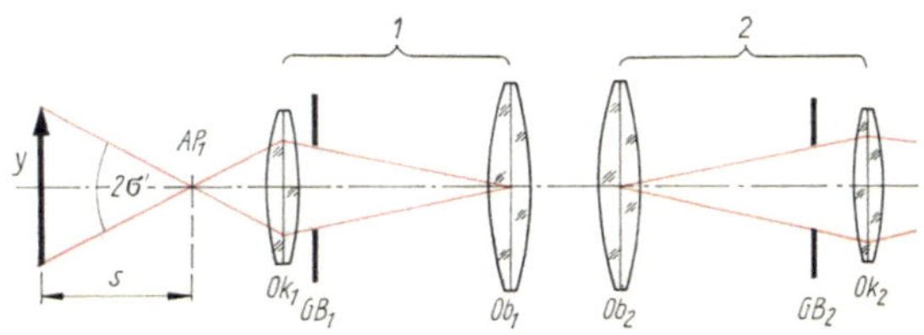

Figure 112 Determination of the apparent field of view of a telescope using a second telescope
1 = 1st telescope; *2* = 2nd telescope

Example: The rod section read with the second telescope through the first one was $y = 2.92$ m. The rod was placed 4.06 m in front of the exit pupil AP_1. Thus from figure 112 it follows that

$$\tan \sigma' = \frac{y}{2s} = \frac{2.92 \text{ m}}{2 \cdot 4.06 \text{ m}} = 0.36; \qquad 2\sigma' = 44.0 \text{ gon}.$$

The apparent field angle amounts to 44 gon. With 1.43 gon for the true field angle, the telescope magnification becomes:

$$\Gamma_F = \frac{\tan \sigma'}{\tan \sigma} = \frac{0.36}{0.012} = 30.$$

Brightness

Under the brightness h of a telescope, one understands the ratio of the amounts of light that hit a unit area on the retina via the telescope or via direct viewing. It depends on the diameter of the objective lens, the telescope magnification, the light loss within the telescope and the diameter of the pupil of the eye. Since the areas of AP and the pupil of the eye are proportional to their diameters d_{AP} and d_P, the brightness becomes for $d_{AP} \leqq d_P$:

$$h = \delta_k \frac{d_{AP}^2}{d_P^2} = \delta_k \frac{d_{EP}^2}{\Gamma_F^2 d_P^2}. \tag{47}$$

The degree of transmission δ_k of an optical system with k surfaces indicates how much of the light entering at EP exits at AP.

Maximum brightness $h = 1$ is obtained according to equation (47) when $\delta_k = 1$ and the exit pupil is smaller than or equal in size to the eye pupil. Enlarging the AP beyond the size of the eye pupil is useless, because then part of the light will never reach the eye.

For telescopic systems, the brightness is generally not given by equation (47) but by d_{AP}^2 (pupil square, often incorrectly referred to as "geometric telescope brightness", which is considered as "relative brightness").

Since surveying instruments are primarily used in daylight when the eye pupil of the observer is about 2 mm, AP's with radii of 1 to 2 mm are used.

More recently, levels have an AP of 1 mm diameter, and theodolites of 1.4 to 1.6 mm. The brightness of such telescopes is 1, respectively 1.96 to 2.56 mm². It is thus for theodolites 2 to 2.5 times as large as for levels.
When light passes through lenses or prisms, a loss of about 5% occurs on each glass/air boundary due to reflections.
Therefore the degree of transmission of an optical system has to be considered with equation (47). This degree of transmission of a single surface depends on the refractive index of the glass and amounts to about 95%. For k glass/air boundary surfaces $\delta_k = 0.95^k$. A system with 10 boundary surfaces has therefore a degree of transmission of $0.95^{10} \approx 0.60$, i.e. a 40% loss of light.
For optical glasses this is improved by coating the surfaces in a vacuum with a thin, transparent film, which reduces the reflectivity and appears violet when hit by light. This is done by vaporization, and the film has a refractive index much smaller than the one of the refracting surface (e.g. $\sqrt{n}$). Sometimes the glass surface itself is changed by etching, submersion or a centrifugal method. The thickness of the film is selected such, that the reflecting partial waves are eliminated or reduced. It's magnitude is in the range of the wavelength of the light.
With such an antireflex film the light loss can be reduced to 1 to 0.5%, which increases the degree of transmission $\delta_k = 0.99^k$ (figure 113). For ten boundary surfaces the light loss amounts to 10% only: Such a telescope gives images which are 50% lighter and have more contrast than a telescope without such a film. It is called T-coat (Zeiss, Jena and Zeiss, Oberkochen for T = transparent), and AR-coat (Kern, Aarau for AR = anti reflecting).

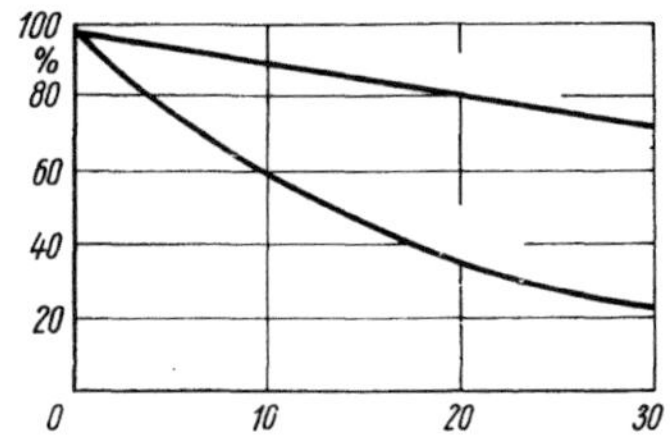

Figure 113 Effect of improvement with antireflex film
1 without film; *2* with film; *3* with film of the optical system; *4* number of planes glass/air

The dusk effect L of a telescope

$$L = c\sqrt{\Gamma_F d_{EP}} \tag{48}$$

has to consider the characteristics of the eye; $c = 1$ has to be used for illuminances of 3×10^{-1} to 3×10^{-4} cd/m² and a diameter of 2 mm for the exit pupil.

Resolution of the Telescope

The resolution of an optical system indicates its capability of imaging two neighbouring object points separately (e.g. resolution of the eye). It depends on the diameter d_{EP} of the entrance pupil and is signified by the smallest resolving angle γ caused by refractions. For $\lambda = 560$ nm, it is the maximum spectral sensitivity of the eye:

$$\gamma = \frac{37 \text{ mgon mm}}{d_{EP}} \tag{49}$$

Thus the resolution is bigger (γ smaller), for bigger d_{EP}. Since $d_{EP} = \Gamma_F d_{AP}$, γ becomes for $d_{AP} = 2$ mm:

$$\gamma = \frac{18.5 \text{ mgon}}{\Gamma_F}. \tag{49a}$$

In order to fully utilize the resolution, Γ_F has to be such that γ corresponds to the resolution of the eye ($\gamma_A = 18.5$ mgon) (see page 39). The usable magnification therefore equals the entrance pupil d_{EP} in millimetres.
For more comfortable viewing, surveying instruments use telescopes with 1.5 to 2 times the amount of usable magnification. With this, one obtains diameters of the exit pupils of 1 to 1.3 mm (see also page 49). For extended objects the normal magnification is most suitable, when the size of the exit pupil corresponds to the size of the pupil of the eye.

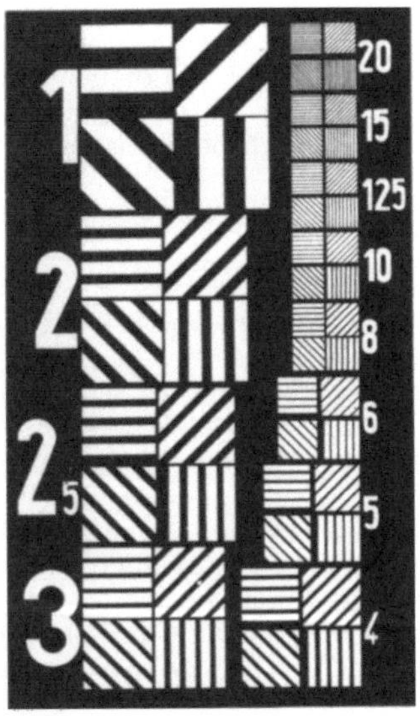

Figure 114 Test board for the determination of the resolution of telescopes

Determination of the Resolution

The resolution can be determined visually by photography or photo electrically. The following method is rather simple: A test board is viewed at a finite distance (figure 114). Lines and separation of its line grids are equal. To also obtain the influence of astigmatism the lines for each value are arranged in four different directions. One obtains the resolution as

$$\gamma = \frac{a}{s}\varrho \tag{50}$$

Figure 115 Siemens-star

where a is the line separation in the grid, when the lines are still separated in all directions, and s is the distance to the test board.
The characteristics of the observing eye have affected this value. To eliminate them, one can place another telescope in front of it if its resolution is smaller than the one of the eye, i.e. its AP is smaller than 1 mm.
The resolution can also be quickly determined with the radial grid method by *Siemens*. The alternately black and white equal sized sectors of the Siemens star (figure 115) correspond to a 5° angle. When testing the objective lens the diameter of the circle of confusion which is not resolved, is determined.

Focal Length

The importance and computation of focal lengths for lenses and compound lenses has been discussed on page 34. Here two methods for the determination of focal lengths will be presented.
To determine the focal length of an objective lens for a telescope with focussing lens, one focusses to infinity (or shortest sighting distance for determining the focal length of the lens system). Then one removes the eye piece and places the telescope in front of the objective lens of a theodolite focussed to infinity (figure 116). In front of the telescope to be tested, a millimetre

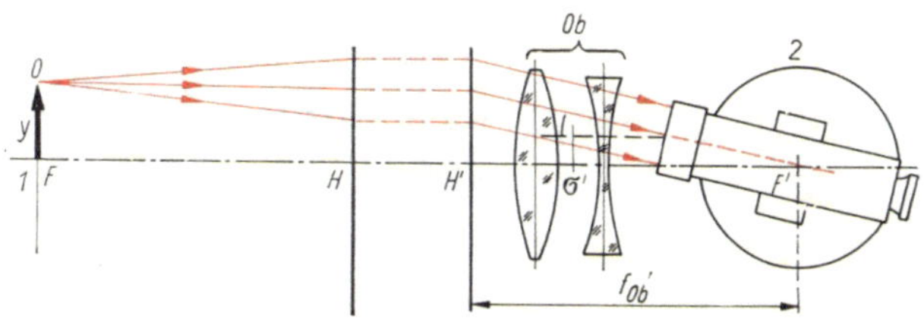

Figure 116 Determination of the focal length of a telescope objective using object size and field angle
1 graduation; *2* theodolite

ruler is shifted until a sharp image appears in the field of view of the theodolite. Then the scale is located in the objective focal plane of the telescope to be tested. After rotating the theodolite around its vertical axis and aiming at the graduation lines, the angles are read at the horizontal circle of the theodolite. According to figure 116 $\tan \sigma' = y/f'_{Ob}$, because σ' determines the direction of a bundle of rays originating in O and exiting from the test telescope, parallel to its axis. Thus

$$f'_{Ob} = \frac{y}{\tan \sigma'} \approx \frac{y}{\sigma'}. \tag{51}$$

Similarly one obtains f, F and H' after reversing the objective lens to be tested. Figure 117 illustrates the arrangement used to determine the equi-

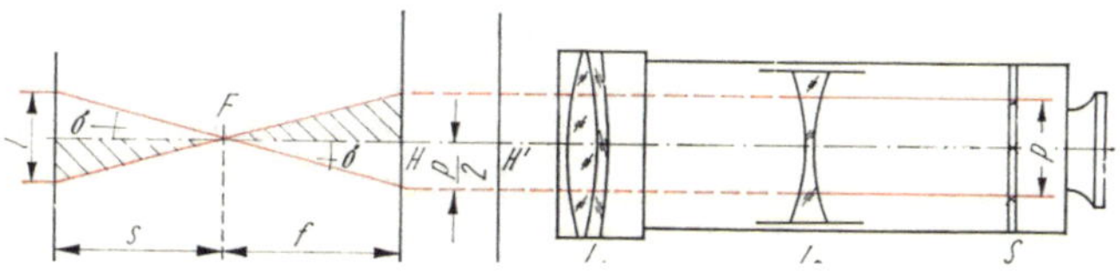

Figure 117 Determination of the equivalent focal length of a lens system with the aid of stadia hairs

valent focal length of a compound lens with the aid of stadia hairs. The rod section l of a distant levelling rod (practically at infinity) appears under the angle σ, thus giving the optically determined distance $s = kl$ (k is the stadia multiplication constant, s the distance between rod and telescope, which can be considered as originating from the focal point for large distances). From figure 117 follows

$$\tan \sigma = \frac{p}{2f} = \frac{l}{2s} = \frac{l}{2kl} \tag{52}$$

and therefore

$$f' = kp. \tag{53}$$

The separation p of the stadia hairs can be measured with a microscope after removing the eye piece. k is equal to 100. Example: The stadia constant has been determined as $k = 100.05 \pm 0.02$, and the separation of the stadia hairs as $p = 2.529$ mm. According to equation (53) the equivalent focal length is $f_{Ob} = 253.2 \pm 0.05$ mm.
The rather short focal lengths of the eye pieces in comparison to the objective lenses can be measured with a microscope after unscrewing of the eye piece. One measures the reduced image l' of a distant rod as generated in its focal plane. The focal length amounts to

$$f'_{Ok} = s\frac{l'}{l}. \tag{54}$$

Imaging Quality

The imaging quality can be determined in the laboratory with the aid of a test board (see figures 114 and 115) in the focal plane of a collimator (see page 53). As a substitute, a sheet with lines or geometric figures can be used in about 10 m from the telescope to be tested.
If perpendicular lines are not sharp at the same time, astigmatism is present. Black figures against a light background which are imaged sharp with open diaphragm and become blurred when the diaphragm is halfway closed (or the eye is moved sideways), indicate spherical aberration. Similarly, chromatic aberration can be tested. If this imaging error is present, the image has coloured edges. During the tests, one has to watch that the images are geometrically similar, i.e. that there is no distortion. A coma can be recognized by one sided halation in spite of central eye position.

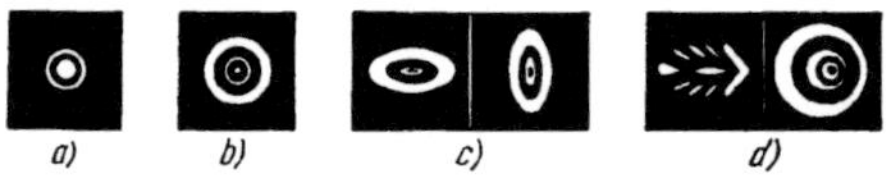

Figure 118 Star images from good and poor objective lenses
a) image of a star at the focal point with a good objective lens
b) star image with spherical aberration
c) with astigmatism
d) with coma

If one checks the imaging quality in the field with the aid of a bright star, its image should consist of a bright spot surrounded by symmetrical circles (figure 118a) for high quality. The brightness of these diffraction circles changes in planes beyond the focal point for a focussing change if spherical aberration is present (figure 118b). With chromatic aberration the images in planes on both sides of the focal point are coloured (figure 84). The rings are distorted regularly for astigmatism (figure 118c), and irregularly for coma (figure 118d); however, this effect can also be caused by incorrect centering of the optical elements.

Aiming Accuracy

The mean square error m_z for aiming or setting of a target with a telescope depends on shape, distance, background and illumination of the target point, meteorological conditions, cross hair, brightness, and to a large extend on telescope magnification, resolution of telescope and eye, and width of the cross hairs.
Since it is practically impossible to obtain a general formula which considers all these influences of the aiming error, telescopes are experimentally compared under controlled conditions.
A formula developed by *Noetzli* in 1915 was later questioned. He proposed

$$m_z = \pm\frac{1.8\text{ mgon}}{\sqrt{\Gamma_F}} \quad \text{to} \quad \pm\frac{2.5\text{ mgon}}{\sqrt{\Gamma_F}}. \tag{55}$$

When setting distant targets by coinciding, one can use

$$m_z = \frac{3\text{ mgon}}{\sqrt{\Gamma_F}}. \tag{56}$$

Yelissejew concluded in 1957 for bisecting a target with a double line:

$$m_z \approx \frac{1.2\text{ mgon}}{\Gamma_F} \quad \text{to} \quad \frac{8\text{ mgon}}{\Gamma_F} \quad \text{as a mean } m_z \approx \frac{4.5\text{ mgon}}{\Gamma_F}. \tag{57}$$

For poor contrast, e.g. target against the sun, or with forest at background, 18.5 or even 37 mgon may have to be used in the numerator.

Often the formula

$$m_z = \frac{c}{\Gamma_F + \Gamma_o} \tag{58}$$

is used, where Γ_o denotes the magnification of the objective lens, but with a ficticious focal length of 10 mm for the eye piece, i.e.

$$\Gamma^o = \frac{f_{Ob}}{10}. \tag{58a}$$

According to *Grödel* $c = 6$ mgon for facourable circumstances, and $c = 12.5$ mgon for natural objects, while *Vogg* suggests to generally use $c = 9$ to 15 mgon.
For a sighting distance s, *Marfenko* found in 1959

$$m_z = \frac{c}{\Gamma_F\sqrt{s}}. \tag{59}$$

Krassowsky lists as maximum error for bisecting

$$m_{z\,\max} = \frac{\alpha}{18} + \Delta \tag{60}$$

when α is the angular separation of the double line (usually 11 to 12 mgon), and Δ the personal error which depends mainly on the brightness of the objective lens, but can change for different directions and thus distort an angle by as much as 0.5 mgon.
The aiming accuracy is also influenced by the thickness d of the cross hair. If the cross hair covers the target, the maximum error is

$$m_{z\,\max} = \frac{d}{f_{Ob}}\varrho. \tag{61}$$

For $d = 8$ µm and $f_{Ob} = 20$ cm it is 2.4 mgon.
Therefore a target can be set more accurately with a double line, than a single one. However, for illuminated targets it is occasionally of advantage to use a single cross hair line.
The exit pupil has a major influence on the aiming accuracy. Best results are obtained if its central part is used because of eye errors of the observer.
If the distant target can hardly be recognized when bisecting, one should use

$$m_z = \frac{20 \text{ to } 37 \text{ mgon}}{\Gamma_F}. \tag{62}$$

The aiming error does not increase if a prism with separating edge is used instead of a cross hair with double line.

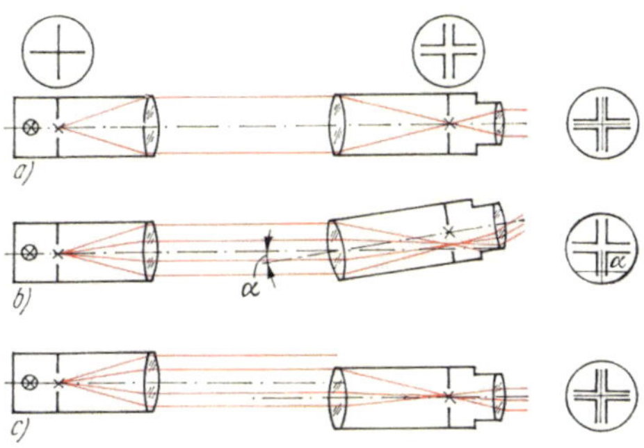

Figure 119 Collimator and telescope
a) identical axes
b) axes not parallel
c) parallel shift of axes

Collimator

In connection with optical instruments, collimators are used to represent an infinitely distant target at a finite distance. They are primarily used for checking and adjusting of optical instruments and for the relative orientation of two optical systems, but can also be utilized to accurately determine small angles. A collimator resembles a telescope, where the eye piece is replaced by a lamp to illuminate the crosshair, a graduated scale or a test plate located in the focal plane of the objective lens. Laboratory collimators have focal lengths of about 500 mm with objective lens diameters of 40 to 60 mm.
An adjusted level, focussed to infinity can also serve as a collimator, if its cross hair is illuminated by a lamp placed behind the eye piece. Rays originating at the centre of the cross hair emerge from the objective lens as parallel rays (figure 119). The theodolites Wild T1, T16, or T2 can also be used as collimators if their eye pieces are exchanged with the eyepiece lamp GEB 32 (figure 120). In comparison to laboratory collimators, it is easier to direct these to the instrument to be tested.
The mutual constancy between collimator and telescope is best utilized if the telescope serves as collimator as for auto collimation. An auto collimator is a telescope, which projects an externally illuminated measuring mark to infinity and images it in its focal plane after reflection on a plane mirror, normal to its collimation axis. The angular difference is twice the amount than when measured with a collimator because of the reflection. An optical beam splitter near the focal point (figure 121) separates the

Figure 120 Wild T 2 with eyepiece lamp GEB 32

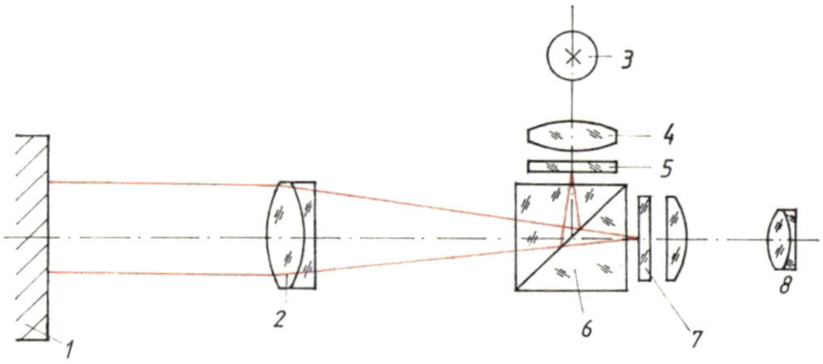

Figure 121 Optical system of an autocollimation telescope
1 autocollimation mirror; *2* objective lens; *3* light source; *4* condensor; *5* reticule; *6* beam splitter cube; *7* partial plate; *8* eye piece

projected and reflected bundles of rays. An auto-collimation system consists of autocollimator and autocollimation mirror. It is used for the adjustment of surveying instruments and telescopes, for measuring the resolution of optical systems, for testing prisms as well as for measuring direction differences when positioning machine parts, and for similar industrial surveying problems. It also is most suitable for checking the proper focussing of a collimator.

1.2. Spirit Levels

Spirit levels are used in surveying instruments to set axes vertical or horizontal as well as occasionally to measure deviations from those positions. Sometimes spirit levels are replaced by compensators operating with a pendulum. The electronic level is also based on the pendulum principle. According to their shapes, one distinguishes between bull's eye levels for rough settings, and tubular levels for more accurate measurements.

1.2.1. Tubular Levels

Design and Operation of Tubular Levels

A tubular level consists of cylindrical glass tube which, on the inside is barrel shaped. It is placed without strain in a metal casing (figure 122). Therefore it is not deformed when the metal casing undergoes thermal or other changes.

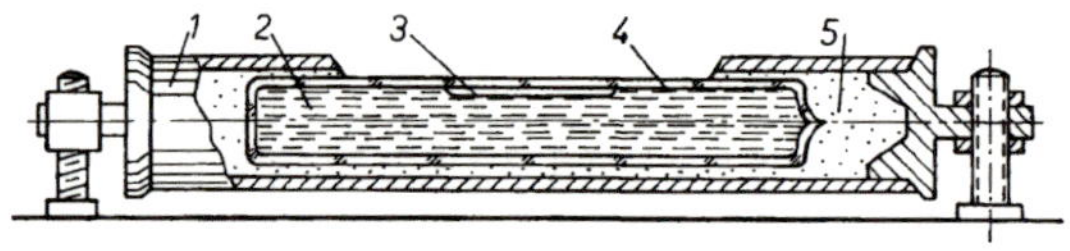

Figure 122 Tubular level
1 housing; *2* ether; *3* bubble; *4* glass tube; *5* wax or plaster

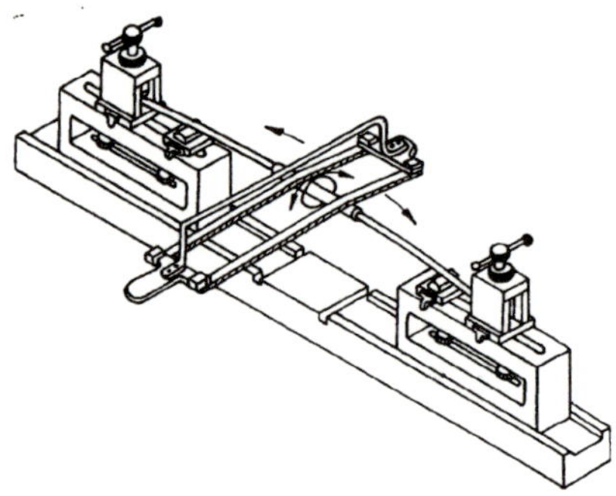

Figure 123 Grinding of tubular level

The quality of the level depends mainly on the grinding of the circular surface. To accomplish this, the tube is placed on a steel mandrel with grinding material, whose surface corresponds to the radius of the circular arc. It is slowly ground by rolling along the mandrel (figure 123). The tube is then filled with a thin liquid with low melting point (below −50°C), usually ether. When only the oblong level bubble filled with vapor of this liquid is left, the ends of the tube are fused together. Temperature changes often affect the vapor such, that some of it changes to the liquid state or vice versa. This changes the size of the bubble. Generally its length should be about $^1/_3$ of the length of the tube at 20°C.
The level bubble always moves to the highest spot of the level. For observation of its position, the outside of the tube carries a graduation with lines perpendicular to the axis of the tube. The separation of two consecutive lines is 2 mm (formally 2.2558 mm = one Parisian line) and is often referred to as Pars.
The bubble axis is the tangent to the inside of the glass tube in longitudinal direction at the centre of the graduation, the normal point. Its projection onto the casing is the setting line. If the centre of the bubble (centre of bubble equals $^1/_2$ of the sum of the readings of both bubble edges) coincides with the normal point, the bubble is level, i.e. the bubble axis is horizontal (figure 124). The separation between normal point and bubble centre for different bubble positions is referred to as bubble deviation, while the position of the bubble centre for horizontal collimation axis or correct vertical axis indicates the centre of run.

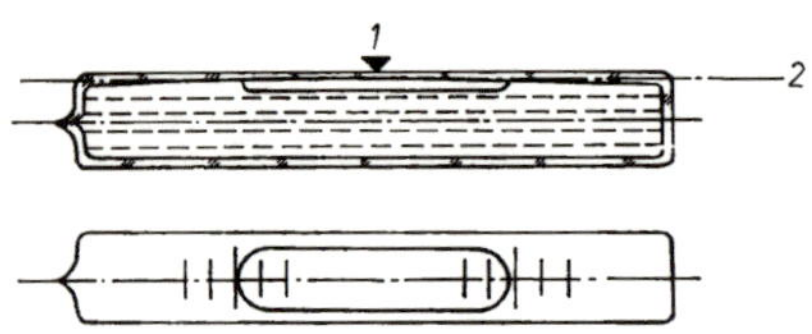

Figure 124 Centered level bubble
1 normal point; *2* bubble axis

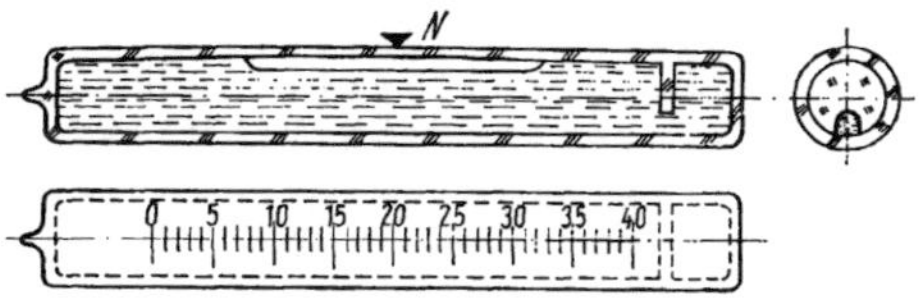

Figure 125 Chambered spirit level

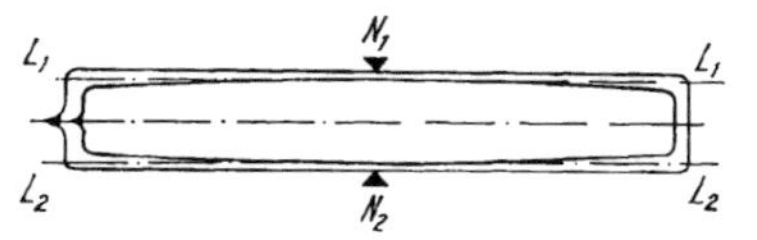

Figure 126 Double- or reversible level

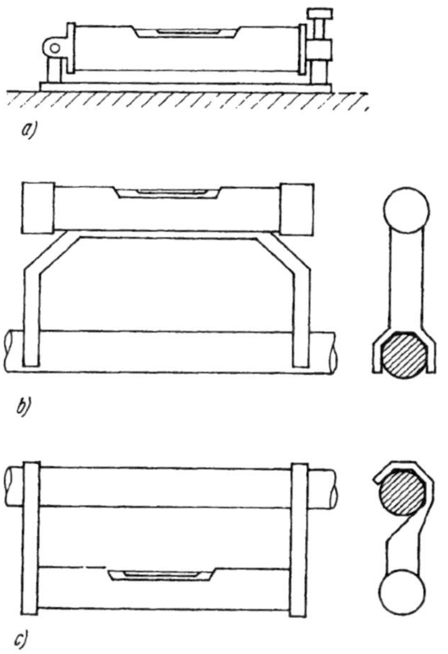

Figure 127 Tubular levels not solidly attached to instrument
a) table level
b) striding level
c) hanging level

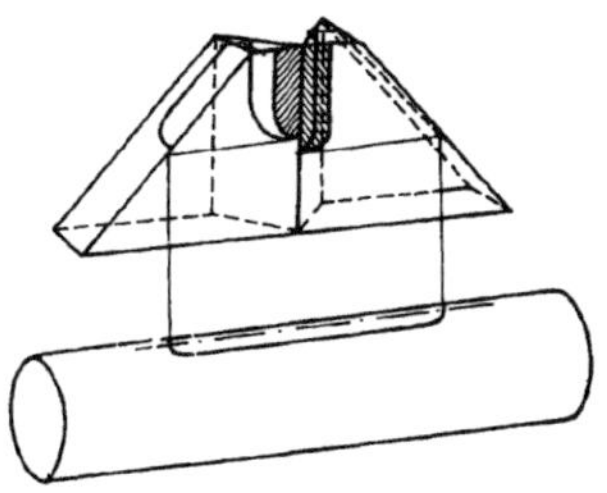

Figure 128 Level bubble with prism system consisting of sphenoids

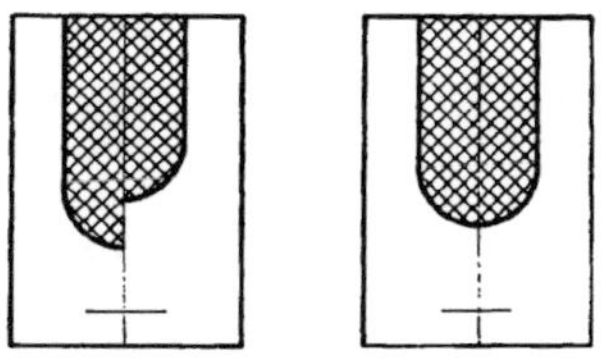

Figure 129 Image of the bubble of a coincidence level
a) not centered
b) centered

Different Kinds of Tubular Levels

There are spirit levels used to set the normal point i.e. to level its bubble axis. These levels have a graduation consisting of two main lines symmetrical to the normal point, and separated by a multiple of 2 mm, approximately equal to the length of the bubble axis at the reference temperature of 20°C. On both sides of these main lines, there are auxiliary lines (figure 124).
Bubble clinometers, used to measure small changes in inclination, have a continuous 2 mm graduation, symmetric to the normal point and covering the whole range suitable for measurements (figure 125).
With chambered spirit levels, the length of the bubble which depends on the temperature of the liquid and can be regulated (figure 125). A glass partition with a small opening at its lower part separates a chamber which is approximately half full of ether. By tilting the level the bubble length can be changed.
Reversible levels are ground on two diametrically opposite sections on the inside of the tube (figure 126). They therefore have two bubble axes which have to be parallel to each other, and can be used in both positions.
Tubular bubbles can be removably attached to an instrument, or fastened permanently. The following ones can be removed and are often used as clinometer levels:

a) the spirit table (figure 127a) to place on a line or plane
b) the striding level (figure 127b) to place on a cylindrical axis
c) the hanging level (figure 127c) to hang onto a cylindrical axis.

The following levels are permanently fastened to theodolites, tacheometers and alidades:

a) the plate level, fastened to the plate of the theodolite either parallel to the horizontal axis (cross bubble) or perpendicular to it (longitudinal bubble). It is utilized to level the instrument, i.e. to get the vertical axis to its correct position
b) the index level, to check the position of the height index with respect to the plumb line
c) the telescope level used to set the collimation axis horizontal, and for levelling instruments
d) the level bubble, fastened parallel to the telescope axis so that it can be levelled.

Level bubbles and index levels are usually made as coincidence bubbles. The bubble is illuminated from below through a mirror which can be rotated. Above the bubble, a prism system consisting of sphenoids (figure 128) is arranged such, that one half each of the bubble ends are imaged beside each other. When the bubble is level, both ends coincide (figure 129). If the bubble is inclined, they move in opposite directions, showing twice the amount of the bubble deviation. Generally, coincidence bubbles do not have a graduation, although occasionally two lines symmetrical to the normal point can be found, which for a properly adjusted prism system appear as one straight line (figure 129). With coincidence bubbles, the observer can set the bubble parallax free without changing his position. Changes in the length of the bubble due to temperature changes have no influence. The setting accuracy is approximately four times that of a regular level, and even higher when magnification during the imaging is used.

Level sensitivity

The level sensitivity p indicates the angular inclination necessary to obtain a shift of the bubble centre by 2 mm. In figure 130 it corresponds to the angle ε between positions L_1L_1 and L_2L_2 of the bubble axis for $a = 2$ mm.
The smaller the sensitivity of a level, the easier it reacts to a change in position. It can be set more accurately, but takes much longer to stabilize. The setting error can be assumed as $0.1p$. High precision levels, i.e. for astronomical instruments, have a sensitivity of 1″ to 2″. Levels for surveying instruments are produced with nominal sensitivities of 10″, 15″, 20″, 30″, 60″ and 120″.
When the temperature rises, the bubble becomes shorter. If the tube is subjected to a temperature increase of 1K the bubble of a 30″ level is shortened by 0.4 mm, a 15″ level by 0.6 mm, and a 5″ by 1.0 mm.

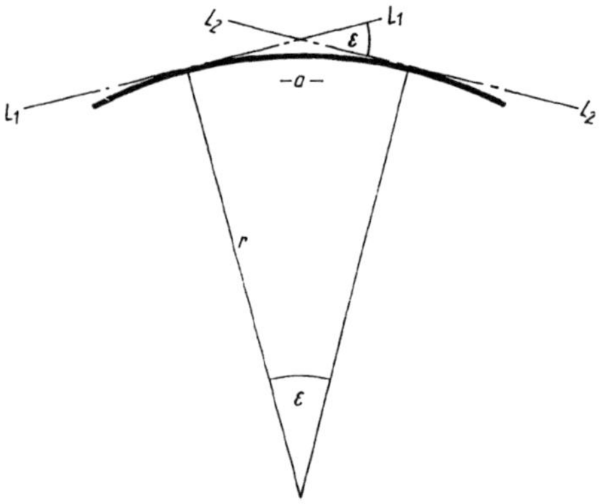

Figure 130 Bubble sensitivity

The large changes of the bubble length for precision levels can be explained by the larger volume of liquid combined with the large expansion coefficient of ether (163×10^{-5}) as compared to glass (3×10^{-5}), and the bubble volume. The level tube is deformed by one-sided heating. When the level is subjected to direct sunlight, the bubble therefore moves towards the sun.

The setting accuracy of the level bubble can be increased up to a factor ten by improving the observation technique. This can be achieved with a coincidence bubble or with magnified imaging of the bubble within the field of view of the telescope eye piece (see figure 483).

In the latter case the graduation is occasionally smaller than 2 mm and then magnified through an optical system. The tilt change of the level corresponding to one graduation line is called graduation value (in millimetres per metre).

The bubble deviation ε, the radius r of the bubble surface and the linear distance a between the bubble ends define their relationship as $\varepsilon = a\varrho/r$. With $a = 2$ mm one can obtain the sensitivity p and the radius r as

$$p = \frac{4'' \cdot 10^5 \text{ mm}}{r} \tag{63}$$

$$r = \frac{4'' \cdot 10^5 \text{ mm}}{p}. \tag{64}$$

The smaller the sensitivity, the larger the radius. It is 13.6 m for a 30″ bubble, 41.2 m for a 10″ bubble and 206 m for a 2″ bubble.

Determination of the Bubble Sensitivity

In addition to the sensitivity, the grinding quality of the tube is also often required. The sensitivity of bubbles, permanently fastened to instruments, can easily be determined with the instrument itself. For more accurate investigations, a level trier is used.

Determination of the Sensitivity with a Levelling Rod

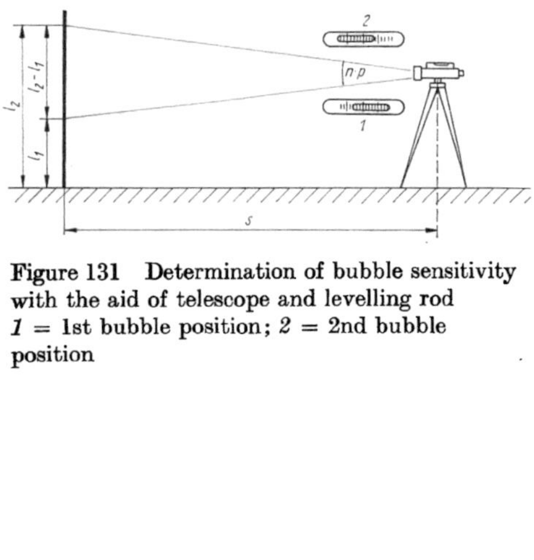

Figure 131 Determination of bubble sensitivity with the aid of telescope and levelling rod
1 = 1st bubble position; *2* = 2nd bubble position

This is a simple method to check telescope—and level bubbles. The instrument is set up at a distance s (20 to 50 m) from a vertical scale, like a levelling rod, such that one of the three foot screws points in direction of the rod (figure 131). Then one reads one bubble end a_1 at the level graduation, and the value l_1 on the levelling rod through the telescope. By rotating the footscrew positioned in direction to the rod, the bubble moves by n graduations, and one observes the bubble end a_2 and the rod reading l_2. The sensitivity of the bubble is then

$$p = \frac{(l_2 - l_1)\,\varrho}{ns}. \tag{65}$$

As a rule, one does not use only two settings, but rather determines the value as the arithmetic mean of a number of readings. Example: The rod readings were $l_1 = 1.812$ m and $l_2 = 1.512$ m. The distance between rod and instrument was $s = 20.63$ m, and the bubble was moved by 10 graduation lines. According to equation (65):

$$p = \frac{(1.812 - 1.512)\text{ m}}{10 \times 20.63\text{ m}} \times 206.265'' = 30''.$$

The sensitivity can be determined more accurately with a scale with mm-graduation. The telescope is aimed such that the horizontal line of the cross hair falls exactly between two graduation lines. In such an arrangement the eye recognizes even slight deviations from the symmetric position, especially with very thin light fringes. Since different bubble lengths may have different adhesion and thus cause inaccuracies, the position of the bubble centre is determined as the mean of the two end readings. Then the horizontal cross hair is moved to the adjacent graduation field, and the bubble position is observed. This process is continued to the edge of the bubble graduation and then reversed. For each bubble position the sensitivity is calculated according to equation (65), with $l_2 - l_1 = 1$ mm. The mean sensitivity is the arithmetic mean of these results.

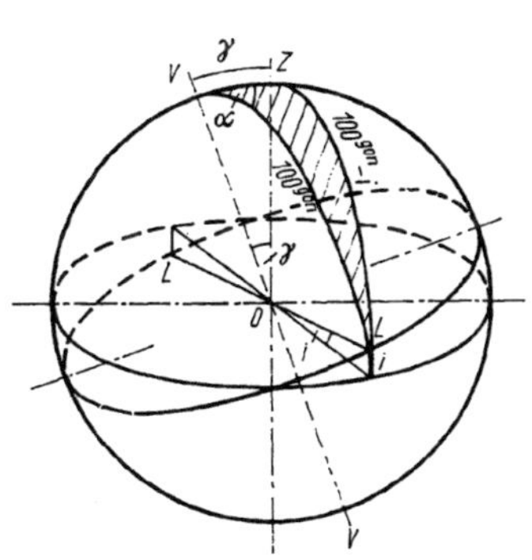

Figure 132 Determination of bubble sensitivity with the aid of the vertical circle

Determination of the Sensitivity Using the Vertical Circle

This procedure by *Comstock*, also called Wisconsin Method (because it was first introduced in Wisconsin, U.S.A.) can be used to check cross levels for theodolites or tacheometers.

It is based on the following theory (figure 132); If the vertical axis VV of the instrument deviates from the plumbline ZO by the angle γ, and if the bubble axis LL is perpendicular to the vertical axis, then the following trigonometric relation in the spherical triangle VZL applies:

$\sin i = \sin \gamma \cos \alpha$.

For two different positions of the instrument plate one obtains

$\sin i_2 - \sin i_1 = \sin \gamma(\cos \alpha_2 - \cos \alpha_1)$.

Since i and γ are small angles, their sines can be replaced by radians. When one considers the angular difference of the angles i as graduation sections multiplied with the bubble sensitivity, one obtains according to the addition theorem:

$$p(i_2 - i_1) = -2\gamma \sin \frac{\alpha_2 + \alpha_1}{2} \sin \frac{\alpha_2 - \alpha_1}{2}.$$

If there is little difference between α_2 and α_1, $(\alpha_2 + \alpha_1)/2 \approx 100$ gon, while $(\alpha_2 - \alpha_1)/2$ is very small, and its sine can be replaced by radians. Therefore

$$p = \frac{\gamma(\alpha_2 - \alpha_1)}{(i_2 - i_1)\,\varrho}. \tag{66}$$

This leads to the following practical application:

The instrument is set up such that one foot screw points towards a well defined target. After levelling of the instrument, the zenith distance for the target is read at the vertical circle with centered index level. By tilting the telescope, the vertical reading is changed, e.g. by 1 gon. Using the foot screw pointing towards the target, the vertical axis of the instrument is tilted by γ until the target comes back to the centre of the cross hairs. Afterwards the bubble axis is positioned approximately normal to the vertical plane through the target such, that bubble is at the edge of the graduation. For the horizontal circle reading an even value is chosen. After reading the bubble ends, one turns the instrument by an even amount and determines the new bubble position. This is continued until the bubble has moved to the other end of the tube. Then the procedure is reversed, and finally γ is checked. If the graduation on the bubble is insufficient, one can attach an auxiliary one.

Example: Determination of the sensitivity of the cross level of the theodolite OT-02 No. 9314. $\gamma = 1°$ (see table 1).

Table 1. Determination of the sensitivity of the cross bubble of the OT-02 No. 9314

Position of Top Part	Bubble reading i forwards		$2[i_n - i_{n-1}]$ (Columns) 2 + 3	Bubble reading i return		$2[i_n - i_{n-1}]$ (Columns) 6 + 7
	left	right		left	right	
1	2	3	4	5	6	7
15° 00′	2.0	13.7		2.1	13.8	
			2.9			2.7
10	3.5	15.1		3.5	15.1	
			3.4			3.3
20	5.2	16.8		5.1	16.8	
			3.0			3.5
30	6.8	18.2		7.0	18.4	
			3.3			3.2
40	8.4	19.9		8.6	20.0	
			3.3			3.0
50	10.0	21.6		10.0	21.6	
			2.9			3.0
16 00	11.5	23.0		11.6	23.0	
			2.9			2.8
10	13.0	24.4		13.0	24.4	
	$2[i_n - i_{n-1}]$		21.7	$2[i_n - i_{n-1}]$		21.5
	$[i_n - i_{n-1}]$ mean		3.1	$[i_n - i_{n-1}]$ mean		3.1

According to equation (66) the sensitivity is

$$p = \frac{1° \cdot 10'}{1.55 \cdot 206.265''} = 6.8''.$$

Determination of the Bubble Sensitivity with the Aid of a Level Trier

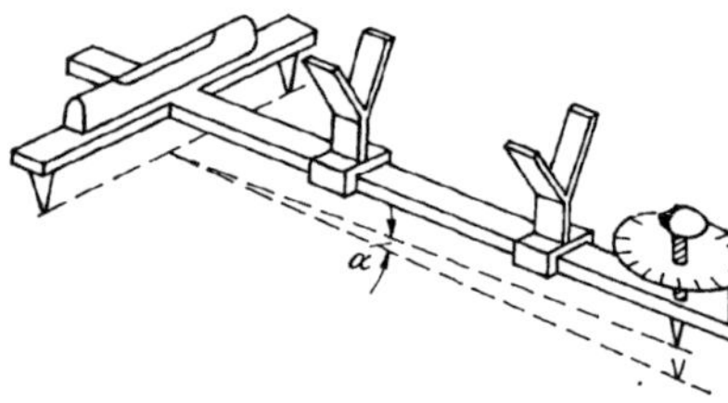

Figure 133 Level trier

A level trier is utilized for highly accurate determination of the sensitivity, as well as for checking the ground arc of the level. A simple design (figure 133) consists of a T-beam which can be tilted with a precision screw, and which has two fork like supports for the level to be tested.
The pitch of the screw is such, that for the effektive length of the trier (between rotational axis of the beam to the centre of the screw) one revolution results in an even value of inclination of the beam. A reading drum at the screw is graduated accordingly. For highest accuracy, more advanced level triers (figure 134) are used, which are set up on a special pillar, isolated from the floor, and free of vibrations. The temperature in the measuring laboratory is kept constant. Usually the part of the level trier which supports the level is covered with a glass box to screen the level bubble from heat radiation, caused for instance by breathing or the human body. For these designs the beam isoften balanced with a horizontal axis and counter weights, so that the sensitive measuring screw does not have to support the weight of the whole beam. The measuring drum is usually subdivided into 120 or 180 parts, with one part representing an inclination of the beam of e.g. 5″. In place of the direct level supports, a plate can be fastened to the beam which then carries the instrument with the bubble to be checked.

Figure 134 Level trier (Freiberger Präzisionsmechanik, out of production)

Figure 135 illustrates the principle of the Hannover level trier as advocated by *Höpcke*. The tilting table of the usual instruments is supplemented with an angular lever with high transmission ratio, while the support axis is replaced by three flat spring joints. The tilting table is connected to the housing with the spring F_3, and the angular lever with the spring F_1, while spring F_2 connects angular lever and tilting table. The tip of the measuring screw presses onto the polished surface of an achat stone, embeded into the angular lever. The measuring range of the instrument is 300″. One interval on the measuring drum corresponds to an inclination of the tilting table of 0.5″. The maximum achievable accuracy with a large level trier is 0.2″. A higher accuracy can only be reached, if the measuring screw is replaced with an interference device, as already done at several institutions. The inclination of the longitudinal axis is then determined from interference lines of equal width.
The checking of a bubble with a level trier uses the following procedure: First the bubble is placed into the supports, or the instrument with the level is set onto the plate. Then the level trier is balanced, and the bubble is brought to its starting position with the measuring screw reading zero.
When testing regular bubbles, the level bubble is moved by rotating the measuring screw for at least two 2 mm intervals in at least two steps once fore- and once backwards. After the bubble has come to rest (after about one minute) the position of the measuring screw and of the centre of the bubble are observed, the latter by reading the bubble ends. The sensitivity is obtained as

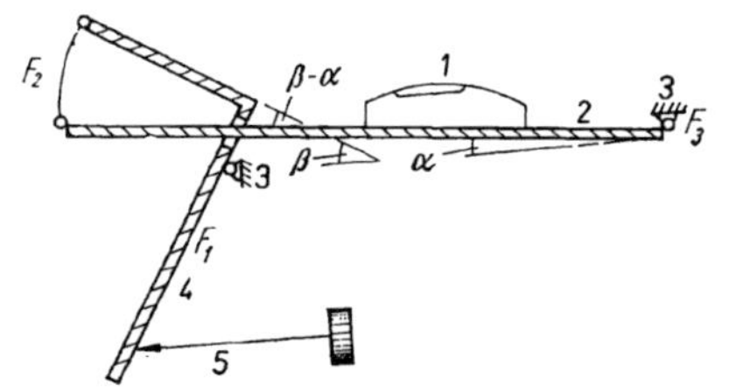

Figure 135 Principle of level trier by *Hoepcke*
1 level; *2* tilting table; *3* housing; *4* angular lever; *5* measuring screw

$$p = \frac{tm}{n} \tag{67}$$

Table 2. Determination of the sensitivity of a levelling bubble

reading at measuring screw	direkt: bubble position lower (l)	direkt: bubble position upper (u)	direkt: length of bubble $(l + u)$	direkt: centre of bubble $(l - u)$	direkt: bubble path $[(l-u)_{n+1} - (l-u)_n]/2$	direkt: $0.4p$ [″]	reverse: bubble position lower (l)	reverse: bubble position upper (u)	reverse: length of bubble $(l + u)$	reverse: centre of bubble $(l - u)$	reverse: bubble path $[(l-u)_{n+1} - (l-u)_n]/2$	reverse: $0.4p$ [″]
1	2	3	4	5	6	7	8	9	10	11	12	13
0	2.5	24.0	26.5	−21.5	4.35	4.60	2.9	23.6	26.5	−20.7	4.65	4.30
20	7.0	19.8	26.8	−12.8	4.55	4.39	7.4	19.1	26.5	−11.4	4.15	4.82
40	11.4	15.1	26.5	− 3.7	3.85	5.19	11.7	14.8	26.5	− 3.1	4.05	4.94
60	15.5	11.5	27.0	+ 4.0	4.50	4.44	15.7	10.7	26.4	+ 5.0	4.35	4.60
80	19.8	6.8	26.6	+13.0	4.25	4.71	19.8	6.6	26.4	+13.2	4.35	4.60
100	24.0	2.5	26.5	+21.5			24.1	2.2	26.3	+21.9		

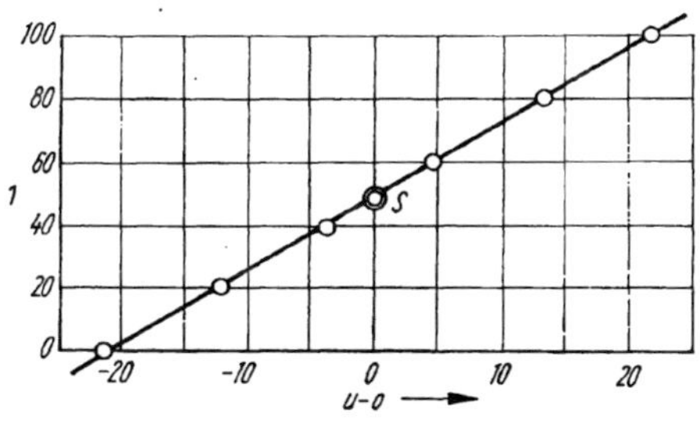

Figure 136 Determination of bubble sensitivity
1 reading on measuring screw

where t is the reading on the measuring drum, m the number of drum intervals, and n the amount of level graduations between the two settings.
The two values for the sensitivity obtained from both steps should not differ by more than 20%.
Measuring levels are treated similarly, however, at least four equal steps are used back and forth. The values for the single steps should not differ by more than 10% from the mean sensitivity. If the readings of the bubble are plotted along the x-axis, and the corresponding inclinations of the level trier normal to it, the slope of the best fitting straight line represents the mean sensitivity of the bubble.
Example: Determine the sensitivity of the level bubble (Ni 004 No. 102579) (see table 2). Since one graduation interval amounts to 0.8 mm for this bubble, $0.4p$ was used. With the 10 values for the bubble movement, the mean value for $0.4p$ resulted in
$200'' : 43.05 = 4.65''$.

Thus the sensitivity of the bubble is
$p = 4.65'' \times 2.5 = 11.6''$
(for graphical solution, see figure 136).

Hirvonen proposed in 1950 a method to determine the sensitivity of a precision bubble which considers all values obtained with the level trier. This method was further developed by *Weise* in 1962. The sensitivity is computed using observations covering the whole range of the graduation as before.
A linear function

$$\bar{y} = a_0 + a_1 x \tag{68}$$

is determined, where x is the inclination of the bubble axis, and $\bar{y}$ the position of the bubble centre after least squares adjustment. The values y to be adjusted are obtained from the forwards readings l, r (resp. o, u) at the left and right, resp. upper and lower bubble edge, and the backwards readings l', r' (resp. o', u') as

$$y_i = (l_i + r_i + l_i' + r_i') : 4 \tag{69a}$$

or, for a coincidence bubble as

$$y_i = (l_i - r_i + l_i' - r_i') : 4 . \tag{69b}$$

The addition constant $a_0 = [y]/n$ is only of importance for the least squares adjustment, which with a_1 provides the sensitivity p directly. Differences in the values Δl, Δr, $\Delta l'$, $\Delta r'$ (resp. Δu, Δo, $\Delta u'$, $\Delta o'$) when reading both bubble edges for steady bubble movement indicate grinding errors, uneven inclination of the bubble axis, or observation errors. Different values on one end in respect to the other indicate grinding or reading errors. If the graduation of the level trier does not match the actual inclination, a factor has to be included; generally however, $\eta = 1$. Without any further explanations, the remaining formulae and an example are presented:

$$b_i = u_i + o_i ; \; b_i' = u_i' + o_i' \tag{70}$$

$$x_1 = \frac{1 - n}{2} ; \; x_2 = \frac{3 - n}{2} ; \ldots \tag{71}$$

(n = number of observed values).

Value of the adjusted function

$$\bar{y}_i = \frac{[y]}{n} + [xy]\, Q x_i . \tag{72}$$

With the Lagrange factor

$$Q = \frac{12}{n(n^2 - 1)} \tag{73}$$

$$v = \bar{y}_i - y_i . \tag{74}$$

The sensitivity results as

$$p = \frac{\Delta\alpha}{[xy]\, Q\eta} \tag{75}$$

with α denoting the position of the measuring screw at the level trier.
The average error is

$$t = \frac{[|v|]}{n} p \tag{76}$$

Table 3. Computation of the bubble sensitivity

No.	α	forwards					return					x	y	$\bar{y}$	v
	[″]	bubble position l	bubble position u	b	Δl	Δu	bubble position l	bubble position u	b'	$\Delta l'$	$\Delta u'$				
1	2	3	4	5	6	7	8	9	10	11	12	13	14	15	16
1	0	2.5	24.0	26.5	1.3	1.2	2.9	23.6	26.5	1.1	1.1	−10	−10.55	−10.34	+0.21
2	5	3.8	22.8	26.6	0.9	1.1	4.0	22.5	26.5	1.1	1.1	−9	−9.38	−9.28	+0.10
3	10	4.7	21.7	26.4	1.1	1.1	5.1	21.4	26.5	1.1	1.1	−8	−8.32	−8.22	+0.10
4	15	5.8	20.6	26.4	1.2	1.1	6.2	20.3	26.5	1.2	1.2	−7	−7.22	−7.16	+0.06
5	20	7.0	19.5	26.5	1.1	1.1	7.4	19.1	26.5	1.1	1.1	−6	−6.05	−6.10	−0.05
6	25	8.1	18.4	26.5	1.1	1.1	8.5	18.0	26.5	1.1	1.1	−5	−4.95	−5.04	−0.09
7	30	9.2	17.3	26.5	1.1	1.1	9.6	16.9	26.5	1.1	1.0	−4	−3.85	−3.98	−0.13
8	35	10.3	16.2	26.5	1.1	1.1	10.7	15.9	26.6	1.0	1.1	−3	−2.78	−2.92	−0.14
9	40	11.4	15.1	26.5	1.0	1.0	11.7	14.8	26.5	1.1	1.1	−2	−1.70	−1.86	−0.16
10	45	12.4	14.1	26.5	1.0	1.0	12.8	13.7	26.5	1.0	1.0	−1	−0.65	−0.80	−0.15
11	50	13.4	13.1	26.5	1.1	1.1	13.8	12.7	26.5	0.9	1.0	0	+0.35	+0.26	−0.09
12	55	14.5	12.0	26.5	1.0	1.0	14.7	11.7	26.4	1.0	1.0	1	+1.38	+1.32	−0.06
13	60	15.5	11.0	26.5	1.1	1.1	15.7	10.7	26.4	1.0	1.0	2	+2.38	+2.38	±0.00
14	65	16.6	9.9	26.5	1.0	1.0	16.7	9.7	26.4	1.0	1.1	3	+3.42	+3.44	+0.02
15	70	17.6	8.9	26.5	1.1	1.1	17.7	8.6	26.3	1.1	1.0	4	+4.45	+4.50	+0.05
16	75	18.7	7.8	26.5	1.1	1.0	18.8	7.6	26.4	1.0	1.0	5	+5.52	+5.56	+0.04
17	80	19.8	6.8	26.6	1.1	1.1	19.8	6.6	26.4	1.0	1.1	6	+6.55	+6.63	+0.08
18	85	20.9	5.7	26.6	1.0	1.1	20.8	5.5	26.3	1.1	1.1	7	+7.62	+7.69	+0.07
19	90	21.9	4.6	26.5	1.0	1.0	21.9	4.4	26.3	1.1	1.1	8	+8.70	+8.74	+0.04
20	95	22.9	3.6	26.5	1.1	1.1	23.0	3.3	26.3	1.1	1.1	9	+9.75	+9.80	+0.05
21	100	24.0	2.5	26.5			24.1	2.2	26.3			10	+10.85	+10.86	+0.01
													$[v] = +5.52$		$[v] = -0.04$

$n = 21;\quad \frac{[y]}{n} = 0.263;\quad [xy] = 816.22;\quad Q = 0.00130;\quad \eta = 1;\quad [|v|] = 1.70$

$\underline{p = 4.71''_7}\quad \underline{t = \pm 0.38''}$

Example: Determination of the sensitivity of a level bubble (Ni 004 No. 102579 from Zeiss, Jena) according to *Hirvonen/Weise* (table 3).
The measurements at the level trier were performed in the usual manner. The measuring screw of the level trier was evenly changed by 5″ between observations. The bubble ends, appearing in the field of view were moved such as to cover the total graduated range. Temperature measurements were taken at the beginning and end. The computation of the level sensitivity is illustrated in table 3 (graduation interval was 0.8 mm).

Adjustment of a Tubular Level

For tabular levels, the bubble axis should be parallel to the setting line, or normal to the vertical axis for levels used to correctly set the vertical axis of an instrument. It is however, possible to measure with bubbles who do not fulfil this requirement, if one considers the centre of run; but work with adjusted levels is much more comfortable. To adjust the levels, they have adjustment screws on one end. This bubble end can be inclined with respect to the setting line with either two pressure screws working in opposite directions, or one pull- and two pressure screws, or one pulling screw in connection with a screw-spring.
Furthermore, levels have occasionally adjustment screws operating sideways to eliminate a cross divergence error. It can be recognized when a bubble moves in opposite directions, if the telescope carrying the bubble is swivelled back and forth. A special adjustment pin or a screwdriver is used to move the adjustment screws.

Adjustment of a Setting Level

For an adjusted setting level, bubble axis and setting line are parallel. If it is placed on a horizontal plane, the bubble is centred in any position of the level. To adjust the level, it is placed perpendicular to the horizontal axis onto an adjustment board, which has two supports at one end, and a setting screw at the other. This board is tilted with the setting screw until the bubble appears centred. When the level is reversed (rotation by 200 gon), the bubble will move, if it is out of adjustment. This bubble deflection corresponds to twice the tilt angle of the adjustment board.

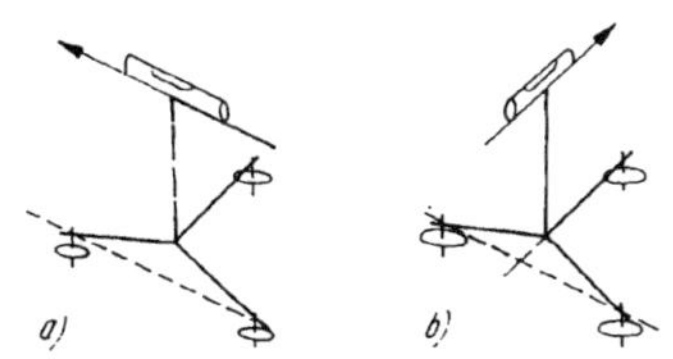

Figure 137 Plumbing of vertical axis

Half this amount is eliminated with the adjustment screws of the level, the other half with the setting screw of the board. Then the bubble axis should be parallel to the setting line. If the bubble moves after reversal, the procedure has to be repeated. The setting line can also be levelled with a non-adjusted level: The centre of run is determined from the two level positions. Then the support is tilted until the bubble centre remains at the centre of run for both positions, and thus has equal separations from the normal point. This also holds for other levels.

Adjustment of a Striding or Hanging Bubble

A striding or hanging bubble is adjusted like a setting level or a plate level. First, however, one has to check whether bubble axis and setting line are in one plane, e.g. whether a cross divergency is apparent. If so, it has to be eliminated with the adjustment screws at the side.

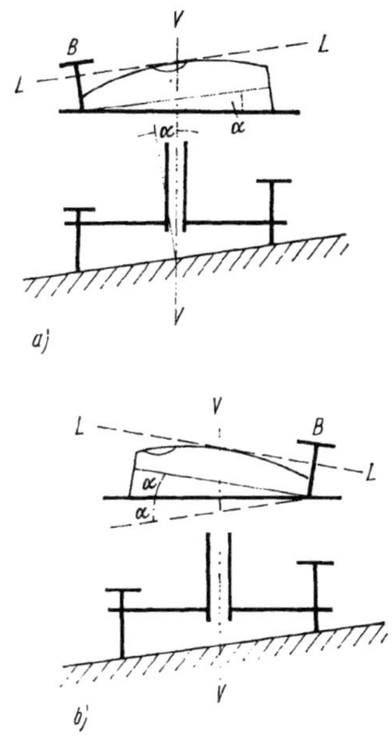

Figure 138 Use of plate level

Adjustment of a Plate Level

The bubble axis of an adjusted plate bubble is normal to the vertical axis of the instrument. The level is adjusted as follows: The bubble is first centered with the level being parallel to two foot screws (figure 137a) of the instrument. Then it is rotated by 100 gon, and centered with the third foot screw (figure 138a). When rotating the foot screws, the bubble moves in direction of the left thumb and the right index finger. One has to always rotate in opposite directions. A deflection of the bubble when turning the level by 200 gon (figure 138b), represents twice the vertical axis error. It is halfways eliminated with the foot screws, and halfways with the adjustment screws of the level.
One also can select another procedure. After the bubble is centered with the level parallel to two foot screws (figure 137a), the level is turned by 200 gon (figure 138b), and half the deflection of the bubble is eliminated with the foot screws. The bubble is now at the centre of run. After turning the level normal to the previous position, the bubble is again brought to the centre of run with the third foot screw (figure 137b). If during slow rotation of this instrument around its vertical axis, the bubble moves from the centre of run, the above procedure has to be repeated. If it remains, the vertical axis is set correctly. With the adjustment screws, the bubble can be set to the normal point and thus is adjusted.

Adjustment of an Index Level

(see section 3.3.2.)

Adjustment of a Telescope Level or Level Bubble

(see section 3.2.4.)

1.2.2. Bull's Eye Levels

The bull's eye level consists of a cylindrical glass container in a metal casing whose lid is ground spherically at the inside (figure 139). Similar to the tubular level, it is filled with a thin liquid, usually alcohol or ether, except for a bubble of a few millimetres' extension. Its bottom is then molted closed. For centering of the bubble, one or more concentric circles around the normal point at 2 mm separation are engraved in the lid. The level axis is normal to the tangent plane in the normal point. The centre of run is the bubble centre for correct vertical axis or horizontal setting plane. The sensitivity of bull's eye levels is usually not better than 8″, and its grinding radius does not exceed 0.8 m. Such levels therefore serve only for rough levelling of a plane. It is advantageous, that this can be accomplished with one set-up. Newer instruments of lower precision occasionally do not have tubular levels anymore but only a bull's eye level with a sensitivity of e.g. 4′, which yields the necessary accuracy for setting the vertical axis correctly.
For an adjusted bull's eye bubble, the centre of run coincides with the normal point. Bull's eye levels are adjusted and checked like setting levels, resp. checked along with the plate bubble. If the latter is centered in all positions, the bull's eye bubble should be also.
Any deviation has to be completely eliminated with the adjustment screw of the bull's eye level.

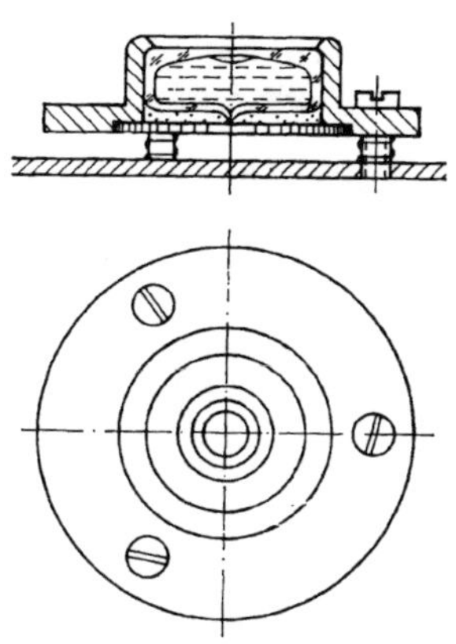

Figure 139 Bull's eye level

1.2.3. Electronic Levels

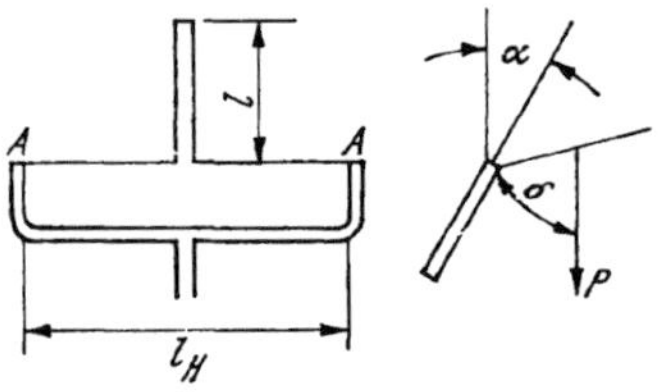

Figure 140 Quartz level

While quartz- and metal levels have not yet reached practical importance, electronic levels are used to determine inclinations with a high accuracy, e.g. for instrument studies or engineering surveys.
Quartz levels are based on the elastic behaviour of a quartz thread which tilts under gravity influence. They are considered astatic (easily movable [Greek]). The level consists of a quartz rod of length l and mass P and a quartz threat stretched between the supports $A - A$ (figure 140). When the level is tilted by α the rod turns the thread and moves by γ from the plumbline. The sum of the moments

$$M_P = \frac{Pl}{2}\sin\gamma\,, \quad -M_k = \frac{4G(\gamma-\alpha)}{l_H} I_P \tag{77}$$

(l_H equals the thread length, G is the module of elasticity, and $I_P = \pi r/2$ is the polar moment of inertia of the thread cross section with diameter $2r$). Results in equilibrium for

$$M_P + M_k = 0\,. \tag{77a}$$

For small values γ and α, the moment equation gives the following:

$$\frac{\gamma}{\alpha} = \frac{8GI_P}{8GI_P - Pll_H}\,. \tag{78}$$

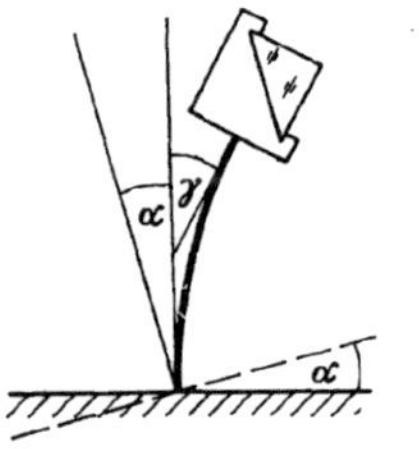

Figure 141 Astatic pendulum

The astatic pendulum is based on the same principle. It consists of a vertical elastic spring, loaded on top with a mass and connected on the bottom to the base with a thin piece of sheet metal (figure 141). When tilting the base by α, the steel spring will swing, because of the imbalance due to the mass, by $n\alpha$. With $k = \gamma/\alpha$, equilibrium is achieved according to equation (78). This spring pendulum is used for gravity determination, and in compensator levels.
Metal levels are based on the principle of a metal plate sliding over another one with a spherical or cylindrical surface with large diameter. Between the plates there is an air cushion of a few micrometres' thickness.
For electronic levels, the movements of a pendulum are transformed into an electric quantity by changing two inductive resistors within a certain arrangement. This quantity is subsequently measured.
During observation, the two Ohm-resistors R_1 and R_2 remain unchanged (figure 142). R_1 can only be changed to adjust the bridge at the zero position of the metal pendulum which swings between the coils R_3 and R_4. If it changes its position, the induction of the coils, and therefore its inductive resistance and the current in the zero branch of the bridge change. The increased current change is shown.

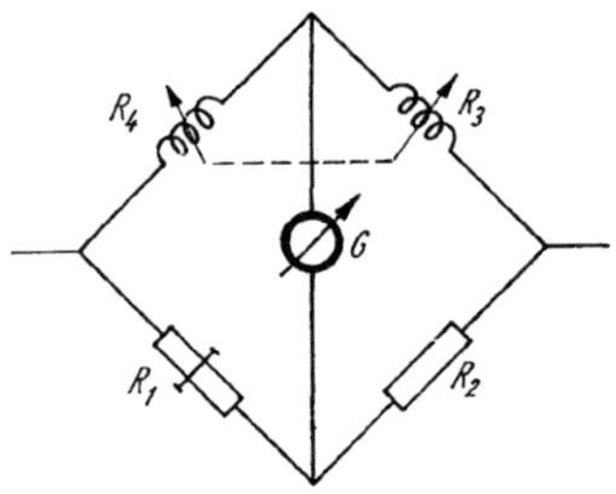

Figure 142 Principle of electronic level

Electronic levels consist of level- and reading unit. The measuring range is at best 2 to 3 gon, the graduation up to 1″.
The electronic level "Talyvel" (figure 143) can be used in three ranges of $\pm 50''$, 100″ and 8′ with graduation of 2″, 4″ and 20″. Batteries (9V) serve as voltage source. The reading unit has a mass of 3.8 kg, the level unit 1.1 kg.
The low frequency A/C voltage generated by the oscillator O supplies the Wheaton measuring bridge R_1, R_2, L_1, L_2 (figure 144). The highly permeable mass K is suspended as a pendulum between the coils L_1 and L_2 (figure 145). In zero position, no A/C voltage arrives at the amplifier V. When tilting the level unit, the position of the mass K in relation to the two coils changes. The changed inductivity generates an A/C voltage at the amplifier V, and a reading at the galvanometer G, which is proportional to the tilt change for small tilts. Thus the tilt change of the level unit can be read at the galvanometer G.

Figure 143 Electronic level "Talyvel" (Rank Precision Industries, England)

The electronic level built by Tese in Remens, Switzerland, has two ranges: ± 0.75 mm/m ($\pm 150''$) and ± 0.15 mm/m ($\pm 30''$). The maximum error of the tilt measurement is less than 0.005 mm/m. The measuring time is 3 s, the mass 4 kg.
Because there is practically no inertia, it can be read immediately, which is an advantage. So is the high measuring accuracy of small angles.
For electrolytic levels, four contacts, arranged in a circle are partly covered by an air bubble. Another current source serves as opposite electrode. Positional changes of the level lead to shifting of the bubble and therefore to changes in the transfer resistance between contact and electrolyt. A level produced by Regelungs- und Meßtechnik Schmidt in Reinbek, F.R.G., has a maximum of 2V supply voltage between two contacts, $\pm 8''$ measuring range, $< 1''$ sensitivity and a mass of 16 g.

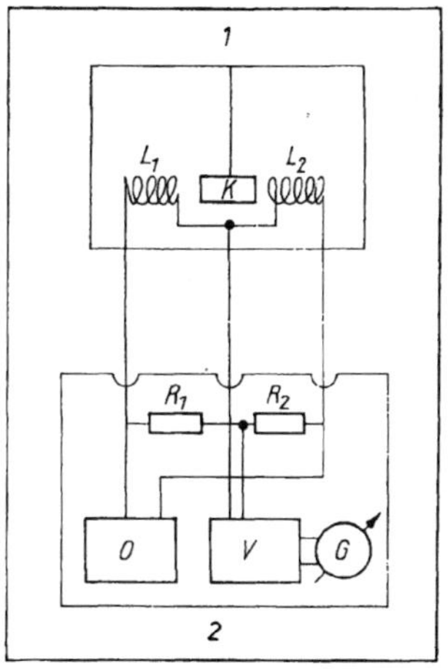

Figure 144 Block diagram of circuits of the Talyvel
1 level unit; *2* indicator unit

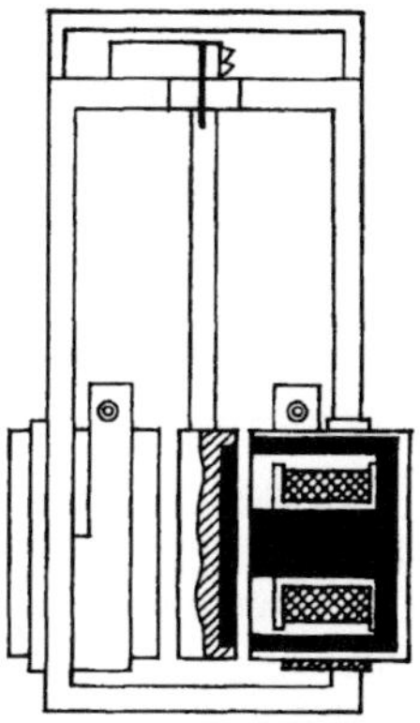

Figure 145 Section through an electronic level

1.2.4. Tilt Compensator

Tilt compensators, short "compensators", are elements which operate automatically under gravity influence. They replace the tabular bubble in levels and optical plummets, as well as the index level in theodolites. The operation of a compensator can be compared to a pendulum with a telescope or reading index fastened in its axis or normal to it. Under gravity influence, the pendulum moves automatically into the plumb line and thus sets telescope or index horizontal or vertical (figure 146a).

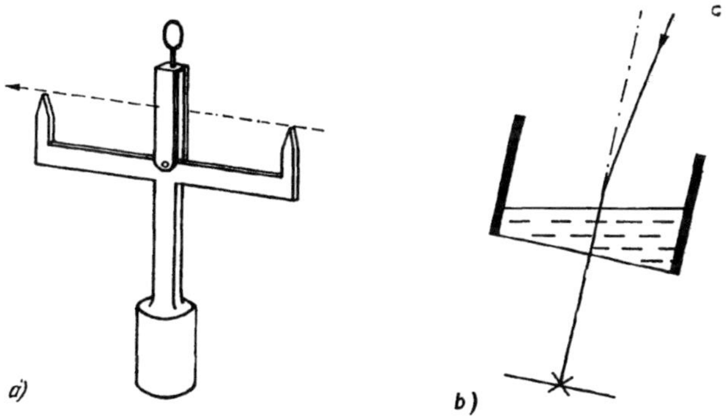

Figure 146 Compensators a) compensator with free swinging pendulum, b) liquid compensator

There are liquid and pendulum compensators. Liquid compensators are based on the characteristic of a liquid to set its surface horizontal when under gravity influence. They can be used as horizontal mirror surface, or—for penetrating rays—as optical wedge with varying refractive angle (figure 146b).

There are mechanical and optical-mechanical pendulum compensators. Compensators have a range of up to $\pm 90'$, and can reach accuracies of $\pm 0.2''$.

2. Instruments for the Determination of Horizontal Angles and Horizontal Projections of Points

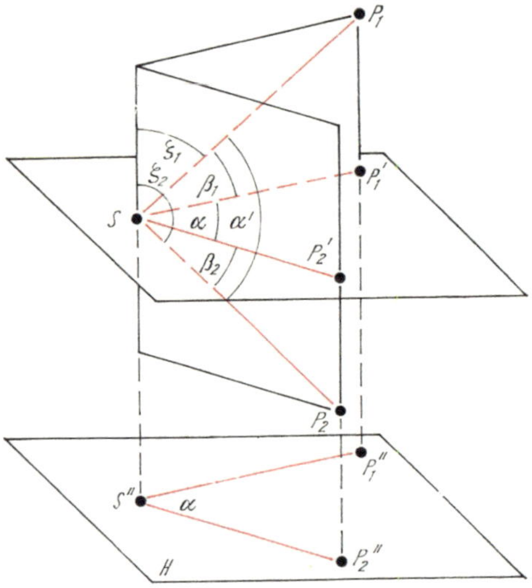

Figure 147 Various types of angles

Depending on their location within a certain plane, one distinguishes the following types of angles (figure 147):
Horizontal Angles (α) lie in a horizontal plane. Their sides S_1' and S_2' are the projections of the rays to the targets P_1 and P_2 onto a plane, parallel to the horizontal reference plane and passing through the station.
Vertical angles lie in a vertical plane. The rays to the target, and the plumbline passing through the station define zenith distances (ζ_1, ζ_2), while the rays to the target and their projection onto the horizontal plane passing through the station result in vertical angles (β_1, β_2). For surveying instruments, primarily zenith distances are read in Central Europe, while vertical angles are preferred in other countries.
Position angles (α') are angles in a random plane. Horizontal angles can be determined either with simple angular instruments, or with the horizontal circle of a theodolite as the difference of two subsequently determined directions. This is the usual procedure in surveying. Right angles or straight angles are staked either with these instruments, or with instruments where the light rays are deflected by a constant angle, due to reflection or refraction. Vertical angles are measured with tilt meters or with the vertical circle of a theodolite. Position angles are not used in surveying. They can be obtained directly with a free hand instrument, i.e. a sextant, or else derived mathematically from a horizontal and the adjacent vertical angles.
The horizontal projection of points is determined with plumbing instruments.

2.1. Instruments for Staking Right and Straight Angles

For the determination of intermediate points on a straight line, the staking of right angles, or the determination of perpendiculars to a line, instruments are used, which permit the staking of right (100 gon) and straight (200 gon) angles. There are diopter –, mirror –, and prism instruments.

2.1.1. Diopter Instruments

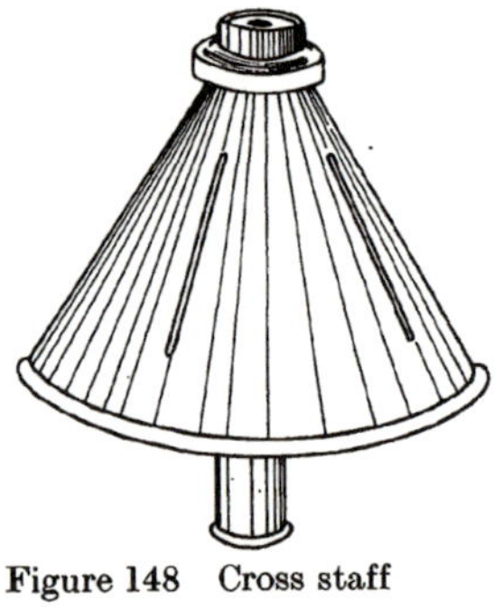
Figure 148 Cross staff

Angle Cross, Cross Staff etc.

Diopter instruments are mainly of historical importance. The sighting devices consist of diopters, which are arranged in a fixed angular configuration. At the angle cross they are located on top of a flat metal ring, while the diopter staff has an octogon or cylindrical prism.
The formerly most common instrument of this type is the cross staff (figure 148). It consists of a blackened conical section fastened to a ground plate with two pairs of slots, about 0.5 mm wide, whose sighting planes are normal to each other. Because it is possible to have inclined sights of up to 35 gon, the cross staff is used in mountainous terrain. A bull's eye bubble is used for levelling the instrument, which is equipped, like all diopter instruments, with a stock as support.
The standard deviation of an angle staked with the cross staff is about 40 mgon. Thus at 50 m distance, a lateral uncertaincy of up to ± 3 cm can be expected.

Checking and Adjusting of Diopter Instruments

To assure proper operation, the instruments have to fulfill the following requirements as demonstrated for the cross staff:

1. The vertical axis has to be vertical when the bull's eye bubble is centered. Check: Use plumbbob in two mutually perpendicular directions.

2. Each pair of slots has to be in a vertical plane when the staff is vertical. Check: A vertical line (e.g. edge of a building) has to appear completely within the view when the staff axis is vertical.
3. Staff axis and line of intersection between the target planes have to coincide. Check: Sight to a point and stake another one at 200 gon. Turn the cross staff by 100 gon and repeat the procedure with the same points. The staked point again has to be on the line of sight.

Errors of a cross staff cannot be adjusted.

2.1.2. Mirror Instruments

Mirror instruments are much more handy than diopter instruments and more comfortable to use when staking fixed angles. They are based on the law of reflection.

Angle Mirror and Mirror Cross

For the angle mirror, (figure 12) φ equals 50 gon, therefore $\delta = 100$ gon, according to equation (7) (figure 50). It is used hand held together with a plumb bob or a plumbing rod to stake right angles or to find the foot of perpendiculars.
The mirror cross consists of two mirrors, arranged on top and normal to the other. With $\varphi = 100$ gon (figure 149), δ becomes 200 gon. This instrument is used to find intermediate points of a straight line.

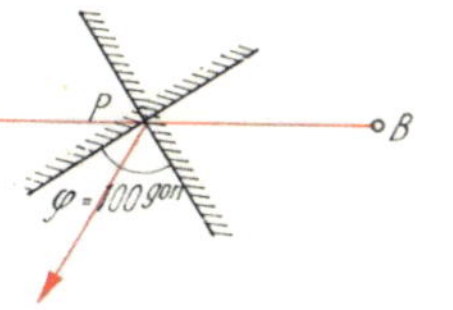

Figure 149 Rays passing through mirror cross

Adjustment of Angle Mirror and Mirror Cross

To check the angle mirror, a point C is lined up between points A and B (figure 150). A right angle in direction D is staked using AC and BC. If D_1 and D_2 coincide at D, the angle mirror is in adjustment. If not, D is to be determined as the mean of D_1 and D_2. Then one of the mirrors is changed with the fastening screws until the angles staked from both sides coincide in direction CD.
When checking the mirror cross, one determines first a point P_1 with point A to the left and B to the right. Then point P_2 is established with point B to the left and point A to the right. The separation P_1P_2 corresponds to twice the error. It is eliminated by setting up the instrument in the middle between P_1 and P_2 and changing the angle between the two mirrors until the images of A and B are on top of each other.

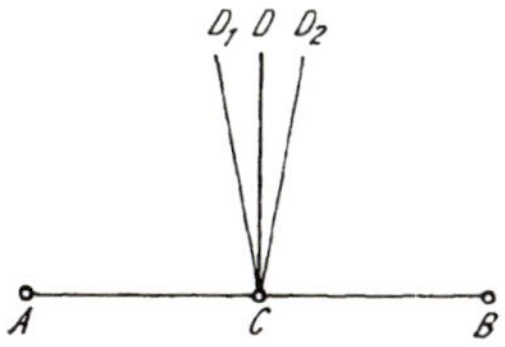

Figure 150 Testing of angle mirror

2.1.3. Prism Instruments

Unlike diopter instruments, where the directions of the two sides of an angle have to be measured consecutively, prism instruments permit the determination of the angle with one sighting. They are therefore used like mirror instruments, namely hand held. Prism instruments are handier than mirror instruments and can be transported more comfortably. Furthermore, simple prisms do not change, therefore no adjustment devices are necessary. The guiding errors of prisms are less than 2 gon. The standard deviation for sighting amounts to 20 to 40 mgon, thus is smaller than for mirror instruments. With this the lateral uncertaincy is ± 2 to ± 3 cm for 50 m. When using plumbing rods it can be reduced by about 50%.
Two reflecting glass surfaces, arranged at a specific angle, are the basis of all angular prisms. The light ray is deflected by twice the amount of the angle between the mirror surfaces because of two reflections. It does not matter, whether the reflections take place at mirror surfaces or whether total reflection occurs on a plain glass surface.
The surfaces of the angular prism where the rays enter or exit have to be normal to each other. Then the refraction of the rays, which also are orthogonal, is the same at both surfaces.
The field of view is relatively small for all prism instruments (about 16 gon), is however, sufficient to clearly see targets like range poles, corners of houses etc.

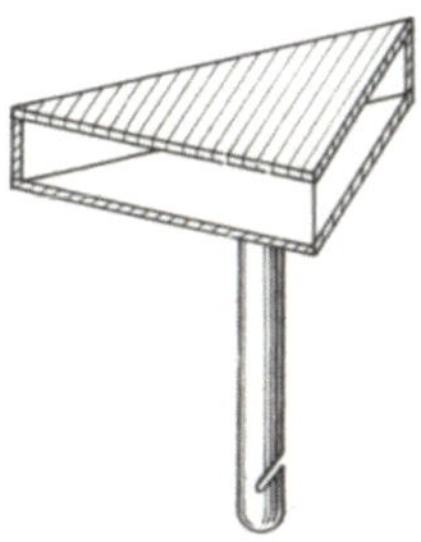
Figure 151 Triangle prism

Simple Angle Prisms

The triangle prism (figure 151) is a rectangular prism with equal sides, whose hypothenusis surface is mirror coated. This prism can be held with a handle attached to the casing, which leaves the sides free. A plumb bob

or a centering rod can be fastened to the handle. Of the two possible images (see figures 64 and 65), only the stationary one is used. If base and top surfaces are mirror coated, inclined sights are possible.
The pentagon prism (figure 67) has several advantages over the triangle prism. There are only stationary images; the images are brighter because light losses are kept small due to the small entrance angles; the vertex of the right angle is located inside the prism; the diameter of the usable bundle of rays is larger. Some types have mirror coated top and bottom surfaces, which permit inclined sights.

Testing of Angle Prisms

Angle prisms are tested on the basis that only one right angle is possible at a point on a straight line. There is no adjustment possibility.

Prism Crosses

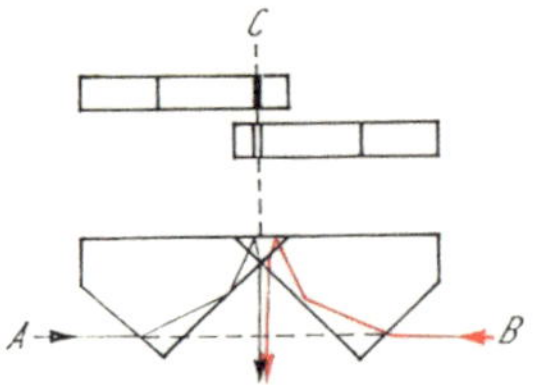

Figure 152 Rays passing through double angle prism consisting of triangular prisms

Prism crosses or double angle prisms (figure 153, 156) consist of two prisms in a casing, arranged on top of each other. These prisms could be triangle-, pentagon — or Wollaston prisms. Each of the two prisms can be used as single prism to stake right angles. When utilizing both prisms, one can place intermediate points on a straight line, or find the foot of normals to a straight line.
Triangle prisms can be placed in different arrangements to each other, e.g. like in figure 152. The prisms of the commonly used double pentagon (figure 154), and of the cross prism (figure 155) with Wollaston prisms (figure 68) are arranged such, that the exit surfaces lie in one plane, and

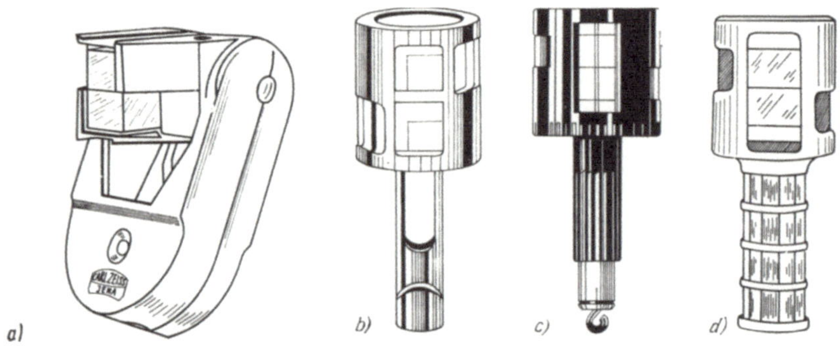

Figure 153 Double prisms a) Zeiss, Jena; b) Zeiss, Oberkochen; c) Wild, Heerbrugg (GDP 2); d) Kern, Aarau; similar ones by Hensoldt, Wetzlar; Lietz, U.S.A.; Rank (SH 4), England; SLOM (SE 5), Paris; Theis, Wolzhausen etc.

that the rays enter and exit normal to the glass surfaces. Figure 157 illustrates a different arrangement with trapezeon prisms with total reflection.
Often the handle below the prisms has an opening to allow straight sights, and sometimes a plane parallel plate is cemented between the prisms for this reason. In order to protect the prisms when not in use, either a turning cover is used (figure 153b, c, d), or else they can be snapped into the casing (figure 153a).
The instrument is held such, that the principal planes of the prisms are horizontal, and that vertical targets appear vertical. The target in direction C (figures 152, 154, 155 and 157) is sighted directly below or above the prisms. If the images of the points A and B coincide vertically with C, the lines connecting A, B and C with the prism are normal to each other. A plumb bob can be fastened through a drill hole in the prism housing. Instead of a plumb bob, a telescopic centering rod can also be used.

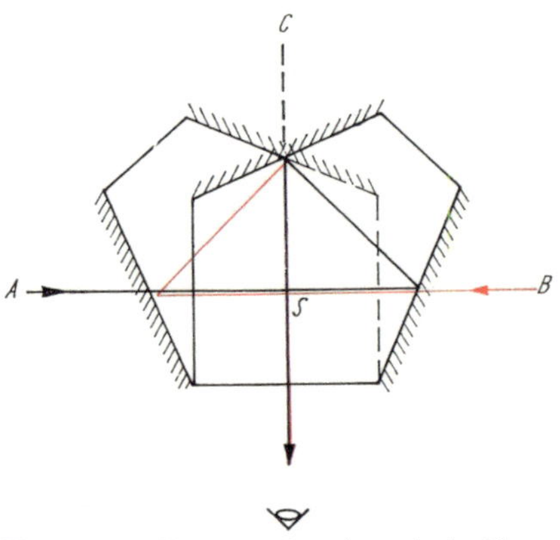

Figure 154 Rays passing through double system

Testing of a Prism Cross

Like for the mirror cross, the test of a prism cross is based on the fact that two points are connected by only one straight line (see figure 150). Adjustments are not necessary, because the production of prism crosses is sufficiently accurate.

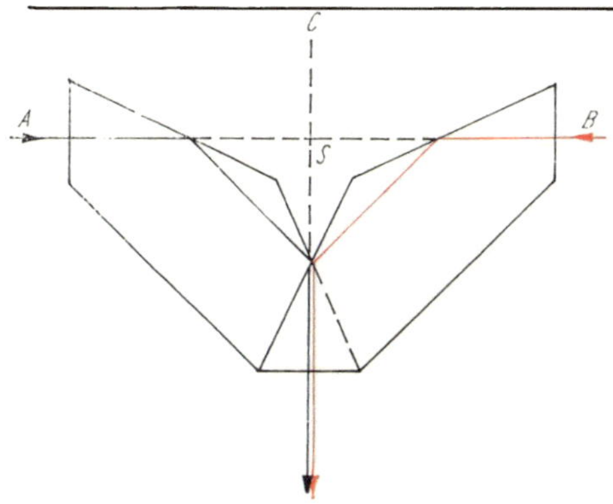

Figure 155 Rays passing through cross prism

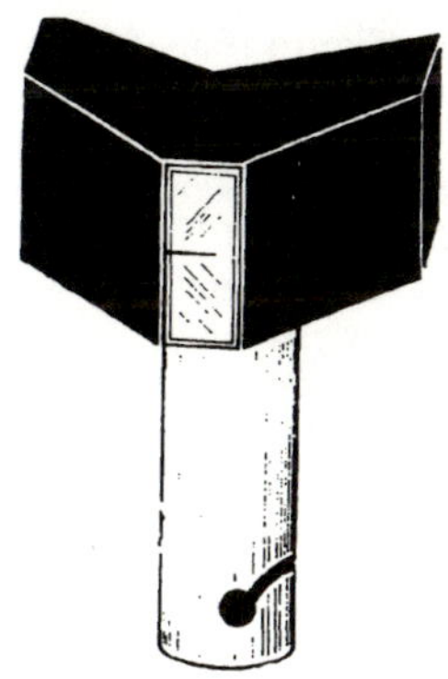

Figure 156 Cross prism (Hensoldt, Wetzlar)

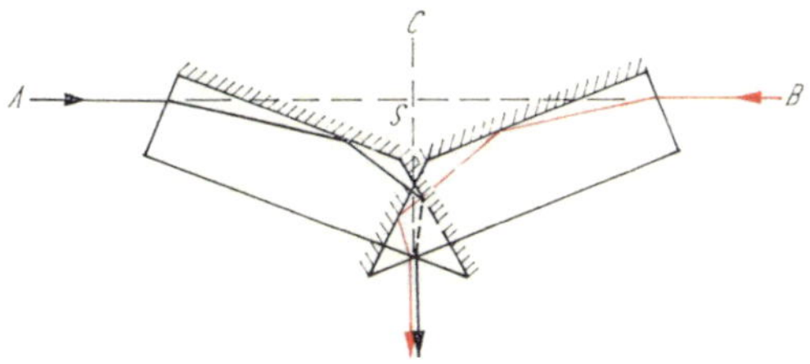

Figure 157 Rays passing through double angle prism SE 3 (SLOM, Paris)

2.2. Simple Instruments to Measure and Stake Horizontal Angles

An instrument for measuring different angles needs the following main features:

1. A graduated circle to read the amount of the measured angle,
2. an axis, perpendicular to the graduated circle, passing through its centre, around which a sighting plane can be rotat‹.l. This plane is generated by diopters or a telescope, and is projected onto the graduated circle for reading.

Goniasmometer and stock compass are simple instruments when compared to the theodolite.

2.2.1. Goniasmometer

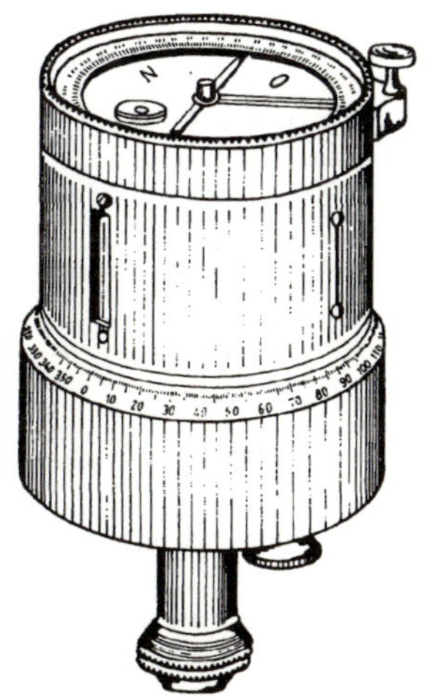

Figure 158 Goniasmometer

Structure of the Goniasmometer

The goniasmometer (figure 158) consists of a conical or cylindrical upper part with slots for sighting, which can be rotated in respect to its base, which is also equipped with slots. The upper part carries an index or a vernier to read directions to a few tenths of a degree on the graduated base. This instrument is used with a rod and levelled with a bull's eye bubble. Often it also has a compass. The goniasmometer is nowadays of little importance.

Testing of the Goniasmometer

The goniasmometer is tested like a cross staff (see section 2.1.1.). However, errors cannot be corrected. Due to its low targetting and reading accuracy. the goniasmometer can only be used for low accuracy measurements.

2.2.2. Stock Compass

Structure and Operation of the Stock Compass

A stock compass is a magnetic compass equipped with a sighting device. It is manufactured either as a self contained instrument or as a supplementary part of other instruments. Magnetic bearings can be measured or transferred to the terrain with the aid of a stock compass.
There are two types of stock compasses used in the surveying field, namely the full circle compass (figure 159), and the orientation compass. The full circle compass is produced either as self contained instrument such as diopter compass (figure 160), telescope compass (figures 164, 165), and as compass

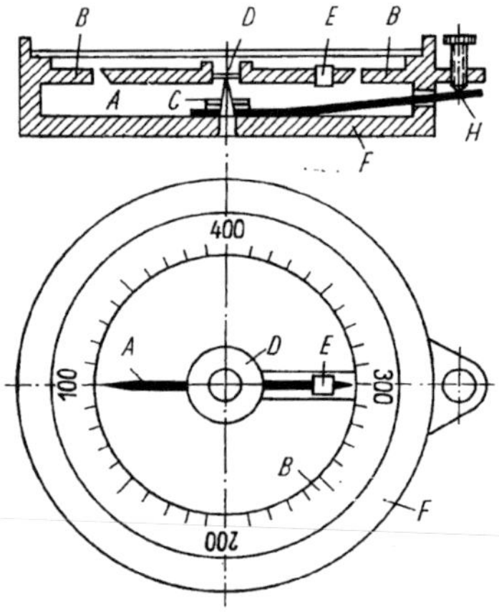

Figure 159 Full circle compass

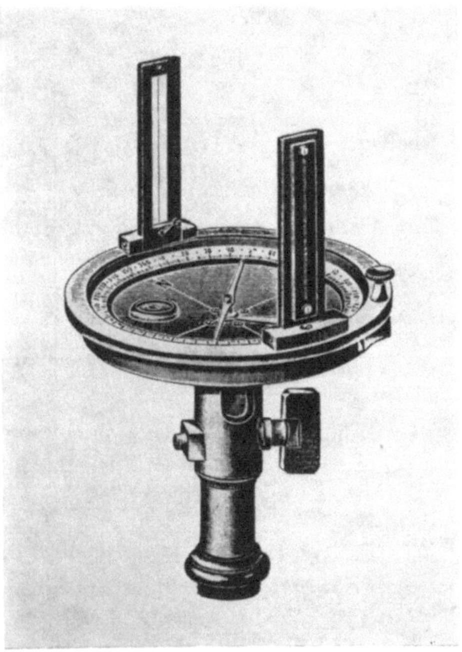

Figure 160 Diopter compass

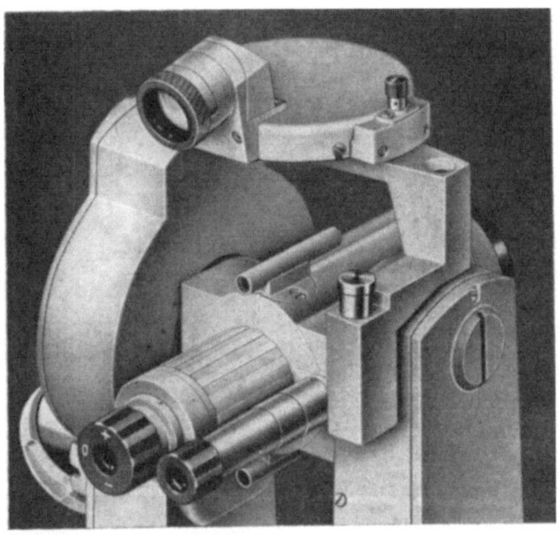

Figure 161 Striding compass

theodolite (figure 166, 167), or else as supplement, such as striding compass (figure 161).
The orientation compass, also a striding instrument, serves only as supplement. It is produced as a box (see figure 171) or as tubal compass (see figure 173).
The most important element of a compass is the magnetic needle. For simple compasses it is punched in rhomb-shape out of sheet metal and simply suspended. For surveyor's compasses it is usually a beam needle (figures 159, 161, 170), suspended on edge. It is manufactured from tungsten- or chrome steel, and has to have high remanence (remaining magnetism after removal of a magnetic field) and high coertion power (magnetic field in opposite orientation to multiply the remanence) with a magnitude field intensity between 50 and 65 A/cm.
The magnetic needle A (figure 159) moves in a precision ground and polished achat-cap D sitting on top of a thin vertical steel tip C. To avoid damage to the of tip due to vibrations, the cap can be lifted off via the lever H of the locking arrangement and pressed against the glass top of the housing. If the instrument is not in use or if one moves around with it, the needle has to be locked, i.e. lifted off the tip and arrested.
The earth magnetism causes not only a deflection of the magnetic needle from geographic north, the magnetic declination, but also an inclination from the horizontal plane. Like for the declination, its value differs in place and time. The needle is kept at equilibrium with a small movable balance piece E. For areas within 500 km radius, one setting of this balance piece is sufficient. The foot of the tip is screwed into the centre of the floor of the cylindrical housing F (10 to 15 cm diameter). The housing contains a graduation B, numbered towards the left. When turning the instrument clockwise, the readings of the needle increase. One immediately obtains the angular value of the magnetic bearing. The circle graduation is usually degrees or half degrees. Tenths of degrees can be estimated. Only a few compasses, whose graduated circle is permanently connected with the magnetic needle, are numbered towards the right. The index is located on the edge of the housing.
Since the needle comes to rest in the magnetic meridian MM, magnetic azimuths μ are measured if the collimation axis coincides with the 0–200 gon line (figure 162). The direction of the magnetic meridian differs from the astronomic meridian AM, which represents the true north-south direction. The deflection of the needle from this direction is the declination δ. It can be positive (towards east), or negative (towards west), depending on the MM in respect to the AM.
As evident from figure 162, there are a number of relations between the declinatoin δ, the meridian convergency γ (angle between astronomic north and grid north, e.g. parallel to x-axis), the needle deflection ν (between grid north and magnetic meridian), the azimuth t (between grid north and direction to the target), and the geodetic azimuth α (between astronomic meridian and direction to the target):

$$\delta = \nu + C \tag{79}$$

$$t = \mu + \nu \tag{80}$$

$$\alpha = \mu + \delta . \tag{81}$$

The magnetic declination changes periodically, daily (depending on the earth rotation), as well as yearly (because of the orbit of the earth around the sun), and progressively over the years, with variations in all cases. Negative declinations are presented as "westerly", and positive ones as "easterly".
The magnetic declination can be determined with the compass, provided the astronomic north is known. For accurate determinations and continuous observations, declinators are used, which are special instruments with recording devices, and which are set up at special observation stations (e.g. in the G.D.R. at the Earth Magnetic Observatory at Niemegk, southwest of Berlin).

Types of Compasses

Full Circle Compasses (table 4)

Diopter compasses (figure 160) have a slot and thread diopter as aiming device. They are fastened to a rod or a simple light tripod and levelled with a bull's eye bubble. For the Schmalcalder compass (figure 163) the aiming

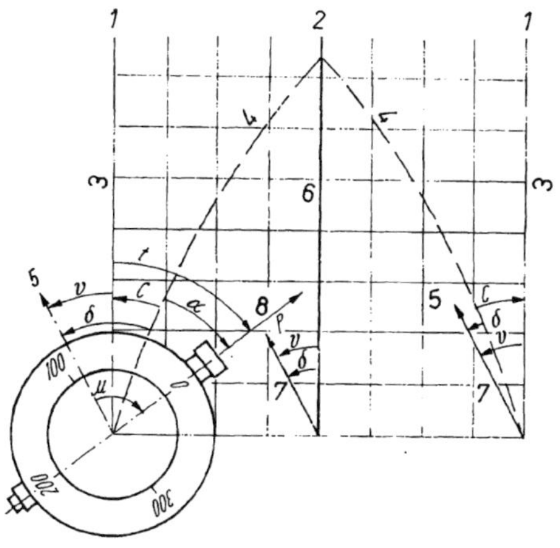

Figure 162 Declination, meridian convergence and needle deflection
1 grid north; *2* geographic north; *3* parallel to the middle meridian; *4* astronomic meridian; *5* magnetic north; *6* middle meridian; *7* magnetic meridian; *8* azimuth

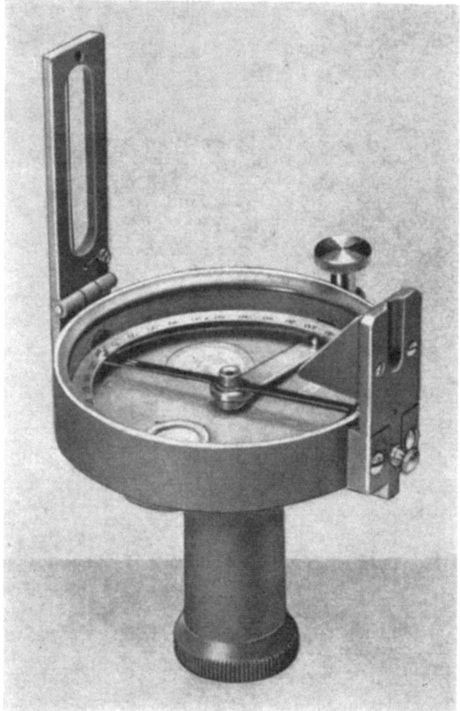

Figure 163 Schmalcalder compass

Figure 164 Telescope compass (Stadia compass BUMON by Breithaupt & Son, Kassel)

Figure 165 Telescope compass (tripod compass Wild B 3)

device is fastened to the magnet needle, instead of the circle graduation. The graduated circle, which is a light ring of carton or aluminium, is fastened to the flat needle in such a way, that its zero mark coincides with the end of the needle pointing north. The needle, loaded with the circle (subdivided to 1°) takes longer to come to rest than a free needle, which is a disadvantage. The marking increases to the right, e.g. by 10°. When aiming with the prism fastened to the eye diopter, one also sees the circle graduation fixed to it. The mirror coating of the prism is interrupted by a thin line to look through, which also serves as reading index. The surface of the prism which faces the graduation, is convex and acts like a magnifying glass, so that the graduation appears in clear view.
Telescope compasses (figures 164, 165) have a telescope instead of the diopters, which permits more accurate sighting. A column, carrying the telescope is fastened to the edge of the plate on which the compass is located. This plate rests on a tribrack, which in turn is supported by a tripod. A bull's eye level is used for levelling. The telescope usually has stadia hairs, and a vertical circle attached to it. Such instruments are called stadia compasses (see section 5.1.2.).
Compass theodolites, also used as stadia compasses are either theodolites with a compass added (figure 166) or theodolites whose graduated circle is rigidly connected to the magnetic needle and thus swings with it (figure 167). For the latter, the needle with circle can be lifted of its pointed centre support and pressed against the platen edge (figure 168). Thus the instrument with the same graduated circle and reading device can be used as a regular theodolite (see section 5.1.1.). Therefore these instruments have better reading devices (figure 169).
The compass is frequently used in mining surveying. The mining compass consists of compass hanger, compass ring and compass box. For the Freiberg design the compass ring is solidly connected to the compass hanger, only the compass box is movable. For the Kassel design (figure 170) both compass box and ring are movable, such that the ring can be snapped into the plane of the hanger for transportation. Compass and clinometer are main tools of the mining surveyor.

Orientation Compasses

If the compass is only used as an attachment to other instruments for orientation purposes, a small section of the compass graduation in a box with a few graduation lines on both sides of the zero index, is sufficient. By attaching such an orientation compass to a theodolite or tacheometer, these instruments can be used as compass instruments.

Figure 166 Compass theodolite (BT I by Ertel, Munich)

Figure 167 Compass theodolite (Wild T 0)

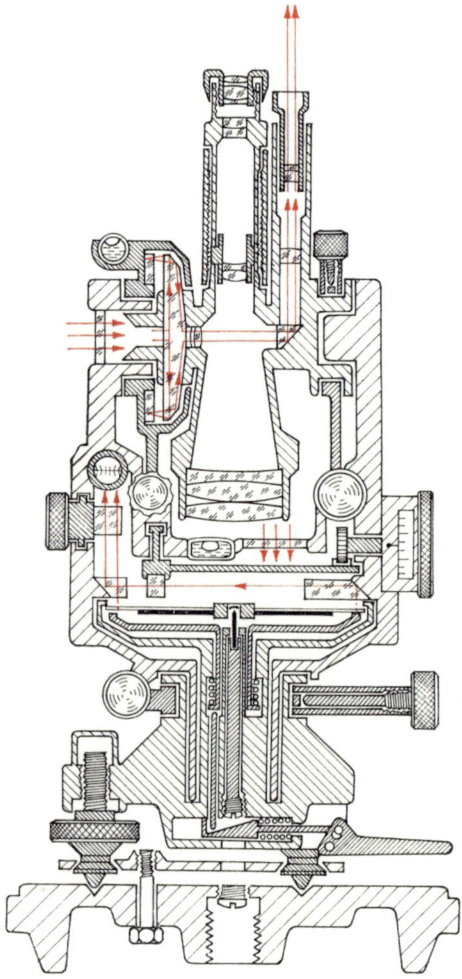

Figure 168 Section through Wild T 0

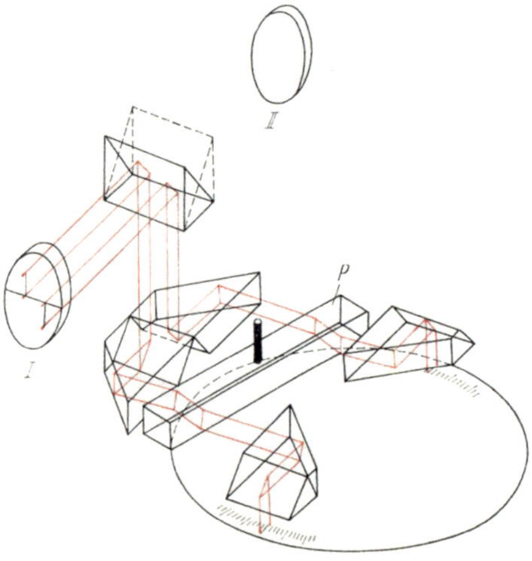

Figure 169 Circle reading for Wild T 0

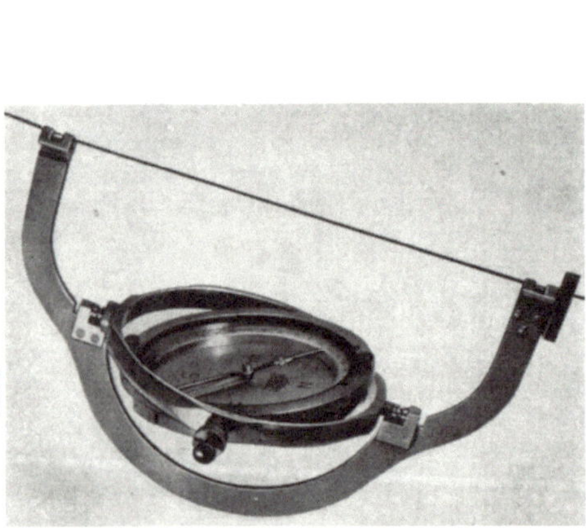

Figure 170 Hanging compass (COMTA by Breithaupt & Son, Kassel)

Table 4. Compass Instruments with Telescope

type	manufacturer	telescope				graduated circles					compass		bubble sensitivity		mass			height	remarks
						diameter		reading device											
								least count											
		Γ_F	diameter of objective lens [mm]	length [mm]	shortest sighting distance [m]	H [mm]	V [mm]	H	V	type	least count	length of needle [mm]	telescope level [″]	bull's eye level [′]	instrument [kg]	container [kg]	tripod [kg]	[mm]	
1	2	3	4	5	6	7	8	9	10	11	12	13	14	15	16	17	18	19	20
BUMON	F. W. Breithaupt & Sohn, Kassel	25	30	168	1.0	100	60	1′ (20 mgon)	10′ (0,2 gon)	*H*: vernier micros. *V*: line micros.	1°	80	30	–	3.2	3.0	3.5	145	60″ cross-level; eccentric telescope; figure 164
TFB	Ertel-Werk, Munich	12	25	160	1.5	85	75	–	1° (1 gon)	index	1°	85	40	12	1.5	1.6	4.0	92	eccentric telescope, spherical support
BT I		18	25	160	1.5	95	70	10′ (0.2 gon)	10′ (0.2 gon)	index and magnifying glass	1°	65	40	30	2.5	1.9	3.9	207	telescope supported on one side; figure 166
8026-30	Lietz Comp., USA	12	20	121	3.7	73	73	5′	1°	*H*: vernier *V*: index	1°	73	–	–	1.6		0.9	140	V-graduation: 50°–0–50°
B 3	Wild Heerbrugg AG	2.5	4.6	56	3.0	65	–	2° (2 gon)	–	diametrical reading estimation with magnifying glass	2° (2 gon)	–	–	8	0.9	0.4	5.0	135	swinging graduated circle with magnet; ball joint support; figure 165
T 0		20	28	135	1.4	68	46	1′ (20 mgon)	20′ (0.2 gon)	*H*: coincidence microm. *V*: diametrical reading estimation	2° (2 gon)	70	60 (Ind.-lib.)	8	2.9	0.8	5.0	240	swinging horizontal circle with magnet; base plate of container is part of instrument; figure 167

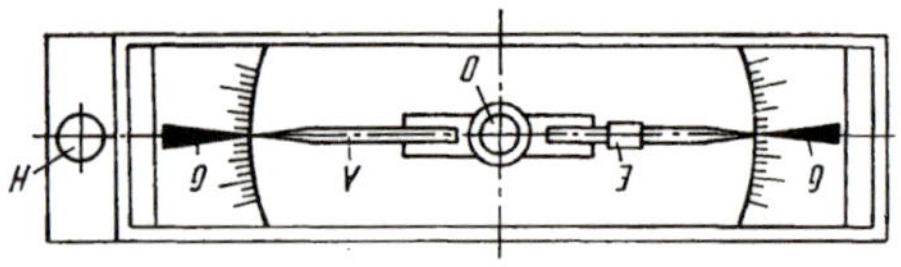

Figure 171 Line-or case compass

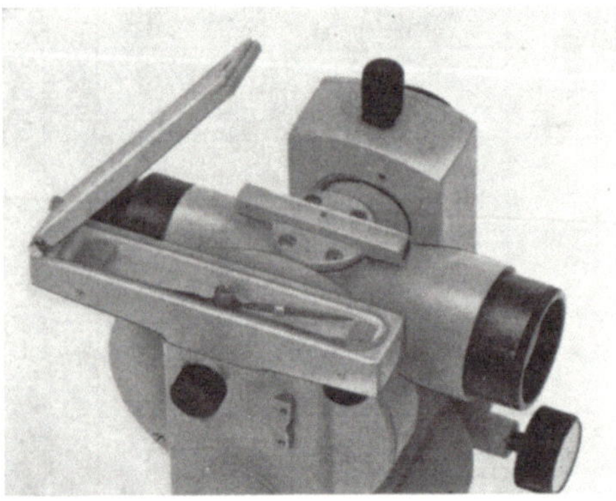

Figure 172 Zeiss, Oberkochen Th 51 with line compass

Figure 173 Tubular compass (Zeiss, Jena)

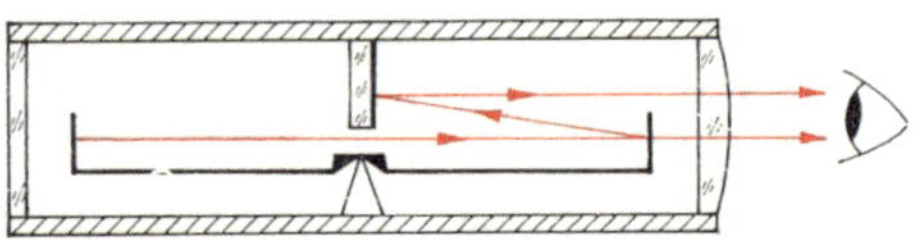

Figure 174 Optical train of tubular compass

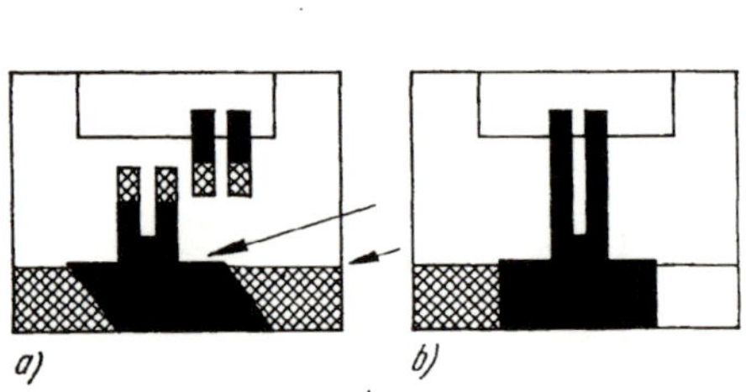

Figure 175 Field of view of tubular compass
a) non coinciding needle ends
b) coinciding needle ends

The line or case compass (figure 171) has small plate segments with index G for the zero direction instead of a graduated circle. Pin and cap D can be protected against damage during transport with an arresting device H. This compass is placed either between the standards or onto the horizontal axis of the theodolite. The case compass OBK (U.S.S.R.), for instance has the following technical data: $\pm 12°$ measuring range, 30′ graduation, 150 mm needle, 180 mm × 47 mm × 25 mm size and 220 g weight. Figure 172 illustrates another form of line compass.

The tubular compass (figure 173) is fastened to one of the standards, or – for some instruments – to the telescope. It neither has graduation nor circle. The ends of the needle are bent perpendicularly upwards. With a mirror placed above the pin one accomplishes, that the ends of the needle appear above each other (figures 174 and 175), and that the paths from the needle ends to the observer are of equal length. The far end of the tubular compass is closed with a ground glass, while the front one has a magnifying glass. When coinciding the far end of the needlo with the mirror image of the near one, the needle points to the magnetic meridian. If one reads the horizontal circle of the theodolite in this position, all further directions are oriented to this direction. The coincidence accuracy is about ± 60 mgon.

This was further developed by Zeiss, Jena (figure 176) where the needle is supported without friction (figures 177, 178) by hanging at a tension tape, which increases the accuracy to ± 20 mgon (figure 179). The needle can be oriented to grid north.

The orientation compass of Zeiss, Oberkochen (figure 180) has a disk-shaped magnet hanging with a spiral spring vertically in a flat housing. A compensation for inclination is not necessary. When the spring is in the plane of the magnetic meridian, it holds the disk in the centre position. When rotating the theodolite by α, the disk strives back towards the centre position. It exerts tension to the spirial spring such, that the magnet disk turns by α in respect to the housing. A mirror attached to the housing, doubles this angle. The full amount of the rotation appears in the field of view of the auto collimation telescope. One orients to magnetic north by setting the movable line centrally between the fixed ones, when turning the theodolite (figure 181).

The "Master Compass" (figure 182), built as attachment by Rank Precision Industries in England is another line compass. It consists of an auto collimation telescope, whose mirror is fixed to a central element.

This element is supported by saphires on top and bottom (for direct and reverse measurements) and has a magnetic bar on both ends. The graduation is read with an index, illuminated from the outside. The vibrations of the magnet are dampened by currents through a copper block.

Testing and Adjusting of the Compass

One can obtain perfect measuring results with a compass if one fulfills certain conditions based on its mechanical design and on physical characteristics of magnetism:

1. The needle shall swing around the centre of the graduation. If the difference of the readings on both ends of the needle is always exactly 200 gon-this condition is fulfilled. However, if an eccentricity of the needle is apparent, the errors can be compensated by taking the mean of the readings on both ends of the needle.
2. The needle shall have sufficient magnetism, and its bearing shall not be damaged. If one moves the needle several times from its position with the aid of a small magnet or an iron object, it has to yield the same reading after coming to rest again.
3. The magnetic axis has to coincide with the needle axis. Especially for rhombic needles, a discrepancy between these axes is possible. This can be determined by placing the needle upside down, provided it has double caps.

To check fully graduated compasses, connected with an angle measuring instrument, one sets the horizontal circle of the instrument to the values 0, 25, 50 gon, ... and reads the compass. The real values have to correspond to the set ones.

Irregularities of the compass, noted by these tests can only be corrected by a mechanic.

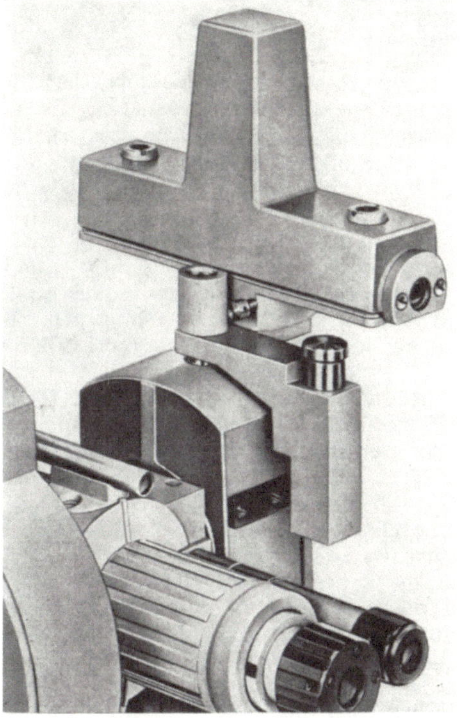

Figure 176 Orientation compass (Zeiss, Jena)
needle length 88 mm, 200 g
dimensions 110 mm × 45 mm × 13 mm

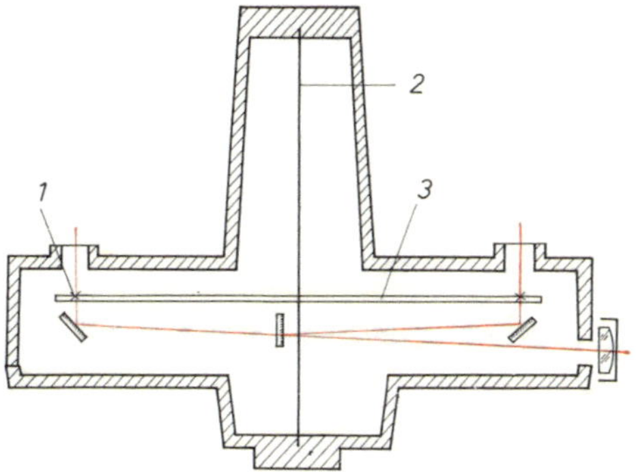

Figure 177 Observation system for orientation compass
1 needle marks; *2* tension tape; *3* compass needle

Figure 178 Internal structure of orientation compass

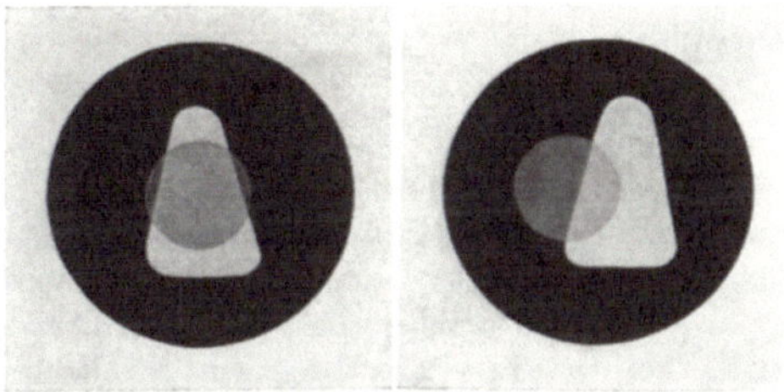

Figure 179 Coincidence setting for orientation compass

Figure 180 Disk compass at Th 4 (Zeiss, Oberkochen)

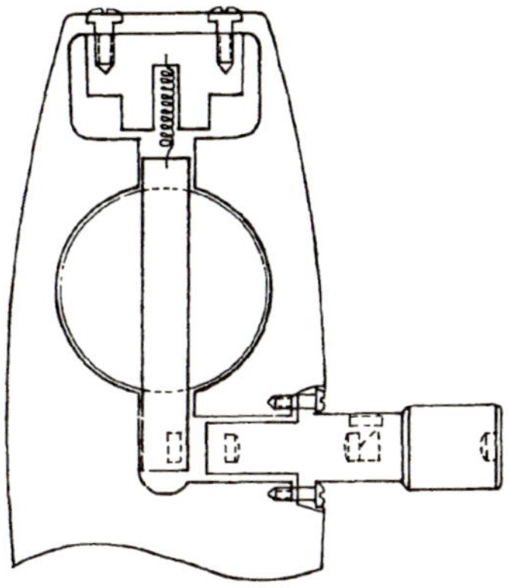

Figure 181 Section through disk compass

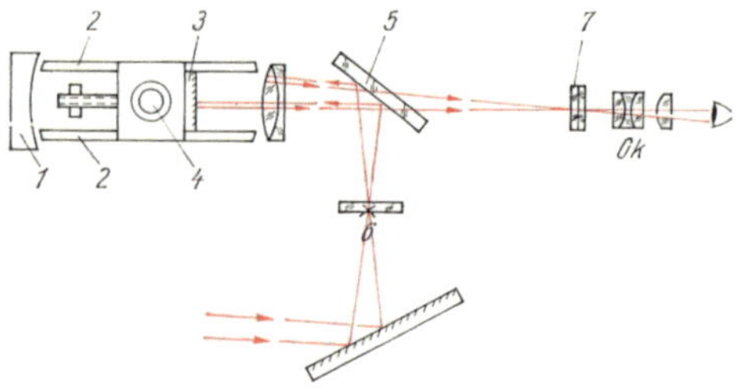

Figure 182 Line compass "Watts Master Compass" (Rank Industries, England). Optical train (from top)
1 copper block dampening; *2* magnet; *3* mirror; *4* central piece with saphire bearing; *5* semipermeable mirror; *6* index; *7* reticule

To the use of Compass Instruments

When using compass instruments, one has to watch especially, that there are no magnetic influences other than the earth magnetism.
The normal magnetic field of the earth is disturbed by local masses (cristaline rocks, iron-, nickel-, cobalt- and other ores). Iron fences, railway tracks, as well as reinforced concrete and hydro lines close to the station also affect the compass needle.
Unlike for other instruments, the clamps of a compass do not consist of steel but of brass or silver. The iron shoes of the tripod legs have little influence on the needle, because they are about 1,5 m below the instrument and arranged symmetrically. When measuring, one has to watch, that persons very close to the instrument have no iron objects (pocket knive, keys etc.) with them.
According to *Ackerl*, one has to keep the following separations, if the deflections should be kept below 0.2 gon: 40 m for hydro masts and steel structures of 20 m height, 15 m for cast iron lantern poles of 5 m height, 10 m for metal fences with iron posts, 20 m for one track railroads, and 100 m for two track railroads.
In addition to periodic changes in declination, irregular variations (magnetic storms) can occur. They cannot usually be noticed and are only recognized when the evaluation shows deviations that cannot be explained. For compass traverses with poor fit, it is advisable to ask a magnetic observatory whether the day in question can be considered magnetically quiet or disturbed. Magnetic measurements, taken on strongly disturbed days, are rarely useful.
An accuracy of $\pm 6'$ can be reached with full circle compasses, and $\pm 1'$ with orientation compasses. At the outset of each job, the magnetic instruments are to be tested. The measurements should be taken with the same instrument. To check each traverse leg, readings should be taken at both ends, direct and reverse. The needle has to be arrested after each reading. When carrying magnetic instruments, one has to be very careful, and avoid shock. It is essential to protect the magnetic needle from heat. It looses its magnetism by shock and heating.

2.3. The Theodolite

The theodolite is an instrument to measure horizontal angles, or horizontal- and vertical angles. Its main parts are a horizontal-, and usually also a vertical graduated circle, a telescope that can be rotated around a horizontal axis and a vertical one, and elements to facilitate circle readings and levelling.
The theodolite also has additional elements with which it can be centered above a point.
For the production of efficient instruments, materials are needed that can meet high demands on stability, despite mechanical, physical and chemical influences. To avoid deformations of instrument parts, materials with small expansion coefficients are used such, that they compensate each other. To eliminate chemical influences, usually non corroding metals and chemically stable glass for optical parts are used. Occasionally, light metal is used to keep the mass small.

Modern theodolites have the following characteristics:

– small size and mass, to ease transportation, set-up and operation; closed in elements, to avoid damage, e.g. by keeping dust and water out;
– comfortable location of the operational elements, such as clamps, screws, level bubbles, reading devices, to that the observer does not have to change his position when levelling, aiming and reading circle or bubble;
– stability of the instrument, so that it is always ready for operation and does not have to be readjusted, even under extreme conditions or after extended use,
– achromatic, distortion free and luminous optics, so that the images are perfectly clear, bright and full of contrast, even under poor illumination;
– suitable instrumental accuracy, so that it can be utilized economically.

2.3.1. Tripod and Support of the Theodolite

The tripod is used to set up the instrument, and should provide a stable, unchanging, and roughly horizontal support. It should be handy to facilitate fast, easy set up without tensions. It consists of three legs and a head.

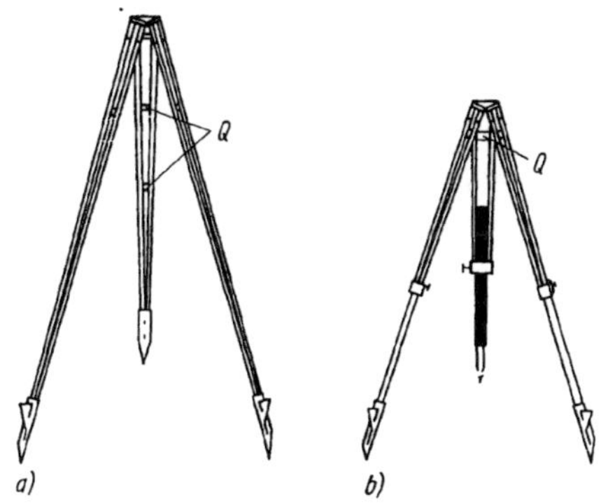

Figure 183 Tripods
a) with rigid legs
b) with telescopic legs

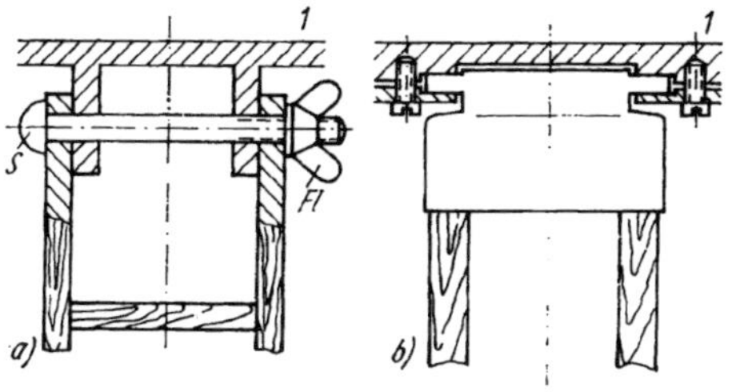

Figure 184 Tripod heads
a) with screw bolts
b) with roller joints
1 tripod plate

The legs are usually made of wood, sometimes of light metal, and consist of two supports with rectangular or circular cross section each, often rigidly connected with cross pieces *Q* (figure 183). At their bottom end, they have metal points with stepping attachments, which can thus be pressed into the ground. For easy transportation and to simplify the set up in uneven terrain, tripods with telescopic legs are produced (figure 183b).
The tripod head for simple instruments consists of a conical plug, while it is a plane plate with connection to the tripod legs for others. The legs are connected such, that they can be rotated in planes perpendicular to the plate and clamped in a desired position. Some tripods have wing nuts (figure 184a) for this purpose, where the bolt *S* is fastened with the wing nut *Fl*. One cannot forget to loosen these wing nuts before moving the tripod, or else the legs as well as the threads of bolts and nuts may get damaged.
In order to avoid tensions when setting up, the nuts should only be tightened after the legs have been stepped into the ground. Most tripods are designed such, that no nuts are used when setting up. For ball bearing tripods, the two supports of the leg are connected on the top with a metal cross piece that is spherical at both sides. These spheres are pressed with screws against the metal plate and into small indentations of the tripod plate. This provides continuous friction, and the tightening and loosening of screws does not become necessary. Cylindrical or conical bearings (figure 184b) are sometimes used instead of ball bearings. The top plate of the tripods for angular measuring instruments has an opening sufficiently large for centering of the instrument.
In England, Japan and U.S.A. one can also find tripods such, that the instrument can be screwed on. For theodolites with optical plummet there are tripods with a shifting plate, where the thread can be shifted in two perpendicular directions without changing the levelling of the instrument (figure 185). The theodolite is screwed onto the tripod head with a thread at its platen. Some theodolites made by Zeiss, Oberkochen, have an orthogonal sliding element between tripod and instrument, which simplifies centering (figure 186).

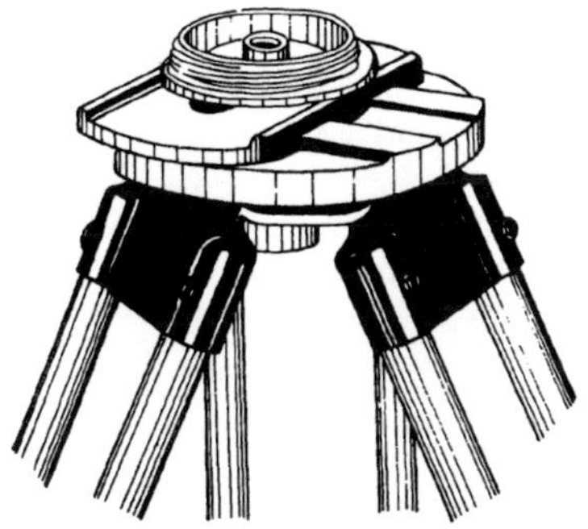

Figure 185 Tripod head with parallel shift plate and thread (Keuffel & Esser, U.S.A.)

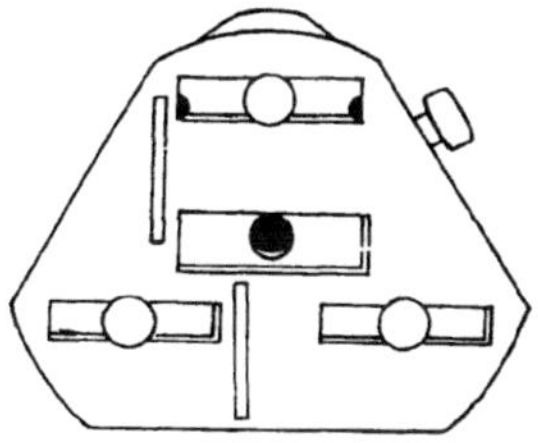

Figure 186 Orthogonal sliding element for Zeiss, Oberkochen theodolites

The solid connection between tripod and instrument with plate-tripods is accomplished by screwing the bolt *A* of the tripod into the spring-plate *Fpl* or the tribrack *G* of the instrument (figure 187). The bolt *A* can be shifted around the opening of the tripod plate *S*, but cannot be lost because of the element *L* which can rotate around the plug *B*. With this, the instrument can be shifted by a few centimetres along the tripod plate in all directions. The bolt *A* is hollow, so that one can look through when using an optical plummet. A plug with a plumb hook can be fastened in it.
Haller's centering tripods facilitate accurate pre-levelling and fast fine levelling (figure 188a). Basically, they consist of a centering rod with the plate for the instrument perpendicular to it.
After placing the centering rod 9 on the ground point, and setting it vertical with the bull's eye level 13, the theodolite, fastened to the recepticle plate 10, is both centered and roughly levelled in one step. The tripod head 11 is machined as a hollow cone, which carries on top and bottom (clamping shell 6) a section of a sphere of different radius but with the same centre. This attachment is mounted such, that it can be shifted on the tripod plate 4 (intermediate shell 5, arresting shell 12).
On the upper sphere section, the recepticle plate is located with the quick lock 1 and the centering plug for the instrument. An extension rod 2 with clamping plate and clamping handle 7 is attached to the recepticle plate.

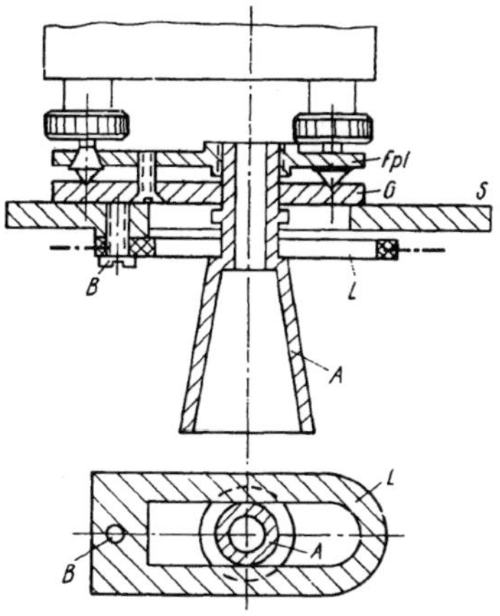

Figure 187 Connection between instrument and tripod via spring plate and tightening screw

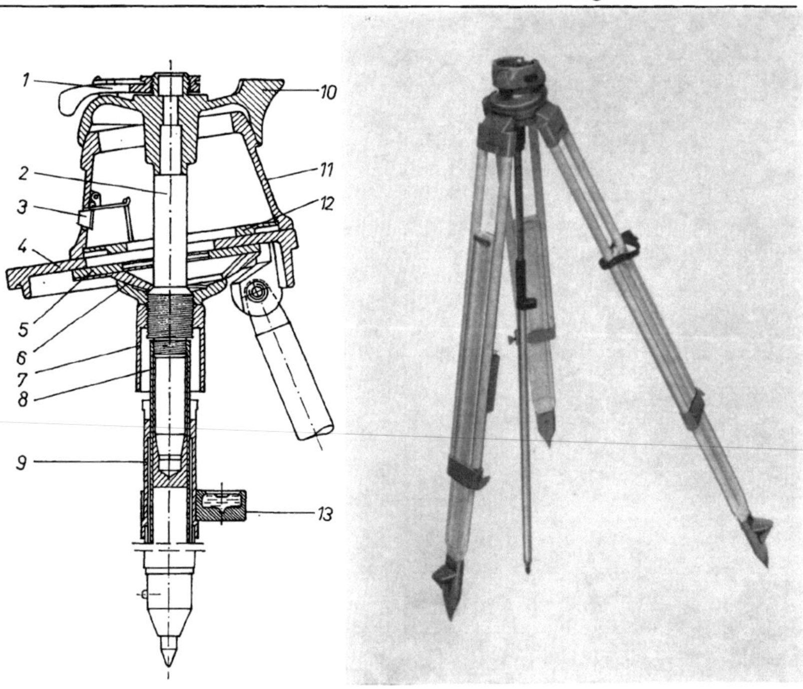

Figure 188 Centering tripods
a) Kern, Aarau
b) Wild, Heerbrugg
1 quick lock; *2* extension rod; *3* push button; *4* tripod plate; *5* intermediate disk; *6* clamp dish; *7* clamp handle; *8* centering rod; *9* centering rod extension; *10* recepticle plate; *11* tripod head; *12* arresting disk; *13* bull's eye level

The clamping handle serves to clamp the recepticle plate to the tripod head, and the tripod head to the tripod plate. The centering rod **8** is screwed to 9 at the clamping handle. When setting up the tripod, one proceeds as usual, places the tip of the centering rod onto the ground point, and steps onto the tripod legs until the bull's eye bubble is near level. By shifting the tripod head (pressure knob 3) the bubble is completely levelled. Then the clamping handle is locked, and the recepticle plate is level to about 2′, if the bull's eye bubble has an accuracy of 10′ (see also figures 307, 321, 328). For the theodolites T1, T16, T2 and RDS, as well as for other tripods, Wild, Heerbrugg, also produces centering rods with spherical section (figure 188b).

2.3.2. Structure of the Theodolite

The theodolite consists of tribrack, bottom, central- and top part. Tribrack and platen are connected by a system of rotational axes – the vertical axis – such, that either only the top or both can be rotated separately, and together.

Tribrack and Bottom Part of the Theodolite

The tribrack is the foot of the instrument. It consists of the tribrack body, the spring plate or spring joints, the base plates, and the foot screws. The spring plate is connected with the base plate, which supports the foot screws (see figure 187). It presses the foot screws onto the ground plate or the tripod. By screwing the spring plate or the base plate onto the tripod, the instrument is fastened to the tripod. For theodolites of low or medium accuracy, the baseplate is occasionally the bottom of the transport container.

The instrument is levelled with the three foot screws of the tribrack (located at 120°), and with the platen level, which usually is part of the top part.

In America and Japan four footscrew instruments are used, with two acting against each other. The base plates of these instruments, like several English ones, have a thread to screw onto the tripod (figures 185, 291, 292).

Foot screws of modern instruments are protected against dust and humidity. It should be possible to turn them easily and steadily, but not too lightly or jerky. One can avoid early wear of the thread by carefully regulating it with the aid of an adjustment nut, sitting on the split foot screw. However, this can only be done after complete removal of the foot screw (figure 189).

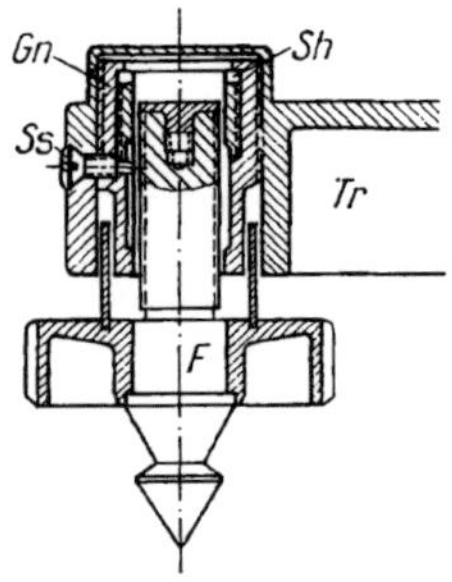

Figure 189 Covered foot screw

The tribracks of theodolites of Zeiss, Jena have the form of figure 190. The vertical footscrews are replaced by knobs with horizontal axes for the instruments of Kern, Aarau (figures 303, 307, 321, 328). The cones of the base plate of the theodolites DKM 1 and 3 fit into the spiral nuts of the knobs and are raised or lowered by their rotation. (figure 191 a). For other instruments, the base ring *B* with the top part is lowered or raised against

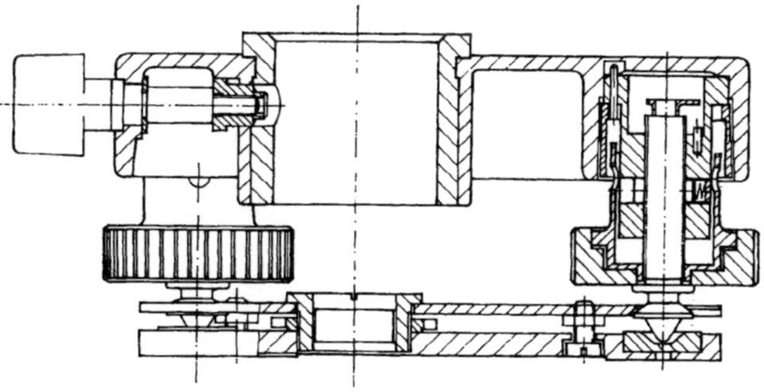

Figure 190 Section through new tripod by Zeiss, Jena

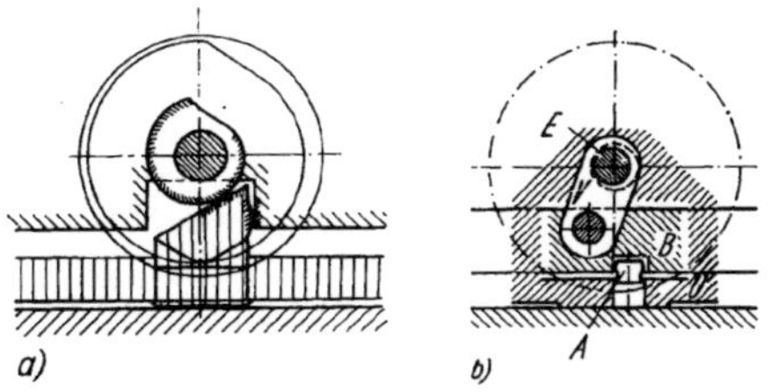

Figure 191 Levelling systems (Kern, Aarau)

the bottom part U via an eccenter E and a connecting piece V (figure 191 b). This is accomplished when turning the knobs.

Due to its mass and inclined position, the instrument always slides to the stopper A. Since the tripod head of the centering tripods is roughly levelled, the shifting range in height is small for both designs.

The bottom part of the theodolites Th2, Th42, Th43, and RTa4 of Zeiss, Oberkochen is manufactured spherically. The theodolite can be tilted around its ball bearing for rough levelling (figure 192). Two knobs are used for fine levelling. The third support is solidly placed, as to fix a membrane, which secures a stiff connection between tripod and instrument. A rectangle slide between spherical footing and tripod plate has the purpose to prevent rotations of the theodolite when centering (see figure 186). The bottom part of the theodolite carries its central and top parts, and is solidly connected to the tribrack during measurements. It contains the supports for the vertical axis, the socket or other elements for connection to the tribrack.

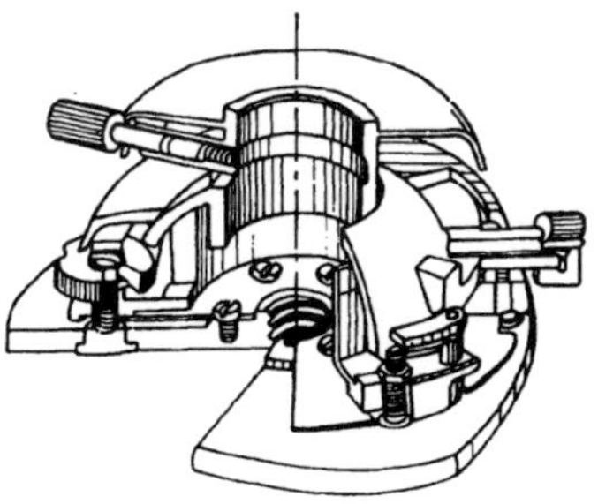

Figure 192 Spherical foot (Zeiss, Oberkochen)

Vertical Axis of the Theodolite

The axis system which connects the top-, centre- and bottom parts of the theodolite is called vertical axis. Its purpose is to support these movable instrument parts and to keep their movements as planned. To keep the mutual position of these parts constant, and to retain their form, the vertical axis is made of special materials. These have high stability and permit easy and perfect operation of the instrument, even at considerable temperature differences. Good stability and easy rotation are usually achieved with rather long axes – generally three to four times as much as the maximum axis diameter. Small play between axis and socket (e.g. 0.5 µm) is also necessary.

Based on their form, one distinguishes conical and cylindrical axes, and ones with disk like grinding surfaces.

Up until a few decades, conical axes were used nearly exclusively. The reason for this is, that they can be manufactured relatively easily, and yield good axes, even at low technical levels. Since the socket has to fit to the plug or vice versa, the fit can be improved by fine grinding, even after wear of the axis. Too much grinding can be corrected by machining the top face S (figure 193).

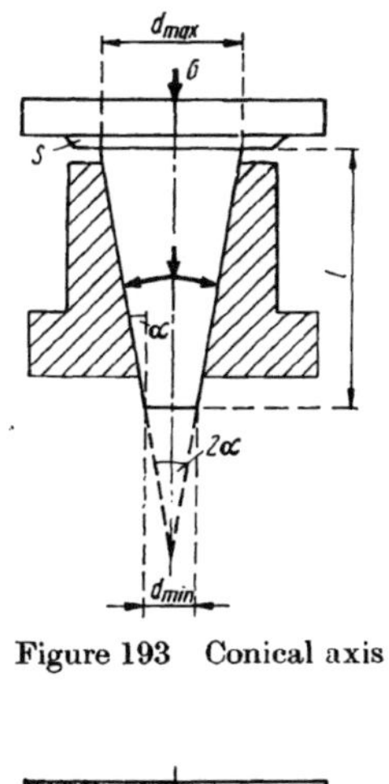

Figure 193 Conical axis

For cylindrical fit, both plug and socket have to be perfectly cylindrical with matching diameters. Optimum centering and minimum wobble of a conical axis is achieved with large axis length l. (e.g. triangulation theodolites with Repsold's system have 140 mm), small diameters d_{max} (about 40 mm and d_{min} of the axis), and small cone angle 2α (usually 4 to 17 gon). Exaggerated friction is caused, if the mass G of the movable top part is directly transferred onto the cone (figure 194). Systematic measuring errors result from both, too much or too little friction. Therefore an oil film of several micrometres' thickness is placed between socket and plug. To further reduce the friction, the surfaces of the containing parts plug and socket are kept as small as possible. This is achieved with 0.3 to 0.5 mm deep cuttings around either the plug or the socket, which leaves only small rings as contacting surfaces. The plugs are placed such, that the mass G is supported mainly by its top face S, while the cone primarily serves for guiding. Too much or too little load on the cone results in sticking or wobbling of the axis. Because of the large mass of instruments with highest accuracy, the load on the axis can be reduced. This is achieved with a special setting, which

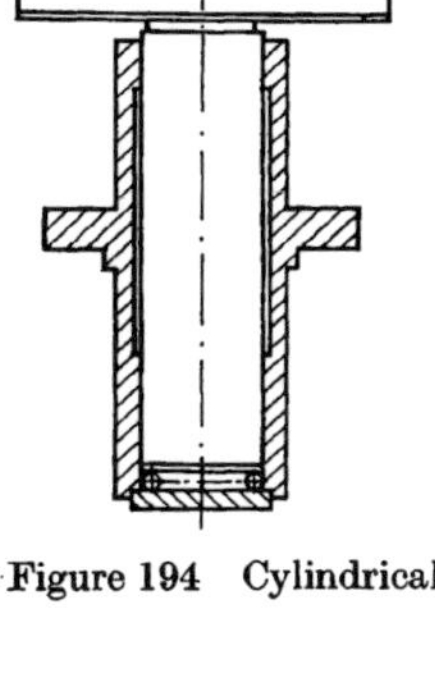

Figure 194 Cylindrical axis on ball bearings

raises or drops the plug in respect to the socket and thus regulates the rotation by changing the space that separates the two. If the plug is an element of the bottom part, the same applies in reversed sense (see figure 201). The load on this regulation device can be as high as 30 kg/cm². One regulates such, that a small amount of play remains. Since the axis basically swims in an oil film, even for optimum regulation, uncontrolable wobble can occur when rotating the top part. This generates measuring errors. Plugs and sockets of conical axes are usually manufactured from different materials, e.g. plug from steel, socket from cast iron or bronze. For cylindrical axes however, both consist of the same material namely of hardened steel. Their production became only possible by the high technical competence, reached during this century. Good, distortion free hardened steels as well as good grinding machines are necessary. The whole mass of the top part is supported either by the top face or bottom face of the plug, or by a sphere attached to it, or by a ball bearing (figure 194).

The cylinder is not affected by forces and is used solely for guiding, provided that the top part is balanced in respect to the vertical axis, and that the latter is really vertical. Cylindrical axes have the advantage, that there is no time consuming axis regulation and oiling. The axis is in good working order without additional care and preparation. It functions over long periods of time without loss in quality, and it can be produced in series. Precision grinding and polishing is a requisite, because the dimension of the plug has to match within 1 μm the dimension of the previously prepared socket. Small deviations of up to 1 μm from the circular cross section cannot usually be avoided. With an oil film of 0.5 μm thickness between plug and socket for ease of rotation, each end of the plug cannot deviate by more than 0.5 μm from its design position. Combined with errors of plug and socket in shape or form, differences of up to ±0.4 μm are practically possible, because the oil film cannot be compressed to zero. These differences lead to errors of up to 0.4 mgon for axes of optical theodolites with 85 mm length and 30 to 50 mm diameter.

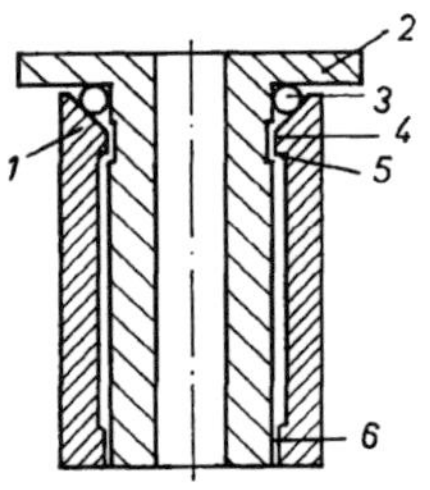

Figure 195 Semi Kinematic axis (Wild T 1, diameter of sphere: 2.5 mm). *1* bush; *2* axis; *3* centeringball bearing; *4* fit; *5* reservoir for lubricant; *6* fit seiding bearing

The friction conditions are improved with reduction of the contact surfaces between axis and socket (figure 194, 195), and with the use of ball bearings, where the friction surfaces remain constant, even after changing the sense of rotation. In the simple cases, the ball bearing serves only to support the top part, while for the semikinematic axes of theodolites produced by Zeiss Jena, Wild Heerbrugg, and Nikon, the radial pressure onto the upper conical extension of the socket is transferred to the ball bearing, e.g. the axis is centered (figure 195). In this case, a narrow guide ring (approximately 5 mm wide) on the lower end of the axis, and the ball bearing on the upper one, provide the guiding mechanism. Similar in principle, but in reversed order, are these elements for the theodolites of MOM, Budapest (figure 196). Axes with extremely precise plain disks as guiding surfaces, which, with a number of precision spheres, form a ball bearing, can be found in theodolites of Kern, Aarau (figure 197) and the precision instruments of the French National Geographic Institute. The ball bearing supports the whole top part and guides it in one plane. For the centering of the axis, a short plug with small diameter within a short socket is used. This axis results in a shortened instrument design.

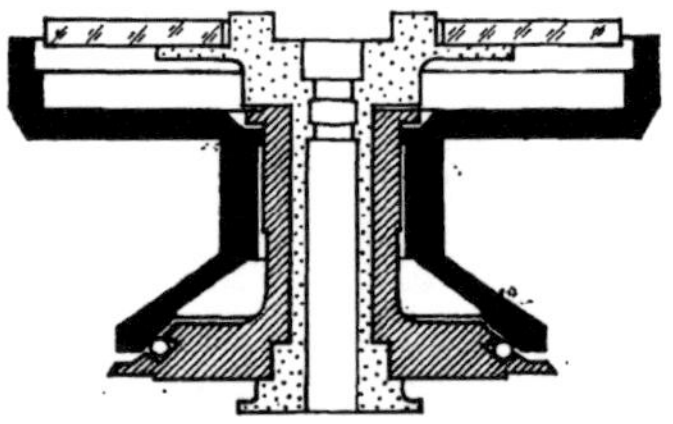

Figure 196 Axis system of the MOM Te-B-Theodolites

Depending on the arrangement of the vertical axis, one distinguishes between single theodolites, repetition theodolites, and theodolites with rotatable circle. Simple theodolites with fixed horizontal circle, are sufficient for simple angle measurements. The angle, consisting of two directions, results in the difference of two circle readings. Besides simple direction measurements, and direction measurements in sets, repetition theodolites are suitable for angle measurements by repetition. To do this, the top part has to be rotated above, as well as together with the central part, after clamping the top to the centre and unclamping the latter. Theodolites with rotatable circle permit successive determination of an angle at different circle positions, in sets, in all combinations or with other methods.

Simple theodolites have a graduated circle solidly connected to the tribrack, and with bottom and centre part thus forming one unit (figure 198). The usually conical axis plug of the top part rests on a support spring within the tribrack socket.

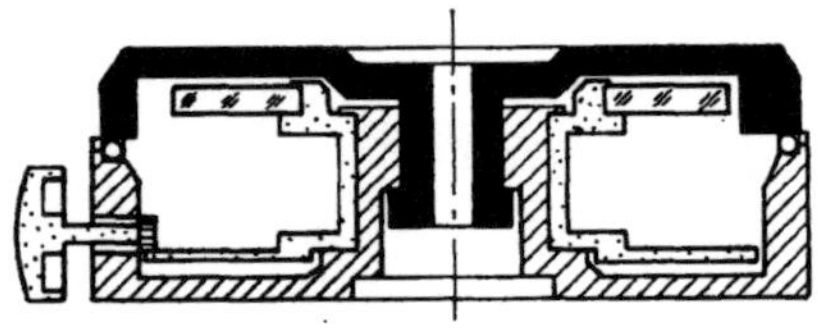

Figure 197 Axis with disk shaped guiding surface and ball bearings (Kern DKM 3, 60 balls with 4 mm ± 0.2 μm diameter)

Repetition theodolites (figures 199 and 200) contain a rotatable centre part, which generates two rotational axes. The centre part can be clamped to the top with a repetition clamp. For the Reichenbach System (figure 199), the centre part extends to the bottom part, and the top part to the hollow axis of the centre one. Centre- and top parts rest on support springs. This system

Figure 198 Simple theodolite (schematic)

Figure 199 Reichenbach's axis system (schematic)

Figure 200 Borda's axis system (schematic)

Figure 201 Repsold's axis system (schematic)

Fig. 198

Fig. 199

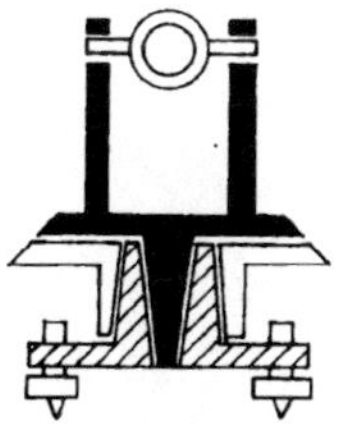
Fig. 200

Fig. 201

has the disadvantage, that the centre part may be dragged along when the top part is rotated. This is avoided for the Borda System (figure 200), where top—and centre parts are arranged on both sides of the bottom part. It is commonly used in repetition theodolites of medium accuracy.
Since angular measurements are usually carried out in sets or in combinations, theodolites of medium and high accuracy do not have the possibility of repetition, but rather a circle drive to change the centre part. Older theodolites of highest accuracy often have the Repsold system (figure 201). The top part (with up to 15 kg mass) can be rotated around the rather large conical plug of the bottom part, which extends upwards. The horizontal circle which can be turned by hand, sits on the plug, just by friction. The socket can be raised or lowered onto the plug with a regulation attachment acting upon the upper end of the plug.
Since the centre and top parts of modern theodolites do not need to be fastened or rotated against each other, their axis systems follow the same principle. The plug of the top part sits in the socket of the bottom part, which is fixed with a pressure disk into a bore hole of the tribrack, or else solidly connected to it. This socket is surrounded by the hollow axis of the centre part, which can be changed with a circle drive (see figures 219, 225, 226).

Centre Part of the Theodolite

The centre part consists of the horizontal circle, and the elements connected to it, which facilitate rotation of the centre part around the vertical axis.
The horizontal circle is produced of materials which resist corrosion and deformations, and which have a clean surface for marking the graduation. Older instruments have metal circles (brass or bronze, even aluminum alloys in the U.S.A.) while modern instruments have glass circles. For metal circles, the graduation is cut directly into the circle in form of lines with a thickness of more than 0.1 mm. Graduations with fine lines however, are marked on small silver strips located along the edge of the circle. For easier attachments of the magnifying glass or the reading microscope, as well as for better illumination for reading, most metal circles are slightly conical (the angle with their base is 9 to 17 gon). The glass circles are 3 to 5 mm thick and placed in a flange, which is fitted to the instrument axis. The graduation is scribed on the plane surface which has to parallel the other one by 3 mgon for accurate instruments. Since graduation errors have less influence for large circle diameters, the circle diameters for high precision instruments are larger (e.g. 27 cm for metal circles) than for instruments of lower accuracy. On the other hand, glass circles have much smaller diameters than metal circles, because much finer graduations can be placed on glass (e.g. with line thicknesses of a few μm). The diameters for glass circles of instruments of lower and medium accuracy amount generally to 70 to 100 mm, for high precision instruments up to 250 mm (see tables 8 to 11).
In Central Europe, the circle graduation is mainly in gon, while degrees are favoured at other places. The graduation unit depends on the diameter of the circle and the type of reading device. It usually is 1 gon, $^1/_2$ gon, $^1/_5$ gon, $^1/_{10}$ gon, resp. 1°, $^1/_2$°, $^1/_3$°, $^1/_6$°, $^1/_{12}$°, $^1/_{15}$°. The lines are numbered such, that the reading is clear-cut (e.g. every 5 or 10 degrees, or every 1 or 2 degrees). Usually increasing numbers appear in the field of view from left to right, or from top to bottom such, that the reading increases for clockwise rotation of the top part. The line thickness depends on the type of reading device, the material of the circle, and the graduation unit. Graduations on silver have line thicknesses of about 0.1 mm for reading with a simple index, and of 0.02 mm to 0.1 mm for vernier reading. Line thicknesses for glass circles amount to 2 to 8 μm.
The graduation is produced either mechanically with an automatic circle graduation machine, or by photo mechanical copying. Mechanically, it is scribed into a waxfilm on top of the glass circle, and then etched and blacken-

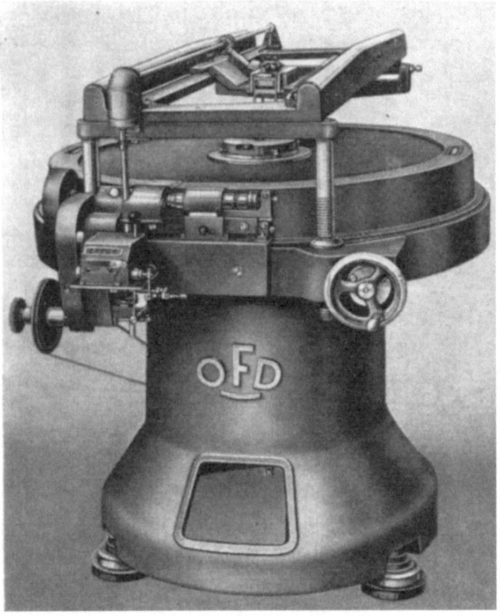

Figure 202 Circle graduation machine
a) Frontal view
b) Engraving element (diameter of master circle 1000 mm)

ed or else directly engraved. The reference circle is a cog-wreath of 50 to 100 cm diameter, screwed onto a round table and covered with a protective ring. This wheel has 1080, 1440 or 2160 cogs, seldom 720, and is rotated through a drive mechanism (figure 202). The table rests on a cast iron column, which guarantees a stable run of the engine, free of shock. An original graduation at the edge of the reference circle serves to control the process of graduation. The drive motor A (figure 203) rotates the disk Hu, which, in turn moves the cog sector Zs back and forth. With this, the cog wheel Zr_1 turns in the same sense. Connected to it, is a locking handle SK, which transports the lock wheel Sr by a predetermined amount of cogs, as set for the desired angular value of the graduation. The transformation wheel $\ddot{U}$ sits on the same axis as the lock wheel, and gears into the cog wheel Zr_2, located on the infinite screw S. This infinite screw—usually a hollow screw (globoid gear) rotates by the same amount as the cog wheel. Since the thread of the hollow wheel forms a globoid with radius of curvature equal to that of the cog wreath, it gears into many cogs of the cog wreath Zk. An erroneous cog, which obtains up to 17 screw passes, has a reduced influence. The rotation of the hollow screw (and with it of the wreath) is interrupted for the time, when the lock handle returns. During this stop, the etching assembly operates on a bridge, supported by three columns with flat thread (see figure 202). The etching tool generates a line on the circle to be graduated. When graduating glass circles, the pressure is very low, and amounts to only 0.03 N. This is, because the graduation of precision theodolites with a 4′ graduation unit contains 5400 lines, and these have a separation of only 0.09 mm for 140 mm circle diameter. Only 3 to 4 lines are marked per minute. The number of the lines to be etched can be set on a counter. As soon as the last line is etched, a mechanism disconnects the gears of the machine, and stops it.
For metal circles, the numbers are engraved with a numbering machine, while an engraving pantograph is used for glass circles (if they are to be engraved) (figure 205). With this engraving pantograph, microscopic engravings in glass can be made (e.g. for crosshair plates).
Widths of 12 μm down to 2, even 0.5 μm can be reached for lines and numbers by a photomechanical process (photo lithographic method). The copied lines have sharp borders and raise the measuring accuracy. The 100 times larger drawed or in the varnish of a foil engraved graduation is reduced with a reproduction camera to the scale 1:1. By contact copying onto a glass body, covered with a film of photosensitive varnish and subsequent developing, a varnish image is produced of a suitable original. This varnish image serves as a mask for metal coating or as an engraving mask to remove structural parts of a metal film, previously attached to the glass surface.

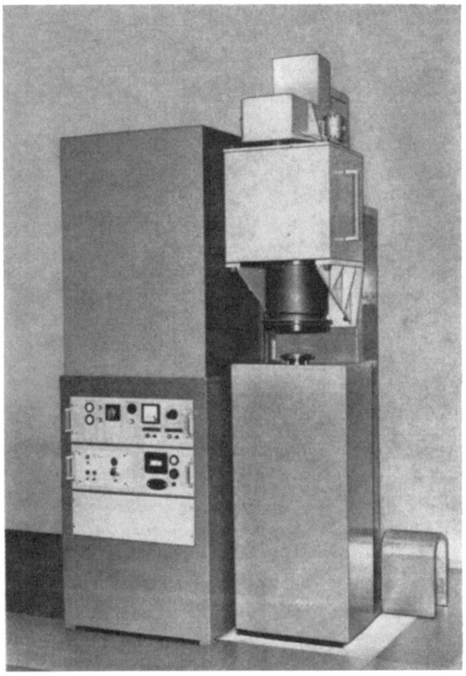

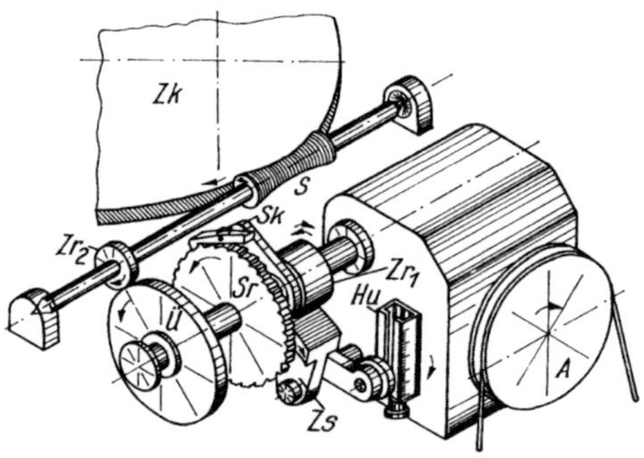

Figure 203 Circle graduation machine with lifting disk and locking handle

Figure 204 Reproduction of graduation via projection illumination

The disadvantages of the photolithographic contact copying process are avoided, when reproducing the graduations by projection illumination (figure 204). The several operations are released automatically after putting on the glass body, covered with a film of photosensitive varnish. The numbers with widths of less than 60 μm are copied with special tools direct on the glass originals. The exposed places are hardened and remain on the glass, while the nonexposed are washed out. The free places are coveed with steams of metals. Then the hardened varnish will be put away from the glass on chemical way, so that the graduation will be visible.
After production, the graduated circles are often tested automatically by photo-electric means, where the difference of certain lines against the reference circle is determined.
The positional errors of the lines cannot exceed 0.5 mgon (1.5″), otherwise the circles are sorted out. The errors are caused by:

Figure 205 Engraving pantograph, height: 500 mm

1. Differences in position between reference circle and circle to be graduated (eccentricity when the rotational centre of the cog wreath does not coincide with the centre of the circle; planar errors, if the two circles do not lie in parallel planes, perpendicular to the machine axis; deformations of the circle to be graduated because of stresses when placed in the casing);
2. Errors in the machine parts (errors in the cog wreath, occured during its production; errors in the hollow screw mechanism);
3. Errors due to wobbling of the axis, and errors of the engraving mechanism;
4. External influences (e.g. temperature changes etc.).

The errors of the first group are periodic, and when plotted give a curve similar to a sine curve. Errors in the hollow screw mechanism cause short periodic errors with amounts of up to three times of the long period ones. They depend on the screw rotation and on the number of cogs in the reference circle. Their period corresponds to one or more screw rotations. The errors of the third and fourth category are mostly random.
One tries to reduce or eliminate the influences of graduation errors either by improving the graduation, or by specific placement of the reading devices and special measuring procedures. Besides various kinds of simple graduations, even with two scales shifted against each other (figure 214d), there are double scales, where two corresponding lines are placed beside each other in different graduation processes (see figure 240a). Graduated circles from Kern, Aarau have two graduations with different diameters ("double circles"), either two equal graduations (figure 206a), or a double line one (figure 206b). However, only one of them is numbered. Other circles have graduations, which are numbered towards the left and the right (see figure 222). The Watts theodolites have a rather complicated graduation (figure 207). The graduation of the self reducing theodolite 4181 of Filotecnika Salmoiraghi, Milano, contains a circular arc, simultaneously engraved with the normal graduation (figure 208).
Graduation surfaces of glass circles, are coated with a silver- and a protective film, where the rays are reflected (figure 225). If the graduation is viewed in penetrating light (figures 219, 226 to 229), it is protected against dust by a circular glass, glued onto it, because imaging of the dust would affect the readings.

Top Part of the Theodolite

The top part of the Theodolite can be rotated against the centre and bottom parts. It consists of housing, standards, telescope, vertical circle, reading devices and levels, resp. tilt compensators.

The Telescope and its Support

The standards bear the horizontal axis of the telescope. This horizontal axis is perpendicular to the optical axis of the telescope. It rests on the standards on two cylindrical contacts of small but equal diameter, which are as far apart as possible. The horizontal axis carries the vertical circle in addition to the telescope. Modern horizontal axes are hollow, so that the optical rays for circle readings can pass through or else that the crosshair can be illuminated at the intersection of collimation- and horizontal axis with a mirror.

Figure 206 Double circle graduations (Kern, Aarau)

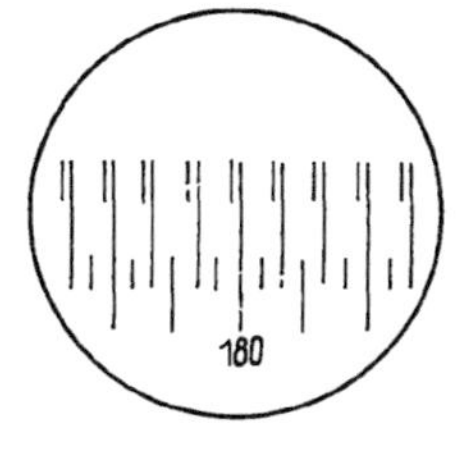

Figure 207 Circle graduations of Watts theodolites ST 200 (Rank Precision Industries, England)

Figure 208 Circle graduation of self reducing theodolite 4181 (Salmoiraghi, Milano)

The standard is designed such that the telescope can be inverted for observations in both positions, usually through both eyepiece and objective lens, otherwise the latter at least. Some time ago, larger instruments were built with an eccentric or reversible telescope (i.e. removing it from the axial bearing and replacing it in reverse). Some modern instruments of high and highest accuracy have mirror telescopes or telescopes with broken path of rays.
The bearings for the horizontal axis are either cylindrical or V-shaped, so that the position of the axis is clearly defined by four points of support. Since they have to support the mass of telescope, axis and all connected parts (e.g. vertical circle) and thus remain under even pressure, the oil film will differ less than for the vertical axis. To provide easy movement for the axis, part of it is supported by relief bearings for telescopes with a large mass.
The horizontal axis should be normal to the vertical axis. When turning the top part around the properly levelled vertical axis, the horizontal axis should describe a horizontal plane. At instruments of older types it can be adjusted, so that horizontal axis tilts can be removed. With cylindrical bearings (e.g. for optical theodolites), the height of a standard can be changed by shifting an eccentric peg after loosening of a screw (figure 209).
Most modern instruments do not permit an adjustment of the horizontal axis, since its bearings are solidly connected with the vertical axis and thus remain stable. The horizontal axis is often made of brass, in order to not disturb a setting compass when used with the instrument.
On the outside of the telescope there are simple sighting devices such as lead notch or a diopter tube (see figures 291 to 330) for rough targetting. It often has a centering mark for placing the instrument below ceiling points. Some theodolites have a telescope level attached to the telescope for levelling the collimation axis.

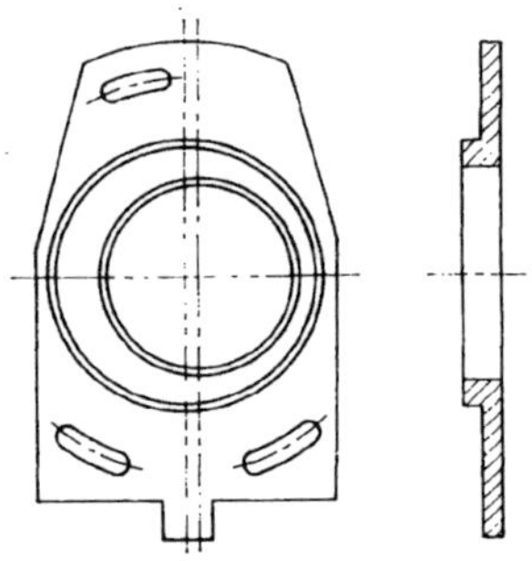

Figure 209 Horizontal axis bearing with eccentric socket

Theodolite Levels

Plate levels are used to level the instrument. To achieve an angular accuracy of 0.1 to 0.2 mgon, the vertical axis of the theodolite has to be plumb within about 1 mgon. This is possible with a 30″ bubble. Usually, a bull's eye bubble is used for rough levelling, and a tubular bubble normal to (longitudinal level) or parallel to (lateral bubble) the horizontal axis serves for fine levelling. Occasionally, both are used (cross bubble). Some low accuracy instruments have only a bull's eye bubble with e.g. 8′ sensitivity plus often a telescope bubble. With a height index bubble, the influence of inclination of the vertical axis can be eliminated. Instruments of medium or high accuracy occasionally have a striding level placed directly onto the horizontal axis. With it, the axis can be levelled, or its tilt during the measurements can be determined, especially for inclined sights, where horizontal axis errors have large influences.

Reading Devices

The amount of the measured angle is determined with the aid of a reading device placed on the top part. One reads directly the graduation line preceeding the index (rough reading), and then determines the separation between it and the index (fine reading). Every reading for a given direction represents the angle between the origin of the graduation and the index of the reading device.
In the simplest case, the fine reading is done by estimation. The accuracy depends on the size of the graduation interval, i.e. the separation of the graduation lines. An experienced observer can quite accurately estimate one tenth of an interval, if the separation is 1.0 or 1.5 mm, and the line thickness 0.1 to 0.15 mm. Considering the graduation unit, these separations would only be suitable for very large circle diameters. Therefore the reading interval is kept smaller, and enlarged optically with a magnifying glass or a microscope.
The reading devices are primarily based on the following principles:

a) Estimation of tenths of the interval (figures 212 and 214)
b) Coincidence of two lines (figures 210, 215, 240, 241)
c) Setting of a line between a double line (see figures 217, 220, 242, 247)

Mechanical reading devices (simple index, vernier), as well as optical readings with screw micrometer, which were used in instruments of high

accuracies are nowadays replaced by optical ones, such as line microscope, scale- and vernier microscope, optical vernier, and optical estimation (double circle reading), simple optical micrometers, and micrometers with automatic means.
While an interval can be read with vernier and magnifying glass to 5 μm and with a screw microscope to 1 μm, optical micrometers permit measurement to 0.1 μm. The magnification depends on the diameter of the circle and its graduation unit. It is generally 3–5 times for direct reading, and 20–50 times for reading microscopes. To eliminate the influence of graduation eccentricity, theodolites of medium and high accuracy have two reading devices separated by 200 gon (the Salmoiraghi 4181 has a 100 gon separation). In addition, the measuring accuracy is increased when reading at two places. For low accuracy theodolites, one reading is usually sufficient. Modern theodolites are designed as optical theodolites, i.e. theodolites whose optical glass circles can be read from the station of the observer, because of special optical systems (reading microscopes). Usually, the readings of both horizontal and vertical circles—either together or one at a time—are imaged in the field of view of the reading eyepiece, placed beside the telescope eyepiece. For better identification, the circle images appear often in different colours. Precision instruments utilize optical micrometers.

Vernier

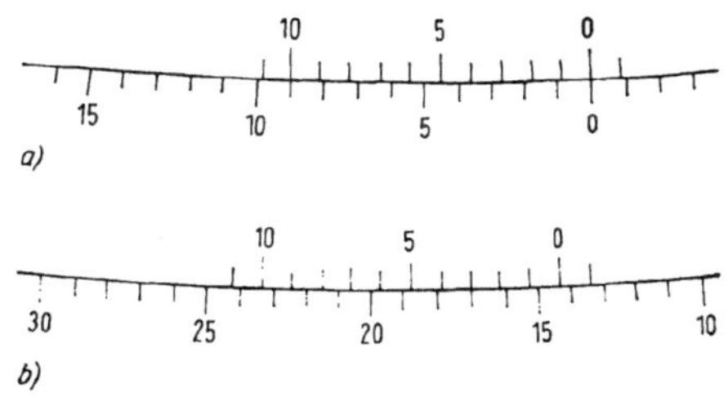

Figure 210 Vernier
a) zero position
b) reading 14.3

The vernier is an auxiliary graduation, placed alongside the main one with the purpose of determining fractures of the graduation unit. n vernier units represent $(n - 1)$ intervals of the main graduation. The following relation exists between the graduation units t_K of the circle and t_A of the vernier:

$$nt_A = (n - 1)\, t_K. \tag{82}$$

The vernier value a is the difference between t_K and t_A;

$$a = t_K - t_A = \frac{t_K}{n}. \tag{83}$$

In zero position of the vernier, its zero line is aligned with a line of the graduation (figures 210a). If 10 vernier units represents $(n - 1) = 9$ graduation intervals with a unit of 1 gon, the vernier value is $a = {}^1/_{10}$ gon. The larger n, the smaller is a. However, for large n, the vernier becomes very long, and reading becomes uncomfortable and difficult. Usually, the vernier value is 5 or 10 mgon resp. 10″, 20″, 30″, or 1′.
The vernier lines are numbered in multiples of a, starting at the zero line and increasing along the main graduation. Where alignment of vernier—and graduation line is achieved, the number at the vernier represents the angular value between the zero line of the vernier and the preceeding graduation line. Therefore, one records the angle of the preceeding graduation line, and then searches for the vernier line, that coincides with a graduation line. The sum of these readings gives the final value, e.g. in figure 210b with $a = 0.1$ and coincidence at the 3rd line we get 14 gon + 3×0.1 gon = 14.3 gon. The vernier of an American transit, illustrated in figure 211 has 30 parts, which correspond to 29 intervals of the ${}^1/_2{}^\circ$ (30′) graduation. Therefore $a = 30'/30 = 1'$. One reads 17° at the circle and 25′ at the vernier, therefore 17° 25′ (resp. 342° 30′ + 5′ = 342° 35′ in opposite direction).
In addition to the coinciding lines, one should compare the two neighboring pairs of lines. At both ends, the vernier usually has an extended graduation (figures 210 and 211). If none of the vernier lines coincides with a graduation line, one uses the mean of two lines which come nearest to coincidence.

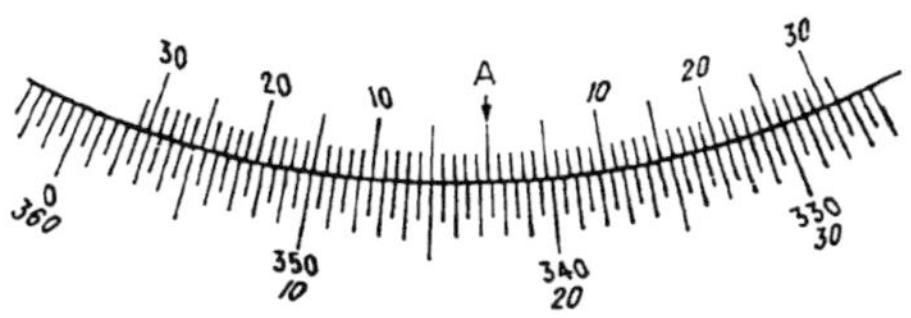

Figure 211 Vernier reading 17°25′ resp. 342°35′ (the numbering towards left omits calculations for staking surveys; it is especially used overseas)

A magnifying glass serves for a more comfortable vernier reading. For focussing, it can be shifted in its casing. It has to be placed such, that the location of coincidence falls into the centre of the field of view. Movable shades help that vernier and graduation are equally illuminated.
The vernier, as described, is additive. The rarely used subtractive vernier has n parts which correspond to $(n + 1)$ graduation intervals. It has to be read opposite to the main graduation.
The standard deviation of a vernier reading can be assumed as $m = \pm 0.3\,a$.
A vernier is tested by coinciding its zero line on various locations of the graduated circle, and then checking its end line, which should also coincide with a graduation line.
The vernier is used with theodolites of medium or low accuracy and metal circle (see tables 8 and 9), as well as for simple angle measuring devices (see figure 158) and for improving rod readings (see figures 611, 614, 616).

Reading Microscopes

Reading microscopes are used to clearly see the graduation intervals, which, for newer instruments, are only hundredths or tenths of a millimetre. In comparison to the magnifying glass, they have the additional advantage, that object and viewing distances become much larger, and thus the eye-piece can be made easier accessible to the eye. Furthermore, a measuring mark, scale, or similar marking can be placed in the plane of the real intermediate image. The separation between graduation and index can be measured very accurately by optical means, without mechanical contact as in the case of the vernier.

To eliminate errors of eccentricity, theodolites of high, often also medium accuracy, have two diametrically opposite reading devices. For optical theodolites, the reading device is located in the hollow axes and standards (see figures 213, 219, 223, 225, 226, 235, 239, 243), while the reading eyepiece is directly beside the telescope eyepiece, or else at the standard, as for some American, Japanese and English theodolites. For instance in the Wild T3 the reading rays of the horizontal circle pass through 390 mm of glass, while the optical path in the MOM TeB23 has 87 optical surfaces. The reading microscopes in the theodolites have the following tasks:

a) Illumination of the graduations

b) Uniting the images of both circles (one reading position each, or two positions on horizontal- and vertical circle, sequentially or at the same time)

c) Imaging of the circle parts so they can be measured.

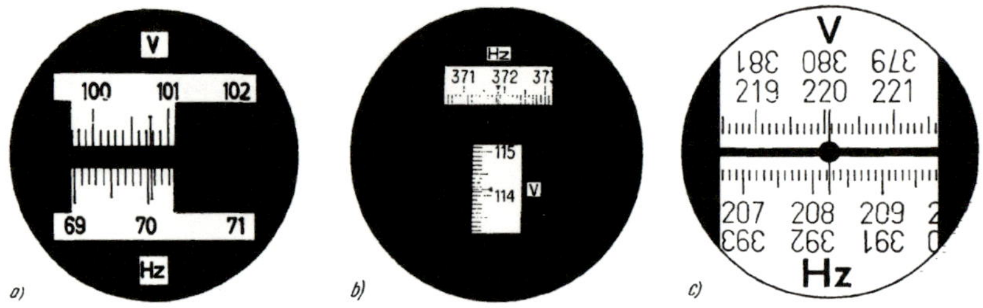

Figure 212 Field of view of line microscopes
a) Zeiss, Oberkochen Th 51 (V: 100.77 gon; H: 70.06 gon) similarly T 30, USSR
b) Wild T 05 (H: 371.85 gon; V: 114.15 gon)
c) Theo 080 A, Zeiss, Jena (V = 220.07 gon, Hz = 208.23 gon)

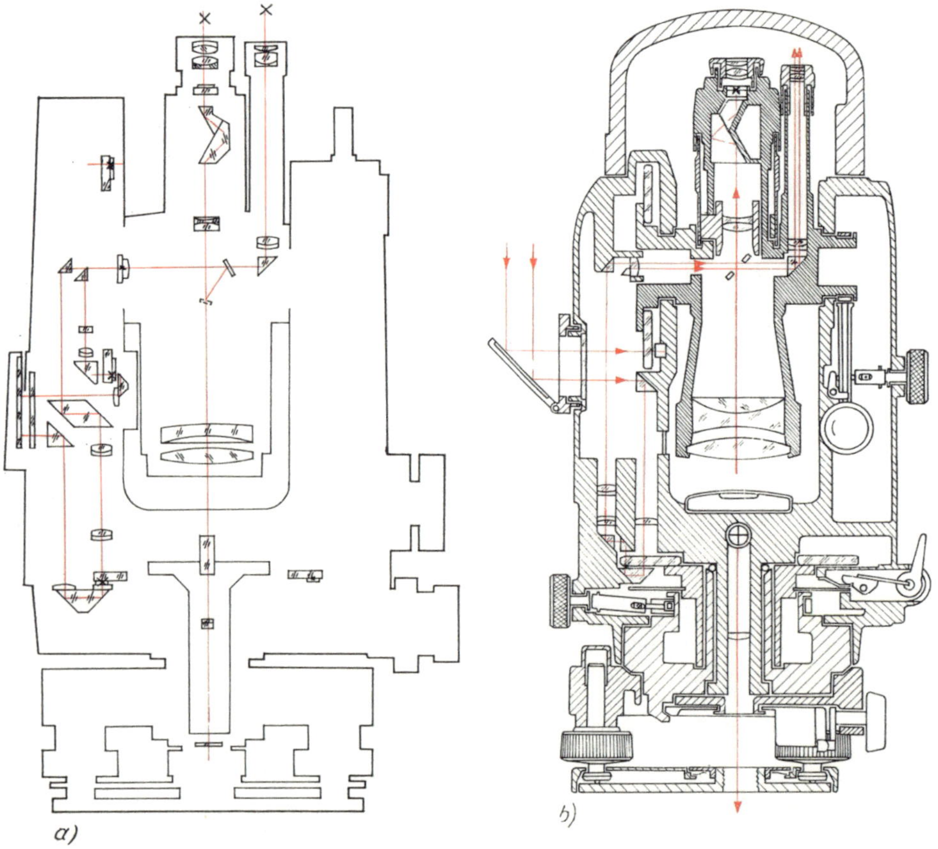

Figure 213 Optical train in theodolites with scale microscopes
a) Theo 020 A, Zeiss, Jena
b) Wild T 16

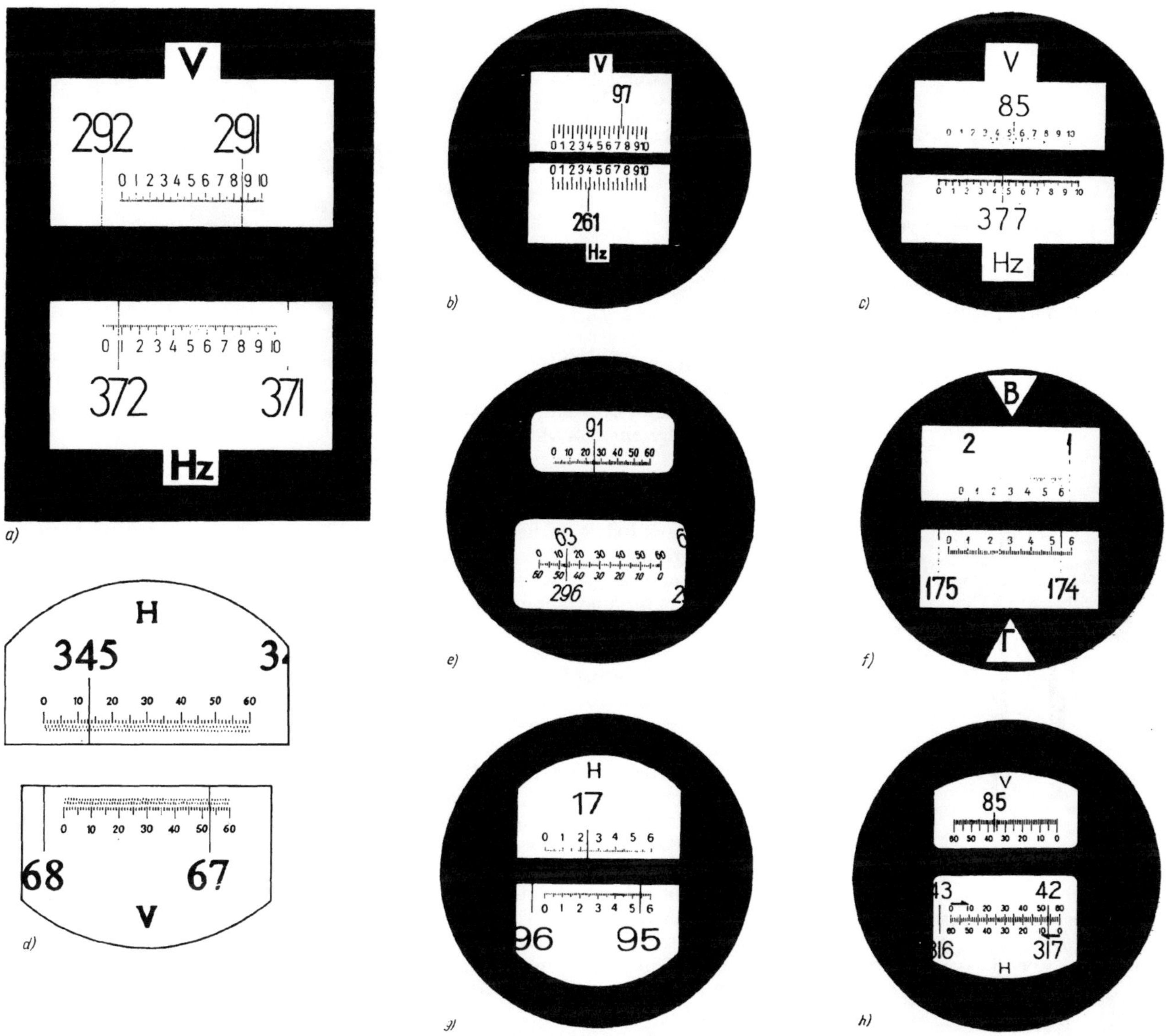

Figure 214 Field of view of scale microscope theodolites
a) Theo 020 A, Zeiss, Jena (*H* (blue) 372.080 gon; *V* (yellow): 291.860 gon)
b) TEKAT, Breithaupt & Son (*H*: 261.39 gon; *V*: 97.73 gon)
c) Th 41 and 42, Zeiss, Oberkochen (*H* (yellow): 377.456 gon; *V* (blue): 85.532 gon)
d) Vickers V 22 (*H*: 345°13′10″; *V*: 67°52′40″)
e) 6020-A1 Dietzgen (*H*: 63°13′.6 resp. 296°46′.4; *V*: 91°25′.4)
f) T-5 and T-10, USSR (*H*: 174°55.0′; *V*: 2°05.0′)
g) MOM Te-D 23 and 43 (*H*: 17°24′; *V*: 95°54′)
h) Lietz T-60 D (*H*: 42°53′.8 resp. 317°0.62′; *V*: 85°36.3′)
i) Wild T 16 (*H* (yellow): 214.935 gon; *V*: 94.065 gon)
k) K 1-S, Kern (*H*: 615.635 gon resp. 84.365 gon; *V*: 105.565 gon)

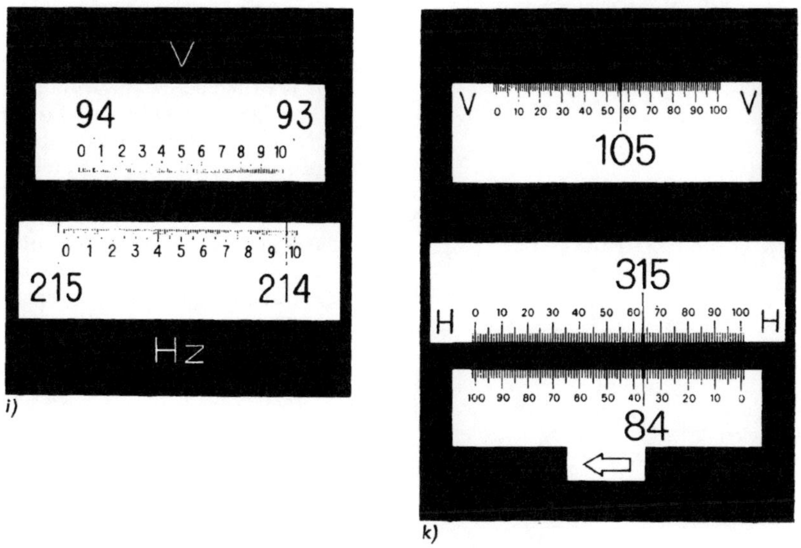

The imaging of the vertical circle reading into the plane of the horizontal circle is mostly enlarged $\Gamma > 1$; the diameters of the vertical circles are smaller than the ones of the horizontal circles. The microscope is adjusted

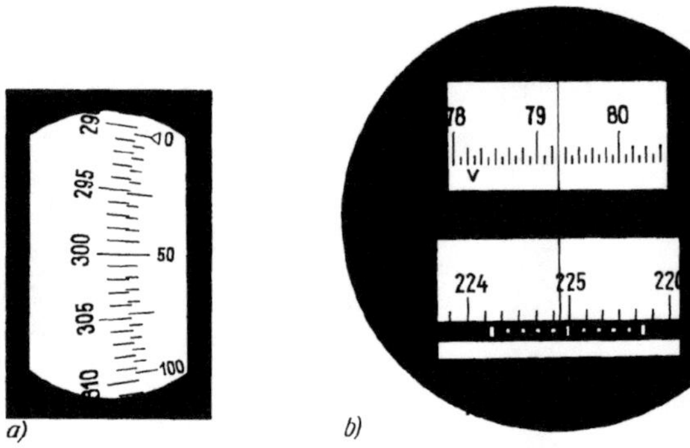

Figure 215 Field of view of vernier microscopes
a) STNO, SLOM (290.50 gon)
b) TG 4b, Officine Galileo (224°54′; line microscope for V: 79°15′)

by the manufacturer such, that there is no parallax between line or index, and image of the graduation. The observer focusses for both line and index or scale by rotating the eyepiece. It is advisable, that all adjustments be done by the manufacturer.

The *line microscope* (sometimes called: estimation microscope) is the simplest of the reading microscopes. The reticule plate of the microscope contains a line as index, which is used for quickly and easily estimating tenths of the graduation intervals (figure 212).
Line microscopes are used in low accuracy theodolites, and to read graduated circles on levelling instruments.

The *scale microscope* has a reticule plate with a scale whose length, when adjusted properly, corresponds to the separation of two neighbouring graduation lines. The numbering, starting at the zero line of the scale, increases in opposite direction to the main graduation.
Optical theodolites of low accuracy usually have a graduation to 1 gon or 1°. The scale has 100 resp. 60 intervals, so that the scale value is 10 mgon resp. 1′, and 1 mgon, resp. 6″ can be estimated. Only one circle part each of the horizontal- and vertical circle is used, with both being imaged simultaneously (figures 213 and 214).
The vernier microscope is a special scale microscope. Instead of the scale, the reticule plate contains a vernier which matches the circle graduation (figure 215).
While vernier microscopes are rarely used, the scale microscope is predominantly used in theodolites of low and medium accuracy, because of the simple and easy reading.

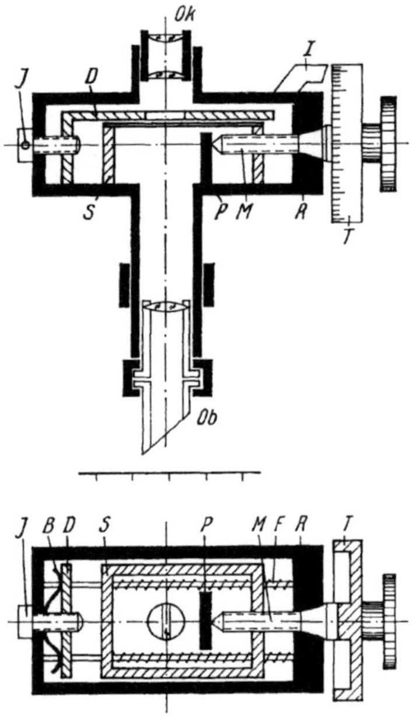

Figure 216 Screw microscope

Screw Microscope. One can obtain smaller scale values than with the scale microscope, if the separation between index line and graduation line is measured with a mechanical or optical micrometer. Until the latter appeared, the screw micrometer was the most accurate reading device, and thus widely used in high accuracy instruments. A microscope equipped with it is called screw microscope.
The distance between the zero mark of the microscope and the image of the preceeding graduation line is determined by the number of revolutions of a precision screw, which moves a double line. The screw micrometer is placed into the microscope tube near the eyepiece (figure 216). The sled S, which carries the reticule with the double line, is located in a rectangular frame R. The sled can be shifted with the micrometer screw M of the frame, which presses against the plate P. To avoid back lash of the screw, two spiral springs F press against the sled from the frame. Complete revolutions of the screw are read at the rake in the cover plate D (figure 217), parts of a revolution of the drum T, which usually contains 30, 60, or 120 intervals (degree graduation), is read with the index I. The zero point of the micrometer is fixed in the rake with a notch. With the adjustment screw J and the spring B, the cover plate can be set such, that the zero points of two diametrically opposite microscopes differ exactly by 200 gon. In order to get a zero reading on the drum, when the double line is set onto the zero point of the rake, the drum, which is held with a spring, can be lifted manually off the index and thus adjusted.
In order to measure the angle between the zero point of the microscope and the preceeding graduation line, the double line is set onto the respective line by turning the micrometer screw (figure 217). To measure one graduation interval (5′), for instance, 2.5 revolutions of the drum are required,

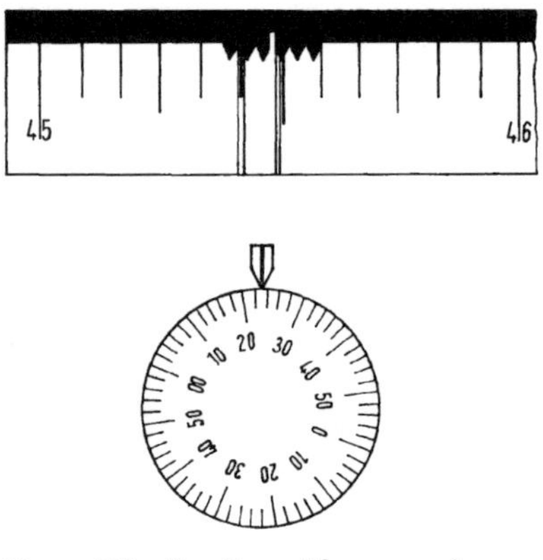

Figure 217 Reading with screw microscope (45°28′23″)

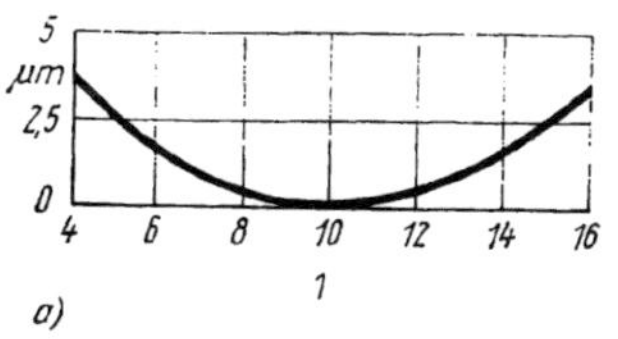

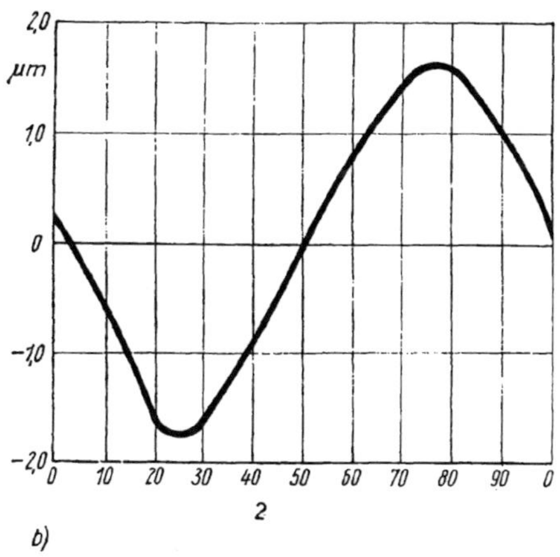

Figure 218 Screw errors
a) accumulative error
b) periodic error
1 revolutions (thread height 0.25 mm)
2 drum readings

which corresponds to 2.5 notches of the rake. Thus, starting from the zero point, each full notch (equals a full revolution) has a value of 2′. Since the drum has 12×5 = 60 intervals, the scale value of the micrometer is 2′ : 60 = 2″. The toal reading is composed of the circle reading (45° 25′) plus the rake (3′) and drum (23″) readings, is therefore 45° 28′ 23″.
The micrometer eyepiece, which can be found on triangulation theodolites and astronomical instruments for improving the aiming accuracy is similarly built. This improvement is based on multiple aiming with the micrometer without changing the telescope setting.
The accuracy of the screw micrometer depends on the precision of its manufacture. The path of the spindle, which turns in the fixed nut of the frame, shall be proportional to the revolution of the screw. Errors in the thread of nut or spindle can cause irregularities. The following errors can occur when measuring with screws (figure 218):

a) Progressive errors because deviations the average screw pitch from its design value correspond to a scale error. This error increases with the measured distance.
b) Periodical errors due to erroneous thread occur with the same amount at each revolution. Its amount changes within each revolution with a sine-function.
c) Irregular errors of the thread because of irregular thread do not usually occur, because the nut covers several threads simultaneously.

The progressive errors have to be considered when measurements occur at different parts of the screw, e.g. for astronomic observations. They

Table 5. Determination of periodic screw errors

screw reading s_n	difference $i_n = s_n - s_{n-1}$	correction $v_n = i_m - i_n$	correction to the reading $\Delta s_n = [v]_1^n$
1	2	3	4
$s_0 \approx 0$			0
	$i_1 = s_1 - s_0$	v_1	
s_1			v_1
	$i_2 = s_2 - s_1$	v_2	
s_2			$\sum_1^2 v$
	$i_3 = s_3 - s_2$	v_3	
s_3			$\sum_1^3 v$
	$i_4 = s_4 - s_3$	v_4	
s_4			$\sum_1^4 v$
	$i_5 = s_5 - s_4$	v_5	
s_5			$\sum_1^5 v$
	$i_6 = s_6 - s_5$	v_6	
$s_6 \approx s_0$			$\sum_1^6 v = 0$
mean	$i_m = \frac{[i]}{n}$	$[v] \approx 0$	

amount to 0.5 to 1.5 μm for 6 to 8 revolutions. By adjusting the microscope, they are eliminated for screw microscopes. They are negligibly small in eyepiece micrometers of triangulation theodolites, because the double line of the sled is only shifted within $^1/_5$ of a revolution for sighting.

Periodical errors, however, have a significant influence on the measuring results. An irregularity of 0.001 mm of the pitch of the screw, for instance, results in a measuring error of 0.1 mgon. Normally, periodical errors do not exceed 0.8 μm. To reduce their influence, screw microscopes are equipped with two double lines for circle readings, whose separation deviates from the apparent size of the graduation interval by half a revolution of the screw. If one matches the two double lines sequentially with the neighbouring graduation lines (to 45° 25′ and 45° 30″ in figure 217), then the mean of these readings eliminates most of the periodical errors. Furthermore, periodical errors can be eliminated by systematically changing the drum when sighting to the first target.

To determine these errors, an interval corresponding to $^1/_4$ to $^1/_8$ of a revolution of the screw is used. Since graduation lines are not available with such separation, one has to substitute signs and numbers in the field of view. The auxiliary interval t is measured several times forwards and backwards with the screw, covering one full revolution (e.g. for $t = 20''$, and a 2′ drum revolution). After each reading, the plate is turned back by t. (table 5).

The most accurate tests of micrometer screws are carried out with precision micrometers or with light interference. On page the run and its determination is treated.

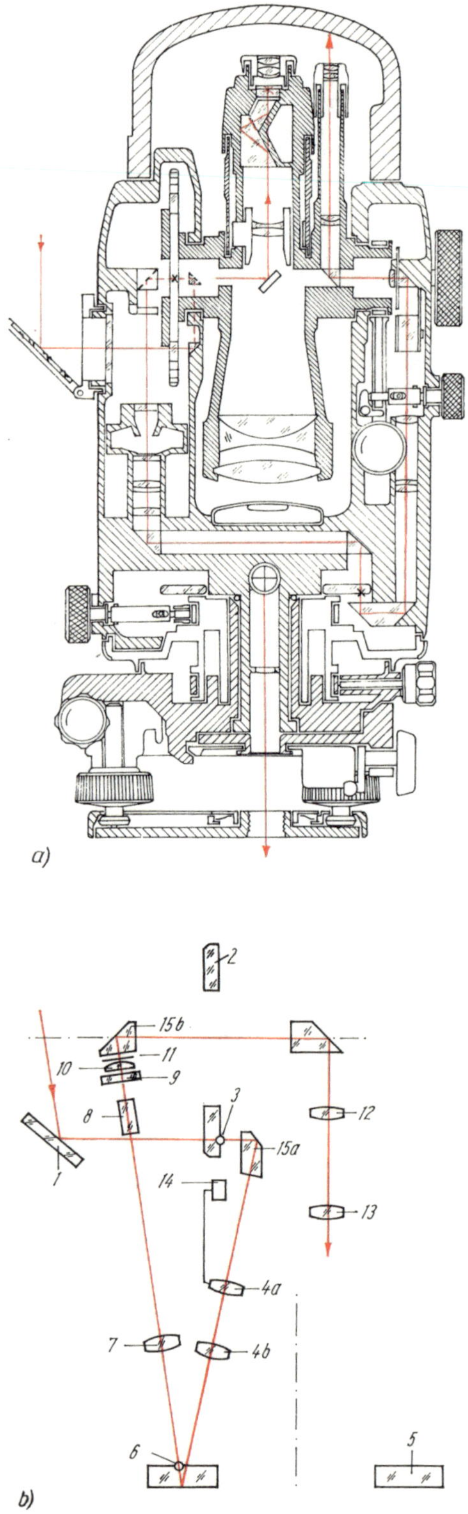

Figure 219 Optical train in theodolites with plan parallel plate micrometers
a) Wild T 1
b) Kern K-1 A

Optical micrometers are the most commonly used and most accurate reading devices for surveying instruments. Their main part consists of an optical system with at least one movable deflection element connected to a scale for the determination of its positional change. Unlike for the screw micrometer, wher the image of the index line is shifted in the reading microscope, a section of the graduated circle is shifted optically in the optical micrometer.

The reading accuracy of optical micrometers can be increased nearly without limit, and its elements can vary according to desired effect and space conditions. Screw errors within the transfer mechanism cause measuring errors ten to one hundred times smaller than for the screw micrometers. Since the micrometer graduation can be edged into glass and can visibly appear in the image plaue of the microscope, it is possible to image it directly beside the circle graduation. Thus there is no need to change the accomodation of the eye when rading circle and micrometer graduation. Then the work becomes less tiring. The whole measuring range of the graduation can be placed on the glass panel, because the latter is magnified by the eye piece. Therefore, unlike for the screw micrometer, resolutions do not have to be counted. The movable deflection element can be built into the path of rays of the reading microscope at any place. The following elements can serve as deflection elements:

a) a plan parallel plate that can be rotated around a fixed axis
b) an optical wedge that can be shifted (shifting wedge)
c) a wedge that can be rotated
d) a lens, that can be shifted or tilted, normal to its optical axis.

Table 6. Maximum deviations of a regular micrometer graduation because of non-linearity of the parallel shift for tangent monitoring

tilt angle	refractive index of the glass	maximum error in % of the total shift
±8°	1.61 1.52	0.01 0.35
±15°	1.59 1.52	0.1 1.0
±25°	1.56 1.60	0.5 1.5
±30°	1.52 1.60	1.0 3.6

For simultaneous reading at two circle positions, normally two deflection elements are used. In reading microscopes of optical theodolites mainly plan parallel plates or shifting wedge micrometers are used. When shifting an optical wedge, the amount of shift is directly proportional to the deflection of the rays. When rotating a plan parallel plate this proportionality is only approximate, which explains the use of a special monitoring device in this case. However, parallel plate micrometers require less space than shifting wedge micrometers.

The Optical Micrometers for a Single Reading contain a parallel plate which deflects a light ray, hitting it under any angle, parallel to its original direction. By rotating the parallel plate, the image of the graduation is shifted until the index line coincides with the previous graduation line. The plate rotation is transmitted via levers to a scale, on which the parallel shift can be read in angular units. A uniform graduation is most suitable for production. This can be achieved by designing the monitoring device such, that it is linear to the tangent of the tilt angle, which in turn is proportional to the parallel shift of the ray in a good approximation. The remaining error, which can be determined by computing the parallel shift for the whole measuring range of the micrometer according to equation (10), can be minimized by proper selection of glass and maximum tilt angle (10° to 15°). This is illustrated in the example in table 6. (from *Haller R.* "Einige konstruktive Möglichkeiten bei der Kreisablesung von Theodoliten und Tachymetern" Zeitschrift für Vermessungswesen 1957, # 1, pp. 12–18 and # 2 pp 41–45). One has to consider, that a 0.5% error causes a reading error of 0.05 mgon. If a non linear graduation is used, based on the actual parallel shift of the ray, glass type and tilt angle can be freely selected.

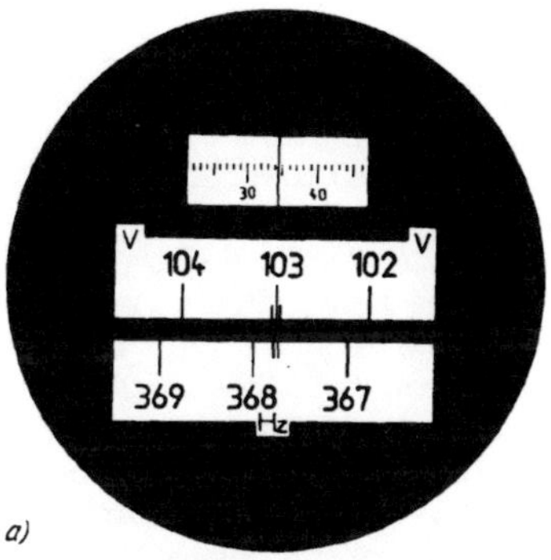

a)

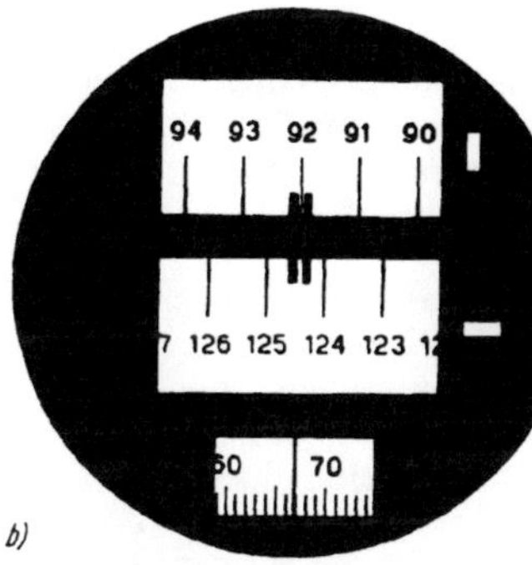

b)

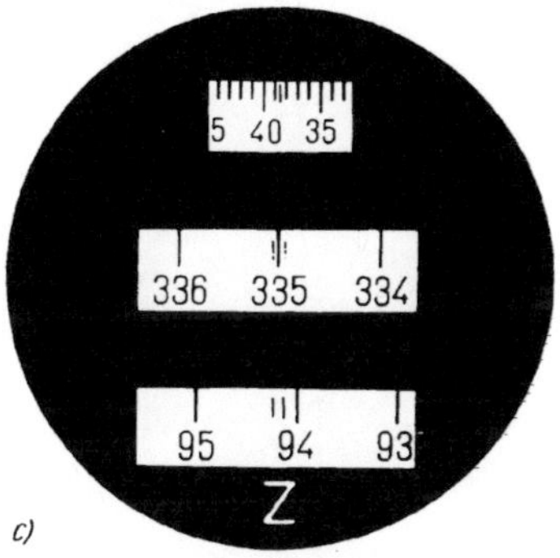

c)

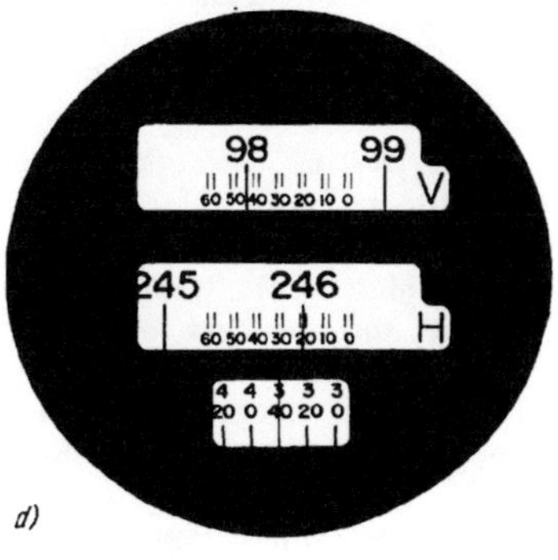

d)

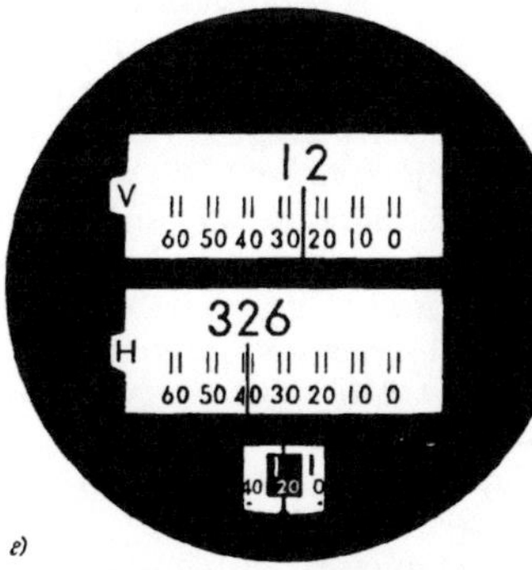

e)

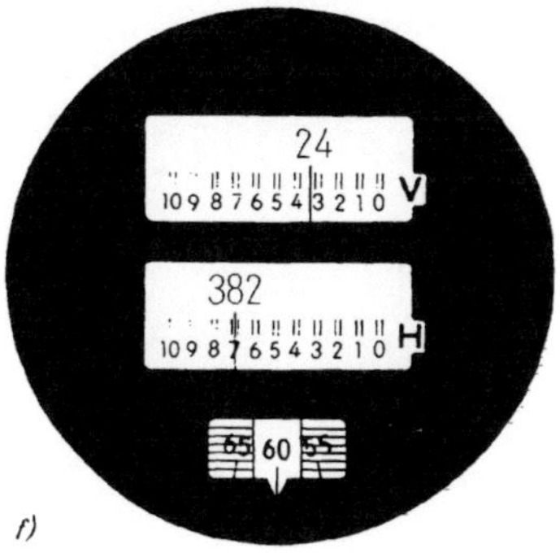

f)

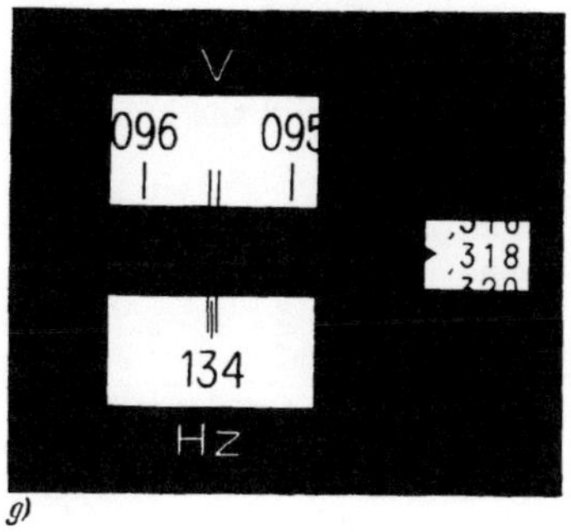

g)

Figure 220 Field of view of microscopes with simple parallel plate micrometer
a) TEAUT, Breithaupt & Son (V: 103.345 gon)
b) MOM Te-D 11 resp. 13, 31, 33 (V: 92.668 gon)
c) TG 3 b, Officine Galileo (H: 335°38′30″)
d) Sokkisha TM-20 (H: 246°23′40″)
e) Topcon TL-20 (326°41′20″)
f) Nikon NT-2 (H: 382.760 gon)
g) Wild T 1 (H (yellow): 134.318 gon)

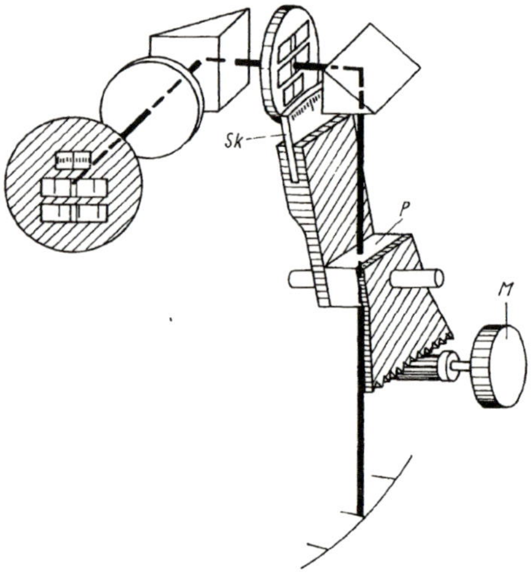

Figure 221 Optical micrometer with tilting plane parallel plate

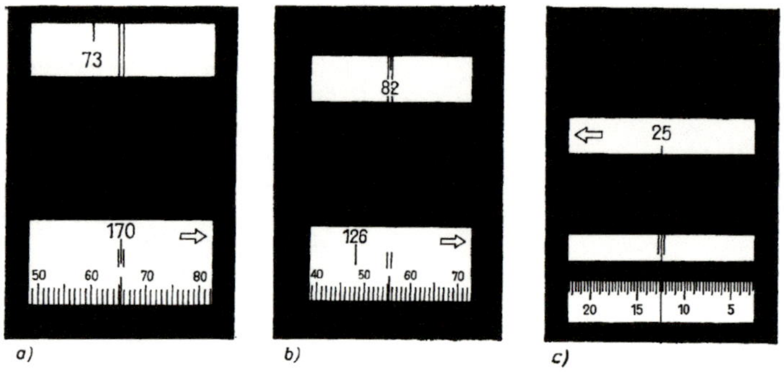

Figure 222 Circle reading for Kern K 1-A
a) graduation to the right H: 170.653 gon
b) V: 82.553 gon
c) graduation to the left: H: 25°12′30″

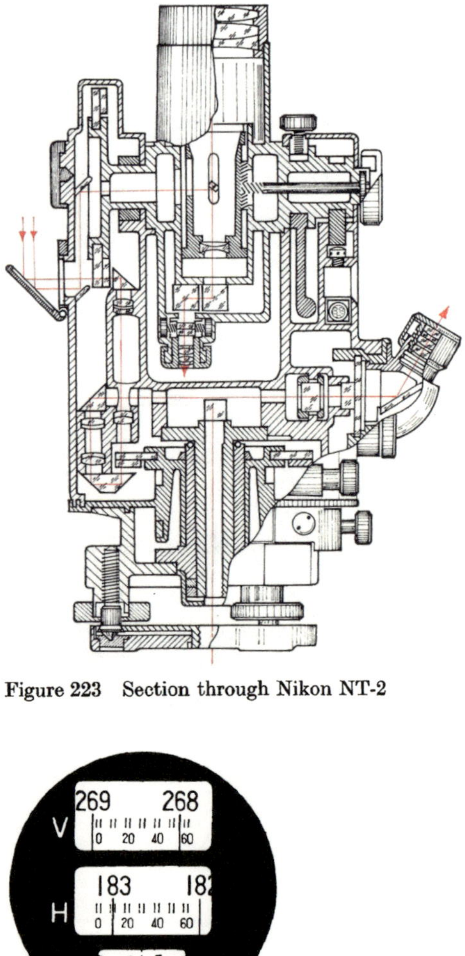

Figure 223 Section through Nikon NT-2

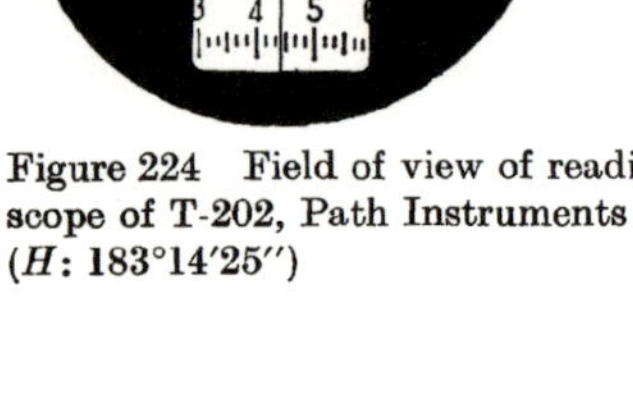

Figure 224 Field of view of reading microscope of T-202, Path Instruments
(H: 183°14′25″)

Figure **219**a shows an example of the path of the rays in an optical theodolite having a reading microscope with simple parallel plate micrometer. The reading position of the vertical circle is illuminated via mirror and prisms, and imaged in the plane of the horizontal circle reading with the aid of prisms, a liquid compensator, and an objective lens. The images of both reading positions are transmitted together into the plane of the diaphragm, using prisms and an objective lens. The image generated there, is deflected with a prism parallel to the collimation axis and viewed with objective lens and eye piece. The objective lenses on both sides of the horizontal circle are double lenses, to facilitate change in magnification and then sharply image not only the lines of the vertical circle in the plane of the horizontal one having equal intervals, but also both circles' position in the diaphragm plane. The diaphragm contains two windows with double lines for the images of horizontal and vertical circles, and one above them for the index line of the micrometer scale (figure 220). The scale, which is fully numbered for some instrument types, is connected to a roughly 6 mm thick parallel plate. The parallel plate P is tilted with the aid of the micrometer screw M until the image of the graduation T is shifted such, that a graduation line is centred between the double line index (figure 221). The scale Sk, which is connected to the parallel plate, moves at the right window and shows the amount of shift at the index (figure 220). The micrometer can be used for either the horizontal or vertical circle reading.

The situation illustrated in figure 219 b is similar. The illumination mirror (*1*) deflects the light rays onto the reading position (*3*) of the vertical circle (*2*). This position is imaged near the reading position (*6*) of the horizontal circle (*5*) using a pendulum element (*4a*) supported at (*14*), and a fixed element (*4b*) of a double objective lens. Both circle positions are imaged with the objective lens (*7*) through the parallel plate (*8*) onto the reticule (*10*) with index. This image is viewed, together with the micrometer scale (*9*), through the objective lens (*12*) and the eyepiece (*13*), after passing through a shiftable diaphragm (*11*) and the deflection prisms (*15b* and *15c*). The graduations of the horizontal circle and the micrometer scale are numbered in both directions. By a switch, either the numbers increasing to the left for horizontal circle and micrometer (figure 222a), or else, the numbers increasing to the right plus the vertical circulation graduation (figure 222c) can be blocked off. With the numbers increasing to the left, no computations are required for staking. Figure 223 illustrates the situation in another theodolite.

If the microscope is not used, one can read it like a scale microscope. On the other hand, it is possible to add an optical micrometer to a scale microscope and measure with it the interval between index and graduation, rather than estimate it (figure 224). In this case, the graduation line is shifted onto the scale index, and the amount of shift is read on the micrometer scale.

Reading Microscopes with Automatic Means for Reading of two Circle Positions are found in theodolites of medium and high accuracy, occassionally even for low accuracy instruments. They use two circle positions to eliminate the influence of graduation eccentricity of the circle. In optical theodolites, two diametrically opposite circle readings are combined optically. In the field of view of the reading eyepiece appears the automatically formed mean of the two single readings. This drastically accelerates and simplifies the reading.

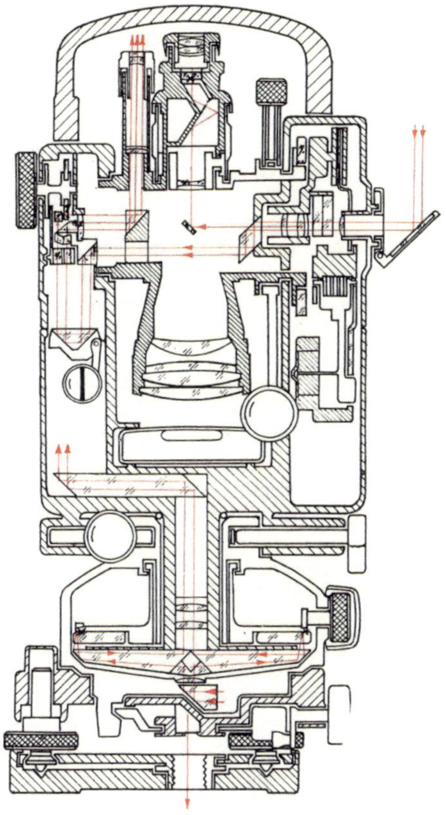

Figure 225 Section through Wild T-2

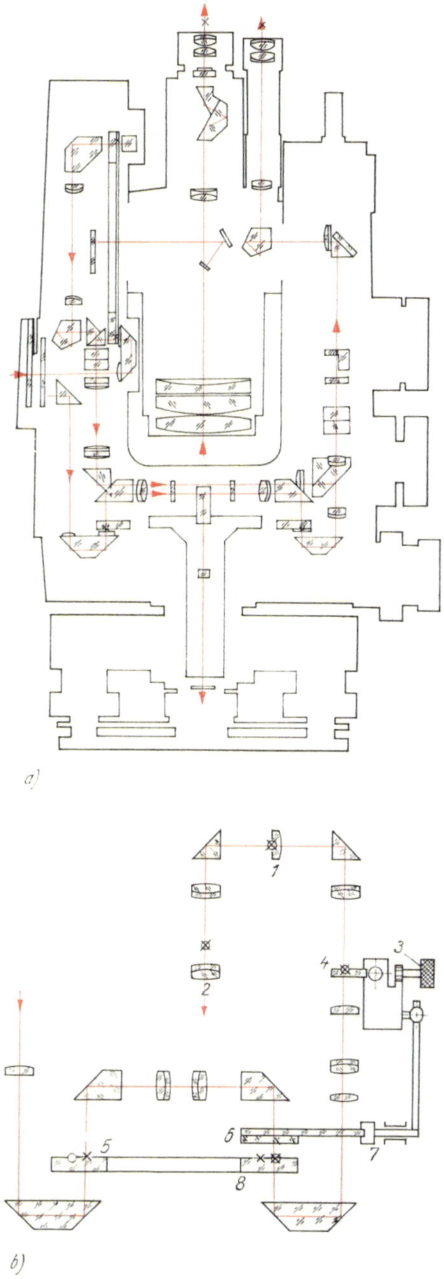

Figure 226 Optical train in theodolites
a) THEO 010 A, Zeiss, Jena
b) DKM 2-A, Kern, Aarau
1 field pupil with 10′ numbers; *2* eye piece; *3* micrometer drive; *4* scale; *5* double lines; *6* parallel plate; *7* support of parallel plate and lever; *8* single lines

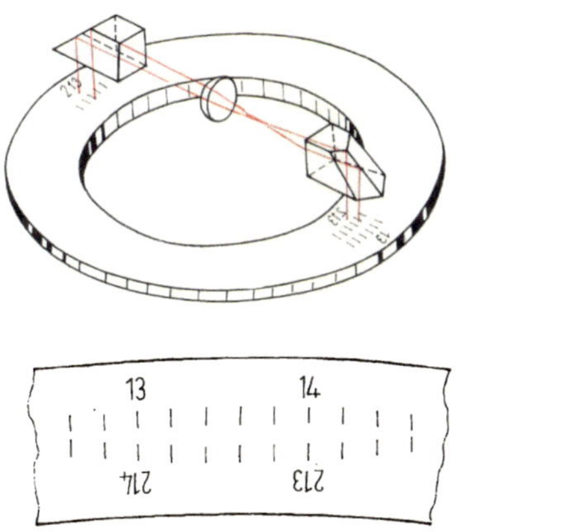

Figure 227 Transfer of circle position into the plane of the diametrically opposite one via deflection prism, objective lens and roof prism

The two circle positions can be combined such, that their images in the microscope

a) touch each other, but run in opposite directions (figures 240, 241)
b) appear above each other and run in opposite directions (figure 242)
c) appear above each other and run in the same direction (figure 247).

The images of the two circle positions can be combined symmetrically, e.g. through the hollow vertical axis and one standard as for the Wild T2 and T3 (figure 225). One can also use both standards, as for the TG 1 from Officine Galileo, Florence, or — as is usually the case — one circle position is imaged onto the second one, and both are then viewed together (figure 226). In the latter case one views through the circle graduation, while for the first case the rays are reflected on the circle surface.

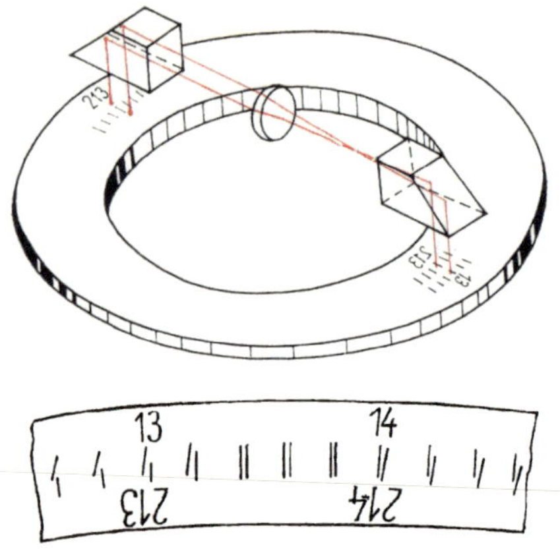

Figure 228 Transfer of circle position into the plane of the diametrically opposite one via two deflection prisms and the objective lens.

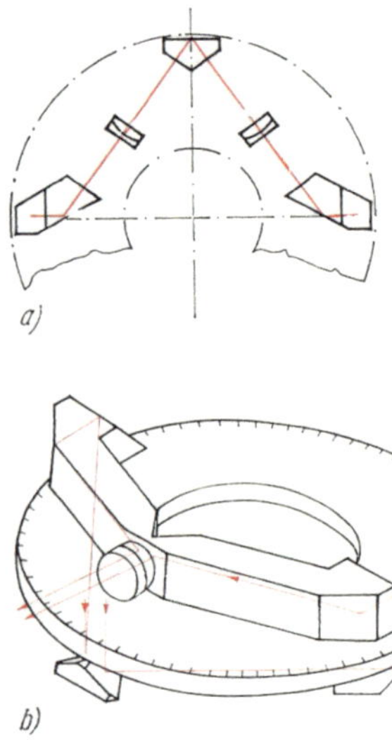

Figure 229 Transfer of circle reading position around the vertical axis
a) MOM Te-B 23
b) Zeiss, Oberkochen Th 2

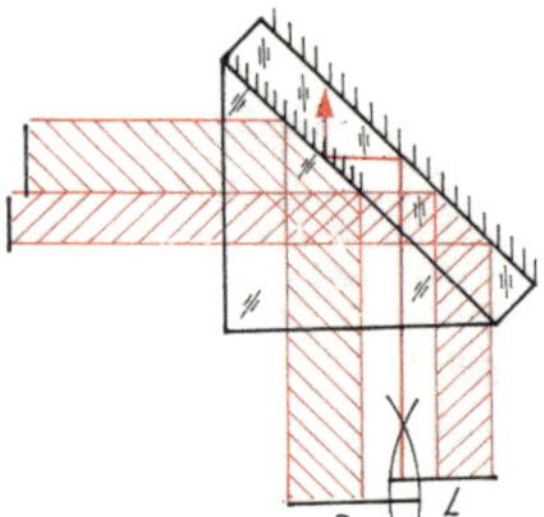

Figure 230 Separation block of an optical coincidence micrometer with prism and wedge

For the circle positions are to appear as running in opposite directions, one of the prisms for transmitting the first circle position onto the second one has to be a roof prism (figure 227).
If, however, diametrically opposite lines are to be combined to double lines running in the same direction, the transmission system contains two simple deflection prisms and an objective lens (figure 228).
The images of both circle positions can also be combined symmetrically around the vertical axis (figure 229).
Circle positions imaged such, that they run in opposite directions, would be separated in the field of view, if they were not brought together with a special system for image separation such, that they are separated by a fine line only. A prism block (figure 230) is most commonly used, which consists of a prism with its hypothenusis surface as semi mirror and of an optical wedge with small prism angle (<9 mgon) and mirror coated outside surface. Reflection on the separation block causes a vertical shift of the rays from the circle position L. The part of the circle section which is not desired, is blocked out by the mirror coating. The bottom edge also serves as separation edge. Since the prism angle is not quite 50 gon, the main rays of both bundles passing through prism and optical wedge are converging in direction of the optical axis of the microscope.
The separation block in other theodolites consists of two prisms cemented together (figure 231). The rays, coming from both circle positions and falling onto the bottom surfaces of the prisms, are shifted and combined by the prisms. Rays passing through the central zone do not reach the field of view. The cemented surface acts as separation surface. Figure 236 illustrates yet another arrangement.
The separation block has the effect that the ends of the lines of both circle graduations hit a common separation line (figure 232). The position, where the difference between the two readings is exactly 200 gon is used as fictitious index, i.e. the indices I_2 and I'_1, coincide. When rotating the upper part or the graduated circle, the images of the two circle positions run against each other, however, the position of I_1 and I_2 remains unchanged. For single readings at two positions one determines first the value of the respective previous graduation line, then the respective interval between graduation line and index, and finally calculates the mean between the two values. In this case, the same value is obtained without single readings and subsequent calculation of the mean, if the value of the graduation line and half the interval between it and the corresponding diametrical line is determined. This interval between the diametrical lines can be estimated (for DK theodolites by Kern, Aarau prior to 1965). More accurate and comfortable readings are obtained, if the interval is measured.
The micrometer in its simplest form, consists of a parallel plate, which is placed behind the first circle position in the path of rays and which can be tilted such that the graduation lines of both circle positions coincide (example: TGlb by Officine Galileo, Florence).
Usually two elements are used to deflect the rays, which means that each has to be tilted by half the amount to reach coincidence, thus resulting in twice the setting accuracy. The two parallel plates, which rotate around a common axis in opposite direction, are placed such, that the rays coming from the two different circle positions pass through a different plate. By tilting the parallel plates with the micrometer drive, diametrical lines are brought to coincidence (figure 233).
The interval between the corresponding diametrical lines as function of the tilt can be read in angular units on a scale connected with the parallel plates. The mechanism which transforms plate tilt into a rotation of the reading scale is quite complicated. Often the parallel plates are monitored with a spiral groove (figure 234). The levers which rotate around two axes and carry one parallel plate each, extend into the spiral groove of a disk,

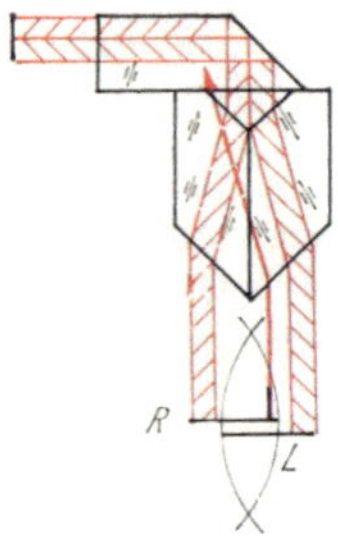

Figure 231 Separation block of an optical coincidence micrometer with two plane parallel plates

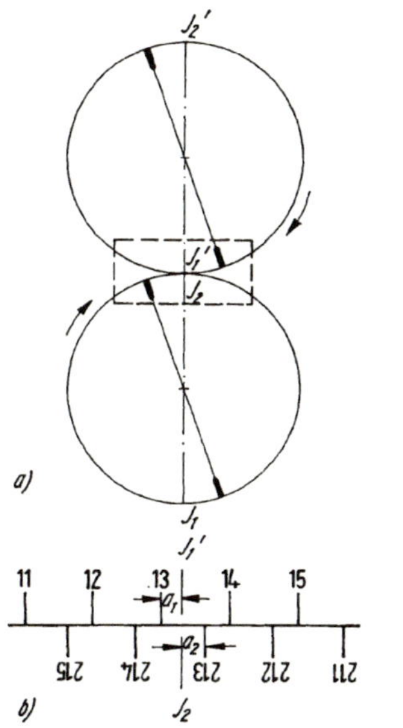

Figure 232 Principle of automatic meaning
a) Schematic representation of contact of the images of both circle positions
b) Field of view (reading 13 gon + $(a_1 + a_2)/2$)

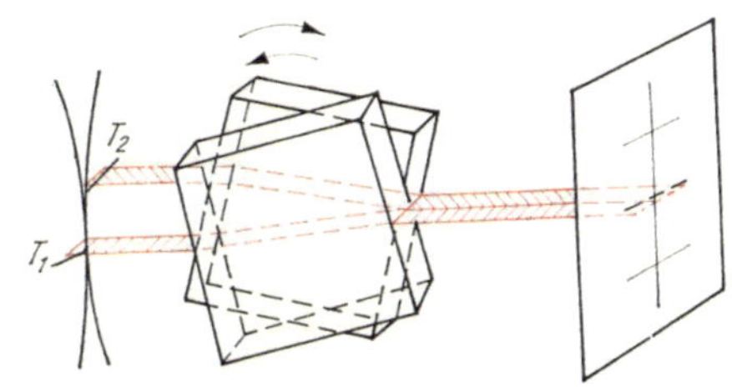

Figure 233 Effect of two parallel plates in optical coincidence micrometers (schematic)

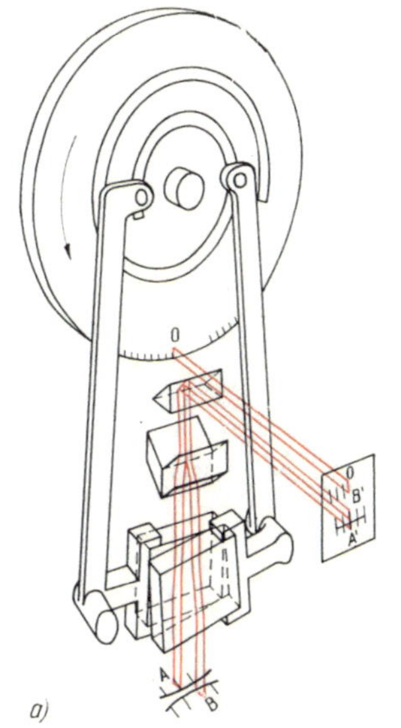

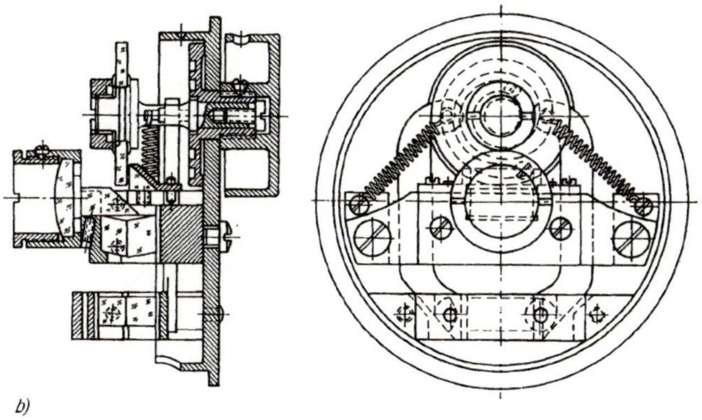

Figure 234 Monitoring of the optical micrometer with the aid of a spiral groove
a) functional scheme
b) arrangement for the Wild T 2 and T 3 (linear separation of the spiral groove: 3 mm, thickness of plane-parallel plates: 4.7 mm, length of lever: 30 mm)

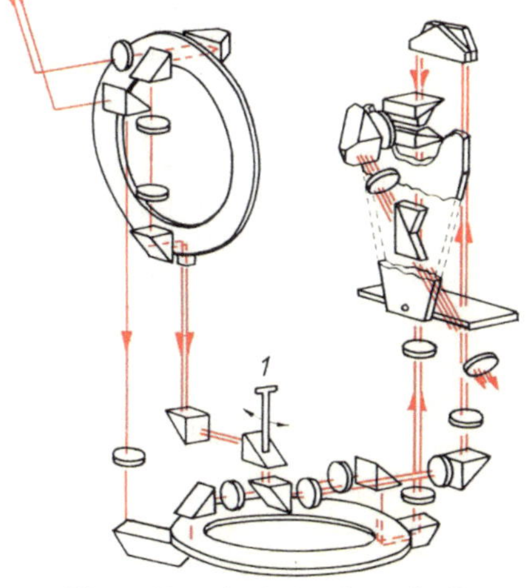

Figure 235 Optical train in the Dietzgen 6020-A2
1 pendulum

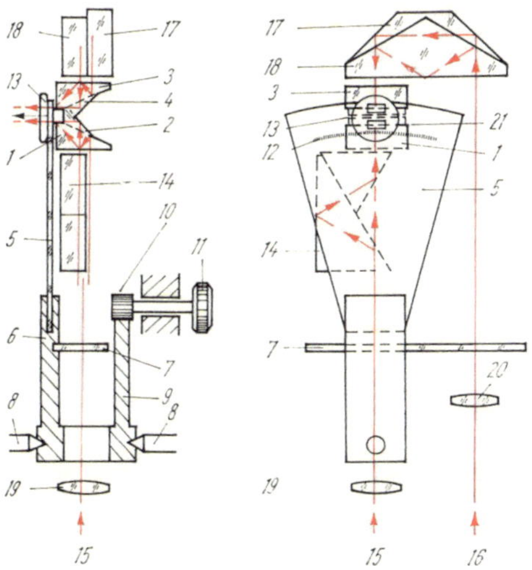

Figure 236 System of the optical micrometer in the Dietzgen 6020-A2

which carries the scale. When rotating the disk, the levers change their position proportional to the rotational angle of the disk. It is also possible to simultaneously shift both diametrical positions in opposite directions with one parallel plate only (figure 235). The first circle position is transferred to the second one as shown in figure 228. The change in direction is generated after the light has passed through the parallel plate. The rays (*15*) of the partial images coming from the horizontal circle, pass through the objective lens (*19*) (figure 236) and the parallel plate (*7*) and are divided afterwards. One partial image is mirror reversed by triple reflection in the

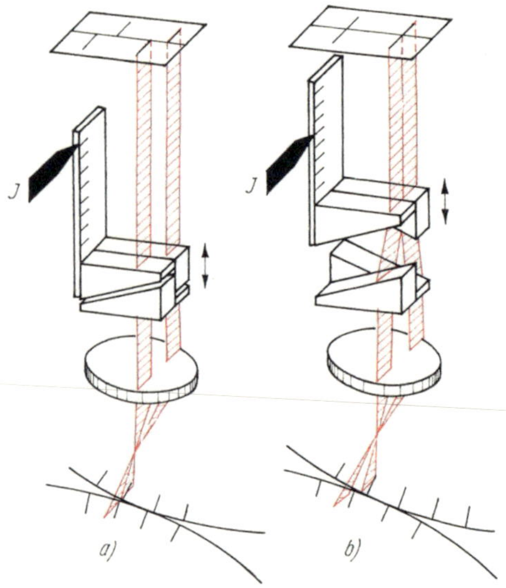

Figure 237 Schematic presentation of an optical coincidence micrometer with shift wedges
a) zero position
b) reading position

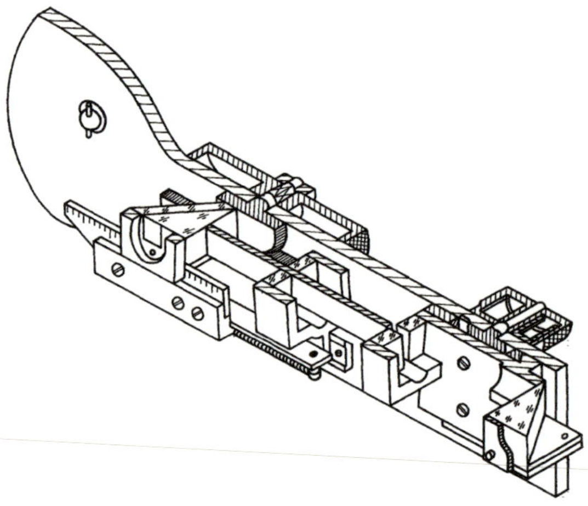

Figure 238 Design of coincidence micrometers with shift wedges

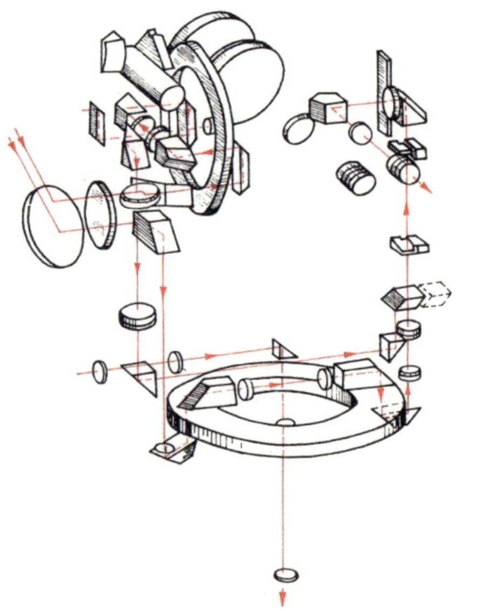

Figure 239 Optical train in an optical theodolite with shift wedges (T-2, USSR)

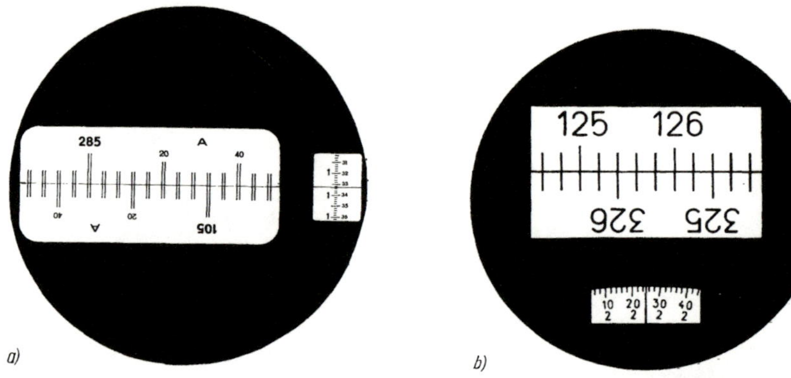

Figure 240 Field of view of reading microscopes with coincidence micrometers
a) Theo 002, Zeiss, Jena (285°17′33.3″)
b) Te-B 23, MOM, Budapest (125.7225 gon)

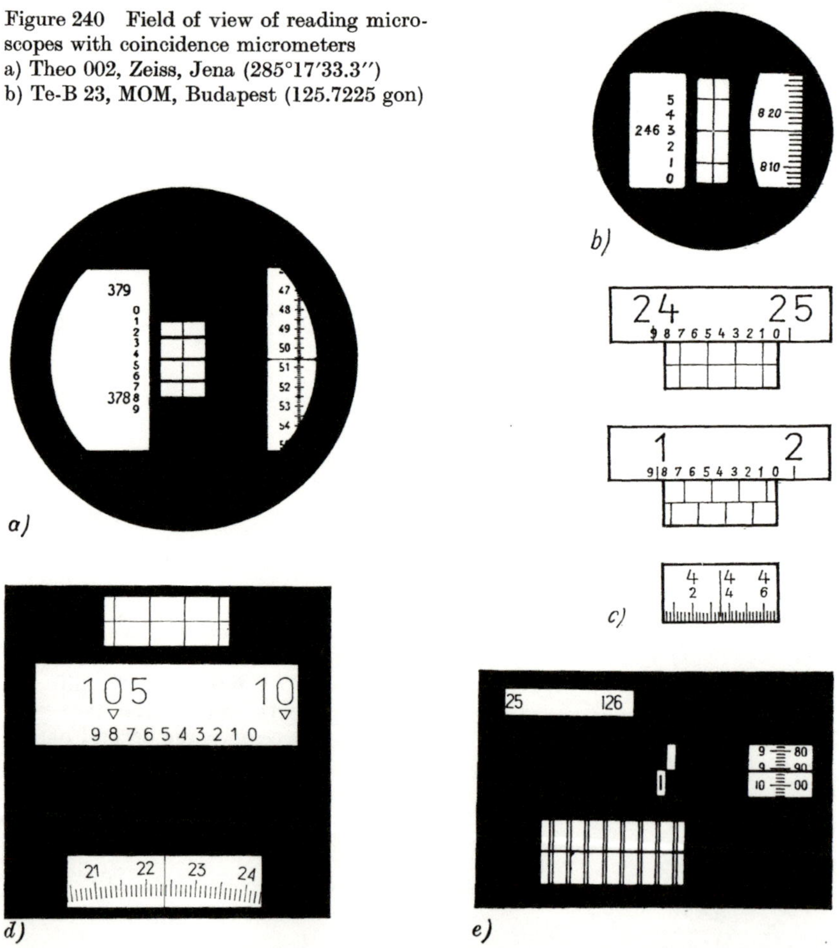

Figure 241 Field of view of reading microscopes with coincidence micrometers and continuous number reading
a) Th 2, Zeiss, Oberkochen (*H* (yellow): 378.8506 gon, *V* (blue))
b) TM-1 A, Lietz, USA (246°38′16.7″)
c) 6020-A2, Dietzgen (24.9435 gon)
d) T 2, Wild, Heerbrugg (*H* (yellow): 105.8224 gon)
e) THEO 010 A, Zeiss, Jena (*H* (blue): 126.1992 gon, *V* (yellow)

reversal prism (*14*), and imaged above the separation edge (*21*) by prism (*1*) via prism (*2*). The other partial image goes directly to prism (*1*), is reversed vertically but not horizontally by reflection at the mirror coated part of the inclined surface, as well as by total reflection at the entrance surface and at the uncoated part of the inclined surface. It is then imaged within the field of view below the separation line. The path of the rays (*16*) coming from the vertical circle is similar. These rays pass through the objective lens (*20*) and the parallel plate (*7*), are then divided with each half passing via prisms (*17*) resp. (*18*) into the prism system (*3*) and (*4*) where the partial images are combined in the same manner as in the prism group (*1*) and (*2*). In the field of view (*13*) the diametrical positions of both circles can be viewed simultaneously. By rotating the driving knob (*11*), the parallel plate (*7*) in its support (*6*) (two pointed screws *8*), is moved via the slot (*10*) and the cog segment (*9*) and, simultaneously with it—without mechanical transmission—the attached scale (*5*). The graduation (*12*) of the scale depends on the actual parallel shift of the ray when tilting the parallel plate. If it is moved together with the parallel plate by the driving knob, the partial images of both circle positions move in opposite directions, so that the graduation lines can be brought to coincidence (figure 241 c).

The shifting wedge micrometer can be easier constructed, and is free of many of the error sources of the parallel plate micrometer (figures 237, 238). Two pairs of optical wedges with equal prism angles are arranged such, that the principal sections are normal to the images of the lines, and that the refracting edges of each pair are opposite to each other. The lower pair is fixed and serves to correct the image quality and to image the diametrical circle sections in the centre of the field of view. The upper pair is connected with the micrometer scale via a rod with cogs. When rotating the micrometer drive, the rotation is transmitted via a cog wheel to the rod, which in turn shifts the wedges and scale measurably, so that the images of the diametrical positions can be coincided. Errors in this mechanism do not influence the results, like when using parallel plates. The larger size, however, is a disadvantage. In figure 239 the path of rays for a theodolite with shifting wedge micrometer is illustrated. With coincidence micrometers, one reads either at an index in the centre of the field of view, or else counts the intervals between corresponding diametrical graduation lines. For each interval, half of its value has to be taken. In figure 240a, the graduation

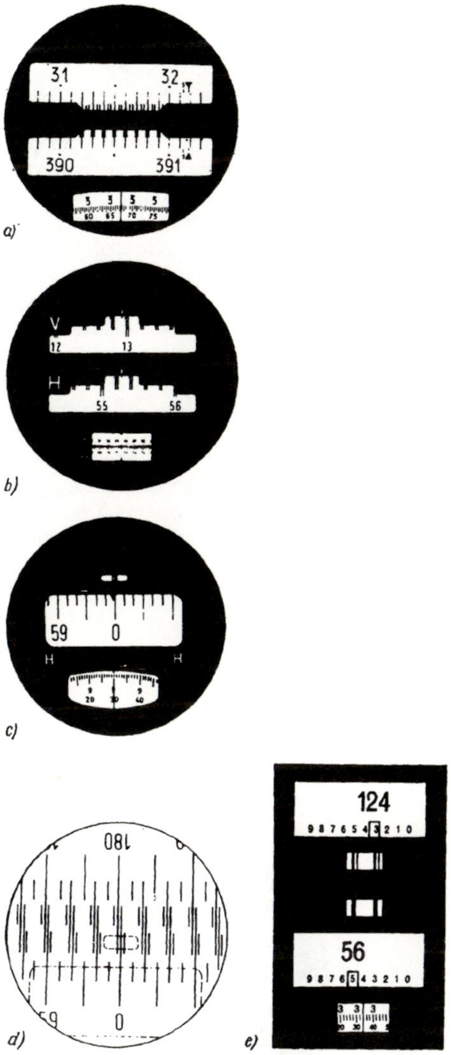

Figure 242 Field of view of reading microscopes with superimposed images running in opposite directions of the diametrical circle positions
a) DKM 3, Kern (*H*: 291.15 gon + 0.03673 gon = 291.18673 gon; compare to figure 206)
b) 4200, Filotecnica Salmoiraghi (*H*: 55.20 gon + 0.0715 gon = 55.2715 gon, *V*: 12.9285 gon)
c) and d) ST 200, Rank Precision Industries (*H*: 395°50′ + 9′30″ = 395°59′30″, comp. to figure 207)
e) DKM-2 A, Kern (*H*: 56.5334 gon)

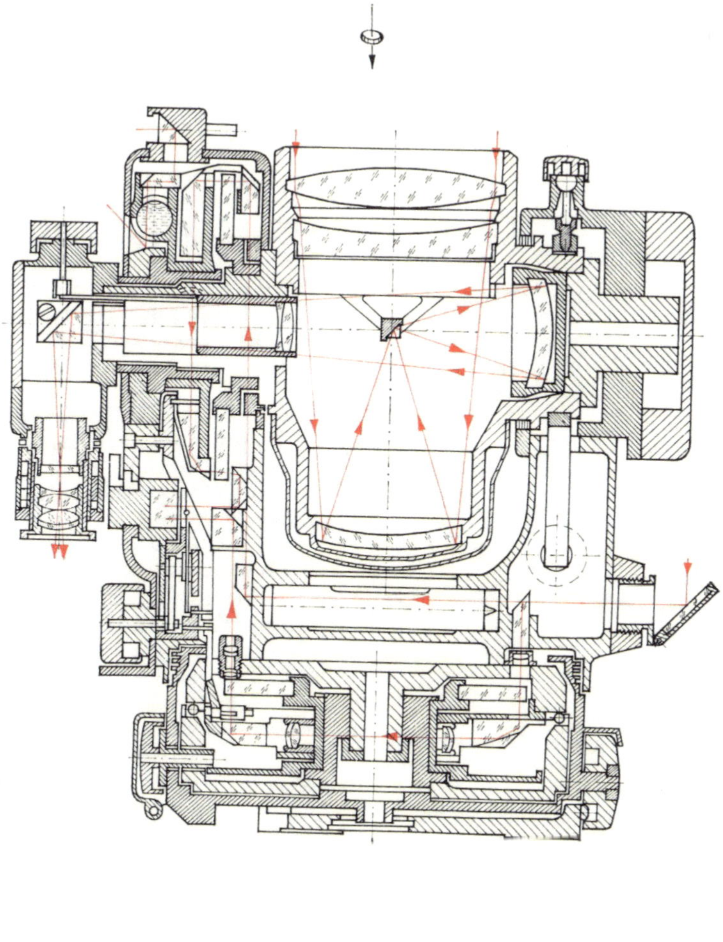

Figure 243 Section through DKM 3 (Kern, Aarau)

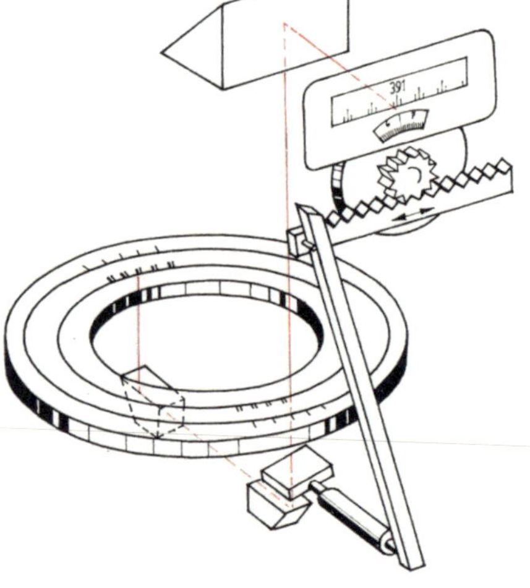

Figure 244 Optical micrometer with plane-parallel plate between both circle positions (DKM 3, schematic)

value of the circle is 4′. The separation of the graduation lines 285° and 105° is 8 intervals; therefore the result is 285° + 8 × 2′ = 285° 16′. The remainder of 1′ 33.3″ is read on the micrometer scale.
In newer theodolites, there is no need to count the intervals. The first six digits of the angle can be read directly (figure 241).
In some designs, the circle sections are not separated by a separation edge as in the coincidence micrometer, but partially or completely superimposed (figures 242, 245). The graduated circles obtain an additional graduation (see figures 206 to 208). The parallel plate of the optical micrometer is placed between the reading positions (figures 243 to 245). The image of the first position can be shifted with it in respect to the second one until the double lines of one position are symmetrical around the single lines of the other (figure 244). By proper selection of line widths and separation of the double lines, together with the movement of the partial images in opposite directions, the setting accuracy can be higher then for plain coincidence reading.
In the Salmoiraghi 4181, the image of the second position, 100 gon apart from the first one (figure 245a), appears as reading index (figure 245b), where a circular arc is used in addition to the graduation lines (see figure 208). The index is therefore not an element of the upper theodolite part, as usual, but of the graduated circle. One can read without and with the micrometer, where the index is shifted between the circle position with the aid of a parallel plate.

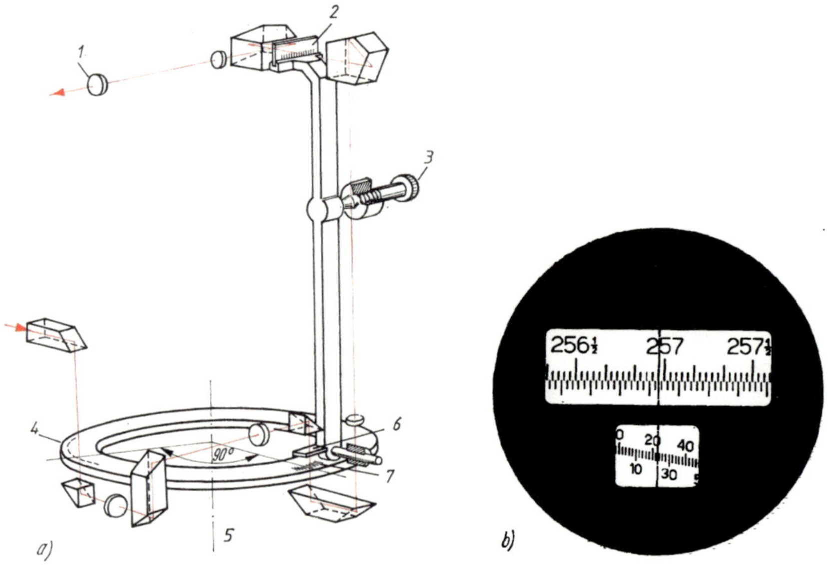

Figure 245 Circle reading for the Salmoiraghi 4181
a) optical train
b) field of view of reading microscope (H: 256°58′23″)
1 reading eye piece; *2* micrometer scale; *3* micrometer knob; *4* reference circle; *5* vertical axis; *6* parallel plate; *7* circle graduation

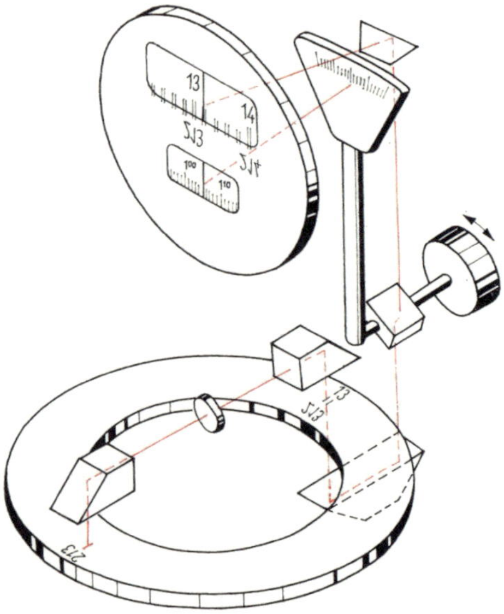

Figure 246 Optical micrometer with plane-parallel plate for centering the double lines consisting of lines of diametrical circle position (schematic)

If the single lines of both circle positions are combined to pairs as in figure 228, only one optical element to deflect rays is required to measure the interval between index and nearest line pair. A parallel plate is mostly used (figure 246). The two lines of the line pairs can either come from two different graduations with equal value (see figures 206a, 247a) or from only one graduation (figure 247c).

Run. The size of the intermediate image, generated by the objective lens of the microscope is designed such, that for a screw micrometer, an even amount of revolutions of the micrometer screw corresponds to one interval of the real intermediate image of the graduated circle. For theodolites with an auxiliary scale in the image plane of the objective lens of the microscope (e.g. scale- or vernier microscopes, scales of optical micrometers) the scale graduation should correspond to the interval of the graduated circle. An error in magnification of the microscope leads to a deviation from this desired value, and is referred to as run.
To determine the run, one sets the zero line of the scale onto a graduation line. The end line of the scale should then coincide with the next graduation line. If not, the difference represents the run. By comparing the length of the scale with various graduation intervals, a mean value for the run is obtained. Generally, the run will not exceed $2\,^0/_{00}$ of the range of the scale. It does not need to be considered for scale microscopes, since the estimation accuracy is of the same magnitude. However, it can have significant effects in reading microscopes with optical micrometers, whose scales, as a rule,

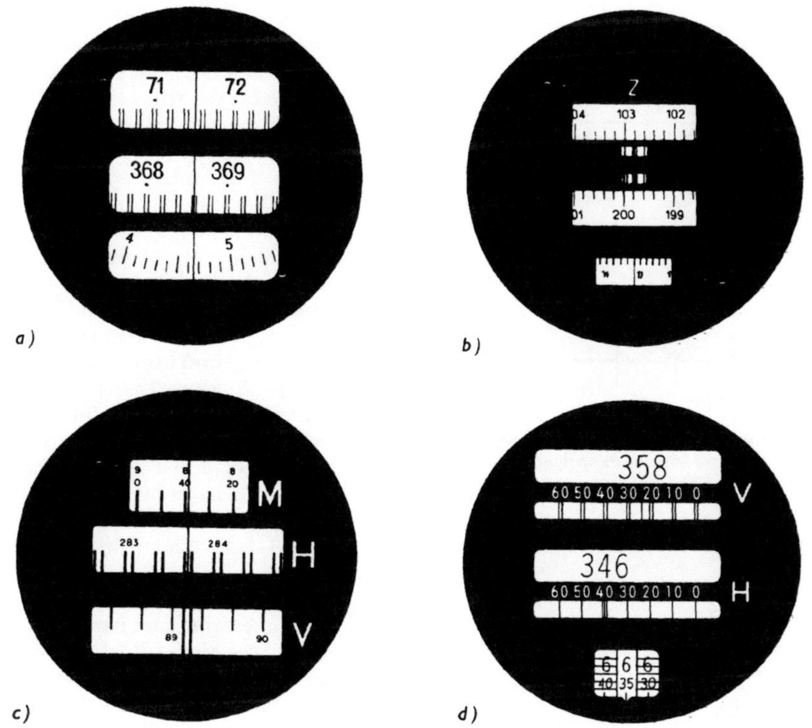

Figure 247 Field of view of reading microscopes with center setting
a) DKM 1, Kern (368.6465 gon)
b) TG 2 b, Officine Galileo (102.9320 gon)
c) Sokkisha TM-10 (283°48′39″)
d) Nikon NT-3 (346°46′35″)

Table 7. Example for run determination

forward measurement						return measurement					
circle position	a 0′ + [″]	b 2′ + [″]	c 2′ + [″]	$r_u = a - b$ [″]	$r_l = a - c$ [″]	circle position	a 0′ + [″]	b 2′ + [″]	c 2′ + [″]	$r_u = a - b$ [″]	$r_l = a - c$ [″]
1	2	3	4	5	6	7	8	9	10	11	12
0° 0′	+0.4	+0.4	+0.2	0.0	+0.2		0.0	0.0	+0.2	0.0	−0.2
	+0.4	+0.2	+0.4	+0.2	0.0	22° 32′	−0.2	−0.2	−0.2	0.0	0.0
			mean	+0.1	+0.1				mean	0.0	−0.1
	+1.2	+1.2	+1.2	0.0	0.0		+0.4	+0.6	+0.6	−0.2	−0.2
45° 06′	+1.0	+1.2	+1.2	−0.2	−0.2	67° 38′	+0.2	+0.4	+0.4	−0.2	−0.2
				−0.1	−0.1					−0.2	−0.2
	+1.0	+0.8	+0.8	+0.2	+0.2		−0.6	−0.4	−0.4	−0.2	−0.2
90° 12′	+0.8	+1.0	+0.6	−0.2	+0.2	112° 44′	−0.2	−0.2	−0.4	0.0	+0.2
				0.0	+0.2					−0.1	0.0
	+0.6	+0.6	+0.6	0.0	0.0		+1.0	+1.0	+1.2	0.0	−0.2
135° 18′	+0.6	+0.6	+0.6	0.0	0.0	157° 50′	+1.0	+1.2	+1.2	−0.2	−0.2
				0.0	0.0					−0.1	−0.2
	+0.8	+1.0	+0.6	−0.2	+0.2		+0.6	+0.4	+0.8	+0.2	−0.2
180° 24′	+1.0	+0.8	+0.6	+0.2	+0.4	202° 56′	+0.8	+0.6	+0.8	+0.2	0.0
				0.0	+0.3					+0.2	−0.1
	+0.2	+0.6	+0.4	−0.4	−0.2		+0.2	−0.2	+0.2	+0.4	0.0
225° 30′	+0.4	+0.4	+0.4	0.0	0.0	247° 02′	+0.2	+0.2	+0.2	0.0	0.0
				−0.2	−0.1					+0.2	0.0
	−0.2	−0.2	+0.4	0.0	−0.6		+0.6	+0.8	+0.4	−0.2	+0.2
270° 36′	+0.2	+0.2	0.0	0.0	+0.2	292° 08′	+0.6	+0.8	+0.6	−0.2	0.0
				0.0	−0.2					−0.2	+0.1
	+0.8	+0.8	+0.6	0.0	+0.2		+0.8	+0.8	+1.2	0.0	−0.4
315° 42′	+0.6	+0.8	+0.8	−0.2	−0.2	337° 14′	+1.0	+0.8	+1.2	+0.2	−0.2
				−0.1	0.0					+0.1	−0.3
			mean	−0.04	+0.03				mean	−0.01	−0.10

$r_{upper} = -0.03''$ $\quad r = -0.04''$

$r_{lower} = -0.04''$ $\quad \Delta r = r_u - r_l = +0.01''$

have several hundred intervals. Therefore, for precision measurements it has to determined carefully, and corrected for mathematically, if its absolute value is more than 0.5 units of the scale. The run correction is $\Delta a = (r/t)\, a$, where t is the total range of the scale, a the reading on the micrometer scale, and r the run. Since transportation of an instrument can result in minute positional changes, the run has to be determined periodically. The adjustment of the optical system of modern instruments to eliminate the run should be left to the manufacturer. It is useful, to eliminate the run during the measurements by spreading the initial readings for sets of angular measurements, equally among the micrometer scale.

Example: Determination of the run of the optical micrometer for the horizontal circle of the Wild T3, No. 7253.

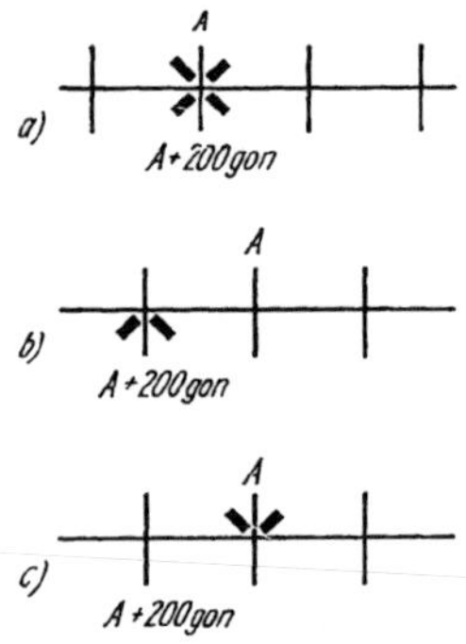

Figure 248 Determination of the run of an optical coincidence micrometer

For optical coincidence micrometers the run is determined separately for the upper and lower image halves, because the imagining rays have different paths. After setting the micrometer scale approximately to zero, the diametrical graduation lines A and $(A + 180°)$ are brought to coincidence with the lateral fine drive (figure 248a). With the graduation line $(A + 180°)$ the interval of the upper half of the image is measured, i.e. one coincides with the micrometer the lines $(A - i)$ and $(A + 180°)$ (figure 248b). At the same time, the interval of the lower half is measured using the graduation line A, e.g. coincide lines A and $(A + 180° - i)$ (figure 248c). The difference between figures 248b and c lies only in the observation of the coincidence on other lines. For each coincidence, the scale is read twice. The determination is carried out back and forth on different circle positions, spread over the whole circle. Furthermore, the minute settings are consistently changed by equal amounts to eliminate the influence of short period circle subdivision errors (table 7). To eliminate circle subdivision errors, one can also measure 10 intervals in sequence at each circle position.

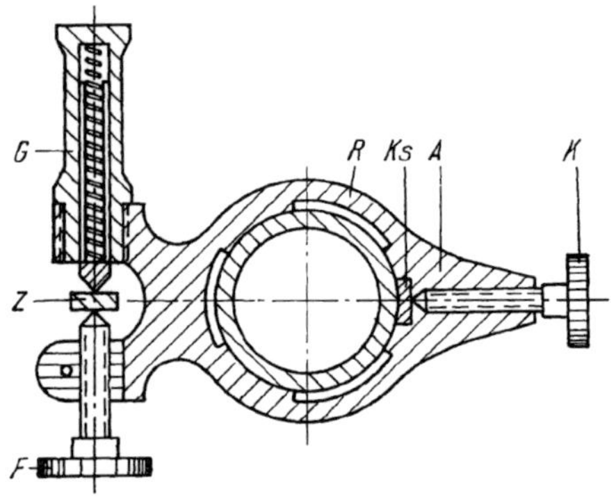

Figure 249 Central clamp

Clamps, Slow Motion Screws, Circle Setting Screw

For precise sighting to a target or at a certain circle position, movable parts of the instruments have to be solidly connected at times and still be changed delicately. The horizontal clamp is used to fix the top part of the theodolite to the bottom one. With the horizontal slow motion screw the top part can be delicately turned in respect to the bottom part. The repetition clamp is utilized for repetition measurements, direction transfers and orientation of the horizontal circle. With it, the centre part is clamped against the top part. The vertical clamp is used to fasten the telescope to the standard. With the vertical slow motion screw the telescope can be delicately tilted, while the vertical index screw is used to delicately change the vertical index according to the index bubble.

Figure 249 illustrates a central clamp as it can be found at low accuracy instruments. By tightening of the clamping screw K the ring R with its extension A is pressed solidly against the adapter of the bottom part via the clamping piece Ks. Since the plug Z of the top part extends between the slow motion screw F and its spring G, the top part is also clamped, can however, be moved with the slow motion screw F.

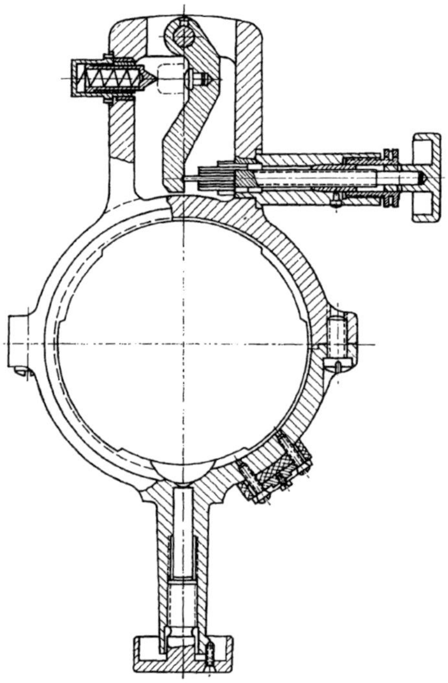

Figure 250 Clamp and slow motion screw with lever transmission (Wild T 3, OT-02)

The setting accuracy of the slow motion screw is increased with a lever transmission (figure 250) for instruments of medium and high accuracy. With clamping screw and clamping piece, the ring is clamped against the adapter, and thus the top part to the bottom part. The slow motion screw presses the lever against the plug of the top part, which presses against the spring, such that by turning of the slow motion screw, the top part is delicately moved. Figure 251 shows another arrangement. Newer designs have clamp and slow motion screw on one axis (see figures 296, 302, 306, 310, 316, 317, 318, 323, 324).

By tightening the clamping screw K, of the type shown in figure 252, the ring R (with extension) is pressed via the clamping piece Ks against the adapter of the top part. The slow motion screw is coaxial to the clamp, and supported independently in the bored hole of the extension. It rests in a ball bearing and is pressed axially into the bearing by a spring. When turning the slow motion screw F, the nut M presses against the lever H. Its movement is transmitted to the spring G of the top part via the plug Bo. Figure 253 shows another design.

The theodolite of the series A of Zeiss, Jena have horizontal- and vertical clamps at the same location, and nearby coaxial slow motion screws for both directions. Of the two pronged clamping levers, the upper one is acting on the horizontal axis, and the lower one on the vertical axis (figure 254). The

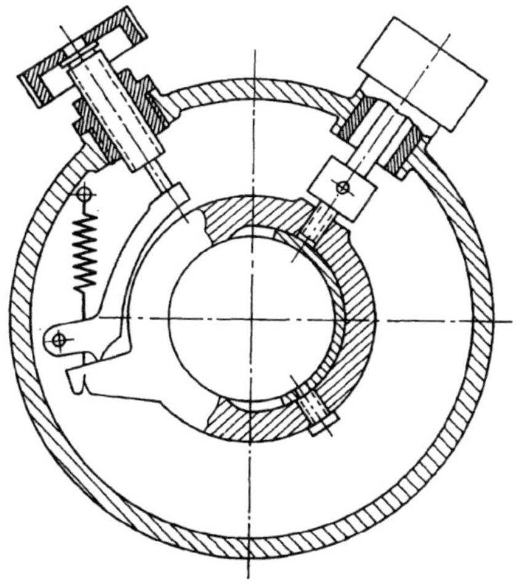

Figure 251 Clamp and slow motion screw with lever transmission (Zeiss, Jena Theo 010)

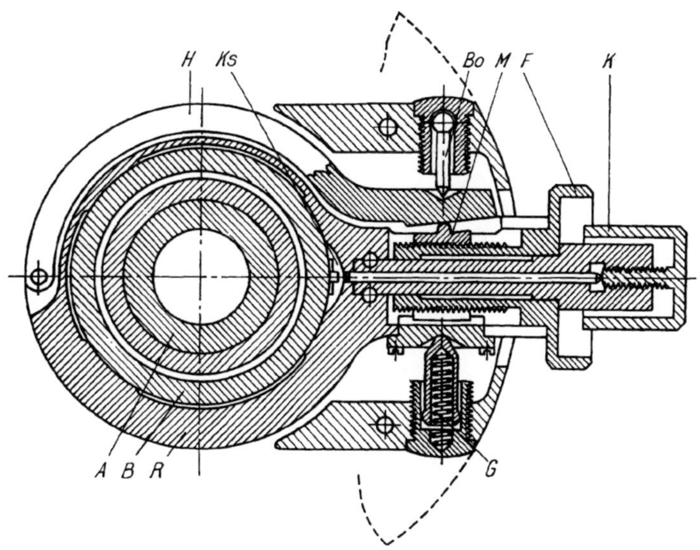

Figure 252 Coaxial clamp and slow motion drive (MOM Te-B 23)

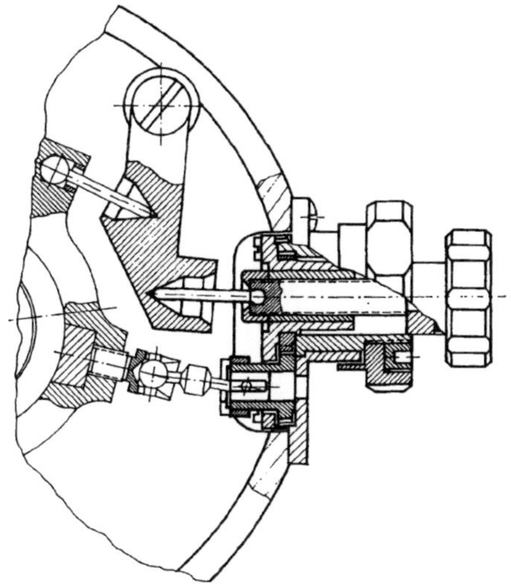

Figure 253 Coaxial clamp and slow motion drive (Breithaupt & Son)

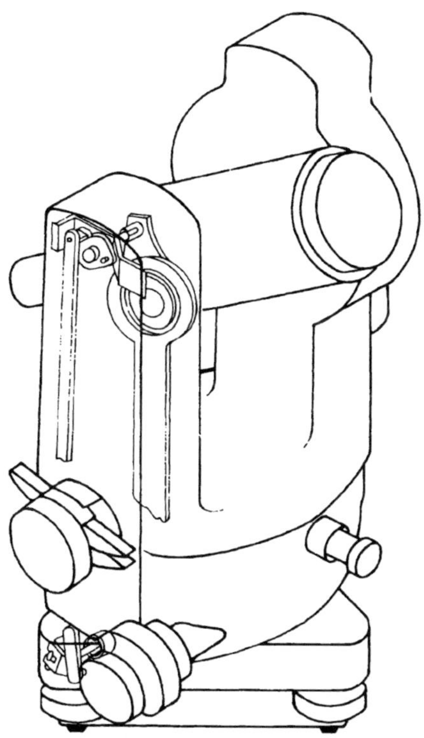

Figure 254 Side- and height clamp with lever rotation centres closely together (Zeiss, Jena)

threads of the slow motion screws act on the respective axes through a lever transmission (figure 255). The exterior thread socket for the vertical slow motion, and the interior thread spindle for the horizontal slow motion are supported by a socket fixed to the instrument such that they can be screwed. Socket and spindle are equipped with driving knobs. By turning the horizontal slow motion screw, the thread spindle is screwed into, or out of the nut. With a swinging bolt it is connected to the transmission lever, which is supported by its short lever arm at the fastening arm. At the end of the fastening arm everything is connected by the power of a spring under pressure of the screw. When turning the vertical slow motion, and thus the thread socket, the nut is shifted back and forth. Its motion is transferred 1 : 1 onto an angle lever with a swinging bolt. With it the direction of movement of the screw is transferred to the telescope lever, positioned in the plane of the horizontal axis. Since it is connected to the angular lever with a swinging bolt, it also follows that movement. Again all parts are connected by the power of a pressure spring, placed at the end of the telescope lever.

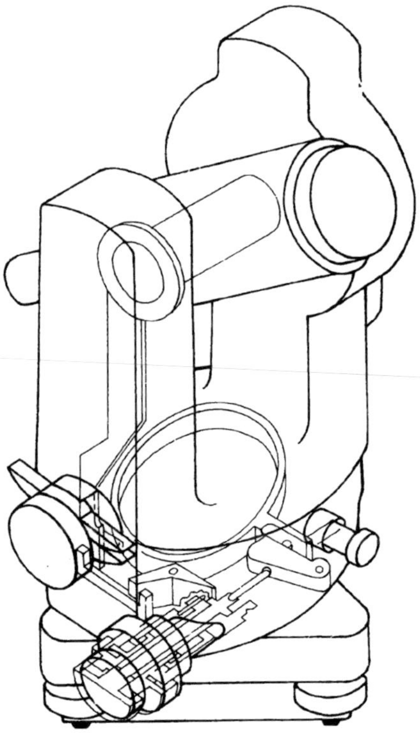

Figure 255 Side-and height clamp with co-axial driving knobs (Zeiss, Jena)

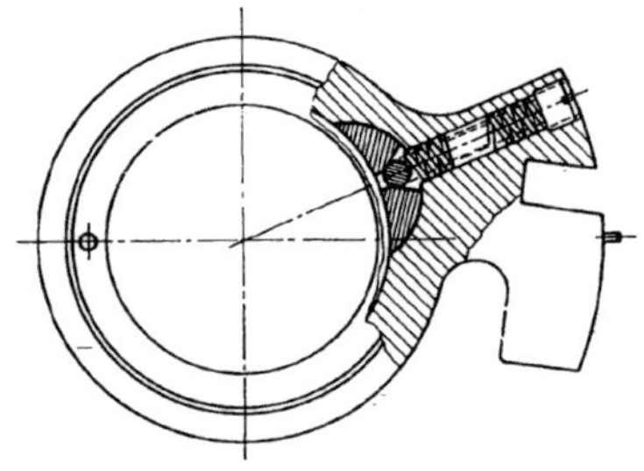

Figure 256 Friction clamp

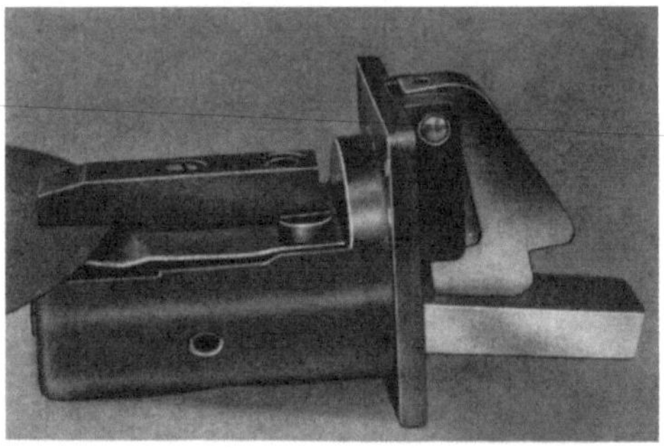

Figure 257 Repetition clamp according to *Mahler*

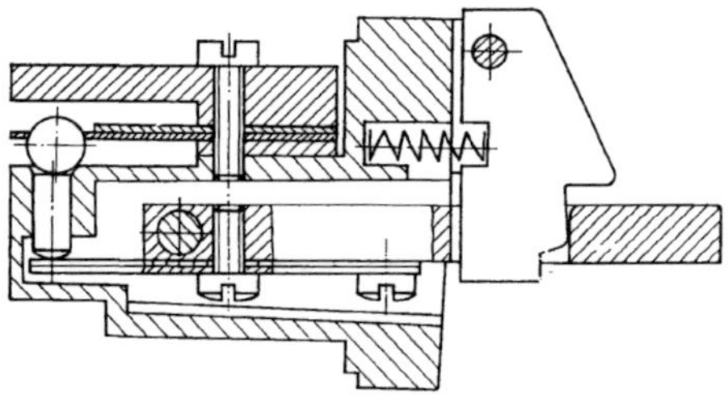

Figure 258 Repetition device of theodolites by Zeiss, Jena

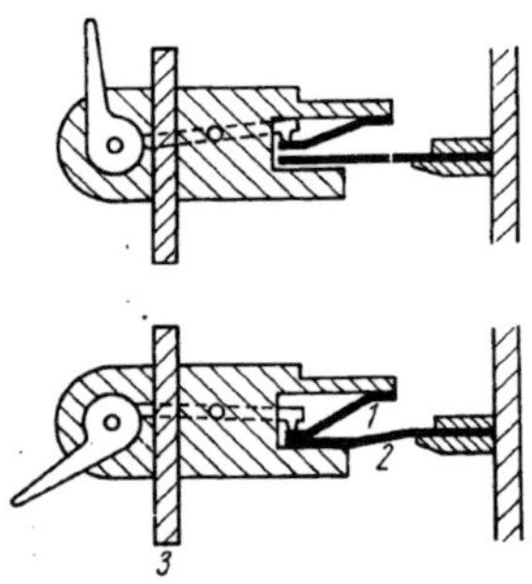

Figure 259 Repetition clamp of the Wild T 16

Occassionally, theodolites have friction clamps and infinite slow motion screws instead of clamping screws or -levers for horizontal- and vertical circles. The two slow motion knobs act on an eccenter via cogwheel gears, whose movement is transferred with a fork to the horizontal axis resp. the horizontal plane. The operational principle of a friction clamp is illustrated in figure 256.
Modern repetition theodolites have clamp and slow motion screw on the top part only, and a repetition clamp (according to *Mahler*) for temporarily connecting centre- and top parts. The axis systems are of the Borda type. The socket of the bottom part can be rigidly connected to the tribrack with the forced centering device. This socket acts as bearing for the top part, and also for the freely rotatable socket of the centre part, which slides along the outside. Once the graduated circle is oriented, it is kept in place by a rippled spring. A flexible disk membrane is rigidly connected to the graduated circle. It can be pressed against the top part with the repetition clamp, so that centre- and top parts are connected and can be rotated together.
At instruments from Zeiss, Jena, the clamping power does not directly act onto the disk membrane, but rather via a flat spring (figures 257, 258). Loosening and clamping of the circle resp. top part are accomplished safely and without errors, if one presses with the finger parallel to the vertical axis and always softly catches with the finger tip the outside lever when it snaps beyond its rest position.
At the Wild T16, a flat spring (*1*) acts under tension upon the membrane (*2*), not at a point, but along its whole width of about 5 mm and thus presses it against the top part (*3*) (figure 259).
Instruments of medium, high, and highest accuracy do not have a repetition clamp, but a circle setting screw to obtain a desired angular reading

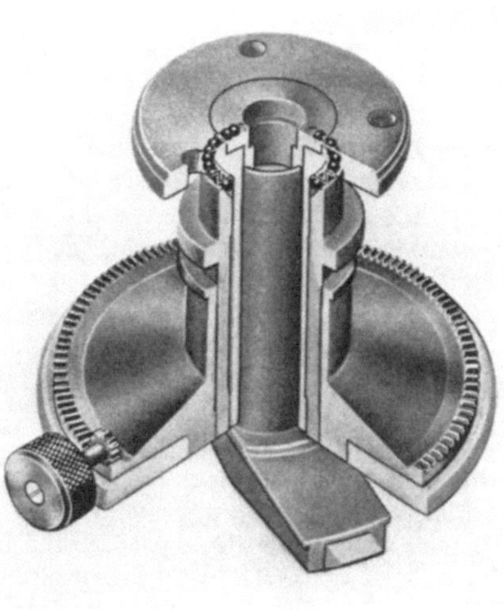

Figure 260 Circle drive (Wild T 2)

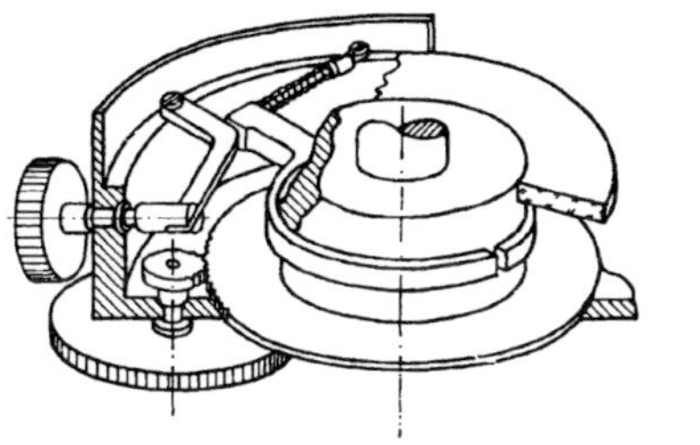

Figure 261 Circle rough-fine drive of the K 1-A, K 1-RA and DKM 2-A (Kern)

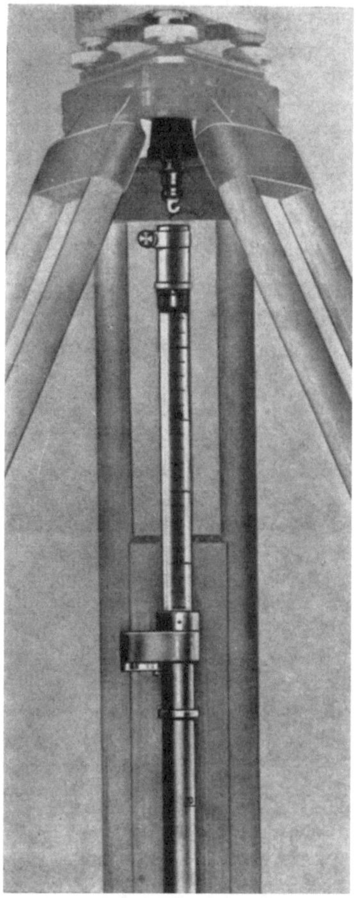

Figure 262 Centering rod

for fixed top- and bottom parts. Then for instance, a disk with cog wheel is rigidly connected to the centre part (see figures 197, 225, 226, 243, 260). It can be moved by turning a knob, which in its rest position is pulled away from the disk by a spring. The circle setting screw is designed as a rough-fine-drive for instruments of Kern and Zeiss, Oberkochen (figure 261), so that the circle can be set to a certain value with coincidence accuracy.

Centering Devices

The theodolite is centered above a ground point with the aid of a plummet. For simple measurements, a plumb bob is sufficient. For more accurate measurements, especially on windy days, a rigid plummet or an optical plummet is preferable.
The rigid plummet, also called centering rod, consists of a telescopic steel pipe with plumbing point and a bull's eye level attached sideways to it (figure 262). Its upper end is inserted into the central clamping screw of the tripod. With a cm-graduation on the pipe the height above ground of the tripod head or the horizontal axis can be read.
When centering, the following procedure is applied: The instrument is first set up roughly above the ground point, then the plumbing point is placed on top of the ground point. After loosening the tripod-instrument connection, the instrument is shifted along the tripod top until the bull's eye level is centered. With the Kern centering tripods, the instrument is also roughly levelled during centering with the rigid plummet.
Optical plummets are either manufactured as separate instruments or included in theodolites. Separate optical plummets are arranged such, that one can center above ground points, or else centre above ground points and below ceiling points (in mining). They consists of a simple little telescope with crosshairs, which can be turned around the vertical axis in a tribrack and which can be levelled with the aid of two level bubbles. Optical plummets with two aiming directions have a broken telescope with horizontal eye piece. They have either two objective lenses on top and bottom, with a deviating prism in between, or else the deviating prism is in front of the telescope (figure 263). Separate optical plummets have forced centering elements so that they can be exchanged with theodolites, targets or subtense bar.

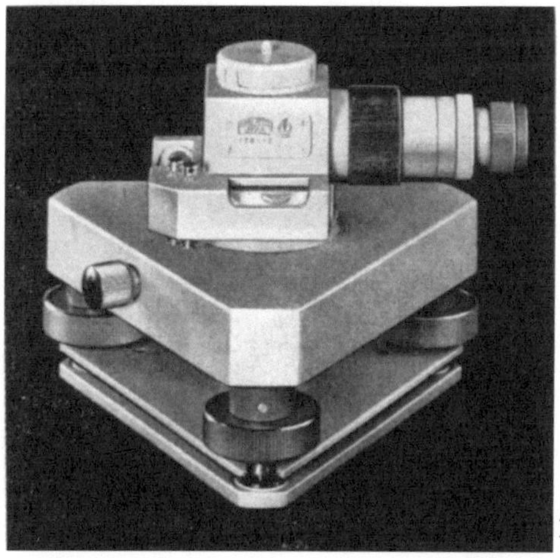

Figure 263 Optical plummet

Optical plummets built into theodolites, consist of a small telescope with reflection prism. They are built into the top part and the hollow vertical axis (occassionally also into targets), or else into the tribrack (Wild T2). After approximately setting up the theodolite above the point, it is accurately centered with the aid of optical plummet and foot screws. Then it is approximately levelled with tripod legs and bull's eye bubble, and accurately levelled with the aid of the platen level and foot screws. Now the centering has to be corrected by shifting of the tripod plate until ground- or ceiling point appear in the cross hair of the optical plummet.
To eliminate a possible collimation error of the plummet in respect to the vertical axis, the top part is rotated by 200 gon, and the instrument is shifted by half of the difference.
When rotating an adjusted optical plummet, the image of the ground point remains at the same distance from the cross hair centre at all directions.

If this is not the case, one has to coincide the cross hair centre with the centre of gravity of the error circle (resp. triangle, for three settings spread equally over the horizon), as marked on the ground. This is done by adjustment of the reticule, carrying the cross hair.
If the plummet is built into the tribrack, a similar procedure is used. In this case, instrument and tribrack are rotated by 133 gon on the locking plate. The plummet in the tribrack can also be tested by laying the instrument horizontally at the edge of a table and then rotating the bottom part. The position of the axis of the optical plummet is marked on a piece of paper, placed at about a 1.5 m distant wall.
Without wind, the centering accuracy for a plumb bob is about 3 to 5 mm, for a rigid plummet about 1 to 2 mm, and for an optical plummet, up to 0.5 mm.
Due to their purpose, cross hair and point can only be seen sharply with the eyepiece within a range of 1 to 2 m for these optical plummets.
A ceiling plummet, which can be placed on top of the telescope has been produced for the Wild Tl, Tl6 and T2 as well as for Breithaupt instruments. For centering the instrument below a ceiling point, or for marking one above the instrument, the telescope is levelled with the aid of vertical circle and index bubble. Then the procedure is similar to the one mentioned for other optical plummets. Another attachment to these theodolites is an objective prism, which can be snapped onto the casing of the objective lens. With this pentagon prism, points can be plumbed up or down. Its main purpose, however, is to transfer given directions onto a different horizon (as needed in mining and for the construction of power stations), and to perform precision shaft plumbing. Special zenith eyepieces, permit upwards plumbing with vertical telescope (see figure 287).

Forced Centering Devices

During forced centering, instrument, target, subtense bar, or optical plummet can be exchanged such, that their vertical axes are forced to be at the same location within a few hundredths or tenths of a millimetre. This is accomplished by mechanical devices, who practically eliminate centering errors.
Every forced centering device consists of a receptacle effectively providing the centering, and a centering element. At least three receptacles are required for measurements; for engineering surveys, often five and more are used.
During measurements the point location is represented by the centering axis. When centering with the Freiberg sphere, or the plate shaped forced centering device together with the Kern centering tripod, it is uniquely defined by the receptacle and valid for all units. Sometimes the centre of the set-up is defined after interaction of receptacle and centering element (three groove forced centering, plug and socket).
There are two construction principles: For forced centering with plug and socket or for the plate type arrangement, the vertical axis is kept vertical, and centred by placing the unit into the receptacle, which has to be levelled only once. For three groove- and Kern forced centering, as well as when using the Freiberg sphere, each unit has to be newly levelled after placing into the receptacle.
Many instruments have forced centering with a cylindrical plug of 34 mm diameter (on the bottom part of the theodolite, subtense bar or target) and a socket (with a drill hole in the tribrack for the optical plummet) (figure 264). The socket contains a free space such that it creates a V-bearing parallel to the axis. The plug is pressed against the edges of this free space, with a clamp which has an angle of 100 gon to the axis (figure 265). Thus a three point support is created. After exchanging the parts, the clamp which can be designed in various ways, is fastened. In instruments from Zeiss, Jena, it presses a simple clamping element directly against the plug, while for Fennel instruments a spring is placed in between. Plug and socket connections of somewhat different design can be found at the TGlb and 2b of Officine Galileo. Instruments from Zeiss Jena can also be forced centered in height with the aid of a clamping ring for the foot screws.
For plate-type forced centering devices, the units are placed with their plane ground base surface onto three points of the bottom surface of the tribrack, which is plate shaped on the top. The centering accuracy is determined by the fit of internal diameter of the tribrack ring and external diameter of the centering flange of the unit (about 0.1 mm). This kind of forced centering is used for the Wild theodolites Tl, TlA, Tl6, and T2, as

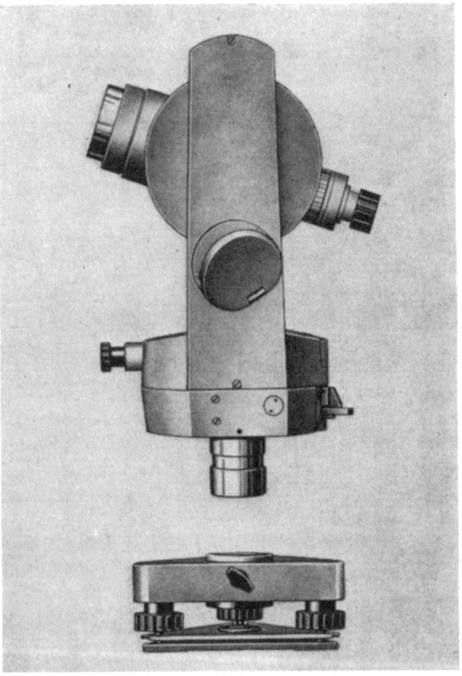

Figure 264 Forced centering with plug and socket (Zeiss, Jena)

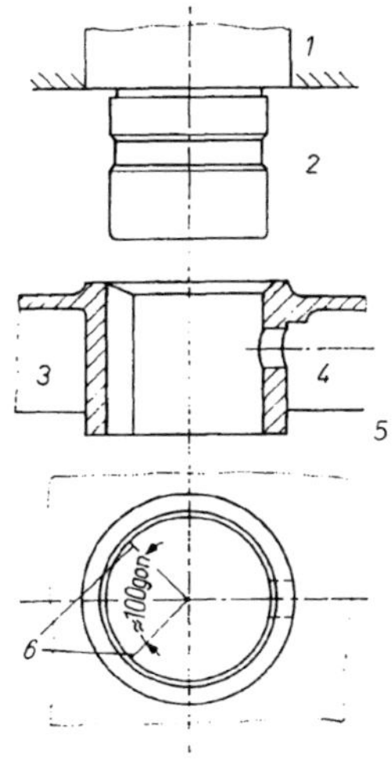

Figure 265 Section through plug and socket; view from top onto socket (34 mm diameter)
1 top part; *2* plug; *3* tribrack; *4* clamp; *5* socket; *6* butting edges

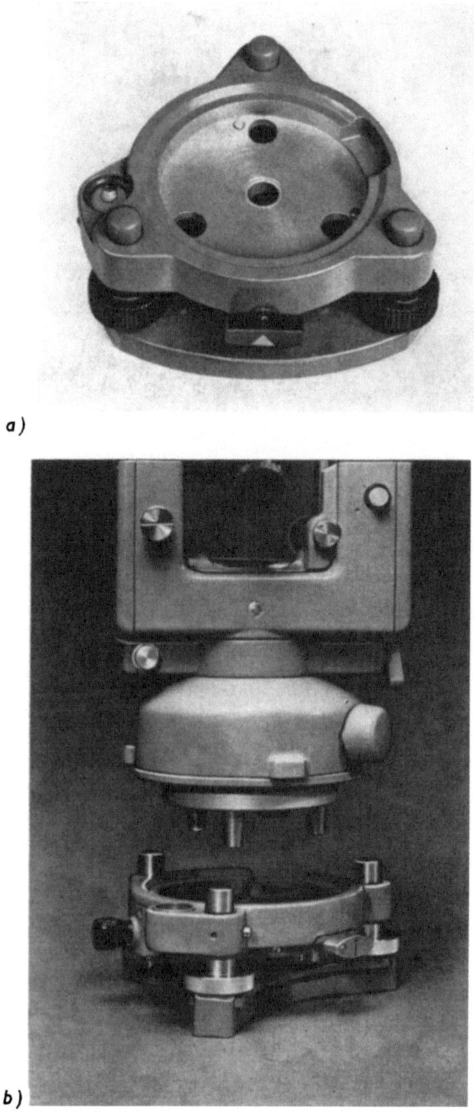

Figure 266 Plate shaped forced centering
a) Wild T 1/Wild T 16
b) Wild T 2

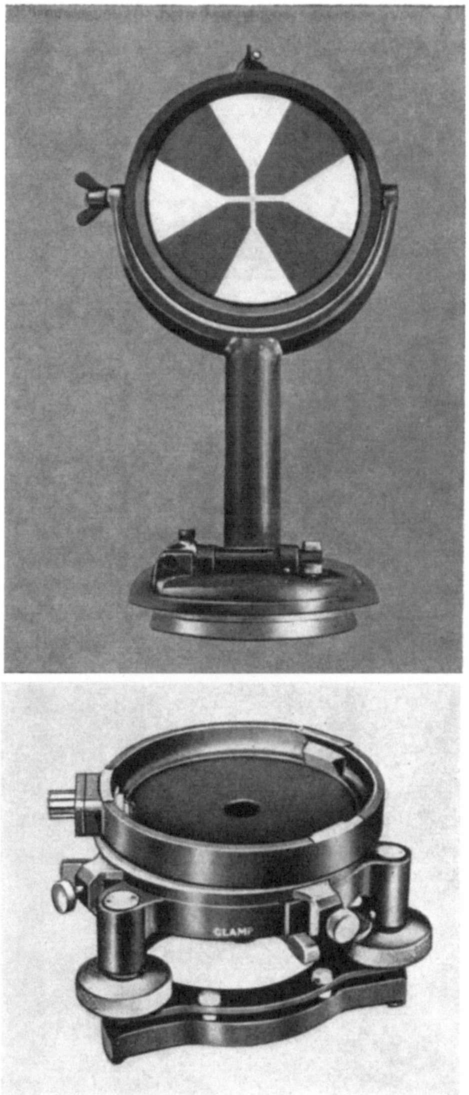

Figure 267 Forced centering for theodolites by Rank Precision Industries (figure) and Vickers

well as for the Wild RDS. For the Wild instruments, the unit is locked into the centering plate of the tribrack with the aid of three pegs of the centering flange (figures 266 and 225). A bull's eye level at the tribrack is used for rough levelling.
For theodolites of the English companies Rank Precision Industries, and Vickers, the locking is accomplished with three points, the receptacles for two of them are fixed at 100 gon separation, while the third one is designed as a movable clamp (figure 267).
In the receptacle plate of the centering tripods of Kern, Aarau (see figure 188a) a centering socket of 12 mm diameter, which uniquely determines the centering axis is arranged coaxially to the centering rod. The units are placed onto three plane receptacles, where the cylindrical peg which acts as centering element fits into a socket on the base plate of the unit. The unit is fastened with the aid of a bayonet lock. Afterwards the levelling has to be improved with the aid of the levelling knobs (see figure 191).
Three grooves, arranged under 133.3 gon in radial directions on a base plate (also used as centering star) are used for three groove forced centering. After rough levelling of the base plate, the instrument is placed with the tips of its foot screws into the grooves, which, however do not uniquely determine the centering axis. The whole tribrack acts as centering element. The centering accuracy depends on how well the ideal position of the foot screws is maintained for all units. This type of forced centering is not often used anymore. There are locking plates for the theodolites Wild T3 and T2.

Figure 268 Forced centering with Freiberg sphere

The three groove forced centering with a centering star is more suitable for pillar observations than when setting up with a tripod.

The Freiberg sphere as part of the Freiberg arrangement is used in mining surveying and for precision engineering surveys (figure 268). The socket is a cylindrical pipe, which is placed into a centering triangle with opening that is fastened to the tripod plate such, that it can be shifted. The spherical centering element with 10 to 20 mm diameter, which is the extension of the vertical axis of the unit, fits into the pipe. Since the Freiberg sphere is a simple forced centering device, especially suitable for pillar observations, some specially built Wild T2 and T3 as well as power dam monitoring equipment by Freiberger Präzisionsmechanik are equipped with spherical forced centering.

There are differences in shape and colour between the targets of different manufacturers (figure 269). Targets usually have a bull's eye bubble for rough levelling, some also have tabular levels (Fennel, Wild). Some targets can be illuminated electrically from behind, so that one can measure at dark or underground.

In addition to the standard targets, Zeiss, Jena and Wild, Heerbrugg produce large targets, which permit electronic distance measurements of up to 8 km.

The targetting equipment of Zeiss, Jena (figure 270) consists of the target and the large target (with or without optical plummet) in red-white colour, which is placed onto the three pegs of the target carrier. Centering is achieved with the centering plug of the target carriers which fits into the inside socket of the target carrier. Tightening of a nut provides solid connection after the target is placed.

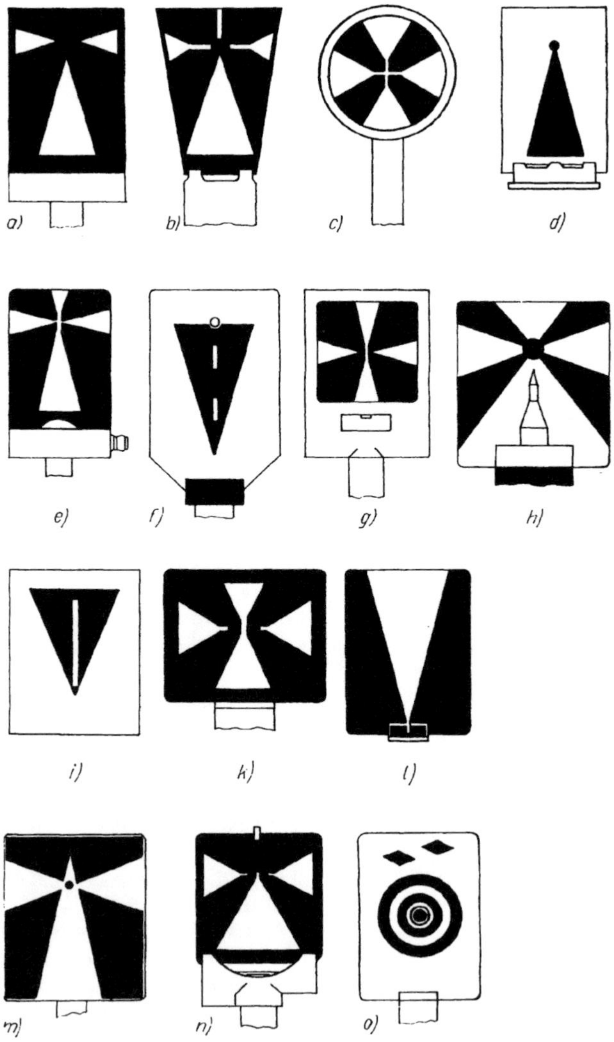

Figure 269 Targets

a) Dietzgen
b) Breithaupt & Son
c) Rank Precision Industries; Stanley; Vickers
d) and e) Fennel
f) Filotecnica Salmoiraghi
g) Kern
h) MOM
i) and k) Officine Galileo
l) Freiberger Präzisionsmechanik
m) Zeiss, Jena
n) Wild, Heerbrugg
o) Zeiss, Oberkochen

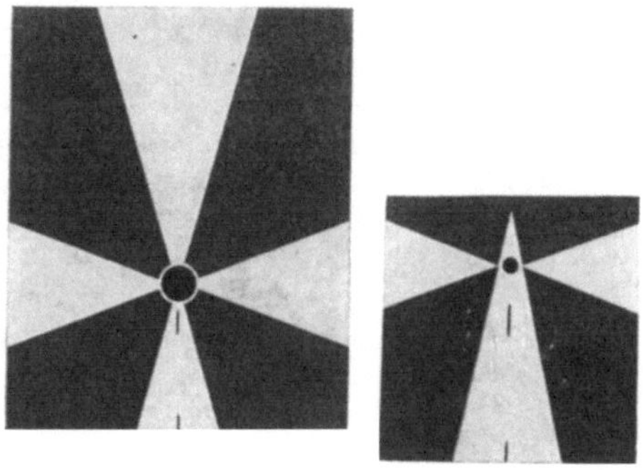

Figure 270 Target and large target board by Zeiss, Jena (red-white)

Figure 271 Large target board by Wild, Heerbrugg on top of normal target (210 mm × 297 mm, black-yellow)

Figure 272 Photographic recording of readings at the Tpr

Figure 273 Ring code section with tracks for 100-, 10-, 1-, 0.1 gon and tact tracks t

The Wild targetting equipment consists of tribrack, carrier (with or without tubular level and optical plummet, also suitable for a prism reflector), and target in yellow-black colours. The large target can be attached to the standard one (figure 271).

Automatic Recording for Angular Measurements

The automatic recording of measured values is a step towards automatization of the whole measuring process

1. Photography of circle reading and subsequent precision measurement of the images

The most suitable times for observations for triangulations of high order are rather short, and can better be utilized, if the amount of time required for circle readings can be reduced. Therefore some theodolites of high and highest accuracy have been equipped with photographic recording devices for the circle reading.
Wild produced a special series of T3 for recording of the horizontal circle reading. Askania produced an instrument with double broken telescope, where sections of the horizontal and vertical circles, the cross level, the face of a watch with second hand, and a small writing table are photographed (figure 272). The recording films are evaluated independently from the field work in an instrument with an optical micrometer, similar to visual reading.
The Theo 002 (figure 330) of Zeiss, Jena is also equipped for photographic recording (with simultaneous visual reading of one graduated circle). This instrument has a broken telescope, eccentrical to the vertical axis, with the eyepiece in direction of the horizontal axis, a pendulum supported mirror to compensate errors in the horizontal- and vertical axes, and an automatic height index (figure 341).

2. Recording by code scanning

Photographic recording of circle readings permits the optimum utilization of the ideal observation times, but just transfers the visual reading from the field to the office. With the analogue-digital conversion by code scanning however, automatic data flow from reading to printing of final results or mapping has become possible. The operation involves coded disks, with which each direction obtains a corresponding uniquely coded signal. The graduation can be scanned galvanically, magnetically or optically.
The code theodolite FLT 3 by Fennel represents a first step. The index position in relation to a coded circle is photographed and subsequently evaluated in the office photoelectrically. The graduated circles have numbering together with the normal line graduation (scale value 0.2 gon) (figure 273).
The angle between two subsequent coincidence positions is 0.1 gon because the diametrical circle positions run in opposite directions. Thus 4000 of such intervals are found on the circle. For identification, the ring code contains four concentric information tracks, with alternately black and white sections of different lengths.
These provide 4000 different black-white combinations for the numbers 0.0 to 399.9. The four information tracks correspond to 100-s, 10-s, 1-s and 0.1-s of degrees. In the middle, two tact-tracks are located, which do not belong to the actual code, but are used in the evaluation device for deciding on how partially shown code-areas on film are to be considered.
Diametral circle positions are combined optically and are photographed on 35 mm film (figure 274) together with the corresponding symbol group (*A*) by exposing a flash. Furthermore, digitally represented key numbers (*c*) are recorded, which are set by the observer for visual control with a twelve digit input key. With the figures 0 to 9 for each digit, one can enter traverse-, station- and target numbers, weather, observer and other data in coded form.
The diametral lines of the circle graduation appear as white marks in black fields (*B*) and serve as analogue indicators for the interpolation between the graduation lines closest to the coincidence.
The vertical circle positions are recorded in a second group with two fields for analogue representation of the graduation lines, and one field for the digital representation of numbers (with combinations between 400.0 to 799.9).

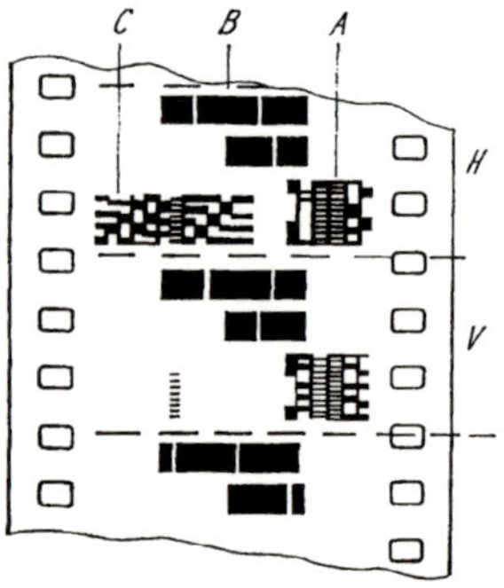

Figure 274 Recording of the horizontal and vertical circle images with coded graduation

The film is developed at an evaluation centre, and evaluated with the film evaluation device ZUSE Z84 in a start-stop-operation, picture by picture (in about $^3/_4$ s per picture). The information of a picture is transformed into 49 characters of a 5 track punched tape. For the complete circle reading to 0.1 mgon, the binary code is transferred to the output tape with the aid of photo reading to 0.1 gon, and the analogue representation of the

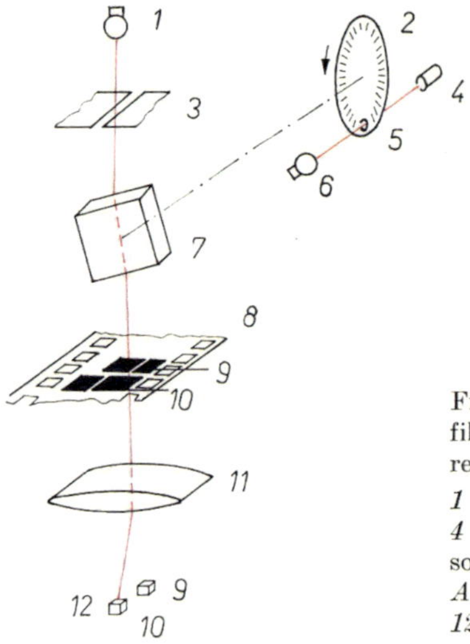

Figure 275 Measuring arrangement of the film evaluation device for evaluating the analog representation of the circle reading
1 light source 1; *2* screen disk; *3* diaphragm; *4* photodiode 2; *5* negative screen; *6* light source 2; *7* prism; *8* recording film; *9* channel *A*; *10* channel *B*; *11* cylindrical collecting lens; *12* photo diodes

graduation lines is interpolated. For this, both images of the scales (*B* in figure 274) are scanned with the aid of a rotating prism by the small light band passing from the light source through a diaphragm (figure 275). The light band is conducted onto two photo diodes via a cylindrical collecting lens. Since the diode *A* obtains light from track *A*, and diode *B* from track *B*, they generate signals as soon as the light reaches the beginning of the tracks or the respective graduation lines. A grid disk (with 18,000 grid lines), connected rigidly to the prism, is used to measure the distance between these signals. With the light source 6, another grid, and the photo diode 4, its angle of rotation is converted to counting impulses at the photo diode 4, which are counted in an electronic counter. In this manner, the 2nd, 3rd, and 4th digits of the reading are obtained. The output tape serves as input to electronic computers. With the code theodolite the same accuracy is reached as with a self reducing tacheometer.

Figure 276 Code disk (Opton Reg Elta 14)

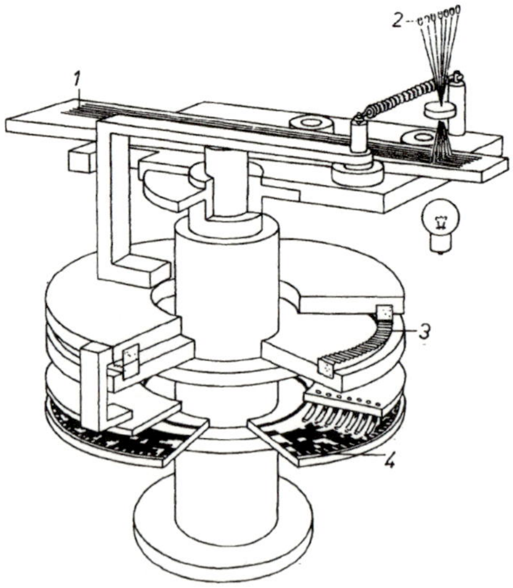

Figure 277 Determination of angles at the Reg Elta 14 shifting wedge with numbers 0–1000 in coded form
2 photo diodes; *3* cog wreath with 400 cogs; *4* code circle with the numbers 0–400

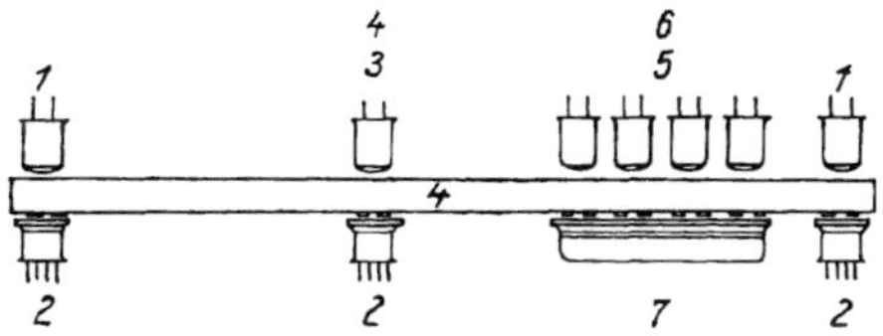

Figure 278 Scanning system of the HP 3820 A
1 scanning track; *2* sensor; *3* illumination; *4* sine track; *5* code disks; *6* code; *7* digital sensors

The angular measurements of the Reg Elta 14 (see figure 759) of Zeiss, Oberkochen are performed by code scanning. The instrument has coded circles (figure 276) with a scale value of 1 gon and micrometers. The code disks are arranged coaxially to cog wheel gears at vertical- and horizontal axes, which serve as limit for scanning of one interval. The instrument thus is designed as cog wheel theodolite. One cog wheel is connected to the bottom part, the other to the top part. When connecting the gears, its cogs lower into the 400 spaces of the lower cog wheel. When the cog wheel gears are connected, the electrical circuit for measuring is closed. The rough reading is thus to 1 gon, while the fine measurement is performed by interpolation of the graduation intervals with a measuring wedge to 0.001 gon. Since the 400 positions of the instrument are accurately defined, a simple mechanical scanning of the BCD-coding is sufficient without additional identifications.

The 10 code tracks are scanned with gold contacts at 4 diametral positions. The rough reading is encoded for punching or display with the aid of a matrix. For the fine measurement, the position of the shifting wedge at the coded scale of the glass circle is scanned by optical electronical means to 1 mgon (figure 277).

In the Hewlett Packard HP 3820 A, the rough reading is obtained with an 8 bit gray code (figure 278) to $^1/_{256}$ (similar to a circle reading of 1°). The fine reading is obtained by interpolation of a sinusoidal trace with 128 periods—where each period is resolved into 1000 parts (figure 279). Thus 4,096,000 increments are generated, i.e. a resolution to 0.1 mgon. Scanning takes place at diametrically opposite circle positions. The results of all three scanning systems are combined in the micro-computer to the final result.

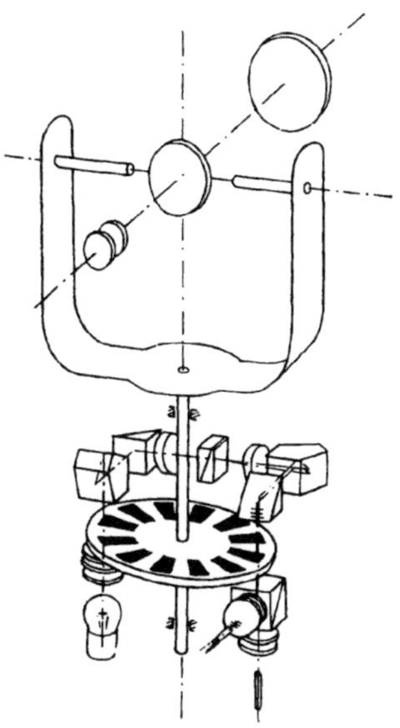

Figure 279 Angular stepping device of a digital theodolite

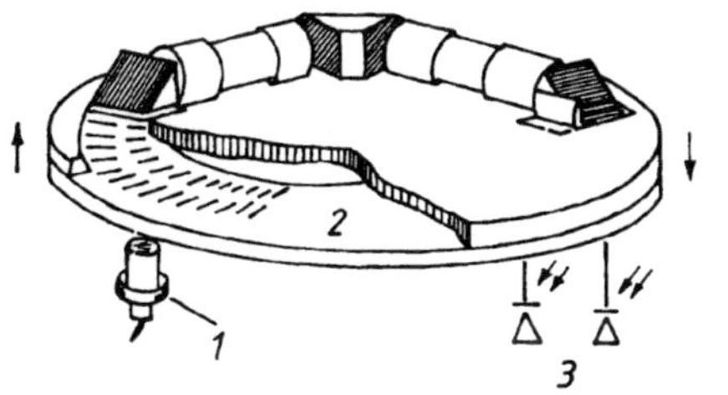

Figure 280 Principle of circle scanning (AGA Geodimeter 700/710)
1 illumination diode; *2* grid disk; *3* receiving diode

3. Recording with the increment- or counting methods

The analogue-digital conversion with the increment- or counting method utilizes line grids, which represent a steadily increasing sequence of equal yes- and no-elements (increments). The number of increments for the reading is coded by a counter. For geodetic accuracies, scanning is carried out optically.

The first digital theodolite based on the increment method, was the DIGIGON by Breithaupt & Son, Kassel. It was not built in series, but demonstrates the functions of this method. An angular stepping device with scanner, which has a circle with grid, solidly connected to the top part are converted into periodic light oscillations. The circle position is illuminated electrically and imaged in opposite sense on top of the diametral position (figure 280). The light emerging behind this superimposed position has twice the modulation period as covered by the number of graduation intervals. If the circle with grid contains 5000 graduation lines, 10,000 periods are generated per revolution, which are converted by two photo diodes into two electrical signals with phase shift. Every period represents an angle of 40 mgon. By interpolation, this value is resolved into 40 impulses, i.e. to 1 mgon. A counter displays directly units of 1 mgon. After interrogation, the angular values can be printed.

The same principle is used in the AGA Geodimeter 710 (figure 760, also AGA Geodimeter 700), as well as in the digital theodolite Vectron by Keuffel & Esser (figure 281). Horizontal- and vertical circles of the Geodimeter consist of two track circles with grid of 70 mm diameter and with 100 lines per millimetre. Each grid track therefore contains 20,000 lines. For angular scanning, the tracks are projected onto the diametral circle position with the aid of a diode, emitting light and with super imposition optics (figure 281). When turning the top part, the superimposed grid tracks move against each other. Because of the slight tilt of the radial grids, two Moire strip systems are generated at the receiving diodes. They result in

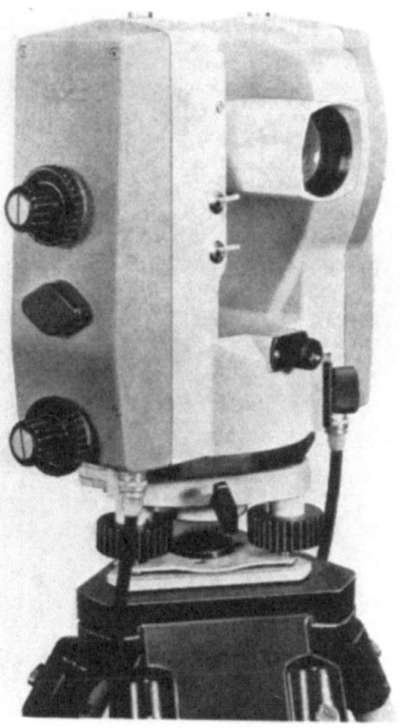

Figure 281 Digital theodolite Vectron (Keuffel and Esser, U.S.A.)

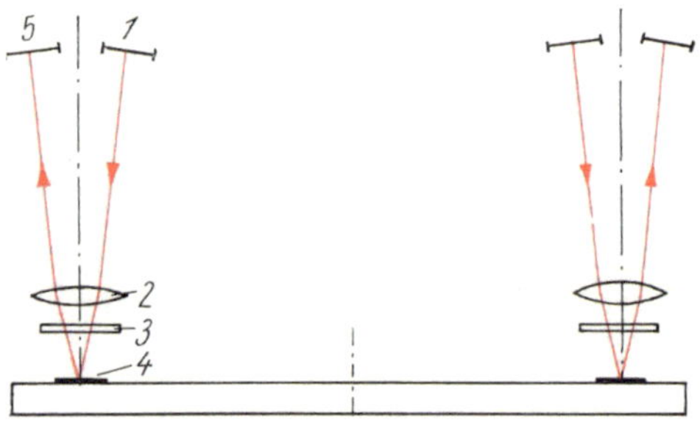

Figure 282 Circle scanning at the Wild TC 1
1 Infrared source; *2* lens; *3* analysator grid; *4* circle with reflecting grid; *5* detector

two nearly sinusoidal light signals, getting light and dark, which form an electro-magnetic field. In addition to the information for angular scanning, they also contain the direction of the rotation. The 40,000 signal periods generated in one revolution due to diametrical circle positions, result in a resolution of 10 mgon. For rough resolution the number of signal periods is counted and stored. The fine resolution is done by interpolation to 0.5 mgon for horizontal-, and 2.5 mgon for vertical angles. With incremental circle scanning, vertical angles can only be measured, if the telescope is set into the zero position before the counting— begins i.e. exactly vertical or horizontal. The electronic reducing tacheometer Wild TC 1 (figure 762) contains a reflecting radial line grid on the horizontal and vertical circles 80 mm diameter. In face of the grid there are arranged diametral devices with transparent cut-outs of grids with the same convergence by the circle moreover a source of infrared radiation and a photodetector. When turning the top part the Moire strips arising by reason of the diffraction move in the plain of the photodetector. They are counted continuously. The interpolation in a Moire period and the automatic means for reading of two diametrical devices give a resolution of 1 mgon. The Vectron (figure 282), consists of precision optical and mechanical parts with electrical scanning knobs and a micro computer with an eight digit display at the front plate. Angular stepping devices are of the incremental type with 0.02 gon as unit.

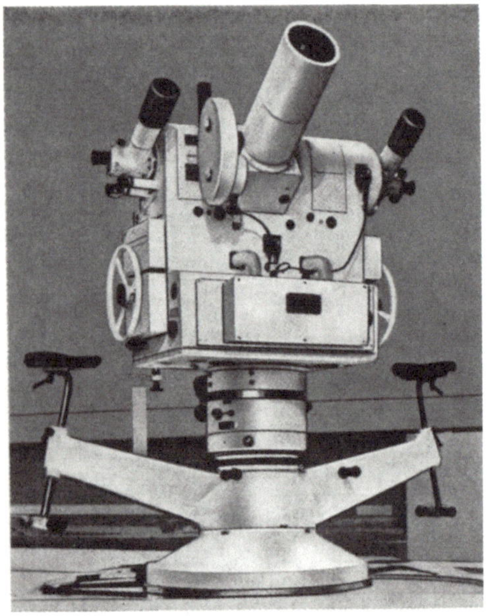

Figure 283 Kino theodolite (EOTSE, Contraves AG, Zurich; mass 1090 kg, height with vertical telescope 2.32 m)

Special Theodolites

Special theodolites are produced for measurements in meteorology, aeronautics and satellite geodesy.
Balloon theodolites are used to measure horizontal- and vertical angles to balloons, to determine wind directions and -velocities in meteorology. Usually, they have one main- and one search telescope, whose magnifications are: 18 times for the GARAT of Breithaupt (no search telescope), 20 × resp. 5 × for the Watts Balloon theodolite (mass: 26.8 kg with container, height 356 mm), and 18 × 40 resp. 3.2 × 10 for the soviet ATK-2 (mass: 4.7 kg). The graduated circles can be read to 0.1° for the Watts theodolite with micrometer, to 0.1° resp. 0.1 gon for the GARAT with vernier, and 1′ for the ATK-2 with scale microscope.
Kinotheodolites are precision instruments for the continuous determination of position and orbit of fast flying objects. This is achieved by multiple intersections, for which the theodolites are synchronized in time, and equipped with movie cameras (5, 10, 20 or 30 images per second) (figure 283). The following of the target takes place with hand wheels or with electrical servomotors. The exchangeable lens- or mirror lens objectives have 30 to 300 cm focal lengths, while the targetting telescope has exchangeable magnifications of 4 × or 5 × and 20 ×. Instruments of this type are the theodolites EOTS (electronic optical tracking system) with the digital model E and the analogue model D, as well as the K400 of Contraves, Zurich.

Theodolite Accessories

For the great variety of theodolite uses, numerous accessories are produced, with exchangeability of instrument and accessory if possible. Among these are optical plummets, fitting into the tribrack, as well as targets and sub-

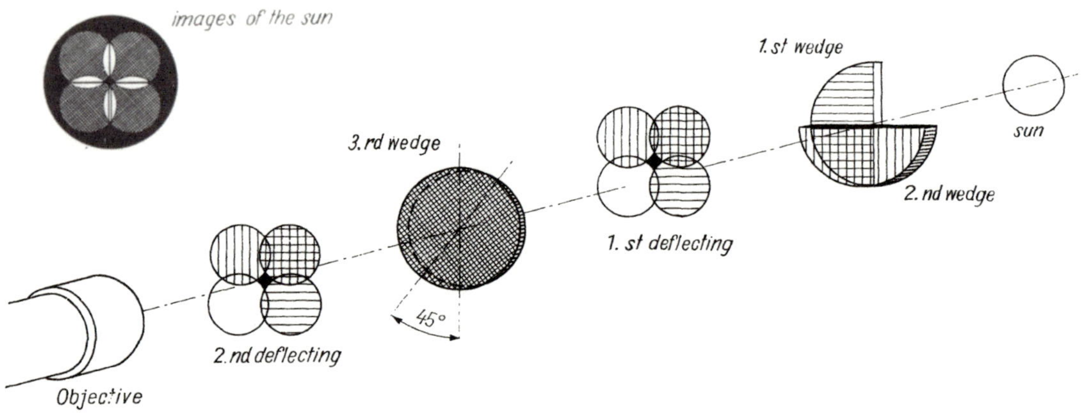

Figure 284 Generation of sun-images with the Wild solar prism

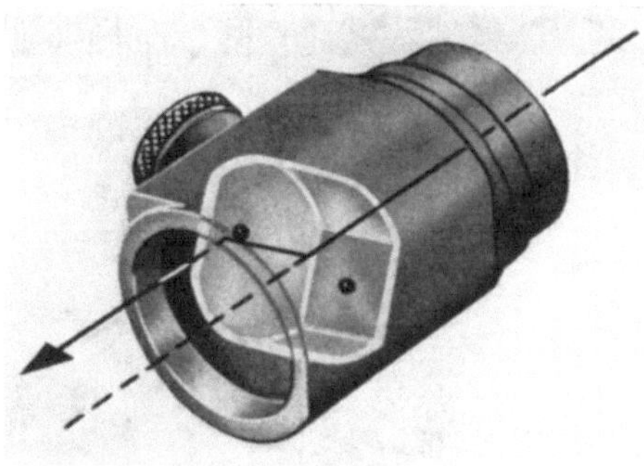
Figure 285 Parallel plates

tense bars; attached to telescope or instrument are gyros, striding- and tubular levels, tubular-, disk- or circular compasses, and telescope attachments for ceiling plummets; in front of the objective lens are distance measuring wedges and objective prisms; and eyepiece attachments such as auto collimation eyepieces, laser eyepieces or eyepiece lamps, which will be dealt with in this book at another section.

Electrical illumination attachments are utilized to illuminate the cross hair (via a reflector inside the telescope that can be rotated) and the circle readings if light conditions are poor. The power sources (batteries) are attached to the standard for short periods of time, and to the tripod for longer periods of time. The illumination mirrors are replaced with lamps.

The *Roelofs* solar prism is fastened in front of the objective lens. It contains three optical wedges, which when pointed to the sun, generate four images of it (figure 284). These are arranged in a square and overlap, except for a quadrilateral in the centre. The telescope is then aimed directly to the centre of the sun, when the cross hair is centred onto this symmetrical quadrilateral. Solar azimuths serve for checking or orienting long traverses, as well as for geographical positioning.

Figure 287 Broken eyepieces

Figure 286 Steep sight prism

Parallel plate micrometers, which are fastened to the telescope objective, are designed to measure small deviations from the collimation axis, e.g. in industrial surveys when aligning axes. Examples are the Wild GPM5 for the Wild T2 with least count 0.1 mm. It is possible to measure with sights less than the shortest focussing distance, if additional lenses are fastened to the objective lens casing.

For inclined sights, eyepiece prisms for telescope and reading microscope can be attached (figure 286). They simplify work with steep sights up to 30 gon. Beyond that, up to the zenith, broken eyepieces (zenith eyepieces, figure 287) are used. To protect the eye for solar observations, a solar filter (e.g. coloured glass with 75% absorption) is placed at the eyepiece. Sun shades in front of the objective lens shield off stray light, which can occur in the telescope when sunlight falls onto it.

Instruments of highest precision are occasionally equipped with a checking telescope at the bottom part, or with a pillar checking instrument, set up below the theodolite. With this telescope, which is equipped with an eyepiece micrometer, and a distant target, the pillar torsion can be determined as a correction (e.g. checking telescope at the T-05, U.S.S.R.; pillar checking instrument for Theo 002 by Zeiss, Jena).

For setting up of a theodolite on top of a pillar, a cast iron base plate with spring loaded centering plug is used (figure 288).

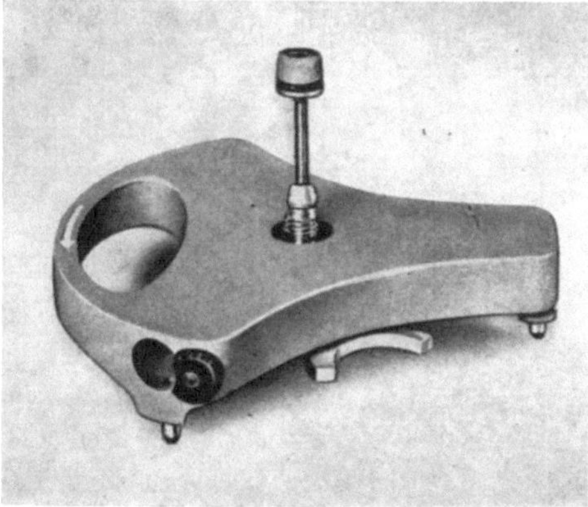
Figure 288 Pillar base plate

For triangulation of highest order, the targets are represented by light signals (figures 289, 290). The instruments are packed in wood-, metal- or plastic containers, which also contain the accessories. For longer transports, e.g. by rail, the instruments with containers are packed into well upholstered crates.

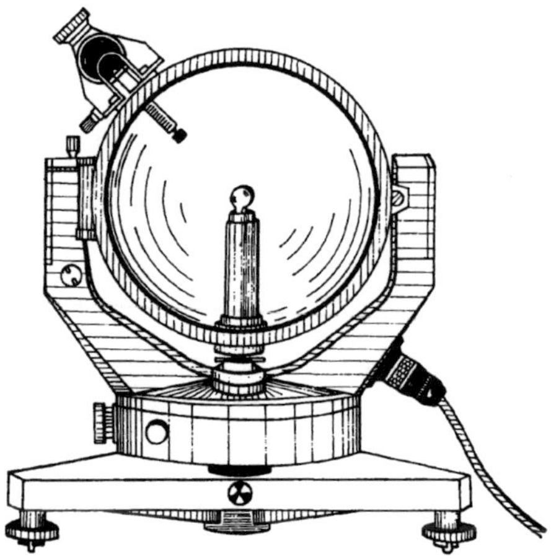

Figure 289 Geodetic signal light TSG 200 (Zeiss, Jena). Range up to 100 km; dimensions 29 cm × 34 cm × 38 cm

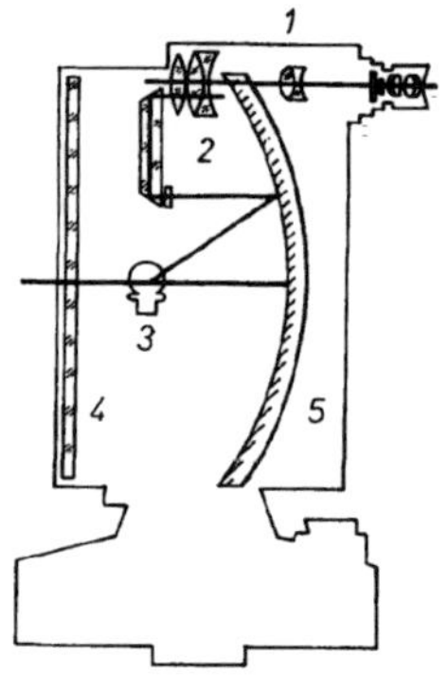

Figure 290 Optical scheme of the TSG 200
1 aiming telescope; *2* triple strip; *3* illumination lamp 6 V, 5 W; *4* cover glass; *5* parabolic mirror with 200 mm diameter

2.3.3. Overview of Different Theodolite Types

There is a great variety in shapes and designations of theodolites. These designations are based on special characteristics, such as type of circle reading, connection between centre part and bottom part, precision of circle reading, purpose, or measuring accuracy.
Based on the type of circle readings, one distinguishes between vernier- and microscope theodolites. Vernier theodolites are still used outside Europe as "transits" (mostly with compass). They are more openly designed (figures 291, 292) in the U.S.A., and more enclosed in Japan (figure 293). These "classical" instruments with metal circles have been completely replaced in Europe, and to a large part outside Europe, by optical theodolites with glass circle and reading microscope. These are attractively designed, often in bright colours, and occasionally arranged in building block style (the use of the same building element for types of different accuracies).
Depending on the connection between central part and bottom part one distinguishes between simple theodolites, repetition theodolites, and theodolites with rotatable circle.
According to the accuracy of circle reading, minute- and second theodolites are listed.
When classifying according to purpose, the different types would be: construction theodolites, engineer's (or stadia) theodolites, and precision theodolites. Construction theodolites are the simplest ones. Engineer's theodolites are mainly used for traversing, small triangulations and staking, while precision theodolites are used for 2nd and 3rd order triangulation and precision staking, and—with special instruments—for 1st order triangulation and other applications of highest accuracy.
When classifying instruments according to purpose, the measuring accuracy becomes the deciding factor when judging the suitability. The criterion is the standard deviation of a direction observed in both telescope positions. In various countries resp. by manufacturers, theodolites are divided into three to six categories, e.g. in U.S.S.R. into technical ($> \pm 10''$), accurate ($\leqq \pm 10''$) and precision theodolites ($\leqq \pm 1''$). The following classification is used for further deliberations:

Theodolites	Standard deviation of a direction observed in two telescope positions
of highest precision	$\leqq 0.2$ mgon
of high precision	$\leqq 0.6$ mgon
of medium precision	$\leqq 2.0$ mgon
of low precision	$\leqq 8.0$ mgon

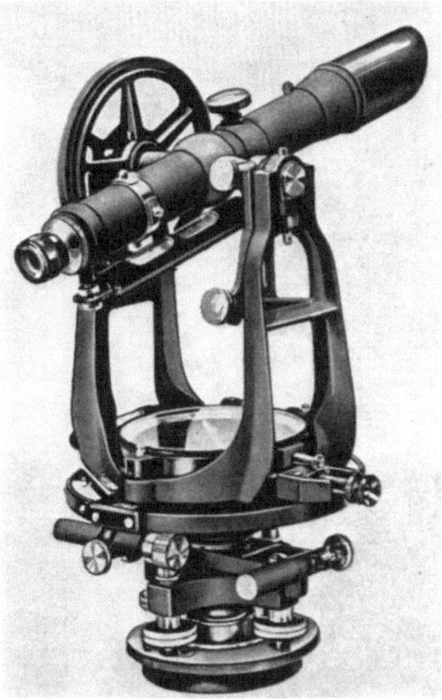

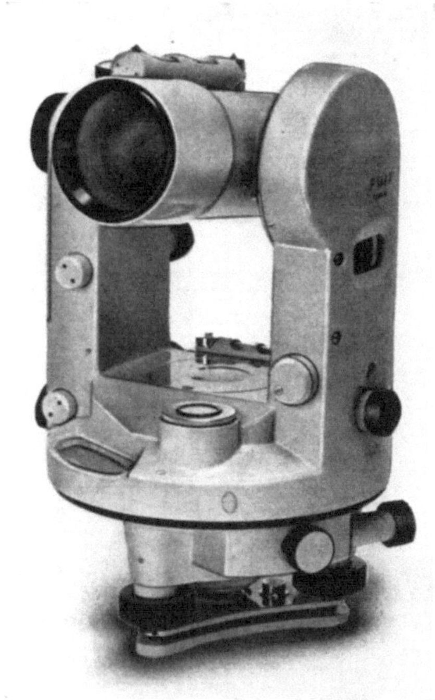

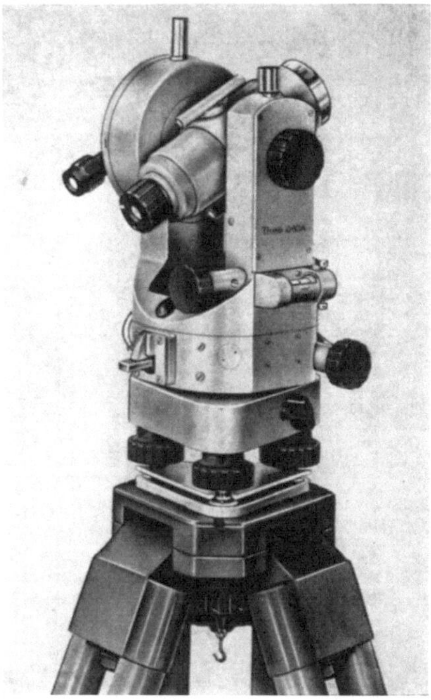

Figure 291 Paragon transit with optical plummet (Keuffel & Esser); height about 35 cm

Figure 292 Hell Gate Precise Theodolite (Gurley); height about 35 cm

Figure 293 NA-2 (Tokyo Sokki); height 250 mm

Figure 294 Construction theodolite K-O-S (Kern), height 315 mm

Figure 295 Micro theodolite THEO 080 A (Zeiss, Jena); height 240 mm

Low Precision Theodolites

Low precision theodolites (table 8, figures 294 to 298) are designed for simple construction measurements, setting out, and small surveys. For this reason, they are small, light, and easy to operate. The diameters of the horizontal circles are about 50 to 80 mm, and 40 to 70 mm for the vertical circles. The telescopes usually have a magnification of 17 to 20 times. Because of the limited accuracy, only simple reading devices are needed, verniers for older models, line-, scale- or vernier microscopes for newer ones, which permit readings to 1 mgon (10′) with minutes being estimated. For this, one circle position is satisfactory, although vernier theodolites have two. For some instruments of this category, the container bottom is used as base plate of the theodolite and is connected to the tribrack with spring leaves (figure 296). Occasionally there are no tubular bubbles, because a bull's eye bubble is sufficient, with respective sensitivity, for levelling the instrument, considering the inherent measuring precision.

Table 8. Low Precision Theodolites (selection)

type	manufacturer	telescope					graduated circles					bubble sensit.				mass			height	remarks
		Γ_F	diameter of objective lens [mm]	field of view [mm]	length [mm]	shortest sighting distance [mm]	diameter H [mm]	diameter V [mm]	reading device, least count H	reading device, least count V	reading device, type	platen level [″]	height index level [″]	telescope level [″]	bull's eye level [′]	instrument [kg]	container [kg]	tripod [kg]	[mm]	
1	2	3	4	5	6	7	8	9	10	11	12	13	14	15	16	17	18	19	20	21
with metal circle																				
GAVEC	Breithaupt, Kassel	15	20	144	145	1.4	100	65	1°	1°	index	45	–	30	10	2.5	2.2	3.5	220	
Transit 50	Brunson, U.S.A.	22	33		260	1.1	160	127	1′	1′	vernier	60	–	35	–	6.8	6.3	5	370	cross-level; compass
with glass circle																				
TEKAT	F.W. Breithaupt & Sohn, Kassel	18	30	90	126	1.1	65	47	5′ (0.05 gon)	5′ (0.05 gon)	scale microscope	–	–	30	4	3.3	2.9	3.5	235	coaxial clamp/fine drive. TEMIN: hanging theodolite; figures 297 and 312
TEMIN																3.2			250	
BTF	Ertel-Werk, Munich	24	30	115	180	1.0	67	50	30′	1°	line microscope	60	–	40	10	2.7	2.0	4.2	220	BTF: metal vertical circle
BTU									5′	5′										BTU: circle illumination with luminence diode
4138-C	Filotecnica Salmoiraghi, Milano	22	27	86	128	1.3	80	57	5′ (0.05 gon)	5′ (0.1 gon)	line microscope	–	–	30	10	2.4	0.9	5.7	220	container base is part of instrument
KO-S	Kern & Co., Aarau	19	24	127	175	0.8	79	69	5′ (0.1 gon)	5′ (0.1 gon)	line microscope	–	–	–	2	4.3	1.8	3.4	240	figure 294
TG 4 bR	Officine Galileo, Florence	22	28	90	115	1.8	49	49	2′ (0.05 gon)	2′ (0.05 gon)	vernier microscope	50	–	30	30	2.0	0.7	4.7	205	container base is part of instrument; reading eyepiece attached to standard
Th 51	Opton Feintechnik GmbH, Oberkochen	20	32	120	137	1.2	70	70	10′ (0.1 gon)	10′ (0.1 gon)	line microscope	45	–	–	–	2.2	2.9	6.5	223	as in figure 296
STNO	SLOM, Paris	22	25	125	150	1.8	42	42	5′ (0.05 gon)	5′ (0.05 gon)	vernier microscope	30	–	25	10	3.5	2.7	4.7	220	telescope supported on one side; SCNO without vertical circle
GT-60	Sokkisha, Tokyo Lietz, U.S.A.	16	21	80	115	1.0	70	70	1′	1′	vernier microscope	137	–	60	6	3.0	2.8	5.0		mining theodolite

Tecogon	Theis & Co., FRG	25	35		160		75	75	10′	10′	line microscope	30	–	–	8	2.2	2.4	5	200	
T-30	U.S.S.R.	20	25	120	140	1.2	72	72	10′	10′	line microscope	45	–	20	–	2.2	1.0	3.8	195	container base is part of instrument; figure 296; T-30 M also as hanging theodolite
T 05	Wild Heerbrugg AG	19	25	134	115	0.8	67	50	5′ (0.1 gon)	5′ (0.1 gon)	line microscope	60	–	(40)	10	3.0	2.2	3.8	222	reading eyepiece attached to standard; figure 298
THEO 080 A	VEB Carl Zeiss JENA	18	32	156	128	0.9	61	61	5′ (0.1 gon)	5′ (0.1 gon)	line microscope	120	–	120	–	2.0	2.2	5.0	250	can also be used as hanging theodolite; figure 295

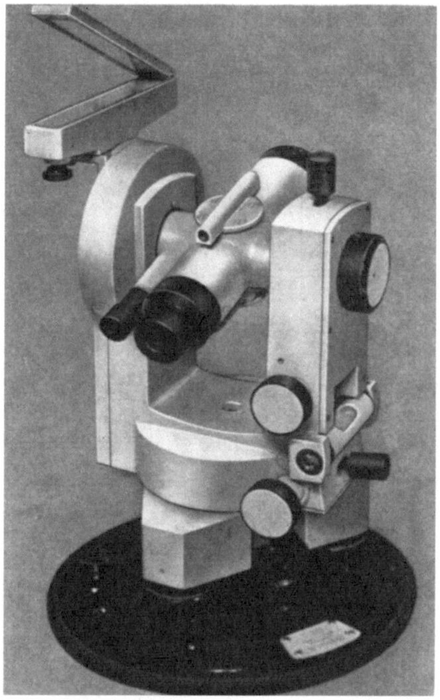

Figure 296 T-30 (U.S.S.R.); height about 24 cm

Figure 297 Small repetition theodolite No. 1103 TEKAT (Breithaupt & Son, Kassel); height 200 mm

Figure 298 Construction theodolite Wild T 05; height 155 mm

Theodolites of Medium Precision

These instruments (table 9, figures 299 to 313) are primarily designed for traversing, small triangulations, setting out surveys and stadia measurements. Based on their purpose, they belong to the most widely used category. In non-European countries, theodolites with metal circle and diametrical vernier readings are still used (figures 291 to 293). The horizontal circles of modern theodolites of this category have a diameter of 70 to 90 mm, with 50 to 80 mm for the vertical circles (mostly with automatic index). They are usually read at only one position, since the influence of eccentricity of circle graduation is kept small. Reading devices consist mainly of scale microscopes and reading microscopes with simple optical micrometer with plane parallel plate. Generally one reads to 10 mgon (1′) or to 3 mgon (10″) with estimation of tenths of the least count. For optical micrometers a 20 × magnification is sufficient, while scale microscopes require up to 70 × magnification. Graduation and scale line have to be sufficiently thin (about 2 to 3 μm). Since this is possible with modern technology, scale microscopes become more and more common because of easier

Table 9. Theodolites of Medium Precision (selection)

type	manufacturer	telescope					graduated circles				
							diameter		reading device		
									least count		
		Γ_F	diameter of objective lens [mm]	field of view [']	length [mm]	shortest sighting distance [m]	H [mm]	V [mm]	H	V	type
1	2	3	4	5	6	7	8	9	10	11	12
with metal circle											
T. V. 2/3	Chasselon, Cachan	24	35	100	175	2.2	108	100	30″	30″	vernier
6140	Dietzgen Corp., U.S.A.	25	35	82	178	1.5	120	68	30″	1′	vernier
6170							110	68	20″	1′	vernier
T-22	Fuji Surv. Instr., Tokyo	27	42	66	181	1.6	125	72	20″	1′	vernier
Transit 50	Gurley, U.S.A.	28/18	38	50/79	300	2.1	160	127	1′/30″ /20″	1′	vernier
Transit 60		28/18	33	60/89	260	1.7					
Transit 110		28/12	29	88/138	216	1.5	130	127	1′	1′	vernier
Constr. Transit		18	30	90	216	1.5	130	102	1′	5′	vernier
Paragon Transit	Keuffel & Esser, U.S.A.	24	36	67	292	1.8	152	127	30″	1′	vernier
Paragon Mount. Trans.		20	33	75	254	1.5	133	127	1′	1′	vernier
H 5	Nippon Kogaku, Tokyo	25	40		167	2.0	120	80	20″	1′	vernier
10-C	Lietz Comp., U.S.A.	28	40	80	175	1.3	110	68	20″	1′	vernier
625		28	38	60	292	1.7	165	127	20″	1′	vernier
550		26	38	70	240	1.9	155	130	20″	1′	vernier
TR-303	Path Instr., U.S.A.	27	42	66	181	1.6	125	80	20″	1′	vernier
BT-20	Sokkisha, Tokyo	28	35	70	170	1.3	120	80	20″	1′	vernier
A	Tokyo Optical	25	40	88	182	1.5	120	80	20″	1′	vernier
NA-2	Tokyo Sokki	25	36	70	193	1.6	120	80	20″	1′	vernier
V 100	Vickers Instr., York	27	38		210	2.0	114	102	20″	20″	vernier
with glass circle											
TEAUT	F. W. Breithaupt & Son, Kassel	30	40	72	150	1.9	73	47	20″ (10 mgon)		opt. microm. (1 par. plate)
US 2	China	26	35		195	2.0	89	89	20″	20″	scale microscope
FT 1 A	Fennel (FRG)	30	40	75	175	1.2	90	70	1′ (10 mgon)	1′	scale microscope
4150-NE	Filotecnica Salmoiraghi, Milano	30	40		172	2.4	90	70	1′ (10 mgon)	1′	scale microscope
4149-A		30	36		172	2.0	90	70	30″ (10 mgon)	30″	scale microscope
6190	Dietzgen Corp., U.S.A.	28	40	80	160	1.5	86	86	1′	1′	scale microscope
6020-AG 6		30	45	90	180	1.5	90	70	1′ (10 mgon)	1′	scale microscope
TH-32	Fuji Surv. Instr. Co. Ltd., Tokyo	25	36	80	108	1.8	92	73	20″	20″	opt. microm. (1 par. plate)
DKM 1	Kern & Co., Aarau	20	30		120	0.9	50	50	10″ (1 mgon)		DK centre setting (1 par. plate)
K 1-A		28	45	90	160	1.8	95	75	20″ (10 mgon)		centre setting (1 par. plate)
K 1-S		30	45	90	160	1.7	89	70	30″ (10 mgon)		scale microscope
T-60 D	Lietz Comp., U.S.A.	28	40	80	175	1.3	95	90	1′	1′	scale microscope

bubble sensitivity					mass				
platen level ["]	height index level ["]	telescope level ["]	bull's eye level [']	automatic height index; range [']	instrument [kg]	container [kg]	tripod [kg]	height [mm]	remarks
13	14	15	16	17	18	19	20	21	22
30	30	–	8	–	4.2	3.0	5.5	200	C. V. 1 without vertical circle
60	–	40	–	–	4.5	4.0	5.0	240	
90	–	40	–	–	4.5	3.5	5.0	240	
90	–	40	10	–	4.6	1.5	5	247	figure 293
80	–	30	–	–	5.5	5.0	5.5	400	cross-level; compass
80	–	40	–	–	5.0	5.0	5.5	400	compass, figure 292
80	–	50	–	–	4.5	4.5	5.5	400	cross-level; compass
80	–	50	–	–	4.5	5.0	5.5	340	cross-level
70	–	30	–	–	8.4	4.5	4.5	310	cross-level; compass, figure 291
75	–	30	–	–	7.3	4.5	4.5	300	transit with 100" platen bubble; 60" telescope bubble
100	–	40	–	–	4.6	3.8	3.9	230	
90	–	40	–	–	5.2	4.1	5.7	310	glass circles, compass
					7.5	5		330	cross-level; compass
					6.0	5		330	
90	–	40	10	–	4.8	2.7	5.7	250	
90	–	60	–	–	5.2	4.1	5.7	270	
100	–	40	–	–	4.5	4.0	3.5	220	cross-level
90	–	40	–	–	5.4	4.1	4.0	320	cross-level, figure 293
45	20	–	–	–	6.2	3.9	6.8	265	
45	–	(30)	5	±8	5.4	3.3	5.0	260	figure 299; coaxial clamps and fine drives FTH 1 without vertical circle
30	30	30	8	–	4.6	4.4	6.5	240	
40	–	30	8	±4	4.0	4.0	7.0	275	coaxial clamps and fine drives
30	30		10		4.6	2.9	7.2	310	
–	–	30	10		4.7	3.1	5.6	300	
90	–	40	–	–	4.9	3.4	4.6	250	type 6191 with optical micrometer (1 plate) 20"
20	–	–	10	±20	4.6	2.4	7.0	310	figure 306, coaxial clamps and fine drives; Ts without vertical circle; Tt with optical micrometer (1 parallel plate) 20"
45	–	–	10		4.6	1.5	5	241	compass
30	30	–	–	–	1.8	1.0	3.4	140	figure 307
30	–	–	–	±10	4.2	2.2	3.4	215	figure 303
30	–	–	–	±20	4.6	2.4	3.4	215	figure 308
30	–	–	10	±5	5.0	3.8	5.9		

Table 9 (continuation)

type	manufacturer	telescope					graduated circles				
							diameter		reading device		
									least count		type
		Γ_F	diameter of objective lens [mm]	field of view [']	length [mm]	shortest sighting distance [m]	H [mm]	V [mm]	H	V	
1	2	3	4	5	6	7	8	9	10	11	12
Te-D 23	MOM, Budapest	26	45	80	143	2.0	77	68	1′ (20 mgon)	1′	scale microscope
Te-C 13									20″	20″	opt. microm. (1 par. plate)
NT-2	Nikon, Tokyo	25	40	96	167	1.8	84	84	20″		opt. microm. (1 par. plate)
TG 3b	Officine Galileo, Florence	22	28	90	115	1.8	49	49	1′ (20 mgon)	1′	opt. microm. (1 par. plate)
TG 2b		29	40	90	177	1.9	80	80	10″ (2 mgon)	10″	optical micrometer
Th 42	Opton Feintechnik GmbH, Oberkochen	30	40	83	155	1.6	98	85	20″ (10 mgon)	20″	scale microscope
T-202	Path Instr., U.S.A.	28	40	64	170	1.8	90	80	10″		opt. microm. (1 par. plate)
T 30 T 6	PZO, Warsaw	26	35	95	160	1.8	94	74	1′ (20 mgon)	1′	scale microscope
TB 1		18	34	95	155	1.9	88	80	2′	2′	scale microscope
ST 310	Rank Prec. Ind. Ltd; England	28	38	65	150	1.8	87	78	10″ (50 mgon)		opt. microm. (1 par. plate)
ST 156		25	38	65	146	1.6	89	64	20″ (10 mgon)		opt. microm. (1 par. plate)
STNO	SLOM, Paris	22	25	127	150	1.8	80	–	1′	–	opt. microm. (1 par. plate)
STO 1		28	40	103	150	1.5	80		1′		opt. microm. (1 par. plate)
TM-10C	Lietz Comp., U.S.A. Sokkisha, Japan	30	40	90	170	1.3	80	70	10″		opt. microm. (1 par. plate)
TM-20C									20″		
T-60 D							95	90	1′	1′	scale microscope
C 15	Stanley & Co., London	20	36		150	1.2	70	64	20″		opt. microm. (1 par. plate)
2 T 5	U.S.S.R.	27	36	90	180	2.0	100	72	1′	1′	scale microscope
T-15		25	34	90	138	1.2	80	72	1′	1′	scale microscope
TD-50	National Instr. and Ophthalm, Glass Ltd., Kalkutta	25	40	90	170	2.5	84	76	1′		optical micrometer
TL-20	Topcon, Japan	28	40	80	160	1.5	86	86	20″		opt. microm. (1 par. plate)
TL-10							90	70	10″		
V 22	Vickers Instr., York	25	38	120	137	1.8	79	64	30″	20″	scale microscope
V 11		15	25	180	127	1.8	70	70	1′ (10 mgon)	5′ (50 mgon)	H: vernier microscope V: line microscope
T 1	Wild Heerbrugg AG	30	42	93	150	1.5	79	79	6″ (2 mgon)		opt. microm. (1 par. plate)
T 16		30	42	93	172	1.7	94	79	1′ (10 mgon)	1′	scale microscope
THEO 020 A	VEB Carl Zeiss JENA	25	36	78	181	1.5	86	86	1′ (10 mgon)	1′	scale microscope

bubble sensitivity					mass				
platen level ["]	height index level ["]	telescope level ["]	bull's eye level [']	automatic height index; range [']	instrument [kg]	container [kg]	tripod [kg]	height [mm]	remarks
13	14	15	16	17	18	19	20	21	22
30	30	–	6	–	5.0	2.5	5.1	310	figure 304; Te-D 13 and 33: optical micrometer Te-D 43: automatic height index ±5′
					4.8	2.8			figure 311. Coaxial clamps and fine drives
100	–	40	–		5.0	4.5	4.2	263	compass; reading eyepiece attached to standard
30	50	30	30	–	2.1	0.7	4.7	205	
30	30	–	30	–	4.8	2.5	6.1	300	
30	–	–	15	±2	4.7	4.2	6.2	290	figure 302; spherical support; rough-fine-focussing
45	–	–	10		5.0	2.0	5.2	273	
30	30	–	10	–	3.9	3.0	5.8	198	TA-6: automatic height index
30	30	–	10	–	3.0	3.0	5.8	181	
30	–				5.4	3.2	5.4	275	reading eyepiece attached to standard
30	30	–	–	–	5.1	3.9	5.4	250	reading eyepiece attached to standard
30	–	40	–	–	3.9	2.3	3.5	190	only one standard; SCNO without vertical circle
30	–	40	–	–	4.8	2.5	4.8	208	figure 309; SCO 1 without vertical circle
30	–	–	10	±5	5.0	3.8	5.9	270	compass; reading eyepiece attached to standard; figure 300. TM 20 E: 40″ telescope bubble; no automatic height index
									TS 20: H = 20″, 40″ – telescope bubble; no automatic height index
60	30	–	–	–	3.2	2.6	6.3	250	
30	15	–	10	±3	3.7	3.5	5.3	335	coaxial clamps and fine drives T-5 K: automatic height index
45	30	30	10		3.1	3.4	5.3	250	T-15 M: mining theodolite
30	–	–	8	–	4.8	2.5	5.1	250	
90	–	40	–	–	4.9	3.4	4.0		90″-cross-level; compass
30	–	–	10	–	4.5	3.7			compass
45	90	–	17	–	5.2	4.7	6.9	200	reading eyepiece attached to standard
45	–	45	20	–	3.0	2.3	6.9	208	reading eyepiece attached to standard
30	–	–	8	±2	5.8	2.8	5.6	265	repetition clamp
30	–	–	8	±5	5.3	2.8	6.5	265	figure 305; repetition clamp
20	–	(20)	8	±5	4.2	4.4	5.8	295	figure 301; repetition clamp; common H/V clamp; coaxial fine drives

Figure 299 Optical repetition theodolite TEAUT (Breithaupt & Son, Kassel); height 330 mm

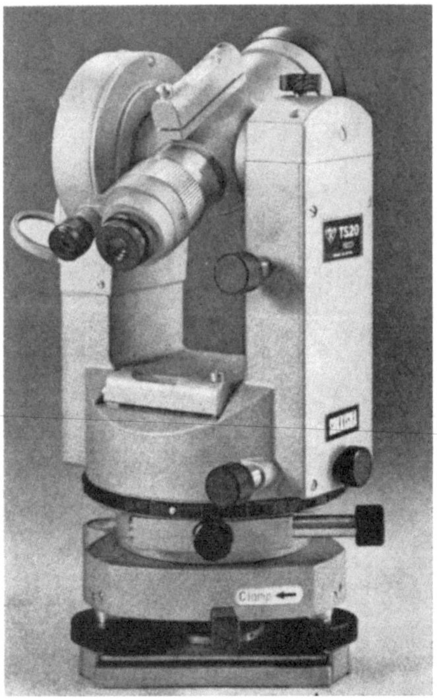

Figure 300 Scale theodolite TS 20 (Sokkisha); height 340 mm

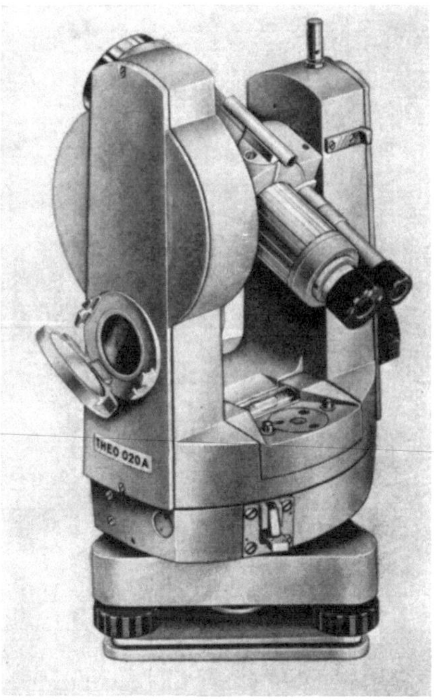

Figure 301 Stadia theodolite THEO 020 A (Zeiss, Jena); height 295 mm

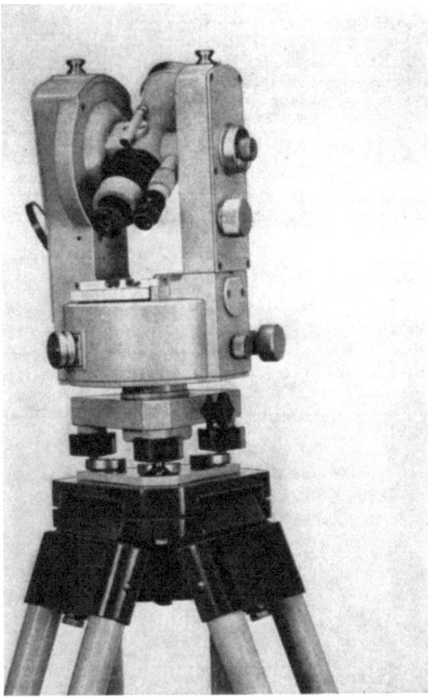

Figure 302 Scale theodolite of medium accuracy Th 41 (Zeiss, Oberkochen); height 268 mm

Figure 303 Engineering theodolite K 1-A (Kern); height 215 mm

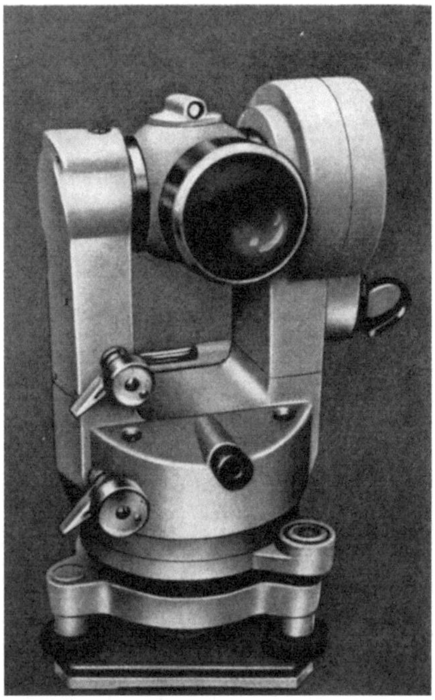

Figure 304 Stadia theodolite MOM Te-D 23 resp. 43; height 230 mm

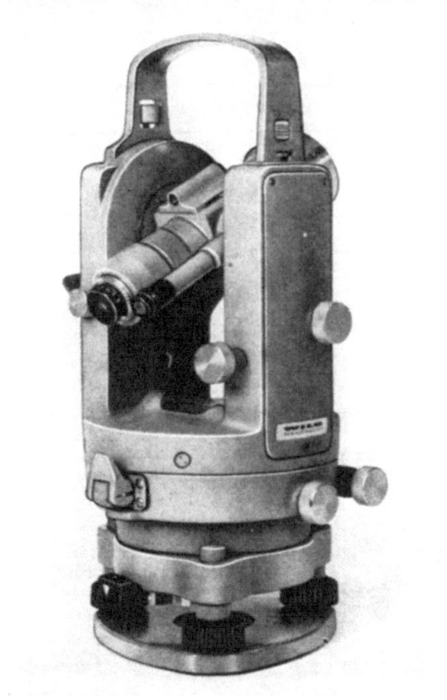
Figure 305 Scale theodolite Wild T 16; height 265 mm

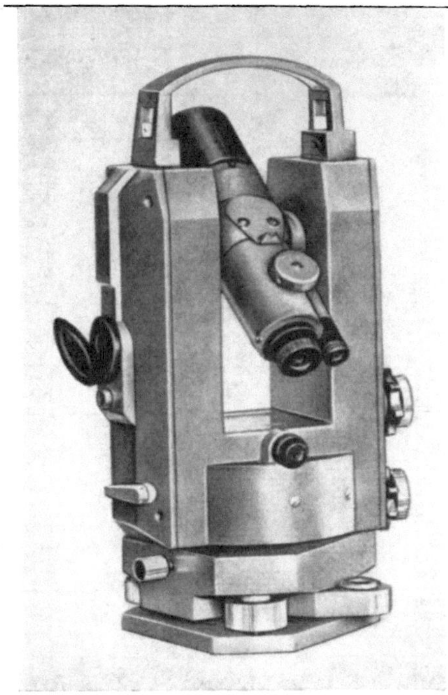
Figure 306 Dietzgen 6020-A 6; height 310 mm

Figure 307 Small double circle triangulation theodolite DKM 1 (Kern); height 140 mm

Figure 308 Theodolite K1-S (Kern); height 228 mm

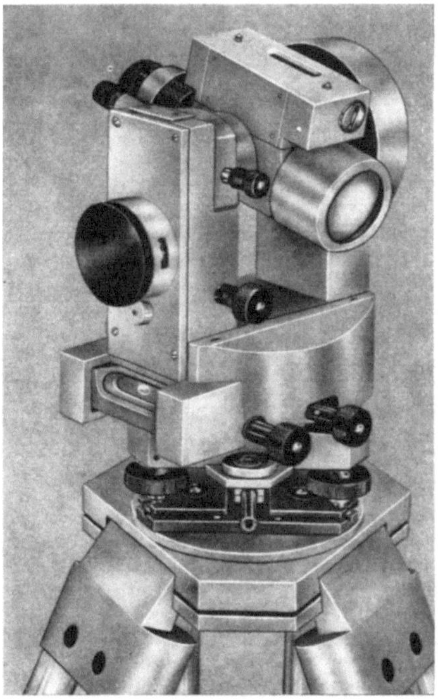
Figure 309 Precision level theodolite SLOM ST 01; height 208 mm

Figure 310 Universal theodolite FT 1 A (Fennel); height 275 mm

Table 10. High Precision Theodolites (selection)

type	manufacturer	telescope					graduated circles				
							diameter		reading device		
		Γ_F	diameter of objective lens [mm]	length [mm]	field of view [']	shortest sighting distance [m]	H [mm]	V [mm]	least count H	least count V	type
1	2	3	4	5	6	7	8	9	10	11	12
A 2	Dietzgen Corp., U.S.A.	30	45	180	90	1.5	90	70	1″ (0.2 mgon)		coincidence micrometer (1 par. plate)
4200	Filotecnica Salmoiraghi, Milano	30	40	172		2.5	90	70	1″ (0.2 mgon)		opt. microm. (1 par. plate)
DKM 2-A	Kern & Co., Aarau	32	45	170	86	1.7	80	74	1″ (0.2 mgon)		opt. microm. (1 par. plate)
TM-1 A	Lietz Comp., U.S.A.	28	40	175	90	2.2	94	80	1″		coincidence microm.
Digital-theodolit	Sokkisha, Tokyo	28	40	175	80	1.3			3″	6″	elect.-mag. scanning
Te-B 23	MOM, Budapest	30	45	143	80	2.0	93	60	1″ (0.2 mgon)		coincidence micrometer (2 par. plates)
NT-3	Nikon, Tokyo	28	40	165	96	2.0	84	84	5″		opt. microm. (1 par. plate)
TG 1b	Officine Galileo, Florence	29	40	177	90	1.9	80	80	1″ (0.2 mgon)		opt. microm. (1 par. plate)
Th 2	Opton Feintechnik GmbH, Oberkochen	30	40	155	82	1.6	100	85	1″ (0.1 mgon)		coincidence micrometer (shifting wedg.)
ST 200	Rank Prec. Ind. Ltd., England	28	41	165	107	1.8	98	76	1″ (0.5 mgon)		opt. microm. (1 par. plate)
2 T 2	U.S.S.R.	27	35	183	90	2.0	100	72	1″		coincidence micrometer (shifting wedg.)
V 406	Vickers Instr. Ltd., York	25	38	159	120	1.8	85	70	1″ (0.1 mgon)		coincidence micrometer (2 par. plates)
T 2	Wild Heerbrugg AG	30	40	150	100	2.2	90	70	1″ (0.1 mgon)		coincidence micrometer (2 par. plates)
THEO 010A	VEB Carl Zeiss JENA	30	40	177	72	1.5	86	86	1″ (0.2 mgon)		coincidence micrometer (shifting wedges)

bubble sensitivity				automatic height index; range	mass			height	remarks
platen level ["]	height index level ["]	telescope level ["]	bull's eye level [']	[']	instrument [kg]	container [kg]	tripod [kg]	[mm]	
13	14	15	16	17	18	19	20	21	22
20	–	–	10	±20	4.6	2.4	5.0	310	coaxial clamps and fine drives; figure 320
20	–	–	10		6.1	5.2	7.2	310	double graduation; figure 322
20	–	–	–	±2	6.2	2.4	5.3	225	*H*- and *V*-rough and fine drive; figure 321
20	–	–	10	±2	6.0	4.0	5.9	280	
90	–	40	10				5.9		*H*/*V* combinations: 15″/30″, 30″/60″, 1′/5′
20	20	–	6	–	5.3	3.7	6.2	310	coaxial clamps and fine drives; figure 317. Te-B3: automatic height index
40	–	30	10	±5	5.0	4.3	4.5	280	reading eyepiece attached to standard
20	30	–	10	–	4.9	2.5	6.1	300	height index bubble within the field of view of the reading eyepiece; fig. 315
20	–	–	10	±2	5.2	4.8	6.2	315	rough-fine focussing; coaxial clamps and fine drives; spherical support; figure 323
20	20	–	–	–	6.3	4.1	6.0	280	reading eyepiece attached to standard; bull's eye level on tripod; figure 314
15	20	–	–	–	5.5	4.5	5.3	298	10″ striding level; coaxial clamps and fine drives; figure 324
20	20	–	20	–	4.8	5.2	6.9	310	figure 319
20	–	–	8	±5	6.0	2.2	6.5		figure 225
20	–	(20)	8		4.5	4.4	5.8	295	common *H* – *V* clamp; coaxial fine drives; figure 316

Figure 311 Stadia theodolite MOM Te-C 13; height 230 mm

Figure 312 Hanging mining repetion theodolite TEMIN (Breithaupt & Son)

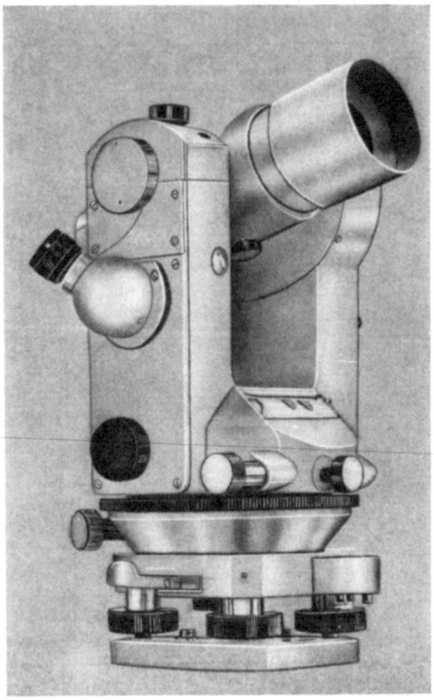

Figure 313 T-202 (Path Instruments Int. Corp.); height 273 mm

Figure 314 ST 200 Microptic one second theodolite (Rank Prec. Ind.); height 280 mm

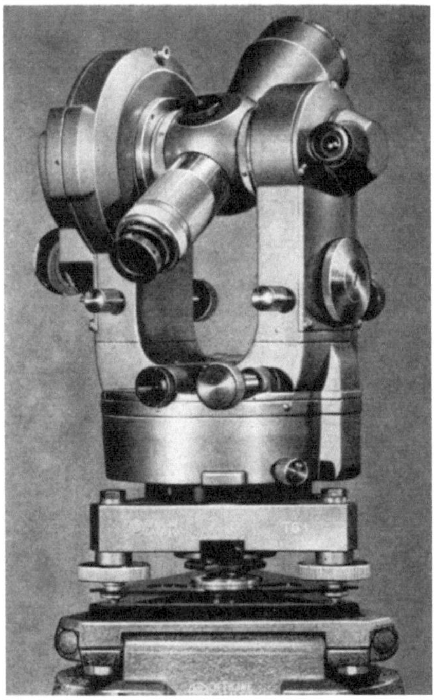

Figure 315 Universal theodolite TG 1 b (Officine Galileo); height 295 mm

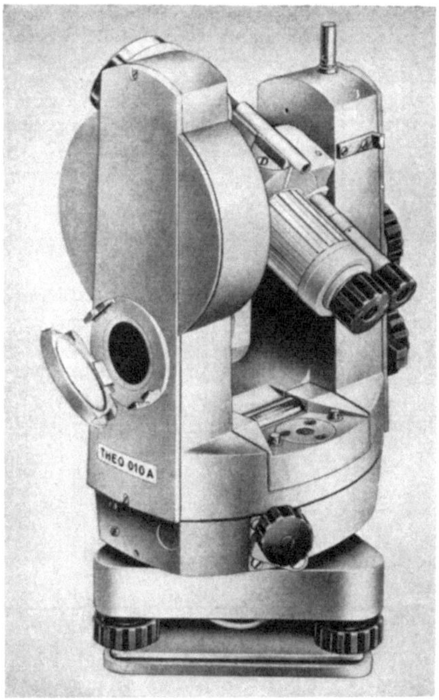

Figure 316 One second theodolite THEO 010 A (Zeiss, Jena); height 295 mm

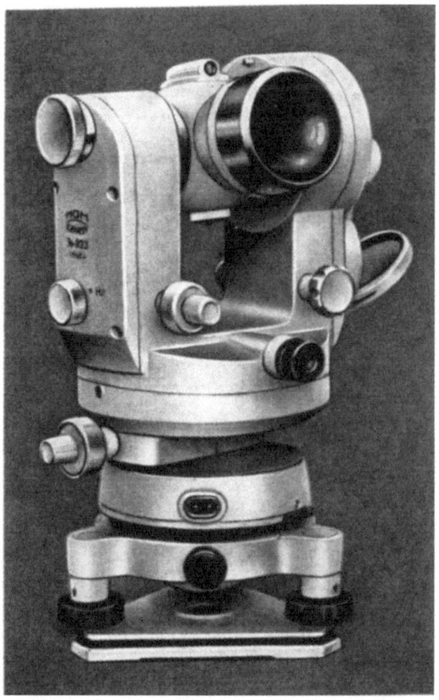

Figure 317 One second theodolite MOM Te-B 23; height 235 mm

Figure 318 Optical Scale theodolite T 5 K (U.S.S.R.); height 285 mm

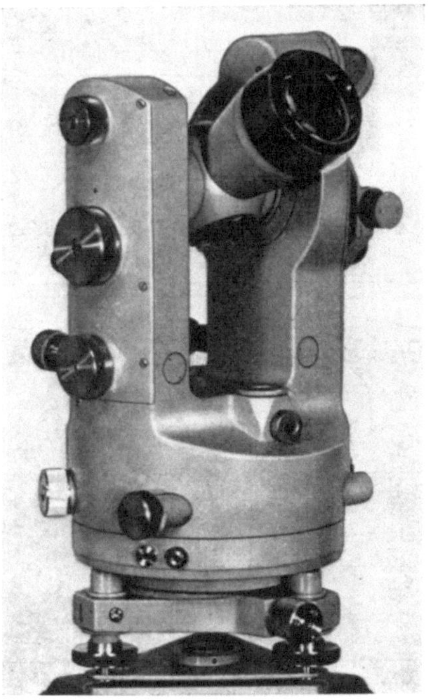

Figure 319 Tavistock theodolite V 406 (Vickers); height 310 mm

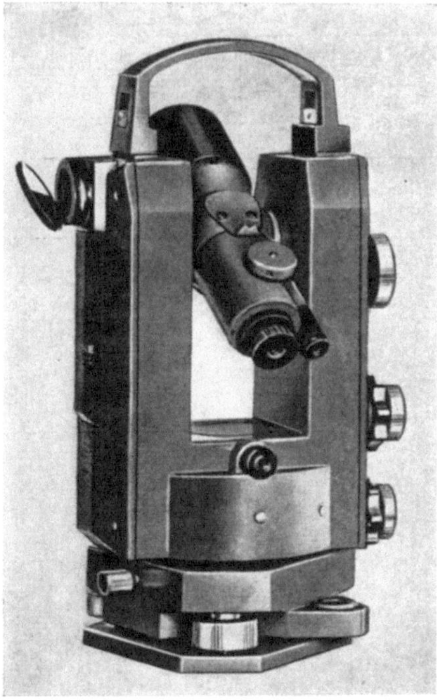

Figure 320 Dietzgen one second theodolite 6020-A 2; height 255 mm

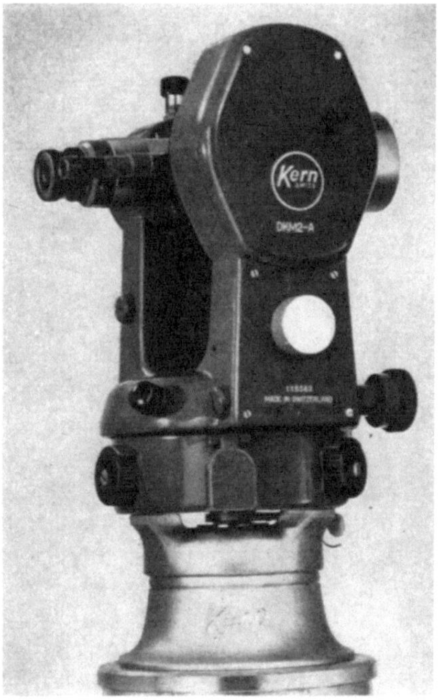

Figure 321 One second theodolite DKM 2-A (Kern); height 240 mm

Figure 322 One second theodolite 4200 (Filotecnica Salmoiraghi); height 310 mm

Table 11. Theodolites of Highest Precision (selection)

type	manufacturer	telescope					graduated circles				
		Γ_F	diameter of objective lens [mm]	length [mm]	field of view [']	shortest sighting distance [m]	diameter H [mm]	diameter V [mm]	reading device: least count H	reading device: least count V	reading device: type
1	2	3	4	5	6	7	8	9	10	11	12
DKM 3	Kern & Co., Aarau	27, 45	72	140	69	19	100	100	0.5'' (0.05 mgon)		double circle setting; optical micrometer (1 parallel plate between the circle positions)
T-1	U.S.S.R.	30, 40	62	255	90	5	135	90	1''		coincidence micrometer (shifting wedges)
T-05		37, 50, 62	64	375	41	5	180	130	1''		coincidence micrometer (shifting wedges)
T 3	Wild Heerbrugg AG	24, 30, 40	60	265	98	4	135	90	0.2'' (0.1 mgon)		coincidence micrometer (2 parallel plates)
T 4		60, 80	70	–	46	100	240	135	0.1''	0.2''	coincidence micrometer (1 pair of parallel plates per circle)
Theo 002	VEB Carl Zeiss JENA	H: 38, 59, 75 V: 32, 50, 65	65	–	40	4000	250	200	0.2''		coincidence micrometer (shifting wedges)

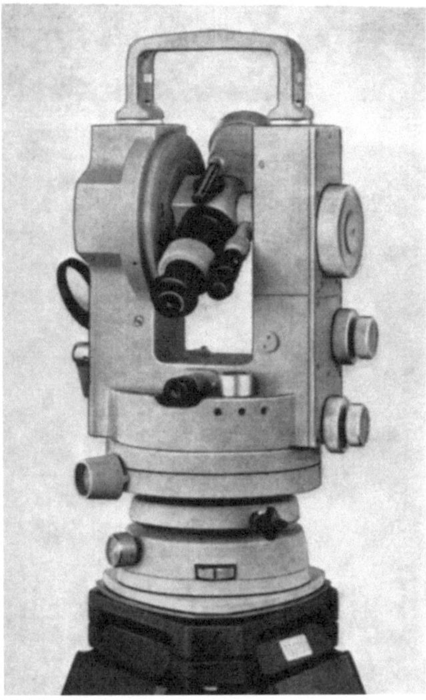

Figure 323 One second theodolite Th 2 (Zeiss, Oberkochen); height 315 mm

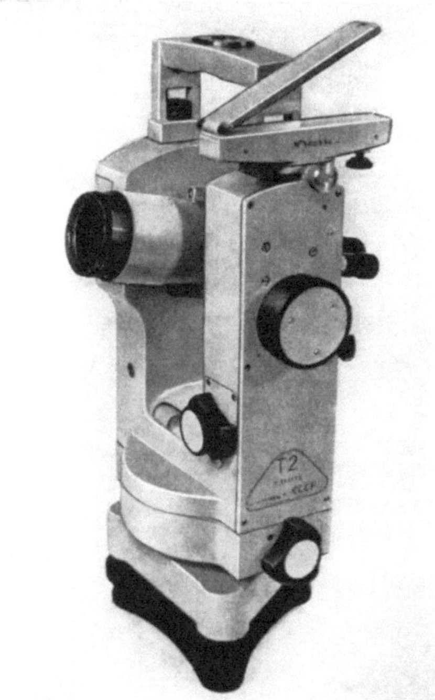

Figure 324 Optical precision theodolite T-2 (U.S.S.R.); height 298 mm

Figure 325 Optical precision theodolite T-1 (U.S.S.R.); height of horizontal axis: 325 mm

bubble sensitivity					mass				
platen level [″]	height index level [″]	telescope level [″]	bull's eye level [′]	automatic height index; range [′]	instrument [kg]	container [kg]	tripod [kg]	height [mm]	remarks
13	14	15	16	17	18	19	20	21	22
10	10	–	–	–	12.2	3.1	7.2	228	mirror telescope (f = 500 mm); levelling knobs; aiming telescope ($\Gamma_F = 11x$); figure 328 Astronomic variant: DKM 3 A
7	12	–	–	–	11.8	5.8	8.0	370	5″ striding level, 4″ Horrebow-level, coaxial clamps and fine drives; figure 325
7	–	–	–	–	19	20.8 + 15.4	–	430	1″ eyepiece micrometer; aiming telescope; telescope: f = 500 mm; figure 326
–	13	–	–	–	11.2	3.7	5.7	350	figure 327. Special model T 3-A for auto collimation and normal observing
–	2	–	4	–	60	50	18	467	broken telescope (f = 578 mm), 1″ hanging level, 1″ Horrebow-level; figure 329
10	–	–	–	±3	60	30	–	530	broken telescope (f = 938 mm); stabilized collimation axis; photographic recording; aiming telescope (Γ_F = 9-, 14-, 18times); figure 330

Figure 326 Optical theodolite of highest accuracy T-05 (U.S.S.R.); height of horizontal axis: 340 mm

Figure 327 Precision theodolite Wild T 3; height of horizontal axis: 287 mm

Figure 328 Large double circle triangulation theodolite DKM 3 (Kern); height 288 mm

Figure 329 Universal instrument Wild T 4; height 467 mm

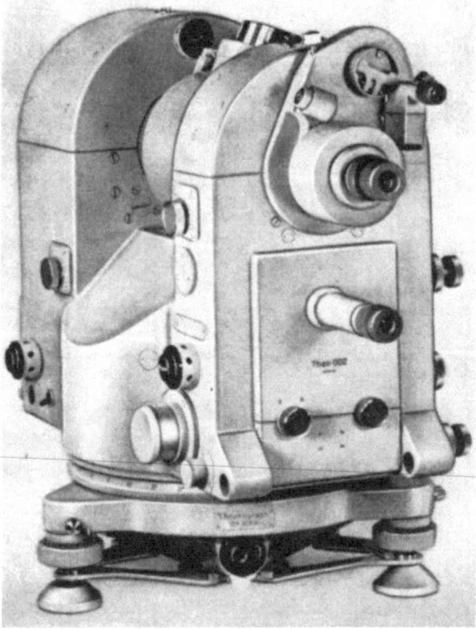

Figure 330 Geodetic-astronomic universal theodolite Theo 002 (Zeiss, Jena); height 530 mm

reading. The telescopes have about 20 × to 25 × magnification. The instruments are suitable for optical distance measurements with stadia hairs and usually permit repetition measurements or orientation of the vertical circle. Clamps and fine motion screws are usually on one axis. For comparison, figure 312 shows a hanging theodolite used in mining surveying.

High Precision Theodolites

High precision theodolites (table 10, figures 314 to 324) are primarily designed for 2nd and 3rd order triangulation, for precise traversing, for highly accurate engineering surveys, and for optical distance measurements with subtense bar. The diameters of the horizontal circles of modern instruments of this category range from 80 to 100 mm, with 70 to 90 mm for the vertical circle (usually with automatic index). To eliminate the influence of eccentricity of the circle graduation, reading takes always place at two circle positions. Older models usually have screw microscopes, even vernier microscopes with optical micrometers, while modern theodolites are equipped with reading microscopes with automatic mean and optical micrometers. Usually 6 of the 7 digits of the circle reading can be directly read. The microscope magnification is 40 × to 50 ×, the telescopes have about 30 × magnification, and 40 mm free objective lens diameter. To set the circle to a desired reading, circle gears are used, occasionally with rough and fine gears.

Theodolites of Highest Precision

Theodolites of highest precision (table 11, figures 325 to 330) are designed for 1st order triangulation, tunnel staking and similar applications. Their horizontal circles have diameters ranging from 100 to 250 mm, with 90 to 150 mm for the vertical circles. Older instruments utilize screw microscopes, while modern theodolites are read with reading microscopes with optical micrometers and automatic meaning. The telescopes, partly designed as broken telescopes (figures 329 and 330) or as mirror telescopes (figure 328), have 30 × to 65 × magnification, and about 60 mm free objective lens diameter. Often different values for magnification are obtained by changing the eye piece. The instruments are heavy and solidly built. The measuring accuracy can be increased by using accessories, e.g. eye piece micrometers, striding level (to consider horizontal axis errors), checking telescope at the tribrack to consider pillar torsion of the signal by sighting to a reference mark with the main telescope for each targetting (figure 326).

2.3.4. Testing and Adjusting of Theodolites

Even when carefully measuring the angles, the results are influenced by errors, caused by irregularities of the instrument (instrumental errors) and the human senses (observation errors), as well as the physical characteristics of the air between instrument and target, which cannot be completely determined. Systematic errors are dangerous, because they can influence the results by one-sided accumulation, while random errors—though not

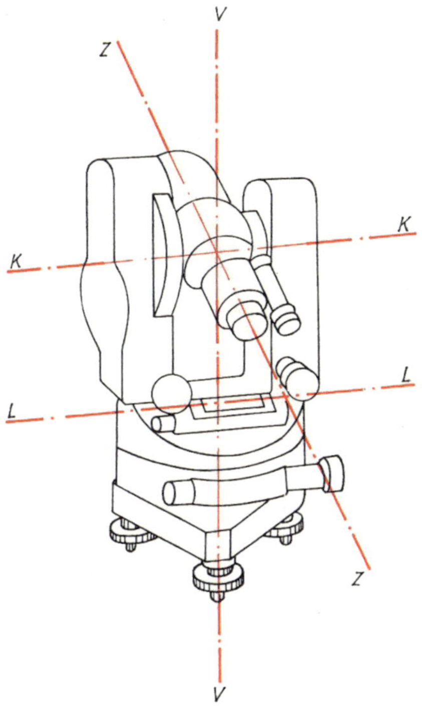

Figure 331 Axes of the theodolite

quite eliminated—can be reduced by using special instruments and procedures as well as by applying least squares adjustments. Therefore, it is very important to know the systematic errors. They can be partly or completely eliminated by special arrangements or with corrections.
Instrumental errors occur, because the design idea can never be mathematically exactly realized in the manufacturing process, or else single elements have slightly moved from their exact position during usage of the instrument. Errors caused by incorrect position of geometrical elements relative to each other (especially axis- and eccentricity errors), as well as errors occurred during graduation of circle or reading scales, can either be eliminated by adjustment or else by special measuring arrangements.
Errors depending on the observer occur when aiming with the telescope and when reading the circles.
The most complicated errors are the ones caused by the physical state of the air (especially refraction and irregular target illumination).
In the following, mainly instrument errors and their elimination are investigated. By considering each error separately, one can assume, that it is small, and that there are no other errors.

Axis Errors

In order to be able to determine horizontal angles with the aid of the horizontal circle, the latter has to be horizontal (and the vertical axis vertical), and the collimation axis has to move within a vertical plane when tilting the telescope. This is the case, when the following holds for the vertical axis (VV), the horizontal axis (KK), the collimation axis (ZZ), and the axes of levels (LL), shown in figure 331 as a cross level:

a) the collimation axis is normal to the horizontal axis ($ZZ \perp KK$)—then the collimation axis moves within a plane when the telescope is tilted—
b) the horizontal axis is normal to the vertical axis ($KK \perp VV$), then the collimating plane is vertical, if
c) the vertical axis is exactly vertical (with $VV \parallel LL$, see section 1.2.1).

If one of these conditions is not fulfilled, a collimation axis-, horizontal axis-, or vertical axis error is present. When determining these errors, one has to watch the mathematical sign.
A secondary horizontal axis error is caused when the horizontal axis wobbles when turning, due to influences on its bearings including thin oil- or grease films. One refers to it as plug-error. A similar error of the vertical axis is called wobble-error. For a good axis, the square sum of the values caused by wobble is a minimum. This error has to conform with the performance of the top part, especially for measurements of highest precision, where secondary horizontal- and vertical axis errors have to be considered.

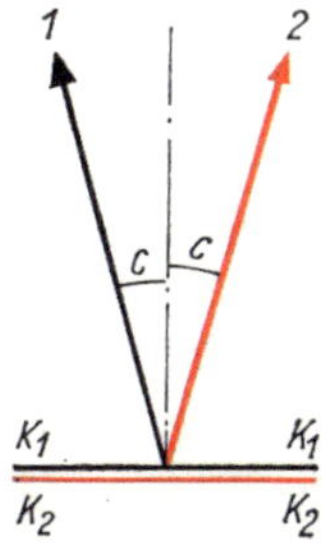

Figure 332 Testing for $ZZ \perp KK$ by plunging the telescope

Collimation Axis Error

Determination of the Collimation Axis Error and its elimination by adjustment

1. Sight to a point at the same height as the horizontal axis, at least 100 m away (direction *1* in figure 332) and read a_1 at the horizontal circle. After plunging the telescope, turn the top part by exactly 200 gon. If the telescope points into direction *2*, the angle between *1* and *2* represents twice the collimation axis error (2c).
This error is eliminated by shifting the crosshair by the amount c, using the adjustment screws which act sideways. This procedure is repeated for checking purposes.
2. The following way of determination is similar to the one mentioned in 1. After sighting point P and reading a_1 at the horizontal circle, the telescope is plunged, one sights again to P and reads a_2 (figure 333). The difference between a_1 and a_2 reduced by 200 gon, is twice the collimation axis error. It is eliminated as mentioned in 1.

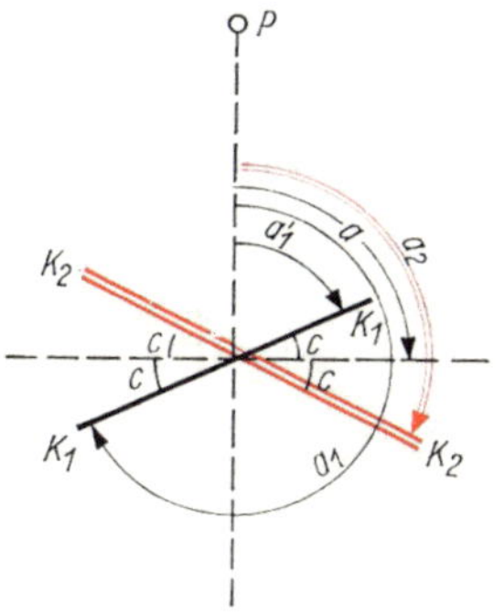

Figure 333 Testing for $ZZ \perp KK$ by horizontal circle readings

Influence of the Collimation Axis Error

The influence (c) of the collimation axis error c on a direction observed in one telescope position when sighting to a point P at a vertical angle β, can be derived with the aid of a semi sphere (figure 334). Because of the error c, the collimation axis is not normal to the horizontal axis KK' and

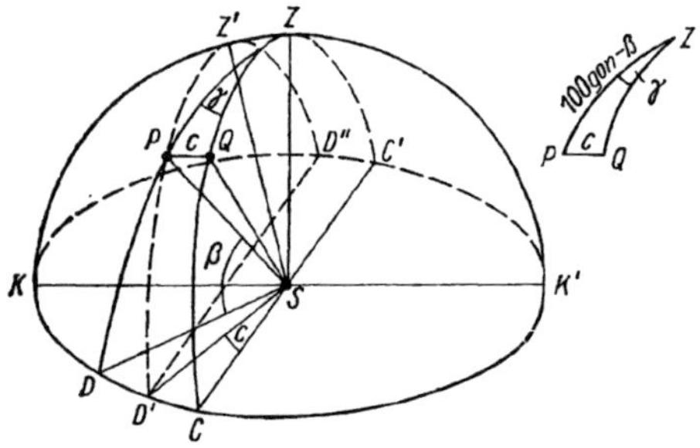

Figure 334 Influence of the collimation axis error

thus does not move within the plane CZC' when the telescope is tilted. It follows a conical path, which intersects the semi-sphere in a circle passing through D', P, Z' and D''. If $Z'S$ were normal to KK', the plane ZQC, normal to the horizontal axis would coincide with the plane ZPD. This is not the case because of c. Instead, ZPD is turned in respect to the normal section by the angle $CSD = (c)$, which is the influence of the collimation axis error c on the value for P to be read at the horizontal circle. To determine (c), one places a great arc $KPQK'$ through P, normal to ZQC. According to the orthogonal spherical triangle: $\sin c = \cos\beta \sin\gamma$ and, since c and γ are small: $c = \gamma \cos\beta$. Since $\gamma = DSC = (c)$, it is obtained as

$$(c) = \frac{c}{\cos\beta} \quad \text{or} \quad (c) = \frac{c}{\sin\zeta} \tag{84}$$

Equation (84) shows, that for horizontal sights, i.e. $\beta = 0$, the influence of the collimation axis is smallest, namely equal to c. For $c = 1.0$ mgon, the following values are obtained for (c):

ζ [gon]	= 100	90	80	70	60	50	40	30	20	10	0
(c) [mgon]	= 1.0	1.01	1.05	1.12	1.24	1.41	1.70	2.20	3.24	6.40	∞

For c the influence $\Delta(c)$ on an angle measured between two points in one telescope position is:

$$\Delta(c) = \frac{c}{\sin\zeta_1} - \frac{c}{\sin\zeta_2} = c\left(\frac{1}{\sin\zeta_1} - \frac{1}{\sin\zeta_2}\right) \tag{85}$$

Therefore the influence of the collimation axis error on an angle measured in one telescope position is smaller, when the difference between the two zenith distances (vertical angles) is small. It is zero if the zenith distances have equal absolute values.

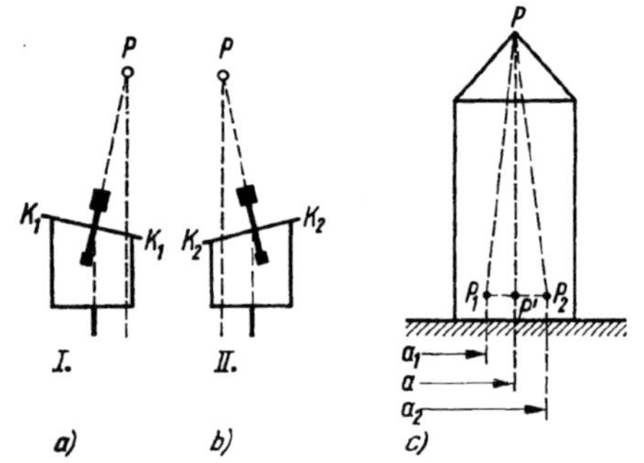

Figure 335 Testing for $KK \perp VV$ by plumbing a high point

Elimination of the Influence of the Collimation Error by Measuring procedure

The influence of the collimation axis error on a single direction, and thus on a horizontal angle is eliminated by observing in two telescope positions. From 1) and 2) above it follows, that for a direction, the Mean of Circle readings from the two telescope positions is free of the influence of the collimation axis error. The difference between the two circle readings always contains twice the collimation axis error. Therefore one can check its magnitude and changes.

Horizontal Axis Error

Determination of the Horizontal Axis Error and its Elimination by Adjustment

1. After the collimation axis error has been eliminated, one sights to a high point P, e.g. the tip of a lightning rod, in both telescope positions and reads the horizontal circle. The difference of the two readings minus 200 gon corresponds to twice the horizontal axis error. To eliminate the horizontal axis error, one sets the telescope to the mean of the circle readings, and raises or lowers the support of the horizontal axis with the adjustment device (figure 335) until P again appears in the centre of the crosshair.
2. After the collimation axis error has been eliminated, one sights to a high point P (figure 335) and reads the value a_1 at P_1 on a ruler placed normal to the collimation axis at the instrument horizon. After plunging the telescope, one sights again to P and reads on the ruler a_2 at P_2. Since $a_2 - a_1$ corresponds to twice the horizontal axis error, one sets the telescope to $a = (a_1 + a_2)/2$ and raises or lowers the support of the horizontal axis until P appears again in the crosshair centre.

For modern instruments with enclosed horizontal axis, this error should only be adjusted by a mechanic. The horizontal axis errors for these instruments may reach up to 3 mgon.

Influence of the Horizontal Axis Error

The directional error (i) caused by the horizontal axis error i when sighting to a point P at a vertical angle β can be derived with the aid of a semi sphere (figure 336). Its base surface is a horizontal plane. K_1K_1' is the correct position (i.e. horizontal) of the horizontal axis, while K_2K_2' is its position

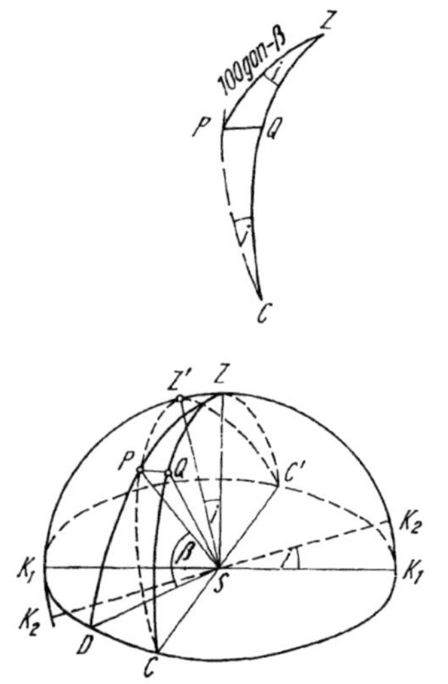

Figure 336 Influence of the horizontal axis error

due to the error i. Therefore the collimation axis moves in the plane $CQZ'C'$ instead of the plane $CQZC'$. Both planes enclose the angle i. When sighting to the point P, the projection falls incorrectly to C instead of to D. The angle $CSD = (i)$ between the vertical planes ZQC and ZPD represents the influence of the horizontal axis error onto the horizontal circle reading when sighting to P.

If one draws a great arc through P normal to ZQC, two rectangular spherical triangles ZQP and CQP are obtained with the right angles at Q. The other angles are: $PCQ = i$, $PZQ = (i)$, $CQ = \beta$ and $PZ = 100\,\text{gon} - \beta$. This leads to $\cos(100\,\text{gon} - PQ) = \sin(100\,\text{gon} - \beta)\sin(i) = \sin PC \sin i$. Since (i) and i are small angles, one obtains with $PC \approx \beta$:

$$(i)\cos\beta = i\sin\beta \text{ and } (i) = i\tan\beta \text{ or } (i) = i\cot\zeta \tag{86}$$

According to equation (86) the influence (i) equals zero for horizontal sights, i.e. for $\zeta = 100$ gon. Therefore, when determining the collimation axis error, a point at the instrument horizon is used, which means that there is no influence of a possible horizontal axis error. For $i = 1.0$ mgon the following influence (i) is obtained:

ζ [gon]	= 100	90	80	70	60	50	40	30	20	10	0
(i) [mgon]	= 0.00	0.16	0.32	0.51	0.73	1.00	1.38	1.96	3.08	6.31	∞

The influence $\Delta(i)$ due to i on an angle measured between two points in one telescope position is

$$\Delta(i) = i\cot\zeta_2 - i\cot\zeta_1 = i\,\frac{\sin(\zeta_2 - \zeta_1)}{\sin\zeta_1 \sin\zeta_2} \tag{87}$$

The smaller the difference between the zenith distances $\zeta_2 - \zeta_1$, the smaller is $\Delta(i)$. For equal zenith distances, the error has no influence on the result.

Determination of the Horizontal Axis Error

The magnitude of the horizontal axis error can be computed with equation (86) if its influence on an inclined sight had been determined with one of the methods mentioned under 1) and 2) (page 128). Based on the determination with ruler (figure 335) one obtains—provided that there is no collimation axis error — $\tan(i) = l/2s$, where s is the distance between instrument and ruler, and $l = P_1P_2$. Since (i) is a small angle, $(i) = l/2s$ sin (0.1 mgon). With the value for (i) from equation (86), one obtains after rearrangement

$$i = \frac{l\tan\zeta}{2s\sin 0.1\,\text{mgon}} \tag{88}$$

For the steep sights, necessary for such investigations, the collimation axis error also has to be considered according to equation (84). The horizontal axis error is obtained with the two circle readings a_2 and a_1 (where $a_1' = a_1 \pm 200$ gon) for a high point observed in both telescope positions

$$i = \left(\frac{a_2 - a_1'}{2} - \frac{c}{\sin\zeta}\right)\tan\zeta \tag{89}$$

It can also be determined with a striding level. In figure 337, represents the centre of the arc of the level, S the centre of run, N the centre of graduation, i the unknown horizontal axis error, φ an unknown error of the level axis, a the separation between centre of run and zero mark, l the centre of the bubble if the zero mark is left, and r the latter for zero mark on the right. With the sensitivity p of the bubble, one obtains for both level positions $pl = pa + i + \varphi$ and $pr = pa - (i - \varphi) = pa - i + \varphi$. Subtracting the second equation from the first one, leads to $p(l - r) = 2i$ and

$$i = p\,\frac{l - r}{2} \tag{90}$$

With the readings l', l'' and r', r'' at the left and right bubble ends, one obtains

$$i = p\,\frac{(l' + l'') - (r' + r'')}{4} \tag{90a}$$

In this manner the horizontal axis error is determined for precision measurements. Equation (86) yields its influence, so that it can be considered,

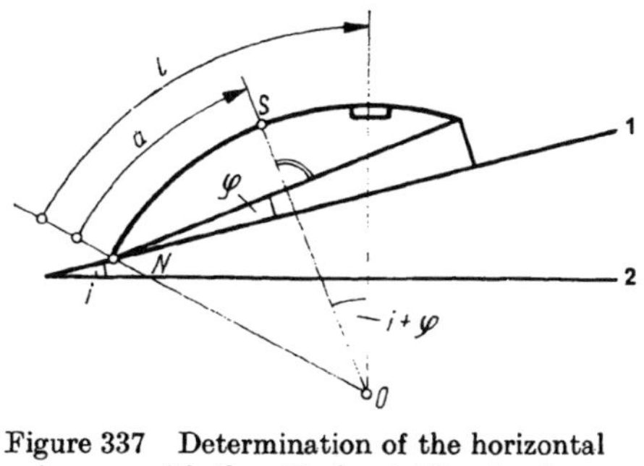

Figure 337 Determination of the horizontal axis error with the aid of a striding level
1 horizontal axis; *2* horizontal plane

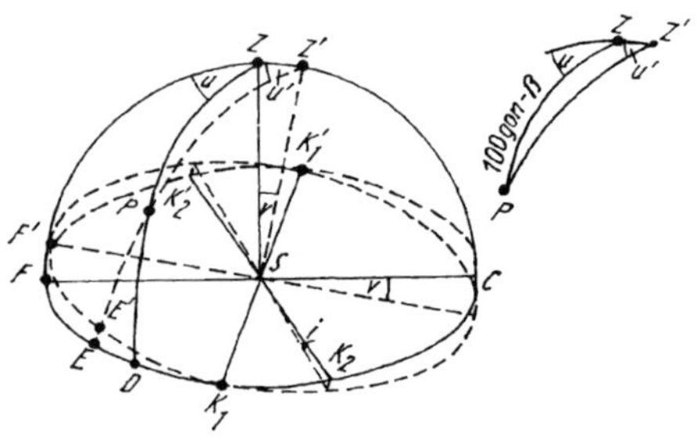

Figure 338 Influence of the vertical axis error

Elimination of the Influence of the Horizontal Axis Error with a Proper Measuring Procedure

Like the influence of the collimation axis error, the influence of the horizontal axis error can be eliminated by measuring in both telescope positions. The mean of the readings in both circle positions is free of the influence of horizontal axis errors.

Vertical Axis Error

The vertical axis error is not an instrumental error but indicates incorrect setting up. It occurs, when the vertical axis is inclineded, but can be avoided by setting the axis correctly vertical.

Influence of the Vertical Axis Error

With the vertical axis error v being the angle between the vertical axis SZ' and the plumbline SZ, the plane of the horizontal circle is also inclined by v against a horizontal plane (figure 338). The great arc passing through Z' and Z; is in a vertical plane. When sighting to a point located on this arc, the horizontal axis K_1K_1' is horizontal, and the circle reading is free of the influence of the vertical axis error. When sighting to a point P located outside this arc, the horizontal axis K_2K_2' is inclined against the horizon by the angle i. On the sphere, the collimation axis follows the great arc $Z'PE$, instead of ZPD, and projects the point P to E instead of to D. The difference between the arcs FD and $F'E'$, which equals the difference of the angles u and u', represents the influence (v) of the vertical axis error v: $(v) = u - u'$. Its amount is obtained in the spherical triangle ZPZ'. According to the cotangent theorem of spherical trigonometry:

$$\tan\beta \sin v = -\cos u \cos v + \sin u \cot u' = \sin u(\cot u' - \cos v \cot u)$$

and, since v is small

$$v \tan\beta = \sin u(\cot u' - \cot u)$$

with $\cot u' - \cot u = \sin(u - u')/\sin u \sin u' \approx (u - u')/\sin u'$
one finally obtains

$$(v) = (u - u') = v \tan\beta \sin u' \text{ or } (v) = v \cot\zeta \sin u' \qquad (91)$$

As evident from equation (91), the influence of the vertical axis error depends on the zenith distance and the angle between the vertical plane through vertical axis and plumb line, as well as the plane through target and vertical axis. It is zero for horizontal sights in any direction. The influence $\Delta(v)$ on an angle measured between two points in one telescope position is

$$\Delta(v) = v(\sin u_2' \cot\zeta_2 - \sin u_1' \cot\zeta_1) \qquad (92)$$

The maximal value of $\Delta(v) = 2v \sin u' \cot\zeta$ is at $u_1' = u_2' + 200$ gon and $\zeta_1 = \zeta_2$ while the minimum, namely zero is at $u_1' = u_2' + 200$ gon and $\zeta_2 = 200\text{ gon} - \zeta_1$. This has to be considered for angular measurements in straight traverses.

Elimination of the Influence of the Vertical Axis Error

Unlike the collimation axis- and horizontal axis errors, the vertical axis error cannot be eliminated by observing in two telescope positions, because it remains constant if the instrument set-up is unchanged. Therefore, when measuring horizontal angles, especially to points outside the instrument horizon, the vertical axis has to be carefully plumbed. The influence of the vertical axis error can be eliminated by adding the following correction to the circle reading

$$(v) = v \sin u' \cot\zeta = i \cot\zeta \qquad (91\text{a})$$

In this formula, $i = v \sin u'$ is the inclination of the horizontal axis, caused by the vertical axis error, and determined with a striding level. When determining i with a striding level, one has to watch for eccentricity of the guide rings, as otherwise $m_i \geqq \pm 1.0$ mgon. Some theodolites have special elements for compensating vertical axis errors.

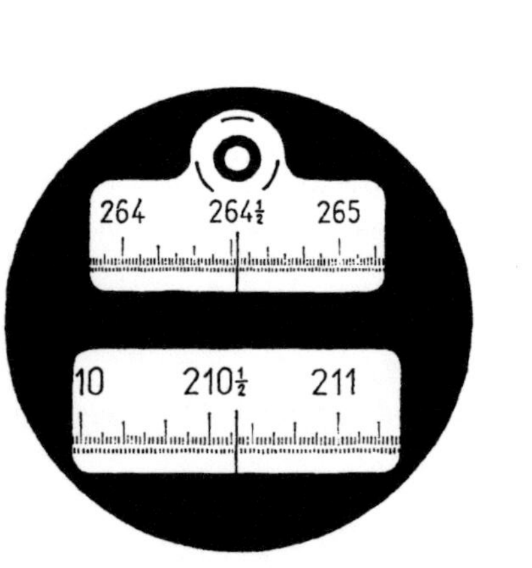

Figure 339 "Optical level" in the field of view of the reading microscope of the theodolite 4149 A (Filotecnica Salmoiraghi)

At the theodolite 4149 by Filotecnica Salmoiraghi, Milano, an "optical level" (figure 339) is linked to the automatic height index. It not only makes it possible to eliminate the component of the vertical axis error in

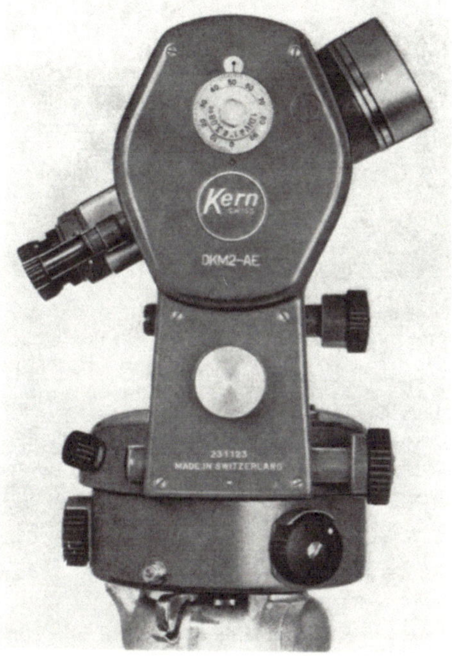

Figure 340 Horizontal axis micrometer of the DKM 2-A (Kern). Frontal view

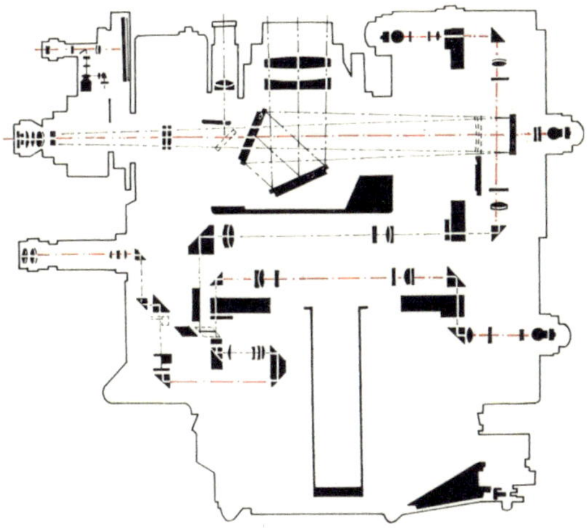

Figure 341 Optical train of the Theo 002 (Zeiss, Jena)

direction to the target, but also to correct the component normal to that direction semi-automatically with a foot screw. In this way the vertical axis can be set correctly to within 5″.

The DKM 2A from Kern, Aarau is built with a horizontal axis micrometer to measure changes of the component of the vertical axis error in direction of the horizontal axis to an accuracy of 1″. The shift of the image of the auxiliary graduation of the vertical circle in respect to the main graduation serves as a measure for this component. This shift is in direction towards, or from the centre of the graduated circle, if there is a component of the vertical axis error in direction of the horizontal axis. It is measured with the aid of a plane parallel plate with central setting. The double line lies in the aperture of the reading microscope, while the single line is imaged there with a special optical system via the fluid compensator of the vertical circle. This image can be centered between the double line, and measured with its own parallel plate (figure 340). A rippled knob with a 90″ scale at the outside of the casing of the vertical circle is used for settings. The practical measuring range is 30″.

In the optical train of the telescope of the Theo 002 (figure 330) a pendulum-tilt compensator is placed, which automatically eliminates the secondary horizontal axis error as well as the influence of the vertical axis error for horizontal angle measurements (figure 341).

Combined Effect of the Collimation Axis-, Horizontal Axis- and Vertical Axis Error

The influences of the collimation axis-, the horizontal axis-, and the vertical axis error are cumulative:

$$(s) = (c) + (i) + (v) \tag{93}$$

The influence $\Delta(s)$ to an angle measured between two points in one telescope position amounts to

$$\Delta(s) = \frac{\sin\zeta_2 - \sin\zeta_1}{\sin\zeta_1 \sin\zeta_2} + i\,\frac{\sin(\zeta_2 - \zeta_1)}{\sin\zeta_1 \sin\zeta_2} + v(\sin u_2' \cot\zeta_2 - \sin u_1' \cot\zeta_1) \tag{94}$$

Equation (94) illustrates, that for horizontal sights the errors of the axes have no influence on the angular measurements. By observing a target in two telescope positions, the collimation axis- and horizontal axis errors are eliminated while the vertical axis error remains. Therefore the correct position of the vertical axis has to be checked frequently.

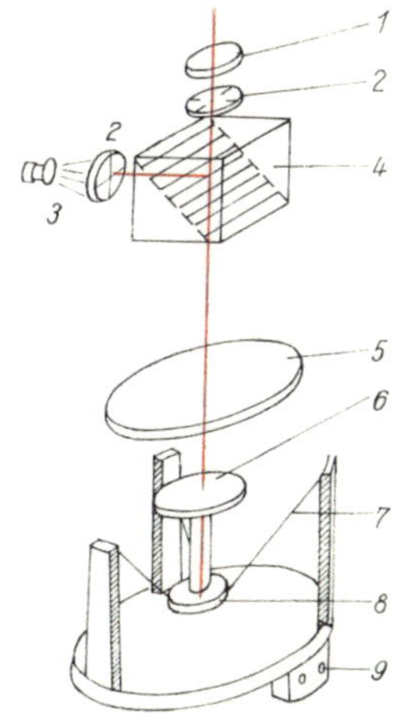

Figure 342 Testing for Wobble error
1 eyepiece; *2* reticule; *3* light source; *4* beam splitter cube; *5* objektive lens; *6* collimator mirror; *7* supporting thread; *8* base of the pendulum; *9* connecting device

Determination of the Wobble Error

The wobble error of the vertical axis can be determined with the aid of precise cross levels (Horrebow levels normal to each other), or by autocollimation, using a mirror placed onto the top part of the instrument. Better results can be obtained, if the collimation mirror is fastened to a pendulum (figure 342). Significant periods of vertical axis wobble lead to conclusions regarding its causes (figure 343). For instance, differences in the

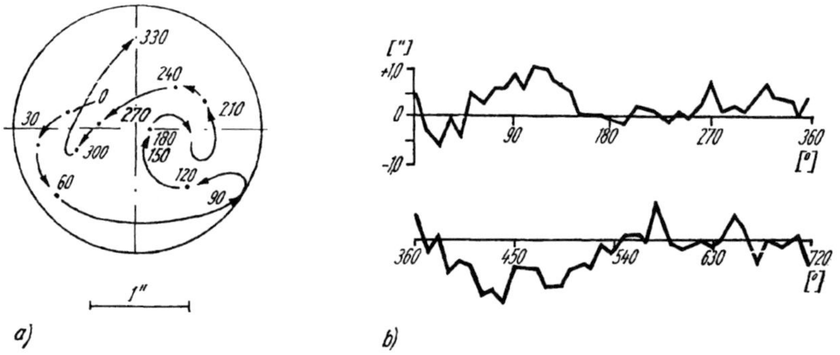

Figure 343 Presentation of the wobble error of the vertical axis
a) of a THEO 020 A (Zeiss, Jena)
b) of a Wild T 3

sphere diameters of semi-kinematic axes cause periods of more than 360°. Periods of 120° are due to poor spherical tracks, while short periodic modulations of are caused by wear of the spheres.
For steep sights with striding level, one has to remember that the influence of the wobble error is not completely eliminated when measuring in two telescope positions.

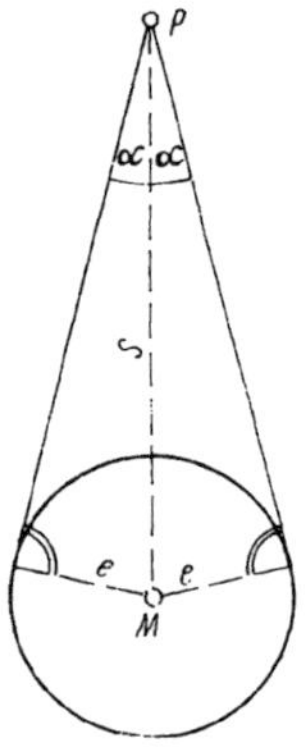

Figure 344 Eccentricity of the collimation axis

Eccentricity Errors

Horizontal angles with the vertex in the rotational centre of the telescope are obtained only, if the vertical axis intersects the collimation axis (figure 147) and if the centre of the graduated circle coincides with the axis of the top part. Deviations from this arrangement lead to eccentricity of the collimation axis (telescope) resp. of the centre of graduation (graduation eccentricity).

Eccentricity of the Collimation Axis

If the collimation axis has an eccentricity (figure 344), it is tangent, when rotated, to a circle with radius e around the graduation centre M. When sighting to a point P in both telescope positions, directional deviations of equal amount and opposite sign occur. The mean of the circle readings in both positions is free of the influence of the eccentricity error.
The eccentricity of the collimation axis causes an error in the measured direction of

$$\alpha_i = \frac{e}{s}\varrho \qquad (95)$$

The larger the sighting distances, the smaller is the influence of the eccentricity. For $s = \infty$, α_∞ becomes 0.
The eccentricity of the collimation axis can be obtained as a by-product when investigating the movement of the focussing lens. Equation (129) is rearranged to

$$c_i = c_\infty + \delta_i + (e/s)\varrho$$

Further rearrangement leads to

$$\frac{c_i}{\varrho} s = \left(\frac{c_\infty}{\varrho} + \frac{\delta_i}{\varrho}\right) s + e \qquad (96)$$

which is a straight line equation. If the error influence δ_i of the movement of the focussing lens is plotted in a rectangular coordinate system, and a best fitting straight line is found by adjustment, then the section e on the coordinate axis corresponds to the eccentricity of the collimation axis. This value e can amount to several tenths of a millimetre.

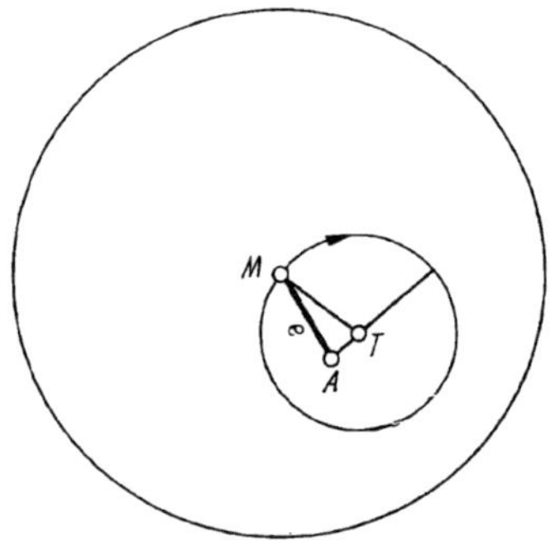

Figure 345 Generation of circle graduation eccentricity

Graduation Eccentricity

Because of inevitable errors during precision mechanical manufacturing and mounting of the theodolite, the vertical axis system and the graduated circle when placing the circle socket cannot be truly centered. The point of rotation T of the graduated circle and the solely geometrically defined centre of graduation M do not coincide, and also are different from the rotational centre of the top part (figure 345). Due to the eccentricity e between graduation centre M and rotational centre A of the top part, the circle readings do not correspond to the actual rotation of the top part,

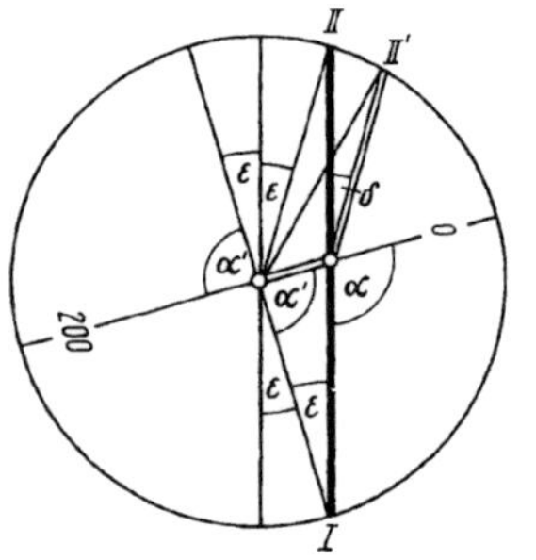

Figure 346 Graduation eccentricity and index bending

but are either too large or too small. The influence of the graduation eccentricity depends on the position of the circle and the direction of the top part.

For instance, if the zero line of the graduated circle coincides with the diameter of the graduation eccentricity M (figure **346**), one reads at reading device I the angle $\alpha' = \alpha - \varepsilon$ instead of the angle between graduation zero line and the sighted direction.

Since the diametrically opposite reading device II provides the angle

$$200 \text{ gon} + \alpha' + 2\varepsilon = \alpha + \varepsilon$$

the mean of both readings is the desired value of α.

If both reading devices are not on one diameter, then the bending angle δ to the second reading device II' is generated. While ε changes as a function of α, δ remains constant and affects all directions in the same manner and thus is eliminated by measuring in two telescope positions (as difference of two directions).

The influence of the graduation eccentricity is therefore eliminated by taking the mean of the readings at two diametrically (or near diametrically) opposite circle readings.

For theodolites with only one circle reading it is only eliminated by taking the mean of the readings in both telescope positions. The influence on the reading of even a small eccentricity can be considerable. For $e = 5\,\mu\text{m}$ and a circle diameter of 90 mm, the maximum influence of the graduation eccentricity (for $\alpha = 100$ gon) is

$$\varepsilon_{\max} = \frac{e\varrho}{r} = \frac{0.005 \cdot 63.7 \text{ mgon} \cdot 10^3 \text{ mm}}{45 \text{ mm}} = 7.1 \text{ mgon}.$$

The influence of the remaining eccentricity can amount to 6 to 12 mgon for theodolites with only one reading position. For these theodolites, the graduated circle should not be moved between the direct and reverse readings, because then the influence of the graduation eccentricity is not eliminated.

The manufacturers attempt to center the axis within 1 μm. Investigations of various theodolites resulted in average values for $\varepsilon_{\max}$ of about 2.0 mgon.

The influence of the eccentricity in various positions of the top part is

$$\varepsilon = \varepsilon_{\max} \cdot \sin \alpha \tag{97}$$

Determination of the Graduation Eccentricity

Although the influence of the graduation eccentricity can easily be eliminated, its magnitude is of interest. For its determination, one sets the reading index I onto any position of the graduated circle and reads the value d' with the index II (figure 347a). Similarly, with II at the initial value, at I, one obtains d'' (figure 347b).

With $\delta \approx \delta'$ one obtains:

$$\begin{aligned} d' &= 200 \text{ gon} + 2\varepsilon + \delta \\ d'' &= 200 \text{ gon} - (2\varepsilon - \delta) \\ \hline d' + d'' &= 400 \text{ gon} + 2\delta \\ d' - d'' &= 4\varepsilon \end{aligned}$$

Thus, depending on the position of the top part, δ and ε_i become

$$\delta = \frac{d' + d''}{2} - 200 \text{ gon}; \; \varepsilon_i = \frac{d' - d''}{4} \tag{98}$$

The diameter which contains M and A is at $\varepsilon_i = 0$ gon. To determine this position, the difference between diametrical circle readings is calculated for various, symmetrically placed positions of the top part. When plotting the d-values as a function of the circle positions, one obtains a sine-curve, whose axis of symmetry is shifted by δ from the abscissa. It intersects the axis of symmetry at the desired point. The linear eccentricity e can be obtained from the graphical plot as:

$$e = \frac{\varepsilon_{\max}\, r}{\varrho} \tag{99}$$

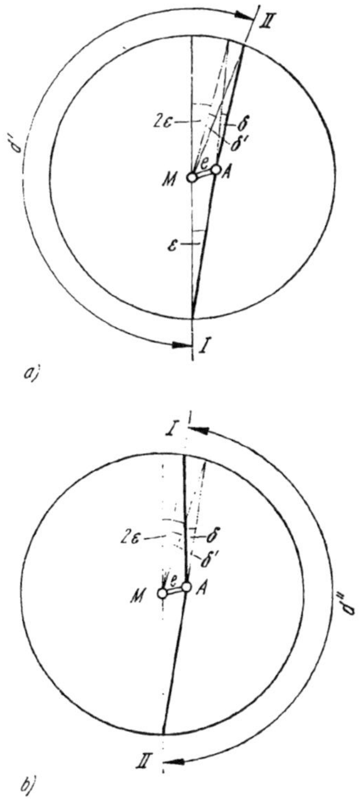

Figure 347 Determination of the graduation eccentricity and of index bending

Since the graduation eccentricity is variable, depending on the circle position, the knowledge of the eccentricity error for the most unfavourable orientation of the circle is necessary in order to obtain a complete insight into the eccentricity situation. For this, the separation between the centre of the graduation M, and the rotational centre I of the graduated circle

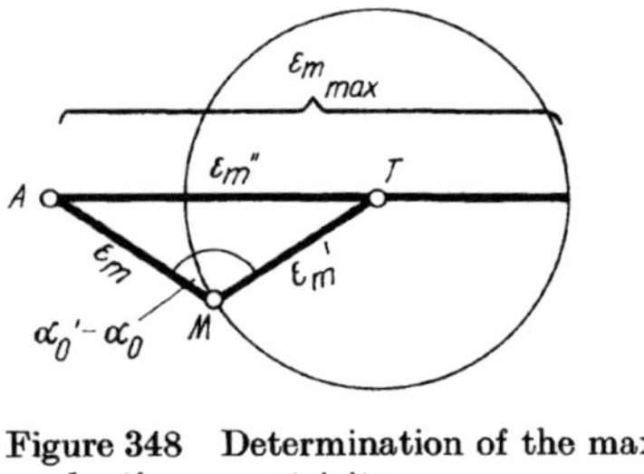

Figure 348 Determination of the maximum graduation eccentricity

is needed in the circle position used for the determination of the eccentricity. The elements are determined as before, however, the top part remains fixed, and the graduated circle is changed symmetrically between readings. At the beginning of the investigation, the zero direction of the circle has to coincide with the one at the beginning of the determination of graduation eccentricity.

With the elements ε_{max} and α_0 of the graduation eccentricity, and ε'_{max} and α'_0 of the circle eccentricity the triangle AMT (figure 348) can be drawn. The most unfavourable circle orientation, as far as effective eccentricity is concerned, is with *A* and *M* on both sides of *T*. The maximum possible graduation eccentricity amounts to

$$\varepsilon_{m_{max}} = \varepsilon'_m + \varepsilon''_m \tag{100}$$

Since mechanical adjustment of two rotational centres (*T* and *A*) is easier than the adjustment of one rotational centre to a geometrically defined graduation point, the separation between *T* and *A* is generally smaller than the ones between *M* and *A*, as well as *M* and *T*.

While the determination of eccentricities for instruments with two readings is usually only of theoretical interest, it is occasionally of practical value for theodolites with only one circle reading. For these instruments, the second reading is replaced for both telescope positions (when determining the circle eccentricity) by observations to a target, or (when determining the graduation eccentricity in the respective circle position) to targets symmetrically distributed over the horizon.

One has to consider the collimation axis error *c*. With the readings a_{I} and a_{II} in both telescope positions, one gets

$$\varepsilon + c = \frac{a_{\mathrm{II}} - a_{\mathrm{I}} \pm 200\ \text{gon}}{2}, \tag{101}$$

and determines *c* either as half the mean of the differences between corresponding readings in two positions, or else from the arithmetic mean

$$c = \frac{(\varepsilon_{max} + c) + (\varepsilon_{min} + c)}{2}. \tag{102}$$

Then ε is determined as difference between the values $(\varepsilon + c)$ on different circle positions and the value *c* as

$$\varepsilon_i = (\varepsilon_i + c) - c. \tag{103}$$

With these values the sine curve is plotted.

Since the setting of targets requires a considerable amount of effort, e.g. every 25 gon around the station, it is easier to proceed as follows:

One sights to a point in direct mode, rotates the top part by e.g. 25 gon and clamps it. Then the forced centering clamp at the tribrack is loosened, and one sights again to the point. Then the tribrack clamp is tightened and the point is precisely sighted. Afterwards the horizontal clamp is opened, and the point is observed in reverse mode. This is repeated for every 25 gon (up to 375 gon). The differences between the circle readings in the two

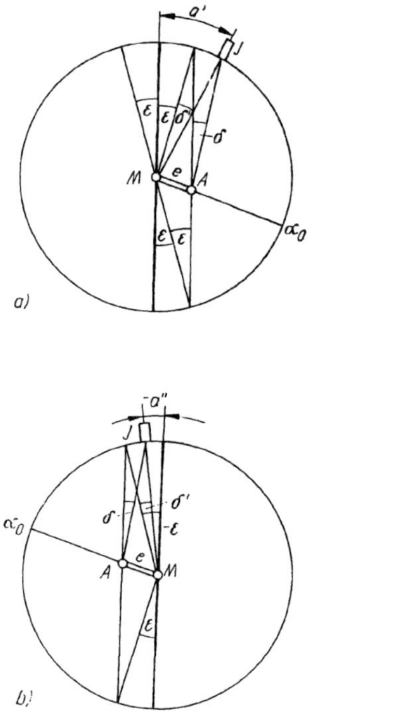

Figure 349 Graduation eccentricity for theodolites with coincidence micrometers

Table 12.

circle position	reading a_1 when lines coincide		reading b_1 for coincidence; line-index		$d_1 = b_1 - a_1$	circle position	reading a_2 when lines coincide		reading b_2 for coincidence; line-index		$d_2 = b_2 - a_2$	$\delta = \frac{d_1 + d_2}{2}$
[°]	[D]	[″]	[D]	[″]	[″]	[°]	[D]	[″]	[D]	[″]	[″]	[″]
1	2	3	4	5	6	7	8	9	10	11	12	13
0	32.4		28.5			180	41.5		35.7			
	.3	64.7	.8	57.3	− 7.4		.4	82.9	.8	71.5	−9.4	−8.4
30	32.2		26.2			210	47.3		43.6			
	.3	64.5	.0	52.2	−12.3		.5	94.8	.8	87.4	−7.4	−9.8
60	48.3		40.9			240	53.8		53.5			
	.5	96.8	.9	81.8	−15.0		.9	107.7	.6	107.1	−0.6	−7.8
90	55.9		46.2			270	35.8		39.4			
	.7	111.6	.1	92.3	−19.3		36.0	71.8	.6	79.0	+7.2	−6.1
120	54.1		43.7			300	37.5		42.2			
	.3	108.4	.4	87.1	−21.3		.6	75.1	.3	84.5	+9.4	−6.0
150	22.6		14.5			330	51.1		54.5			
	.7	45.3	.4	28.9	−16.4		.3	102.4	.3	108.8	+6.4	−5.0
note: D = double seconds											mean	−7.2

telescope positions represent the differences between two diametrical values of a diameter.

For theodolites, where the image of the graduation lines of one reading position is transferred into the plane of the other one to form a double line image, the eccentricity has the effect, that the two lines of a double line, coming from diametrically opposite circle positions have different separations.

For theodolites with coincidence micrometers, the distances between the images and the index on the separation prism are different. The eccentricity is determined as follows. At positions, spread evenly over the circle, e.g. every 25 gon, the lines of the diametrical graduation positions are first coincided (figure 349a). Then the line of one graduation position is coincided with the index (or another fixed mark appearing in the field of view, such as a grain of dust, fault in the reticule etc.). The difference of the two readings results in $a' = \varepsilon + \delta$. In the diametrical positions of the top part (figure 349b), one similarly obtains $-a'' = \varepsilon - \delta$. Therefore

$$\delta = \frac{a' + a''}{2} \qquad (104)$$

$$\varepsilon = \frac{a' - a''}{2} \qquad (105)$$

Example: Determine the graduation eccentricity of the theodolite Wild T3, No. 7253 (table 12).

The separation between diameter and index is $\delta = 7.8''$. With the ε-values, the graph shown in figure 350 was plotted. The angular eccentricity amounts to $\varepsilon_{max} = 14.2''$, $\alpha_0 = 199°$.

The values obtained when determining graduation eccentricity were:

$\delta = 12.1''$, $\varepsilon'_{max} = 9.8''$, $\alpha_0 = 210°$.

Therefore, the maximum possible graduation eccentricity is 14.8'', which represents a linear amount of 5 μm according to equation (99).

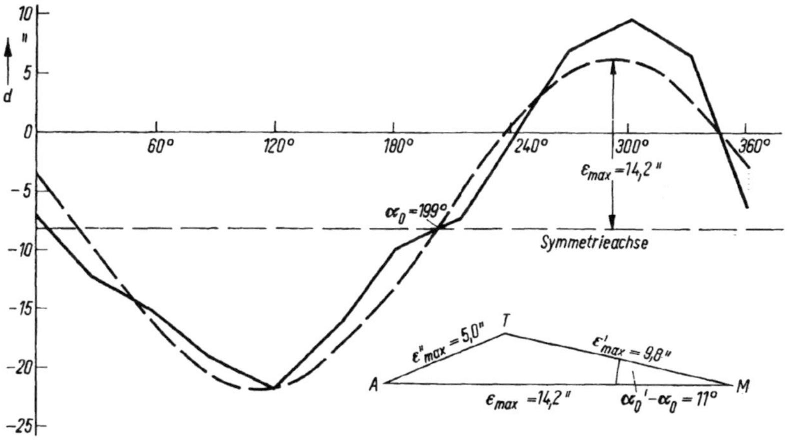

Figure 350 Presentation of graduation eccentricity (Shown are d_1 and d_2 from table 12)

Graduation Errors and their Determination

The accuracy of a circle graduation is apparent by the magnitude of its graduation errors. For modern instruments, graduation errors are the main source of differences between sets of directions, while targetting- and reading errors are smaller.

The causes for the total graduation error, composed of an irregular or random part, and a regular or periodic part, were presented on page 81. The regular errors are largely eliminated by changing the circle reading symmetrically between sets of horizontal directions, while the random errors of all lines cannot be determined exactly. Therefore, the graduation study shall give an overview of the quality of the graduation with a diagram, or with an average value of the graduation error. For precise measurements with the fixed vertical circle, the readings have to be corrected for systematic graduation errors.

The errors of the graduation lines for instance, can be determined by sequentially measuring all graduation intervals with an optical micrometer.

Because of the run, their sum has to be adjusted to 400 gon. The remaining differences to the nominal value represent errors of the graduation line, influenced however, by eccentricity-, and observational errors.
Since such an investigation is rather comprehensive and time consuming, and since precise instruments have readings on two diametrically opposite circle positions, one does not determine the errors of all graduation lines, but rather the diameter errors, which are the mean of the errors of two diametrical lines. This is done symmetrically over the full circle. Therefore, the circles have to be studied between 0 and 200 gon only. For double circle theodolites with simple graduations only, the investigation has to be extended to the full circle, because in reverse telescope position, different graduation lines are used than in the direct mode.
For practical surveying, studies of operational graduated circles within the theodolite are of importance, not when they are dismantled. Eccentricity measurements provide a rough idea on the graduation errors. More accurate and generally used methods utilize the direct observation of a test angle. This test angle is obtained by

a) sighting with the theodolite to two targets or collimators (sighting method)
b) placing a surface mirror or prism in front of the objective lens such, that it covers half of the field of view. The test angle is then fixed once directly, and once by refraction with mirror normal, and the collimation axis (reflection method, especially suited to check instruments with fixed vertical circle).
c) fixed stops in a special pot like device, into which the theodolite is placed (stop method; no sighting or targetting necessary).

While sighting- and reflection methods are about equivalent as far as accuracy, time requirement and stress on the observer are concerned—with the reflection method, the test angles are established faster, and, for suitable distances, only simple requirements are placed on the targets—the stop method is generally less accurate. For practical purposes, usually the sighting method is utilized, especially since the necessary accessories are easily obtained.
Wild's Method. If the test angle is approximately 40 gon, the measured angle is influenced by for instance, the random observational error λ_l and the graduation errors t_l and t_r. With the graduation errors t_0, t_{40}, t_{80} ... of the diameter 0 gon, 40 gon, 80 gon ... and the random observation errors λ_1, λ_2, λ_3 ... the true values α_s of the measured angles $\alpha_i \approx 40$ gon are obtained as

$$\begin{aligned}
&\text{between } 0 \text{ and } 40 \text{ gon}: &&= \alpha_1 + t_{40} - t_0 + \lambda_1\\
&\text{between } 40 \text{ and } 80 \text{ gon}: &&= \alpha_2 + t_{80} - t_{40} + \lambda_2\\
&\text{between } 80 \text{ and } 120 \text{ gon}: &&= \alpha_3 + t_{120} - t_{80} + \lambda_3\\
&\text{between } 120 \text{ and } 160 \text{ gon}: &&= \alpha_4 + t_{160} - t_{120} + \lambda_4\\
&\text{between } 160 \text{ and } 200 \text{ gon}: &&= \alpha_5 + t_0 - t_{160} + \lambda_5
\end{aligned} \tag{106}$$

The probable value of the angle is

$$\alpha_s = \frac{1}{5}\sum_{i=1}^{5} \alpha_i \tag{107}$$

The graduation errors fall out in the mean, because the same lines are used when changing the circle, and the measuring series returns to its initial point. A comparison between the angles and the mean results in corrections $(v)_i$ for each angle:

$$\begin{aligned}
\alpha_s - \alpha_1 &= (v)_1 = t_{40} - t_0\\
\alpha_s - \alpha_2 &= (v)_2 = t_{80} - t_{40}\\
&\cdots\cdots\cdots\cdots
\end{aligned} \tag{108}$$

When t_0 is referred to 0 gon, the reduced diameter errors are

$$\begin{aligned}
t_0 &= 0 \text{ gon}\\
t_1 &= (v)_1\\
t_2 &= (v)_1 + (v)_2\\
t_3 &= (v)_1 + (v)_2 + (v)_3\\
&\cdots\cdots\cdots\cdots
\end{aligned} \tag{109}$$

The x-axis of the graphical representation of these values passes through the origin and has a different position for different initial points. In order to be able to refer several series of measurements to one axis, the values t

Table 13. Determination of circle subdivision errors according to Wild

circle position	reading								$\alpha_i = \frac{\alpha_i^I + \alpha_i^{II}}{2}$	$(v)_i =$ $\alpha_s - \alpha_i$	t_i	(t_i)	v_i	$\Delta =$ $\alpha_i^I - \alpha_i^{II}$
	direct telescope position				reverse telescope position									
	a_1 [D]	a_2 [D]	$a_m = a_1 + a_2$ [″]	a_i^I [″]	a_1 [D]	a_2 [D]	$a_m = a_1 + a_2$ [″]	α_i^{II} [″]	36° + [″]	[″]	[″]	[″]	[″]	[″]
1	2	3	4	5	6	7	8	9	10	11	12	13	14	15
1st series of measurements														
0° 00′	11.3	11.1	22.4		11.8	11.9	23.7							
36° 00′	17.5	17.6	35.1	12.7	19.6	19.6	39.2	13.5	13.10	+0.64	±0.00	−0.09	+0.87	−0.8
36° 00′	7.4	7.8	15.2		13.2	13.3	26.5							
72° 00′	14.3	14.6	28.9	13.7	20.4	20.4	40.8	14.3	14.00	−0.26	+0.64	+0.55	−0.03	−0.6
72° 00′	7.4	7.4	14.8		11.2	11.1	22.3							
108° 00′	14.4	14.1	28.5	13.7	19.0	19.0	38.0	15.7	14.70	−0.96	+0.38	+0.29	−0.73	−2.0
108° 00′	8.2	7.9	16.1		13.4	13.9	27.3							
144° 00′	14.3	14.4	28.7	12.6	20.5	20.5	41.0	13.7	13.15	+0.59	−0.58	−0.67	+0.82	−1.1
144° 00′	6.8	6.9	13.7		12.8	12.4	25.2							
180° 00′	13.5	13.3	26.8	13.1	19.6	20.0	39.6	14.4	13.75	−0.01	+0.01	−0.08	+0.22	−1.3
									Σ 68.70 mean α_s = 13.74	±0.00	+0.45 t_s = +0.09	±0.00	[vv] = 2.01	[ΔΔ] = 7.90

M amounts to 36° 00′ 13.97″

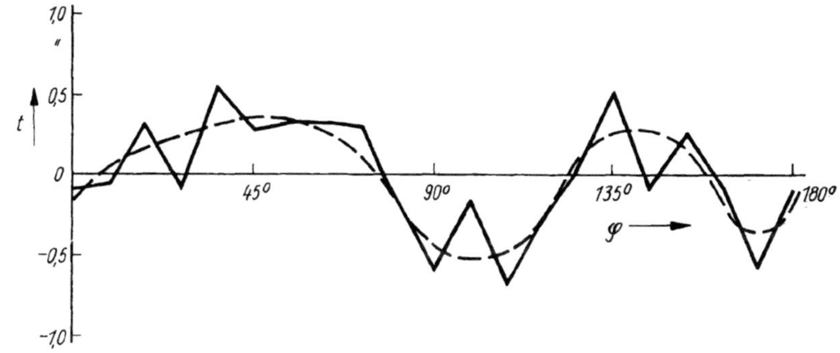

Figure 351 Presentation of diameter errors of the horizontal circle, determined with the *Wild* method

have to be determined:

$$t_s = \frac{1}{5} \sum_{i=1}^{5} t_i \tag{110}$$

and the differences to the mean have to be found

$$(t)_i = t_i - t_s. \tag{111}$$

Thus, with additional series, starting at other initial points (e.g. 10 gon, 20 gon, 30 gon), diameter errors can be determined for 10 to 10 gon and represented graphically as a curve.

Example: Determination of the graduation errors of the horizontal circle of the theodolite Wild T3, No. 33258 with photographic recording.

The theodolite was set up on a pillar such that an angle of 36° ± 1′ was obtained with two collimators placed at the horizon. The angles were measured in direct mode, subsequently at the circle position 0°, 36°, 72°, ... 144°, and in reverse mode backwards, starting at the right target with 180°, 144°, 108° ... Similarly, a second series with 9°, 45°, 81° ... a third series with 18°, 54°, 90°, ... and a fourth series with 27°, 63°, 99°, ... was observed. For each sighting, the reading consisted of two coincidences (double seconds). The evaluation was carried out according to equations (107) to (111). Table 13 contains the first series.

The best fitting curve of the diameter errors shows the characteristics of the regular errors (figure 351). The distances to the polygon connecting actual values can be considered as being random errors. The mean diameter error τ_1 is obtained from the diameter errors with

$$\tau_1 = \pm \sqrt{\frac{[(t)_i (t)_i]}{n}} \tag{112}$$

as being $\pm 0.35''$. It can also be determined from the standard deviation m_1 (which includes observation- and graduation errors) of an angle measured at different circle parts in both telescope positions. Then the standard deviation m_2 (which includes observation errors only, since the influence of the graduation errors remains constant) of an angle measured at the same circle position has to be removed. This angle is calculated from differences of the readings in both telescope positions. Therefore,

$$\tau_2 = \pm \sqrt{\frac{m_1^2 - m_2^2}{2}} \tag{113}$$

where

$$m_1 = \pm \sqrt{\frac{[v_i v_i]}{n-1}} \quad \text{and} \quad m_2 = \pm \sqrt{\frac{[\Delta_i \Delta_i]}{4n}}$$

with $v_i = \alpha_i - \alpha_M$ and $\alpha_M = \frac{[\alpha_s]}{n}$ where s is the number of series of measurements, and $\Delta_i = \alpha_i^{\mathrm{I}} - \alpha_i^{\mathrm{II}}$.

For the case at hand

$$m_1 = \pm \sqrt{\frac{6.74}{19}} = \pm 0.60''$$

and

$$m_2 = \pm 0.12'' \quad \text{lead to} \quad \tau_2 = \pm \sqrt{\frac{0.60^2 - 0.12^2}{2}} = \pm 0.42''.$$

With Wild's method one obtains a good idea on the quality of a graduated circle with just about 2 hours' measuring and 2 hours' evaluation time. The accuracy can be increased, for instance, by measuring two sets at each circle position.

Heuvelink's Method. Often a more detailed analysis is desired for graduated circles of triangulation theodolites. For the Heuvelink method, which requires about one day each for measurements and evaluation, a test angle, given by two targets—collimators or physical stops—is measured. This test angle may be $\beta = 50$ gon and is measured at $n = 10$ circle positions (every 20 gon) or $\beta = 45°$, $n = 9$ each in $s = 4$ series and two sets (one set: left—right, right—left). Differences between the measured angle and its mean value provide information on the graduation errors. For instruments with two reading positions, diameter errors are obtained, while line errors are obtained by instruments with only one reading position. The method is arranged fore and back such, that observation- and graduation errors are obtained separately, since the same angle is measured several times with unchanged circle position. The means p_1 and p_2 at one circle position are combined to the total mean p for each series, and these means p of each series are combined to give the most probable value α_s of the angle. The total mean square error M of the p-values, which contains observation- and graduation errors, is obtained from the probable observation corrections $v_i = \alpha_s - p$ as

$$M^2 = \sqrt{\frac{[v_i v_i]}{n-s}} \tag{114}$$

For the total mean t, one obtains the mean square graduation error

$$t = \pm \sqrt{M^2 - m^2} \tag{115}$$

by separating the observation error

$$m = \sqrt{\frac{\mu^2}{2}} \tag{116}$$

The mean square error of the observation of one direction

$$\mu^2 = \frac{1}{4}\left(\frac{[dd]}{2n} - \frac{[d]^2}{4n^2}\right) \tag{117}$$

is computed from the differences d of the measured angles in one face in back and fore measurement (differences of half sets). The total mean square graduation error τ of a direction (a diameter) can be directly determined from equation (115). It consists of a random and a systematic component.

The systematic graduation error $\Delta\varphi$ at the circle position φ can be presented by a Fourier Series:

$$\Delta\varphi = a\sin(2\varphi + A) + b\sin(4\varphi + B) + c\sin(6\varphi + C) + \dots \tag{118}$$

By reading at two diametrically opposite circle positions, uneven multiples disappear. The unknowns of this series are computed as follows:

$$\tan(i\beta + K) = \frac{-[(p-\alpha)\sin 2i\varphi]}{[(p-\alpha)\cos 2i\varphi]}\,; i = 1, 2, 3; \quad K = A, B, C \tag{119}$$

$$K_i = \frac{-[(p-\alpha)\sin 2i\varphi]}{n\sin i\beta\sin(i\beta + K)} = \frac{[(p-\alpha)\cos 2i\varphi]}{n\sin i\,\beta\cos(i\beta + K)}\,; \tag{120}$$

$i = 1, 2, 3$; $K = A, B, C$; $K_i = a, b, c$; β are the mean values of the s values of α.

By substituting these values into equation (118), the systematic graduation errors can be computed for each circle position and presented graphically.

Since the influence of the graduation error remains constant for the values p, determined by repeated measurements of the same angle at the same position, the following connection exists between the mean square error of the graduation τ of an angle p, and the total mean square graduation error τ of a direction:

$$2\tau^2 = t^2 = M^2 - m^2 = M^2 - \frac{\mu^2}{2}$$

Therefore,

$$\tau^2 = \frac{M^2}{2} - \frac{\mu^2}{4} \tag{121}$$

Step by step the periodic part of the τ-values can be separated according to the number of terms of the Fourier Series, such that only random parts τ', τ'', τ''' remain:

$$\tau' = \sqrt{\frac{M'^2}{2} - \frac{\mu^2}{4}}, \quad \text{where } M' = \sqrt{\frac{[v'v']}{n - s - 2}}$$

$$\tau'' = \sqrt{\frac{M''^2}{2} - \frac{\mu^2}{4}}, \quad \text{where } M'' = \sqrt{\frac{[v''v'']}{n - s - 4}} \tag{122}$$

where

$$[v'v'] = [(p-\alpha)^2] - \frac{2}{n}[(p-\alpha)\sin 2\varphi]^2 - \frac{2}{n}[(p-\alpha)\cos 2\varphi]^2$$

$$[v''v''] = [v'v'] - \frac{2}{n}[(p-\alpha)\sin 4\varphi]^2 - \frac{2}{n}[(p-\alpha)\cos 4\varphi]^2.$$

Table 14. Determination of the horizontal circle of the Wild T 3 theodolite No. 35 258 with photographic recording, according to the Heuvelink method (1st series)

series	circle position	set number	direct			reverse					set number	direct			reverse						
			l_1' [″]	r_1' [″]	$p_1' = r_1' - l_1'$	l_1'' [″]	r_1'' [″]	$p_1'' = r_1'' - l_1''$	$p_1 = {}^1/_2(p_1' + p_1'')$	$d_1 = p_1' - p_1''$		l_2' [″]	r_2' [″]	$p_2' = r_2' - l_2'$	l_2'' [″]	r_2'' [″]	$p_2'' = r_2'' - l_2''$	$p_2 = {}^1/_2(p_2' + p_2'')$	$d_2 = p_2' - p_2''$	$p = {}^1/_2(p_1 + p_2)$	$(p - a)$
1	2	3	4	5	6	7	8	9	10	11	12	13	14	15	16	17	18	19	20	21	22
1	0°	1	14.6	30.3	15.7	14.9	30.5	15.6	15.65	+0.1	18	10.1	25.2	15.1	10.0	26.2	16.2	15.65	−0.9	15.65	−0.07
	20°	2	9.9	26.0	16.1	9.3	27.0	17.7	16.90	−1.6	17	19.1	32.9	13.8	16.9	33.7	16.8	15.30	−3.0	16.10	+0.38
	40°	3	12.9	30.9	18.0	13.5	30.4	16.9	17.45	+1.1	16	21.1	37.9	16.8	22.1	38.1	16.0	16.40	+0.8	16.92	+1.20
	60	4	14.6	31.2	16.6	14.1	30.0	15.9	16.25	+0.7	15	25.5	41.3	15.8	25.4	43.2	17.8	16.80	−2.0	16.58	+0.86
	80°	5	21.7	37.5	15.8	23.1	37.5	14.4	15.10	+1.4	14	20.8	34.2	13.4	21.1	35.6	14.5	13.95	−1.1	14.52	−1.20
	100°	6	19.2	34.6	15.4	17.3	34.1	16.8	16.10	−1.4	13	23.3	39.4	16.1	24.8	40.6	15.8	15.95	+0.3	16.02	+0.30
	120°	7	21.3	38.0	16.7	22.2	38.9	16.7	16.70	0.0	12	25.3	41.5	16.2	24.6	42.3	17.7	16.95	−1.5	16.82	+1.10
	140°	8	24.5	39.0	14.5	24.4	39.2	14.8	14.65	−0.3	11	22.9	38.4	15.5	23.8	38.3	14.5	15.00	+1.0	14.82	+0.90
	160°	9	17.8	32.0	14.2	19.0	33.3	14.3	14.25	−0.1	10	24.0	38.1	14.1	25.3	39.0	13.7	13.90	+0.4	14.08	+1.64
		$\sum$	156.5	299.5	143.0	157.8	300.9	143.1	143.05	−0.1		192.1	228.9	136.8	194.0	337.0	143.0	139.90	6.0	141.51	0.00

$a_1 = 45°\,00'\,15.72''$

Table 15. Computation of A, B, C and a, b, c using equations (120) and (121)

series	$[(p-\alpha)\sin 2\varphi]$	$[(p-\alpha)\cos 2\varphi]$	$[(p-\alpha)\sin 4\varphi]$	$[(p-\alpha)\cos 4\varphi]$	$[(p-\alpha)\sin 6\varphi]$	$[(p-\alpha)\cos 6\varphi]$
1	+2.647″	+1.117″	+3.881″	−2.339″	−1.367″	+2.820″
2	+0.821″	+0.940″	−0.151″	−1.812″	+0.864″	−1.010″
3	+1.256″	−3.227″	−0.577″	−2.004″	−0.110″	+1.865″
4	+1.199″	−0.728″	+2.158″	−3.703″	−0.290″	+0.716″
$\sum$	+5.923″	−1.898″	+5.311″	−9.858″	−0.903″	+4.391″

$\tan(\beta + A) = +3.121;\quad \beta + A = 72.23°;\quad A = 27°;\quad a = +0.24$

$\tan(2\beta + B) = +0.5385;\quad 2\beta + B = 28.30°;\quad B = 298°;\quad b = +0.31$

$\tan(3\beta + C) = +0.206;\quad 3\beta + C = 191.63°;\quad C = 57°;\quad c = +0.18$

$\Delta\varphi = +0.24''\sin(2\varphi + 27°) + 0.31''\sin(4\varphi + 298°) + 0.18''\sin(6\varphi + 57°)$

Table 16. Determination of periodic circle graduation errors, according to equation (119):

$\varphi°$	$2\varphi + 27°$	$\sin(2\varphi + 27°)$	$4\varphi + 298°$	$\sin(4\varphi + 298°)$	$6\varphi + 57°$	$\sin(6\varphi + 57°)$	1. Gl.	2. Gl.	3. Gl.	$\Delta\varphi''$
0	27	+0.454	298	−0.883	57	+0.839	+0.109	−0.274	+0.151	−0.01
5	37	+0.602	318	−0.669	87	+0.999	+0.144	−0.207	+0.180	−0.12
10	47	+0.731	338	−0.375	117	+0.891	+0.176	−0.116	+0.160	+0.22
15	57	+0.839	358	−0.035	147	+0.545	+0.201	−0.011	+0.098	+0.29
20	67	+0.920	18	+0.309	177	+0.052	+0.221	+0.096	+0.009	+0.33
25	77	+0.974	38	+0.616	207	−0.454	+0.234	+0.191	−0.082	+0.34
30	87	+0.999	58	+0.848	237	−0.839	+0.240	+0.263	−0.151	+0.35
35	97	+0.992	78	+0.978	267	−0.999	+0.238	+0.303	−0.180	+0.36
40	107	+0.956	98	+0.990	297	−0.891	+0.229	+0.307	−0.160	+0.38
45	117	+0.891	118	+0.883	327	−0.545	+0.214	+0.274	−0.098	+0.39
50	127	+0.799	138	+0.669	357	−0.052	+0.192	+0.207	−0.009	+0.39
55	137	+0.682	158	+0.375	27	+0.454	+0.164	+0.116	+0.082	+0.36
60	147	+0.545	178	+0.035			+0.131	+0.011	+0.151	+0.29
65	157	+0.391	198	−0.309			+0.094	−0.096	+0.180	+0.18
70	167	+0.225	218	−0.616			+0.054	−0.191	+0.160	+0.02
75	177	+0.052	238	−0.848			+0.012	−0.263	+0.098	−0.15
80	187	−0.122	258	−0.978			−0.029	−0.303	+0.009	−0.32
85	197	−0.292	278	−0.990			−0.070	−0.307	−0.082	−0.46
90	207	−0.454					−0.109	−0.274	−0.151	−0.53
95	217	−0.602					−0.144	−0.207	−0.180	−0.53
100	227	−0.731					−0.176	−0.116	−0.160	−0.45
105	237	−0.839					−0.201	−0.011	−0.098	−0.31
110	247	−0.920					−0.221	+0.096	−0.009	−0.13
115	257	−0.974					−0.234	+0.191	+0.082	+0.04
120	267	−0.999					−0.240	+0.263	+0.151	+0.17
125	277	−0.992					−0.238	+0.303	+0.180	+0.24
130	287	−0.956					−0.229	+0.307	+0.160	+0.24
135	297	−0.891					−0.214	+0.274	+0.098	+0.16
140	307	−0.799					−0.192	+0.207	+0.009	+0.02
145	317	−0.682					−0.164	+0.116	−0.082	−0.13
150	327	−0.545					−0.131	+0.011	−0.151	−0.27
155	337	−0.391					−0.094	−0.096	−0.180	−0.37
160	347	−0.225					−0.054	−0.191	−0.160	−0.40
165	357	−0.052					−0.012	−0.263	−0.098	−0.37
170	7	+0.122					+0.029	−0.303	−0.009	−0.28
175	17	+0.292					+0.070	−0.307	+0.082	−0.16

graphical representation see figure 352

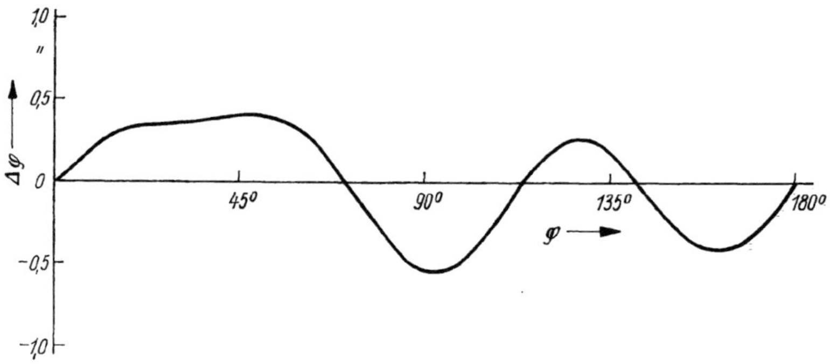

Figure 352 Diameter errors of a horizontal circle, determined with the *Heuvelink* method

Table 17. Computation of standard graduation errors using equations (114) to (119), (121) and (122), $n = 36$; $n = 72$; $s = 4$

series	$[(p-\alpha)^2]$	$[dd]$	$[d]$	n_α	$\frac{[dd]}{n_\alpha}$	$\gamma^2 = \frac{[d]^2}{n_\alpha^2}$	μ^2	μ
1	2	3	4	5	6	7	8	9
1	8.6065	25.75	−4.3	18	1.431	0.057	0.343	±0.59″
2	1.8643	17.94	−6.4	18	0.996	0.126	0.218	±0.46″
3	3.8795	12.42	−3.4	18	0.690	0.036	0.163	±0.40″
4	7.8211	13.12	−3.2	18	0.729	0.032	0.174	±0.42″
$\sum$	22.1714	69.23	−17.3	mean	0.962	0.063	0.224	±0.47″

$m^2 = 0{,}1130$

$[vv] = 22.1714, \quad M^2 = \frac{[vv]}{n-s} = \frac{2{,}1714}{36-4} = 0.6929, \quad M = \pm 0.83''$

$[v'v'] = 20.0223, M'^2 = 0.6674, [v'', v''] = 13.0564, M''^2 = 0.4663, [v'''v'''] = 11.9399,$
$M'''^2 = 0.4592$

$2\tau^2 = M^2 - m^2 = 0.5799, \ \tau = \pm 0.54'', \ 2\tau'^2 = 0.5544, \ \tau' = \pm 0.53''$

$2\tau''^2 = 0.3533, \ \tau'' = \pm 0.42'', \ 2\tau'''^2 = 0.3462, \ \tau''' = \pm 0.42''$

standard deviation of the constants

$$m_A = \frac{M\varrho}{a\sqrt{n}} = \pm 33°, \quad m_a = \frac{M}{\sqrt{n}} = \pm 0.13''; \quad m_B = \frac{M\varrho}{b\sqrt{2n}} = \pm 18°,$$

$$m_b = \frac{M}{\sqrt{2n}} = \pm 0.10''; \quad m_C = \frac{M\varrho}{c\sqrt{n}} = \pm 44°, \quad m_c = \frac{M}{\sqrt{n}} = \pm 0.13''$$

standard deviation of the correction

$$m_{\Delta\varphi} = M\sqrt{\frac{5}{2n}} = \pm 0{,}22''$$

While μ represents a measure of the quality of the observation, τ', τ'', τ''' signify the quality of the graduation.

Example: Investigation of the horizontal circle of the theodolite Wild T3, No. 33258 with photographic recording, using the Heuvelink Method, and a test angle of 45° (tables 14 to 17, figure 352). The total mean square graduation error is thus $\pm 0.54''$, and the random mean square graduation error $\pm 0.42''$.

Yelissejew Method ("Method of symmetric connections")

Since, in the opinion of the Soviet surveyors, the Heuvelink Method does not yield the necessary accuracy for determining systematic errors for theodolites used in Ist and IInd order triangulation, a simplified "Bruns' rosette method" according to *Yelissejew* is used in the Soviet Union. Without considering the theory of these methods, measuring procedure and evaluation are explained.

If diameter errors of e.g. every 3° are to be investigated, one measures angles of 36°, 45°, and 60° in rosettes, and in two sets between two collimators with the following circle positions:

$$\begin{array}{lll} 0° + \gamma & 0° + \gamma & 0° + \gamma \\ 36° + \gamma & 45° + \gamma & 60° + \gamma \\ 72° + \gamma & 90° + \gamma & 120° + \gamma \\ 108° + \gamma & 135° + \gamma & \\ 144° + \gamma & & \end{array} \qquad (123)$$

For γ, one has to set 0°, 3°, 6° ... 177°. For each of the 60 rosettes, the mean value is computed, e.g. for the 60° angle and $\gamma = 3°$.

$$\alpha_s = \frac{\alpha_3 + \alpha_{63} + \alpha_{123}}{3} \qquad (124)$$

The subscript indicates the starting positons. Furthermore, the differences

$$v_3^6 = \alpha_s - \alpha_3, \ v_{63}^6 = \alpha_s - \alpha_{63}, \ v_{123}^6 = \alpha_s - \alpha_{123} \qquad (125)$$

are determined, for which the superscript "6" indicates, how often the respective angle can be found in a full circle. With these values the corrections (γ) for the diameters are found, which represent the mean of the diameter corrections for the graduation line γ and for the following line:

$$\omega_\gamma^{10} = \frac{-2v_\gamma^{10} - v_{\gamma+36}^{10} + v_{\gamma+108}^{10} + 2v_{\gamma+144}^{10}}{5}$$

$$\omega_\gamma^{8} = \frac{-2v_\gamma^{8} - v_{\gamma+45}^{8} + v_{\gamma+135}^{8}}{4}; \quad \omega_\gamma^{6} = \frac{-2v_\gamma^{6} - v_{\gamma+60}^{6}}{3}$$

$$\omega_\gamma^{6.10} = \frac{\omega_\gamma^{6} + \omega_{\gamma+36}^{6} + \omega_{\gamma+72}^{6} + \omega_{\gamma+108}^{6} + \omega_{\gamma+144}^{6}}{5}$$

$$\omega_\gamma^{10.6} = \frac{\omega_\gamma^{10} + \omega_{\gamma+60}^{10} + \omega_{\gamma+120}^{10}}{3}$$

$$\omega_\gamma^{30} = 0.4\,\omega_\gamma^{10} + 0.4\,\omega_\gamma^{6.10} + 0.6\,\omega_\gamma^{6} + 0.6\,\omega_\gamma^{10.6}$$

$$\omega_\gamma^{8.30} = \frac{\omega_\gamma^{8} + \omega_{\gamma+12}^{8} + \omega_{\gamma+24}^{8} + \ldots + \omega_{\gamma+156}^{8} + \omega_{\gamma+168}^{8}}{15}$$

$$\omega_\gamma^{30.8} = \frac{\omega_\gamma^{30} + \omega_{\gamma+45}^{30} + \omega_{\gamma+90}^{30} + \omega_{\gamma+135}^{30}}{4}$$

$$(\gamma) = \frac{1}{3}\omega_\gamma^{8} + \frac{1}{3}\omega_\gamma^{30.8} + \frac{2}{3}\omega_\gamma^{80} + \frac{2}{3}\omega_\gamma^{8.30}. \qquad (126)$$

check:

$$\Sigma\omega_\gamma^{10} = \Sigma\omega_\gamma^{8} = \Sigma\omega_\gamma^{6} = \Sigma\omega_\gamma^{30} = \Sigma(\gamma) = 0.$$

If the diameter errors are to be determined for every 4°, it is sufficient to measure angles of 40° and 36°, while for 5° to 5° measurements of angles of 40° and 45°, or 20° and 45° are suitable. The fact that the diameter error is directly determined, is of advantage. For measurement and evaluation, approximately 1.5 times the amount of time is required than for the Heuvelink Method.

Roelofs' Method. For this method a bundle of rays with four rays at 0, 15, 37, and 90 gon is measured, such that the angles formed of 15, 22, 37, 53, 75 and 90 gon are approximately equal to the angles, which proved to be most suitable for the determination of two each of the 14 coefficients of the Fourier Series when using a single angle method. A much higher accuracy is obtained than with the two angle method. The 14 coefficients are determined with approximately the same accuracy, and practically without correlations. All graduation lines of the circle are utilized during measurements.

Test Instruments for Graduated Circles. For testing of graduated circles in surveying instruments, test instruments are manufactured by F. W. Breithaupt & Son, Kassel. The PRUTE instrument (figure 353) consists of a base plate, which can be levelled, and two collimators, forming the test angle, which can be selected between 30° and 50°.

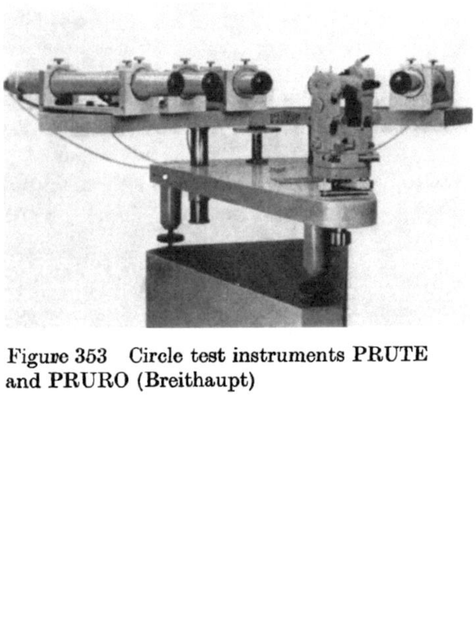

Figure 353 Circle test instruments PRUTE and PRURO (Breithaupt)

PRURO contains four collimators which form angles of 0, 15, 37 and 90 gon (Roelofs' method). The collimators have a 60 mm aperture and 600 mm focal length. By utilizing electronic data handling, the times required for the evaluation can be significantly reduced.

The measuring accuracy for circle graduations has to be better than 10^{-7}, independent of the radius of the graduated circle. Normal methods cover only 1% to 2% of the graduation lines, with measuring accuracies of about 5×10^{-7}.

With the automatic graduation test instrument (ATP), which is used for official state surveying in the G.D.R., accuracies of $\pm 1 \times 10^{-8}$m (0.02″ for 200 mm diameter) are obtained. The instrument consists basically of a heavy metal cast body, which bears an axis with double conical support. A round table is connected to this vertical axis. It contains a centering device, which accepts the test instrument and serves to level and adjust the graduation concentrically to the axis system. Above this rotational system, a stable bridge is placed. It carries four photo-electrical microscopes arranged along the circumference of the circle, two each diametrically opposite. These can be shifted in radial direction and thus adjusted to the diameter of the test instruments. One of them serves as principal microscope. Because of the high sensitivity of the measuring systems of 10^{-9} m, only a standard wave (spectral lamp) can be used as norm. A calibration device with inductive capabilities is placed in between, and compared with an interference comparator. By storing the measurements on punched tape a continual data flow is obtained.

Table 18. Determination of the short periodic errors of the horizontal circle of the theodolite OT 02 No. 9335

interval [']	a ['']	b 2' + ['']	c 2' + ['']	$a-b$ ['']	$a-c$ ['']	$\frac{v_0}{2}$ ['']	$\frac{v_u}{2}$ ['']	v ['']
1	2	3	4	5	6	7	8	9
0 … 4	+0.2	−0.4	−0.2	+0.6	+0.4			
	0.0	−0.2	0.0	+0.2	0.0			
			mean	+0.4	+0.2	−0.46	−0.55	−1.01
4 … 8	+0.2	+0.4	+0.6	−0.2	−0.4			
	0.0	+0.2	+0.8	−0.2	−0.8			
				−0.2	−0.6	+0.14	+0.25	+0.39
8 … 12	0.0	+0.8	+0.8	−0.8	−0.8			
	0.0	+0.8	+1.0	−0.8	−1.0			
				−0.8	−0.9	+0.74	+0.55	+1.29
12 … 16	+0.4	+0.4	0.0	0.0	+0.4			
	+0.2	0.0	0.0	+0.2	+0.2			
				+0.1	+0.3	−0.16	−0.65	−0.81
16 … 20	+0.4	0.0	+0.6	+0.4	−0.2			
	+0.6	+0.2	+0.8	+0.4	−0.2			
				+0.4	−0.2	−0.46	+0.15	−0.61
20 … 24	0.0	+0.6	+1.0	−0.6	−1.0			
	+0.2	+0.8	+1.0	−0.6	−0.8			
				−0.6	−0.9	+0.54	+0.55	+1.09
24 … 28	+0.2	0.0	+0.6	+0.2	−0.4			
	0.0	+0.2	+0.8	−0.2	−0.8			
				0.0	−0.6	−0.06	+0.25	+0.19
28 … 32	0.0	−0.4	−0.4	+0.4	+0.4			
	−0.2	−0.6	−0.2	+0.4	0.0			
				+0.4	+0.2	−0.46	−0.55	−1.01
32 … 36	0.0	+0.4	+0.6	−0.4	−0.6			
	+0.2	+0.4	+0.6	−0.2	−0.4			
				−0.3	−0.5	+0.24	+0.15	+0.39
36 … 40	−0.4	−0.6	+0.2	+0.2	−0.6			
	−0.6	−0.4	+0.2	−0.2	−1.0			
				0.0	−0.8	−0.06	+0.45	+0.39
40 … 44	0.0	0.0	+0.4	0.0	−0.4			
	+0.2	−0.2	+0.6	+0.4	−0.4			
	+0.2			+0.2	−0.4	−0.26	+0.05	−0.21
44 … 48	0.0	−0.4	0.0	+0.4	0.0			
	0.0	−0.2	+0.2	+0.2	−0.2			
				+0.3	−0.1	−0.36	−0.25	−0.61
48 … 52	0.0	+0.6	+0.8	−0.6	−0.8			
	0.0	+0.6	+1.0	−0.6	−1.0			
				−0.6	−0.9	+0.54	+0.55	+1.09
52 … 56	+0.2	+0.6	+0.8	−0.4	−0.6			
	+0.2	+0.8	+1.0	−0.6	−0.8			
				−0.5	−0.7	+0.44	+0.35	+0.79
56 … 60	−0.6	−0.8	−1.2	+0.2	+0.6			
	−0.6	−1.0	−1.2	+0.4	+0.6			
				+0.3	+0.6	−0.36	−0.95	−1.31
			mean	−0.06	−0.35	$\sum$=0.00	+0.05	+0.05

Elimination of the influence of long period graduation errors with the measuring procedure.
The influence of long period graduation errors is largely eliminated by multiple measurements of the same angle at various positions, evenly spread over the circle.

Determination of Short Period Graduation Errors. For angular measurements of high accuracy, short period graduation errors (error of neighbouring graduation lines) have to be considered, whose amount may surpass the one for long period errors. Small angles are not influenced by long period errors, but by short period ones. They are caused primarily by errors of the hollow screw. Its period is generally the smallest common multiple of the value $360 \times 60'/z$ (with z being the number of cogs of the main cog wheel), of the scale value i of the circle, and of the number n of the transfers of the cog sector onto one revolution of the transfer wheel. Since the errors of the sector are relatively small, n can eventually be neglected. For $z = 1440$ and a 5′ graduation, the period is 15′, while for a 4′ graduation it is 60′.
To investigate short period errors, one measures for instance, consecutively all intervals of a circle sector of 1 to 3° similarly as for the run determination (see page 96). To eliminate random errors, this is repeated at other circle positions, e.g. for 10′ theodolites between 0 and 1°, 30 and 31°, 60 and 61° ... 150 and 151°, or for 20″ theodolites between 0 and 2°, 30 and 32° etc. By separating the run, e.g. by comparing the single values with the mean of each series, the amounts of the short period errors are obtained.
Example: Determine the short period errors of the horizontal circle of the theodolite OT-02 No. 9335, 3rd series, circle sector 60 to 61° (table 18).
The short period errors of the circle positions, appearing at the top and bottom within the field of view are:

$$v_o = 2[(a - b)_m - (a - b)_i]; \; v_u = 2[(a - c)_m - (a - c)_i] \tag{127}$$

The differences of the corrections of neighbouring diameters are

$$v = \frac{v_o + v_u}{2} \tag{128}$$

With the corrections from the measuring series, the short period errors are computed (table 19) and plotted. The tendency of the short period

Table 19. Determination of the short periodic errors of the horizontal circle of the theodolite OT-02 No. 9335 (Computations)

circle line	corrections of the intervals at the circle positions [″]							short periodic errors [″]	
[′]	0°	30°	60°	90°	120°	150°	mean	preliminary	final
1	2	3	4	5	6	7	8	9	10
0								0.00	+0.14
	−0.37	−0.66	−1.01	−1.38	−0.40	−0.13	−0.66		
4								−0.66	−0.52
	+0.08	+0.48	+0.39	+0.72	−0.25	−0.28	+0.19		
8								−0.47	−0.33
	+0.98	+0.54	+1.29	+0.97	+0.40	+1.07	+0.88		
12								+0.41	+0.55
	−0.02	+0.38	−0.81	+0.17	−0.25	+0.26	−0.04		
16								+0.37	+0.51
	−0.27	−0.52	−0.61	−0.98	−0.95	−1.03	−0.73		
20								−0.36	−0.22
	−0.07	+0.08	+1.09	−0.18	−0.15	+0.02	+0.13		
24								−0.23	−0.09
	+0.03	+0.84	+0.19	+0.77	+0.15	+0.67	+0.44		
28								+0.21	+0.35
	−0.62	−0.91	−1.01	−0.08	−0.40	+0.37	−0.44		
32								−0.23	−0.09
	−0.02	−0.46	+0.39	+0.02	−0.26	−1.23	−0.26		
36								−0.49	−0.35
	−0.32	+0.28	+0.39	+0.47	+0.05	+0.37	+0.21		
40								−0.28	−0.14
	−0.02	−0.76	−0.21	+0.67	+0.70	+0.02	+0.07		
44								−0.21	−0.07
	−0.57	−0.22	−0.61	−0.93	+0.14	+0.12	−0.34		
48								−0.55	−0.41
	+0.40	+0.45	+1.09	+0.17	+0.15	+0.27	+0.24		
52								−0.13	+0.01
	+0.58	+0.74	+0.79	+0.62	+0.60	+0.67	+0.67		
56								+0.54	−0.68
	−0.07	−0.42	−1.31	−0.83	−0.75	+1.13	−0.50		
60								(+0.04)	(+0.18)
							$\sum$ mean	−2.08 −0.14	

errors for theodolites of the type OT-02, whose period is about 16′, can be recognized in figure 354. Considering the long period errors of the first and last line, the long period errors of the graduation lines in between can be interpolated.

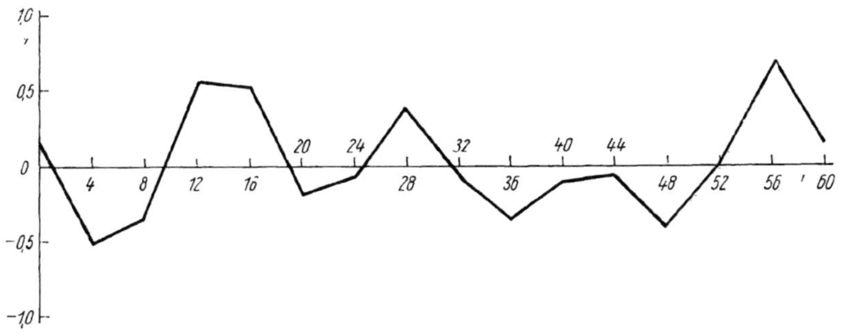

Figure 354 Presentation of short period graduation errors

Elimination of the Influence of Short Period Errors with the Measuring Procedure. The influence of the short period errors on angular measurements can be eliminated, when the sets of directions are started at subdivision lines of a degree which are equally spread over the period of the short period errors.

Errors of the Reading Device

For the investigation of the most important errors, namely screw-errors and run, the reader is referred to pages 87 and 96. Investigation of the telescope (see section 1.1.2).

Investigation of the movement of the focussing lens. The direction of the horizontal component of the collimation axis of a telescope with focussing lens can be obtained by determining the collimation error for various distances to object points. The target is sighted in both telescope positions, and the difference between the circle readings is calculated. Here the telescope eccentricity has to be considered.
The focussing error δ_i, caused by the movement of the focussing lens is obtained as

$$\delta_i = c_i - c_\infty - \alpha_i \tag{129}$$

where c_∞ is the actual collimation error for infinite focus, c_i the collimation error for the target i, and α_i the angle caused by the telescope eccentricity for the respective distance, as computed with equation (95).
Since the influence of the movement of the focussing lens on the collimation axis is much larger for short sights than for long ones, one selects close targets in denser sequence when studying this error, e.g. ∞, 80 m, 50 m, 30 m, 20 m, 10 m, 7 m, 5 m, 3 m, 2 m, 1.5 m, 1.2 m.
The direction of the vertical component of the collimation axis can be obtained similarly by determining the index error.

Elimination of the influence of the movement of the focussing lens. Since the experimental arrangement is similar to that for the determination of the collimation error, it is obvious, that the influence of this error is eliminated by using the mean of the readings in both telescope positions.

Determination of the Sensitivity of the Level Bubbles (see section 1.2.1).

Errors of the Observer

Determination of the Reading Error. The mean square reading error can be determined by multiple readings (in practical situations usually twice) of the same circle position at various circle- and index positions. For theodolites with coincidence micrometer for instance, one coincides the lines of diametrically opposite circle positions every 20 gon. With the differences d of the two micrometer readings, the standard deviation of one coincidence is

$$m_a = \pm\sqrt{\frac{[dd]}{2n}} \tag{130}$$

Determination of the Sighting Error. The mean square error of one sighting can be determined by multiple (e.g. 10 times) sighting of a point with two circle readings each. With the differences v between the combined mean and the mean of the readings for each sighting, the following standard deviation is obtained:

$$m_{z,a} = \pm \sqrt{\frac{[vv]}{n-1}} \tag{131}$$

which contains the reading error [equation (130)]. The standard deviation m_z of one sighting is obtained by separating these errors:

$$m_z = \pm \sqrt{m_{z,a}^2 - \frac{m_a^2}{2}} \tag{132}$$

Determination of the Standard Deviation of one Direction. The standard deviation of a direction observed in both telescope positions can be determined by observing four well identified targets, equally spread over the horizon in four complete sets. The weather conditions should be as follows: air temperature, +15°C ± 10K, light air movement, no rain, light wind (≦4 m/s). For an expected standard deviation of more than 2 mgon, the distance to the target should be less than 1 km, for m_r = 0.6 to 2 mgon, 1 to 3 km, and for m_r < 0.6 mgon, 3 to 5 km.
In the second case, the measurements should be repeated by the same observer at another day, and in the third case at two other days with different light conditions.
The means of the observations in both telescope positions are reduced in each set to the first direction. The values obtained are to be meaned from all four sets $(r_i) + [r_i]/4$.

Table 20. Determination of the standard deviation of a direction

set	difference				sum (across)
	Δ_1	Δ_2	Δ_3	Δ_4	
1	2	3	4	5	6
1	0	$r_{2,1} - (r_2)$	$r_{3,1} - (r_3)$	$r_{4,1} - (r_4)$	$[\Delta_i]$
2	0	$r_{2,2} - (r_2)$	$r_{3,2} - (r_3)$	$r_{4,2} - (r_4)$	$[\Delta_i]_2$
3	0	$r_{2,3} - (r_2)$	$r_{3,3} - (r_3)$	$r_{4,3} - (r_4)$	$[\Delta_i]_3$
4	0	$r_{2,4} - (r_2)$	$r_{3,4} - (r_3)$	$r_{4,4} - (r_4)$	$[\Delta_i]_4$
	0	$[\Delta_2\Delta_2]$	$[\Delta_3\Delta_3]$	$[\Delta_4\Delta_4]$	$[\Delta_i\Delta_i]$

With the differences between reduced directions and respective mean value (table 20), and the sums $[\Delta_i]$ of the differences, calculated for each set separately, and with the sum of the squares of the differences $[\Delta_i\Delta_i]$ for all sets, the sum of the squares of the corrections $[v_r v_r]$ is obtained:

$$[v_r v_r] = [\Delta_i\Delta_i] - \frac{1}{4}([\Delta_i]_1^2 + [\Delta_i]_2^2 + [\Delta_i]_3^2 + [\Delta_i]_4^2) \tag{133}$$

This gives the standard deviation m_r ofa direction measured in both telescope positions

$$m_r = \pm \sqrt{\frac{[v_r v_r]}{(s-1)(n-1)}}, \tag{134}$$

where s is the number of sets, and n the number of directions. If the measurements are repeated, the quadratic mean of the standard deviations for the different days is to be computed.

2.4. Gyro Instruments

With the gyro instrument, geographic north can be directly determined. The orientation of the gyro is more accurate than the magnetic orientation and is faster than astronomical azimuth determination. Surveying gyros are especially suitable for directional orientation of long or open traverses, orientation and direction transfer in mining, and direction measurements in engineering surveys. Two directions are required to determine the position of a point by resection. Traverses can be measured with compass instruments also.

2.4.1. Fundamentals of Gyro Instruments

Any solid body, undergoing any rotation can generally be considered as being a gyro. Its fundamental characteristics are its inertia and its precession (directional change due to external moments acting on its rotational axis). According to possible rotations around all three axes, around two axes, or only one axis, one distinguishes gyros with three, two or one degree of freedom.

For geodetic purposes, the declination gyro is the most important one. Its rotational freedom around the horizontal y-axis is either taken away or else reduced by tying the rotational- or impulse axis to the horizontal plane with a ballast below the centre of gravity. It can rotate freely around two axes only, namely the horizontal gyro axis HH and the vertical axis VV (figure 355).

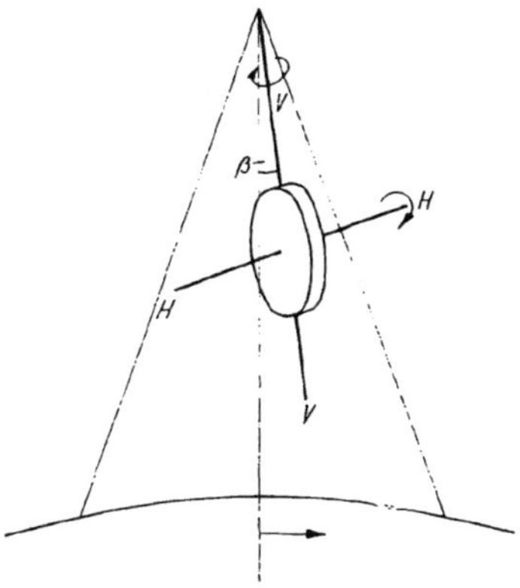

Figure 355 Principle of effect of a gyro with two degrees of freedom

Due to its inertia, the fast rotating gyro tries to retain its direction in space. The earth's rotation however, causes a deviation of the center of gravity from the plumbine of the supporting point, which is counteracted by the gravity moment of the ballast piece. The gyro is forced into a rotation around the plumb direction (precession) such, that the gyro axis swings into the geographic north direction. Then the axial vector I_ω of the rotational impulse of the gyro lies in the same meridian plane as the rotational vector ω_E of the earth.

In screw position of these vectors, a precession moment is present at the gyro

$$M' = I_\omega \, \omega_E \cos \varphi \tag{135}$$

because only its horizontal component $\omega_E \cos \varphi$ acts onto a horizontally bound gyro at a geographic latitude φ. If the gyro axis (rotational vector I_ω of the gyro) forms the angle α with the meridian direction (figure 356), the directing force R is generated, which directs the gyro with precession velocity into the meridian. With $R/M' = \sin \alpha$ and equation (135), the directing force becomes

$$R = I_\omega \, \omega_E \cos \varphi \sin \alpha = I_\omega \, \omega' \tag{136}$$

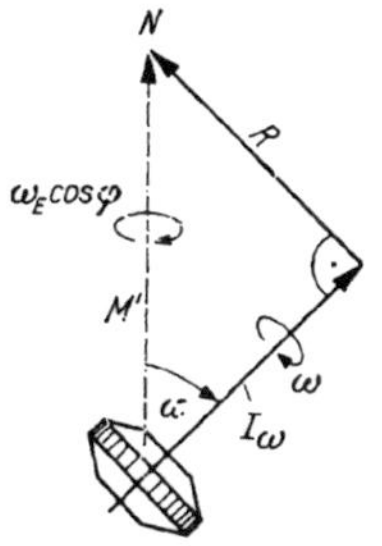

Figure 356 Diagram of directive forces

The larger the deviation α from the meridian, the larger the precession velocity:

$$\omega' = \omega_E \cos \varphi \sin \alpha \tag{137}$$

The maximum directing force is reached when the meridian axis is in East-West direction. There is no directing force in the meridian, because it is the gyro's normal position ($\alpha = 0$). Equation (136) also shows, that the directing force is at a maximum at the equator ($\cos \varphi = 1$), and that the gyro operates best and fastest there. It is reduced for higher latitudes, and the gyro reacts more slowly, while it does not work at all at the pole. When increasing the rotational impulse I_ω by enlarging either the rotational speed or the inertial moment, a large directional force R is obtained. An enlargement of the inertial moment—the body of the gyro—is undesirable, because the mass to be transported becomes larger, and energy losses will occur due to friction. An increase in the number of revolutions is limited by the strength of the material.

2.4.2. Structure of the Surveying Gyro

The precession of declination gyros is utilized for the determination of the North direction with

1. pendulum gyros
2. rate-gyros
3. pendulum supported rate-gyros.

The most important one is the pendulum gyro (figure 357). As surveying gyro, it consists of:

a) the directing part, whose elements participate in swinging around the north direction
b) the part, taking the direction, which is connected to a theodolite (gyro theodolite), or mounted on top of a theodolite and
c) the instrument support.

The principal part of the directing part is the actual gyro, formed by the rotor of a non synchronized D/C motor.

The stator lies within in the runner. Frequencies of 333 to 500 Hz lead to 20,000 to 30,000 revolutions per minute. The gyro is supported at its innermost gimbal joint (the frame or ring of the cardanic support, here gyro cap)

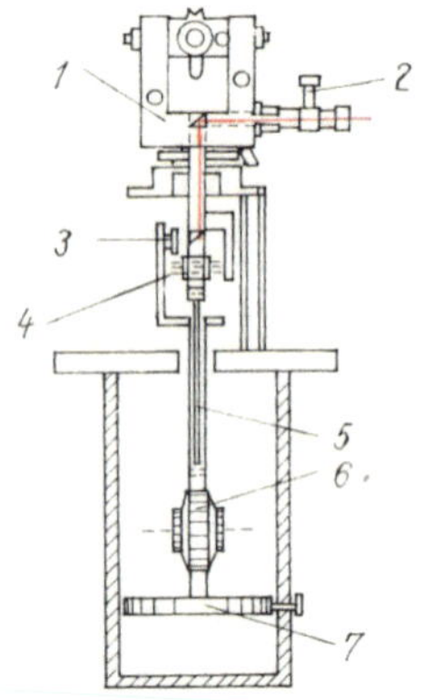

Figure 357 Schematic structure of a pendulum gyro
1 theodolite; *2* autocollimation telescope; *3* autocollimation mirror; *4* current spirales; *5* carrying ribbon; *6* gyro motor; *7* dampening

like a torsion pendulum with horizontal axis. It is thus elastically bound to the local plumb line, and swings around the meridian because of the earth's rotation. This dampened oscillation is followed with an autocollimation device. The mean position is determined from its turning points. Its period is

$$T = 2\pi \sqrt{\frac{I_\omega}{Pc(\omega_E \cos\varphi + D_B/I_\omega)}}\ [\mathrm{s}] \tag{138}$$

where P is the mass of the oscillating system, c the separation between centre of gravity and support, and D_B the return force of the supporting ribbon. The direction transfer is carried out with the attached theodolite.

There are two types of gyros hanging on ribbons. The period of a gyro with liquid support is relatively long—because of the bearing liquid—while its mass is large.

The unsupported gyro, oscillating in air, is more suitable for geodetic purposes. It hangs with its whole mass on a strong, thin and elastig metal ribbon. With electronic (e.g. *Lear*) or manual tracking (e.g. *Fennel*) of the upper ribbon clamp, the torsion of the ribbon can be eliminated.

The carrying ribbon connects the directing part with the one for direction transfer, e.g. the top part of the theodolite. In U.S.S.R., gyros with liquid support and peg bearings are produced.

The electrically bound rate-gyro (figure 358), used in Great Britain and U.S.A., provides the North direction by the precession of the gyro, instead of the oscillation around the meridian. The motor is mechanically fastened at a special position in respect to the instrument container. The horizontal component of the earth rotation, acting on the y-axis (input axis) causes a precession around the z-axis (output axis), which generates an A/C voltage in a sensor. This is amplified and converted to D/C, and fed to a motor with opposite rotation, which acts on the output axis and cancels the directing force of the gyro. Polarity and intensity of the necessary current can be read on a display instrument. This instrument has to be turned in its frame until it indicates zero. Then the gyro axis is in the meridian. A second measurement is taken with the gyro rotating in opposite direction.

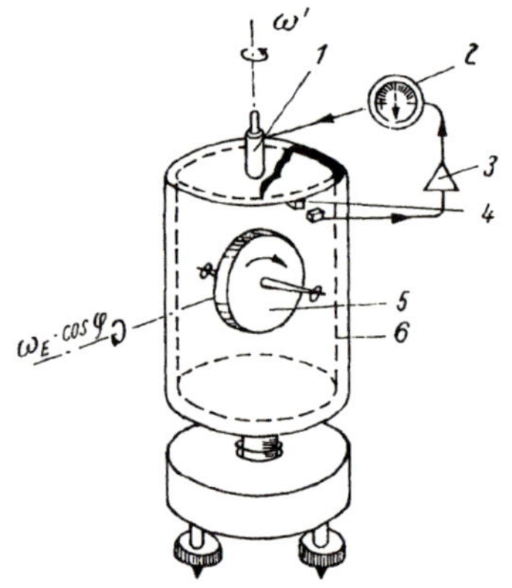

Figure 358 Rate gyro
1 motor acting in opposite position; *2* null indicator; *3* amplifier; *4* sensor; *5* gyro motor; *6* gyro housing

Due to the rigid tie of the z-axis, levelling errors of the instrument act as errors $\Delta A = \delta \tan\varphi$ of the north direction. Furthermore, errors in the directing system can be caused by external vibrations. Therefore, a combination of the pendulum and the rate-gyro was developed (figure 359).

A gyro hanging on a ribbon in a frame is arranged such, that its output axis positions itself into the plumbline. A motor, acting in opposite direction, reduces the precession of the gyro. Its input signal is also transferred to, the moment giving device of a second gyro, mounted in the frame with horizontal rotation axis. This gyro reacts with a precession around the x-axis to the input moment around the y-axis. This precession is also sensed and monitors a motor on the outside casing, which moves the frame until the gyro stands in the meridian.

At the laser gyro (figure 360), a helium-neon-laser emits infrared coherent light from both ends of a laser tube into both directions of an optical frame. Its corners contain mirrors and prisms. In the zero position of the frame, the frequencies of both rays are the same, but differ when moved, e.g. due to the earth rotation. The change in frequency depends on the rotational velocity.

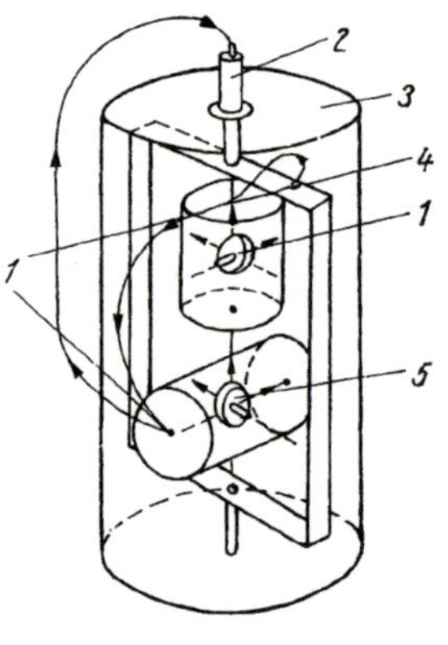

Figure 359 Rate gyro system according to *Kearfott*
1 sensor; *2* azimuth following motor; *3* outside casing; *4* frame; *5* azimuth stabilisation (2nd gyro)

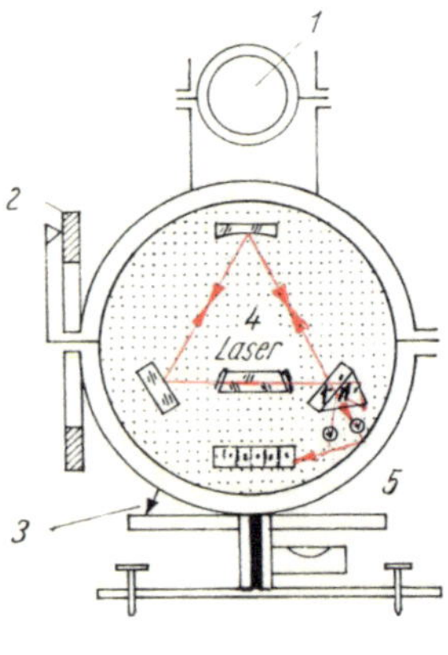

Figure 360 Laser gyro
1 telescope; *2* vertical circle; *3* reading horizontal circle; *4* laser; *5* casing of the laser gyro

2.4.3. Overview of Surveying Gyros

The equipment of ribbon supported surveying gyros, internationally referred to as gyrotheodolites (figure 361) has a mass of 30 to 70 kg. In about 30 to 60 minutes, the north direction is given to within ±3 to 5 mgon. By utilizing integrated circuits, printed on plates, and D/C motors, instruments with smaller mass could be built (figures 362 to 364), e.g. instruments where transformation and source of the current are one unit (MW 50 by WBK Bochum, ±1′; MVT-3 by VNIMI, Leningrad, ±30″). Recently, surveying gyros with automatic north setting have been developed. For this, the signal determining the north direction is either generated by a gyro dampened with air bearing support (north directing turning gyro NSK Teldex, Heidelberg), or else taken form a gyro, whose cap is bound electronically to the housing (ribbon supported bound gyro by Bodenseewerk Gerätetechnik, Überlingen, FRG). The accuracy amounts to ±3 to 5′.

Gyro attachments of about 2 kg mass can be fastened with a bridge onto theodolites, and are supplied by battery and converter unit via cables or at MOM, Budapest, through the connection plug of the theodolite. The north direction is obtained within ±10 mgon in 20 to 30 minutes (figures 365 to 367). These attachments are best suited for theodolites of medium accuracy.

Direction Gyros (figures 368 and 369) are in between gyro attachment and compass, as far as accuracy is concerned. They are a combination of simple graduated circle and goniometer with a gyro, as used in gyro attachments.

The gyro attachment Wild GAK 1 (figures 365 to 366) is operated by a fully transistorized converter with output of 115 V A/C and 40 cycles. Converter and 12 V battery are placed inside a handy housing. The battery lasts for

Table 21. Surveying Gyros (selection)

type	manufacturer	type of gyro	swinging system; half period T/2 [s]	warm up time [min]	measuring duration [min]	mass: instrument [kg]	mass: complete equipment [kg]	standard deviation of one measurement [″]	remarks
1	2	3	4	5	6	7	8	9	10
ABLE	Autonics, U.S.A.	rate	–	12	20	20	45	35	manual setting to precision moment zero
PIM	English Electric	rate		15	40		60	30	manual setting to precision moment zero
Gyromat	WBK, Bochum (FRG)	ribbon in air	171	0.3	7	2	16	3	
MW 50a	Giesenberg and Schmieder (FRG)	ribbon in air	90	2		8.1	23	60	
Gytar	Kearfott (U.S.A.)	2 rate	–	6	35		50	20	steering with azimuth gyro
Gi-B 1	MOM, Budapest	ribbon in air	270	2	35	11	55	15	figure 361
Gi-C 13		ribbon in air	220	1.5	25	4.1	21	25	figure 363
MW-2	U.S.S.R.	plug bearing, supported		14	45	23	55	35	
AG		liquid		12	40	22	93	60	
MVT 2		ribbon	120	7	15	16	33	30	automatic following
ARK 1	Wild Heerbrugg AG	ribbon in air	210	1.5	10	5.5	27.4	20	direction gyro; figure 368
gyro attachment									
GP-1	Lietz Comp. (U.S.A.) Sokkisha (Japan)	ribbon in air	240	1.5		4	17.3	20	
Orientor	U.S.A.				35		45	45	
MT-1	U.S.S.R.	ribbon in air		11	40	16	45	20	automatic following
GAK 1	Wild Heerbrugg AG	ribbon in air	240	1.5	20	1.8	13.3	20	figure 365

6 hours' measurements. The gyro oscillations are dampened with three vertically and axially acting sensor pins, which are automatically activated shortly before the circle is set free. In this manner, a suitable amplitude of the gyro oscillations of 3 to 6 gon can be achieved.
The direction gyro Wild ARK 1 (figures 368 and 369) has a telescope with 5 times magnification, upright image, and 10° field of view. The horizontal

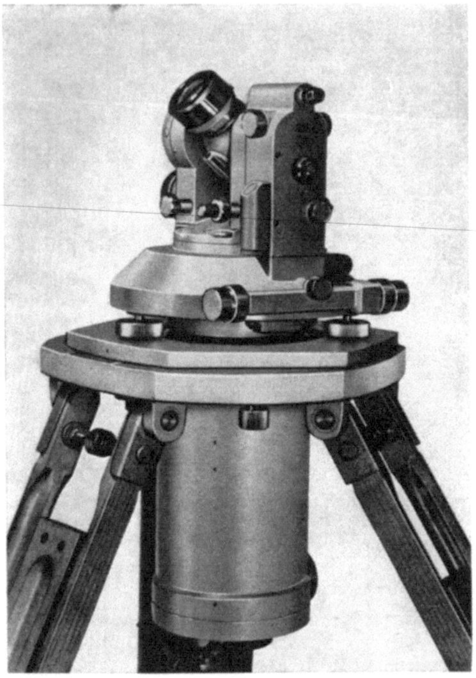

Figure 361 Gyro theodolite Gi-B 1 (MOM, Budapest)

Figure 362 Gyro theodolite KT 3 by FENNEL (not produced)

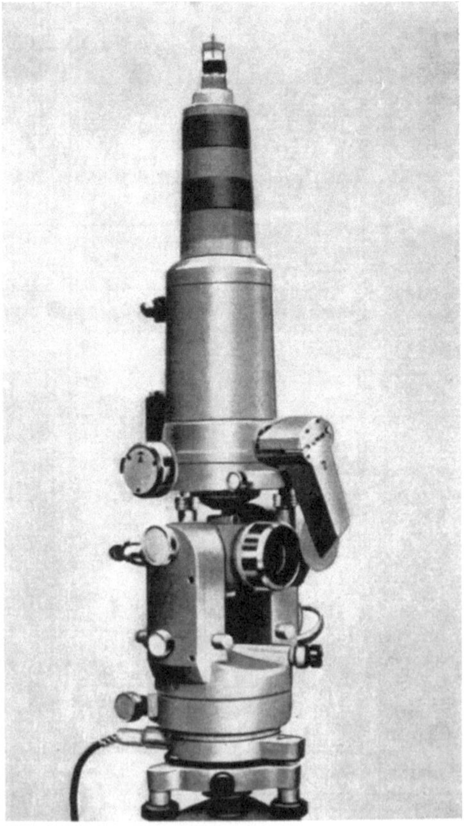

Figure 363 Gyro theodolite Gi-C 11 (MOM, Budapest)

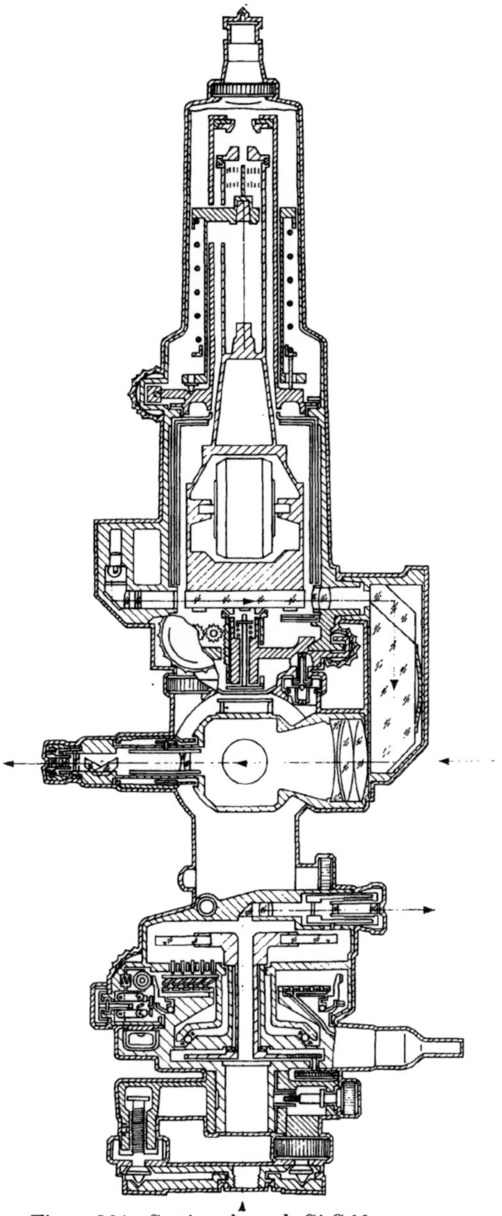

Figure 364 Section through Gi-C 11

Figure 365 Gyro attachment Wild GAK 1 on T 16 with converter GKK 1 (height of GAK 1: 340 mm)

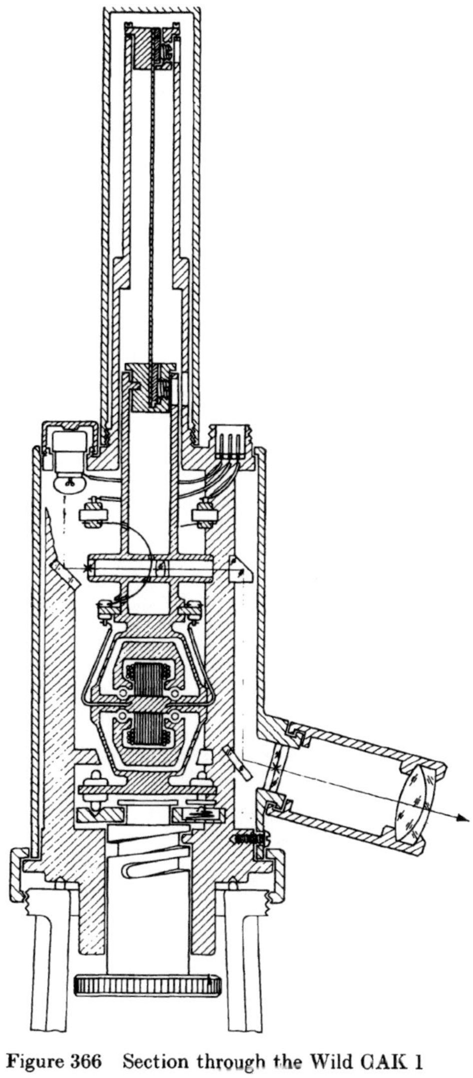

Figure 366 Section through the Wild GAK 1

Figure 368 Direction gyro Wild ARK 1

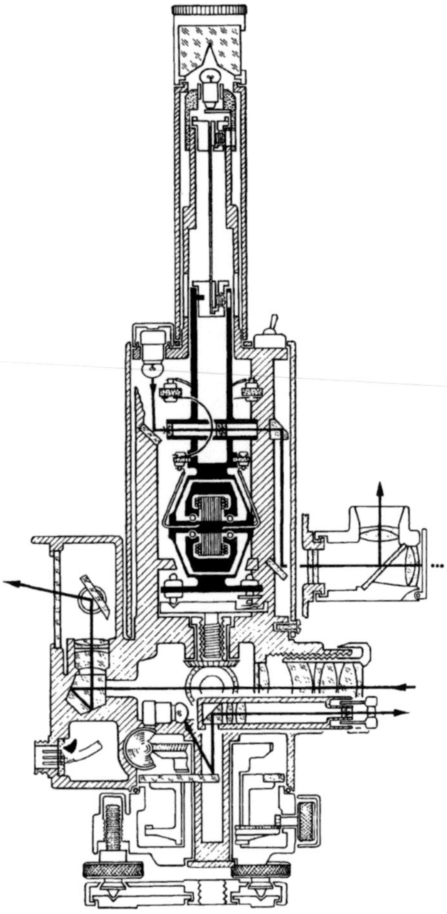

Figure 369 Section through the Wild ARK 1

Figure 367 Circle scale and gyro mark (light index) of the Wild GAK 1

circle is subdivided into 640 intervals to 10% and can be read to 1% with a scale. After dividing the circle reading by 16, the angular value is obtained in gon. The instrument has an infinite tangent screw.

The surveying gyros can be classified as

- gyrotheodolites of high accuracy ($\leqq \pm 2''$)
- gyrotheodolites of medium accuracy (± 10 to $15''$)
- gyrotheodolites of low accuracy ($\pm 1'$)
- gyro compass for mining applications (± 5 to $10'$).

Instruments to be used underground have to be protected for fire-damp. Since efforts in time and equipment are considerable for high accuracy measurements, surveying gyros only gained acceptance in mining and for some engineering surveys, but not in general surveying. They are also widely used for military purposes, because the low accuracies achieved with handy instruments satisfy these needs.

2.4.4. Measurements with Surveying Gyros

The oscillating gyro does not exactly position itself into the north-south direction, but performs a weakly dampened rotational oscillation symmetrically to the meridian direction, because of its inertia. The direction of the gyro axis is projected as a wandering mark onto a scale with the aid of a light index, and observed with a magnifying glass (figure 367). Before measuring, the top part of the instrument is manually pre-oriented to a given target with a compass or by reading the turning points. Measurements can be performed with:

1. the turning point method (by following)
2. the amplitude method (without following)
3. the passage method, or
4. a combination of these (figure 370).

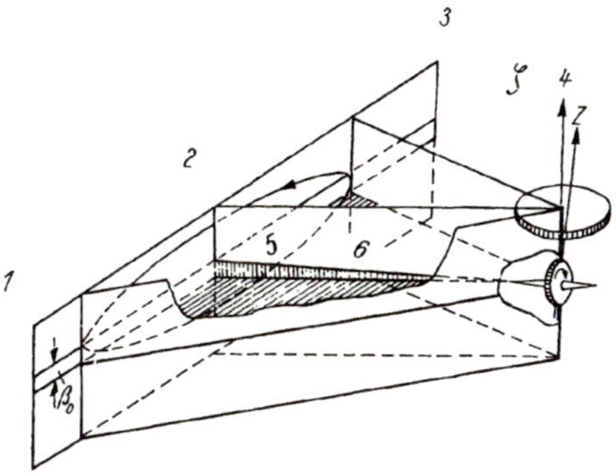

Figure 370 Combination method
1 west; *2* north; *3* east; *4* φ (plumb line;)
5 horizon; *6* equilibrium

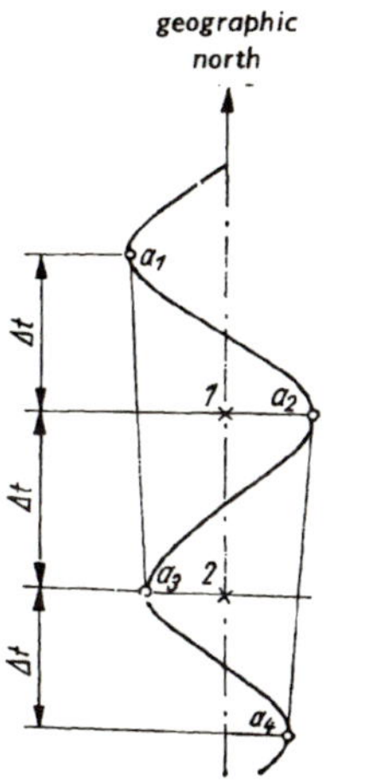

Figure 371 Schuler mean

When measuring turning points, the wandering mark is always kept in the centre of the index by following with the tangent screw. At the turning points, the gyro appears to rest for a short time. In this position, the horizontal circle is read. Using at least three observed turning points (figure 371), the circle reading for geographic north can be computed with the Schuler-mean:

$$N = \frac{1}{2}\left(\frac{a_1 + a_2}{2} + a_2\right) = \frac{1}{2}\left(\frac{a_2 + a_4}{2} + a_3\right) \text{etc.} \tag{139}$$

According to *Halmos*, four turning points would be the optimum. External influences or the heating of the motor, exceed the uncertainties of reading and computation. With automatic following, manual following errors, which cause accuracy losses of up to 50%, can be avoided. When the gyro is properly dampened and adequately pre-oriented, it is sufficient to read the turning points at the amplitude scale (amplitude method).

The passage method is most widely used, especially with gyro attachments. Here the times of passing of the wandering mark (or the zero point of the reticule) through the centre of the index are recorded with a stop watch, and the amplitude is read at an auxiliary scale. The correction to the orientation is proportional to the amplitude and to the time difference on both sides of the index. The measurement of time can be done automatically to ± 0.01 s $\triangleq \pm 1''$ with photo transistors.

Since only relative measurements are possible with a gyro (the gyro axis is not observed directly), its calibrated value has to be determined periodically with a calibration line of known azimuth. The calibrated value is the angle between the normal to the mirror and the gyro axis, plus a constant, dependent on the instrument type. It is obtained as the difference between the display of the gyro and the azimuth of the calibration line.

The errors of surveying gyros can be divided into three groups. The display error depends on the quality of the oscillation observations and on the observation conditions. The instrument error is characteristic for the variance of a single measurement from the mean of several measurements. It encompasses instrument-, setting-, and reading errors, and is determined with several measurements without new centering or transporting.

The orientation error signifies the variance of a single measurement from the mean, and is determined from repeated measurements under different conditions at various known orientation lines.

2.5. Optical Precision Plummets

Recently, engineering surveys have increasing requirements for precision plumbing in connection with the construction of skyscrapers, television towers, hydro masts, bridges, power dams, and -stations, etc., when monitoring elevators and pipelines, for deformation measurements, as well as in mining.

For this, precision plummets are used, which, compared to optical plummets, are equipped with telescopes with larger magnification and focussing lens, designed for longer distances, and with more accurate levelling devices thus that higher accuracies for longer distances can be achieved (table 22). The

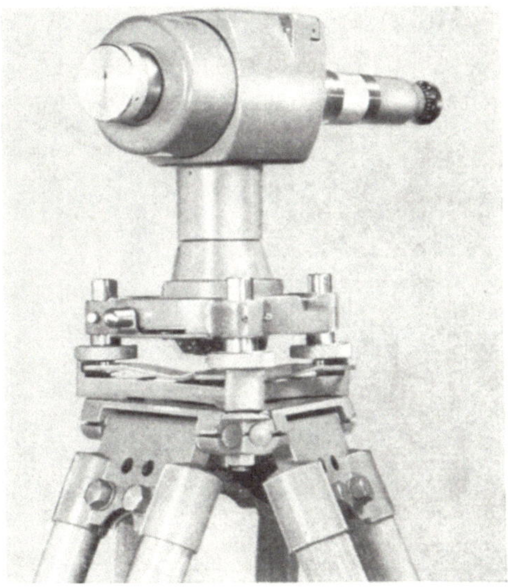

Figure 372 Section through the Wild ZNL

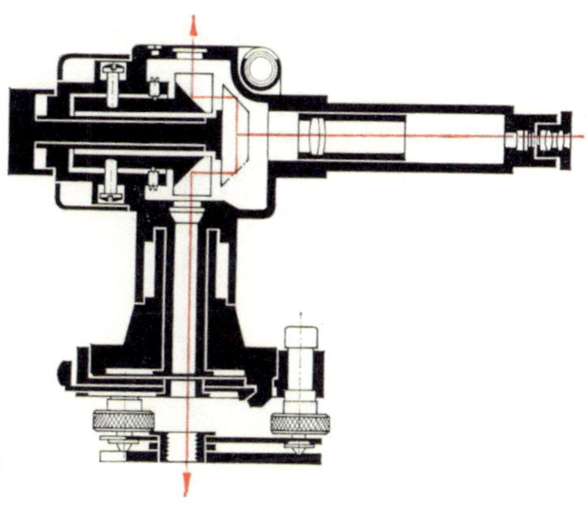

Figure 373 Zenith and nadir plummet Wild ZNL, heigth 170 mm

Table 22. Optical Precision Plummets

type	manufacturer	telescope			bubble sensitivity [″]	length [mm]	height [mm]	mass		maximum range [m]	accuracy of one plumbing	remarks
		Γ_F	diameter of objective lens [mm]	shortest sighting distance [m]				instrument [kg]	container [kg]			
1	2	3	4	5	6	7	8	9	10	11	12	13
with level bubble												
TELIM	F. W. Breit-haupt & Son, Kassel	42	50	2.0	20	230	350	7.1	6.5	2 bis 600	1 : 100,000	cross-level; figure 375
OL	Kern & Co., Aarau	22.5	30	0.8	20	135	160	3.7	1.8	100	1 : 50,000	figure 374
GLQ	Wild Heer-brugg AG	40	60	4.6	–					5 bis 500	1 : 300,000	mercury horizon
ZBL		5	8	0.3	60	160	100	1.5	0.5	20	1 : 10,000	
ZNL		10	12	0.6	20	240	170	2.6	2.2	2 bis 100	1 : 30,000	figure 372, 373
with tilt compensator												
OZP	U.S.S.R.	31	34	3	–	350	150	5.0		3 bis 500	1 : 10,000	
Auto-plumb	Rank Precision Industr. Ltd., England	30 17	40 25	1.8 0.9	– 45	195	215			150	1 : 200,000	top line: zenith plummet bottom line: nadir plummet figure 379, 380 20′ bull's eye level
PZL 100	VEB Carl Zeiss JENA	31.5	40	2.2	–	–	335	4.8	2.8	100	1 : 100,000	figure 376, 377; 8′ bull's eye level
ZL	Wild Heer-brugg AG	24	36	0.9	–			3.1	2.2	100	1 : 200,000	figure 378; 4′ – bull's eye level NL = nadir plummet

Figure 374 Optical precision plummet Kern OL (height 160 mm)

Figure 375 Optical plummet TELIM with electrical illumination (Breithaupt & Son); height 350 mm

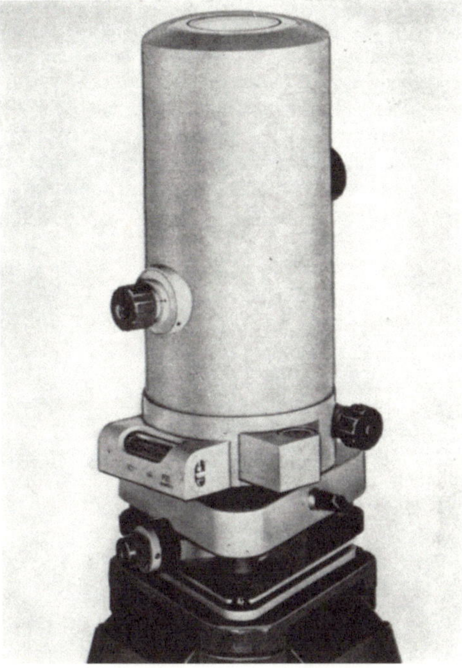

Figure 376 Precision plummet PZL (Zeiss, Jena) height 355 mm

telescopes of the optical plummets are either supported at the horizontal axis – then the measurements are carried out in two vertical planes normal to each other — or else are arranged vertically. Depending on the direction of sight, one distinguishes between nadir- and zenith plummets. The collimation axis can be set vertical with the aid of a liquid horizon (mercury or oil for auto-collimation observations for nadir sights), with a spirit bubble, or with a compensator.

Like the plummet DOPLO from Breithaupt, the zenith- and nadir plummet Wild ZNL (figure 372) contains a telescope with a prism that can be switched for nadir- or zenith observations (figure 373). The optical precision plummet Kern OL (figure 374) has a second telescope instead of the reversing prism. The instrument has regular foot screws, and can be used with other tripods as well. The instrument TELIM (figure 375) by Breithaupt contains telescope placed in the tribrack, which can be rotated around the vertical axis.

The precision plummet PZL 100 (figure 376) by Zeiss, Jena, has a compensator and a 100 gon notch and was developed from the Ni 007 (see figures 377, 467). The collimation axis of the zenith plummet ZL (figure 378) by Wild also sets automatically. Watt's Autoplumb (figure 379) contains a compen-

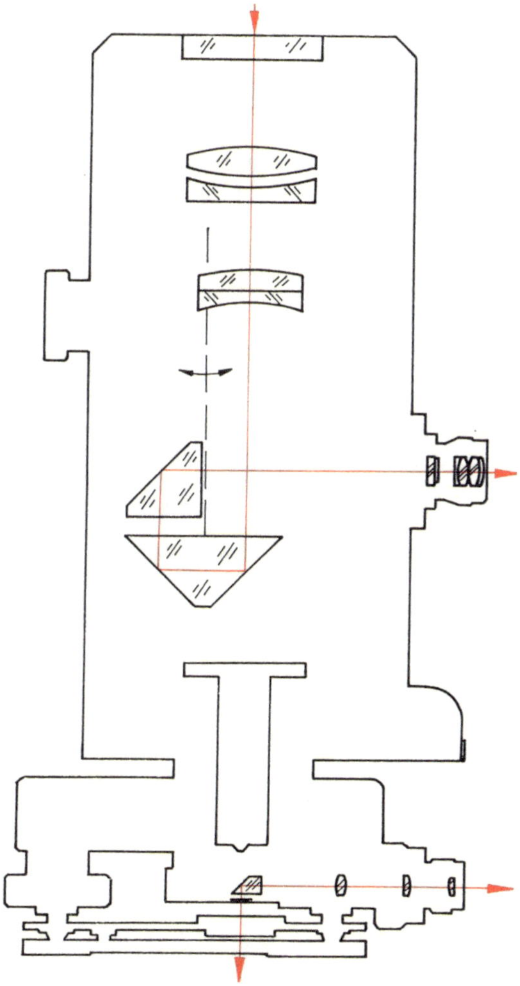

Figure 377 Optical train in PZL

Figure 378 Automatical zenith plummet ZL (Wild)

Figure 379 Watts Autoplumb (height 215 mm)

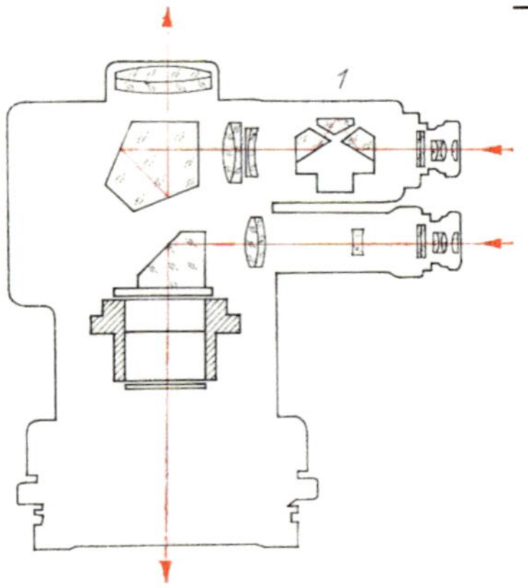

Figure 380 Optical system of the Watts Autoplumb
1 compensator

sator in the zenith telescope, which is identical to the one in the compensator level Autoset (see figure 455) and which automatically sets the collimation axis vertical (figure 380). The amount of deviations from the plumb line can be measured with a micrometer. The lower telescope is used for nadir sights. The Soviet "Zenit-OZP" contains a lens compensator. By switching a pentagon prism in front of the telescope, nadir- or zenith sights are obtained. Similarly, the Ni 2 (see figures 470, 471) by Zeiss,Oberkochen, can be utilized as a precision plummet by placing a prism in front of the objective lens.

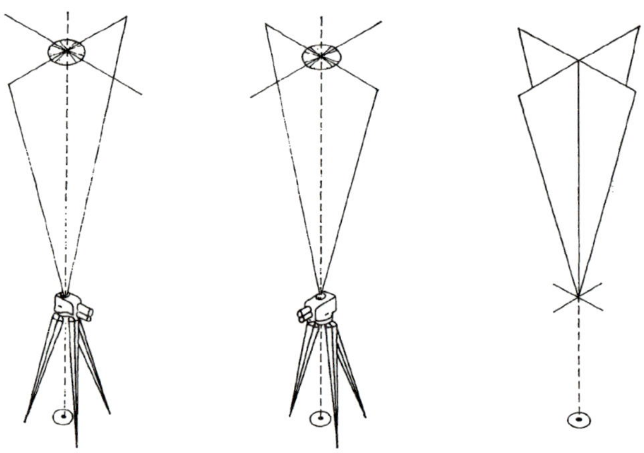

Figure 381 Plumbing with compensator plummets

Optical precision plummets are operated like simple optical plummets. The vertical axis VV, the collimation axis ZZ and the bubble axis LL should be arranged as follows: $LL \perp VV$, and $ZZ \parallel VV$ ($ZZ \,|||\, VV$!). The collimation error, in this case the angular deviation between collimation axis and vertical axis, is composed of the eccentricity e which is small and can be neglected (e.g. for the Kern OL it is less than 0.02 mm), the primary collimation axis error c, which is the angle between vertical- and collimation axis at ∞-focus, and the secondary collimation axis error c', which is a rotation of the collimation axis depending on the focus setting. Its influence is eliminated by observing the target in two diametrically opposite positions of the top part. Since compensator plummets only provide one vertical plane which contains the plumbline, two sights normal to each other are required for unique plumbing (figure 381). The results obtained with this method contain residual levelling errors as well as the influence of the wobble of the vertical axis. The mean of plumbings in four by 100 gon different positions of the top part is practically free of these influences.

Figure 382 Alignment instrument with striding level (Freiberger Präzisionsmechanik); length 593 mm

Plumbing can be accomplished by estimating intervals when transferring coordinates or when measuring deviations from the plumbline, by reading at graduated targets, or by shifting targets placed on cross rails until they are centered. Recently, optical precision plummets have occasionally be replaced by gas lasers (see section 2.6.) for work on high constructions. A disk with a cross mark is placed on the working platform on which the laser ray coming from vertically below is imaged as a light dot.

2.6. Alignment Instruments

Figure 383 Equipment for deformation surveys of power dams NITAL (Breithaupt & Son)

Decades ago, instruments measuring small deviations from a straight line were only used in first order control surveys. With the development of engineering surveys, these instruments gained importance for alignment- and deformation measurements of engineering structures as well as for industrial surveys.

Autocollimation devices and alignment instruments are well suited to monitor buildings for horizontal changes under varius loads and at different temperatures. As examples are mentioned: Deformation surveys of power dams where the amounts are obtained by measuring deviations from a straight line and comparing them to the results of previous measurements; alignments of machine parts along straight lines in industry. Since the deviations are small, the measuring accuracy has to be high.

Alignment instruments (figure 382) do not have horizontal- or vertical circles, but have a tiltable telescope with high magnificatior. With the instrument set up solidly, first a fixed reference mark at the end of the line is sighted. Intermediate points represented by movable targets, placed on forced centered rails normal to the line, are then aligned in the vertical plane

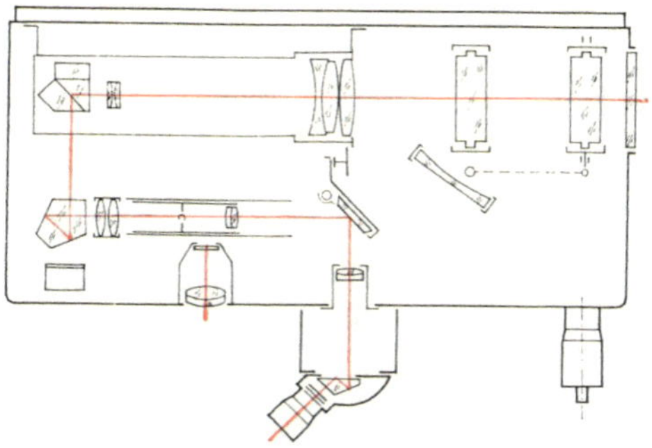

Figure 384 Alignment telescope FF 1 (Zeiss, Jena)

containing the station and the reference mark. The amount of deviation from the line is then measured.

The equipment for deformation surveys of power dams NITAL by Breithaupt consits of a telescope (42 × magnification, 50 mm diameter of the objective lens, ±30° tilting angle), a fiscal reference mark and movable target (figure 383).

Alignment instruments, similar to the ones used in surveying, are common in industrial surveying, e.g. in mechanical engineering, for the placement of large machine parts and in instrument construction. They are always used in connection with a reference point, which may be a fixed part of the test object or a special target. The alignment telescope FF 1 (figure 384) is based on the principle of double image measurements. Therefore inevitable errors in the guidance of the focussing lens do not affect the measurements. The images of the target move against each other, which leads to twice the setting accuracy. The measuring range is ±1 mm, the least count 1 μm, and the range with lens attachment is 200 m.

Autocollimation devices are used primarily in mechanical engineering and in laboratories for the alignment of equipment and machinery. These devices consist of autocollimator and autocollimation mirror. Depending on how the image shift is measured, one distinguishes between optical and electro-optical autocollimators. It is visually observed for optical autocollimators, and measured photo-electrically for electro-optical ones.

Optical autocollimators are usually 300 to 600 mm long, have a focal lenght of about 500 mm, with the diameter of the objective lens being 40 to 60 mm, and a mass of 3 to 5 kg. The measuring range is 10 to 30′, even less for some.

To shorten the instrument, the path of light is ocassionally bent (figure 385), or else a microscope is used instead of the objective lens (figure 386). Measuring screws or optical micrometer serve as measuring devices. The least count is between 0.1 and 2″. For optical systems without telelens, the microscope may be placed in the optical axis with the cross hairs moved sideways (figures 385 and 386) and vice versa.

For systems with telelens, cross hair and micrometer are usually placed centrically so that the cross hair and its image appear simultaneously in the field of view (figure 386). Because of its high setting accuracy, the autocollimation method has gained special importance in industrial surveying. Therefore some manufacturers produce autocollimation eyepieces for their theodolites (figure 387), so that the telescope can be used as autocollimator.

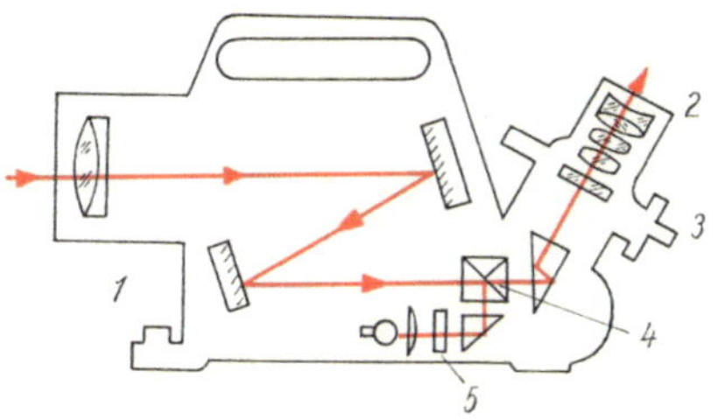

Figure 385 Optical system of the autocollimation telescope by Leitz, Wetzlar
1 objective lens; *2* eyepiece; *3* micrometer; *4* beam splitter cube; *3* cross hair

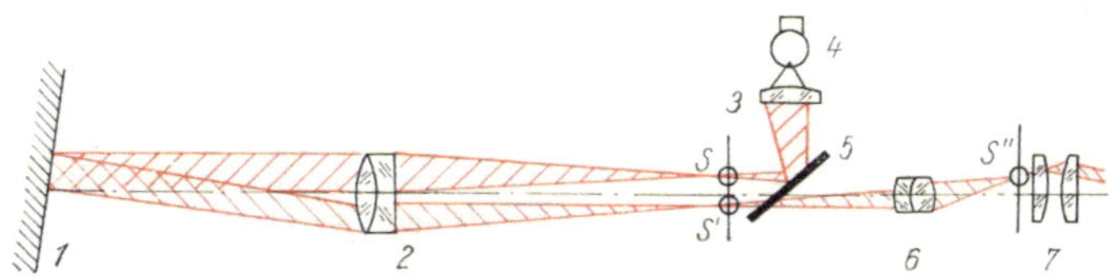

Figure 386 Optical system of the Microptic autocollimator (Rank Prec. Ind.)
1 mirror; *2* objective lens of telescope; *3* cross hair; *4* light source; *5* splitter surface; *6* objective lens of microscope; *7* micrometer

Wild produces the autocollimation eyepiece GOA (Figure 388) for the theodolites T 1, T 16, T 2 and T 3, which can easily be exchanged with the regular telescope eyepiece. It contains a beam splitter prism instead of the mirror (figure 387).

Similar devices are manufactured by Zeiss, Jena for the THEO 010A, and by Zeiss, Oberkochen for Ni 1 and Ni 2, 41 to 43. They are used by Rank, England for the line compass (figure 182) and in orientation magnetometers in mining.

In U.S.S.R., the autocollimation eyepiece by *Monchenko* (figure 389) is used. A crosshair is engraved at the contact surface where the prisms are cemented together. In their eyepiece, two images appear of it, namely one image created by rays passing through the cross hair plane to the mirror and being reflected to the eye piece, and the other one created by parts of the rays passing through both prisms and being reflected at the mirror coated surfaces, as well as at the back of the cemented surface.

For electro-optical autocollimators, the subjective reading is replaced by an objective, automatic setting and reading. Thus subjective errors are practically avoided.

Electro-optical autocollimators contain measuring devices with motor driven movable parts. The mobile part of the micrometer is rotated via a servo motor until coincidence between measuring mark and its image is sensed by

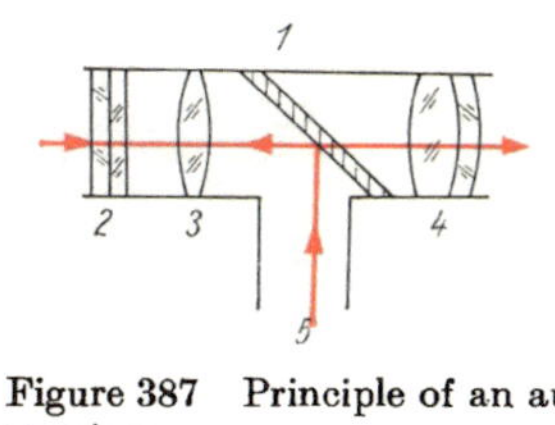

Figure 387 Principle of an autocollimation eyepiece
1 semipermeant mirror; *2* reticule; *3* field lens; *4* eye lens; *5* illumination plug

Figure 388 Wild T 2 with autocollimation eyepiece GOA

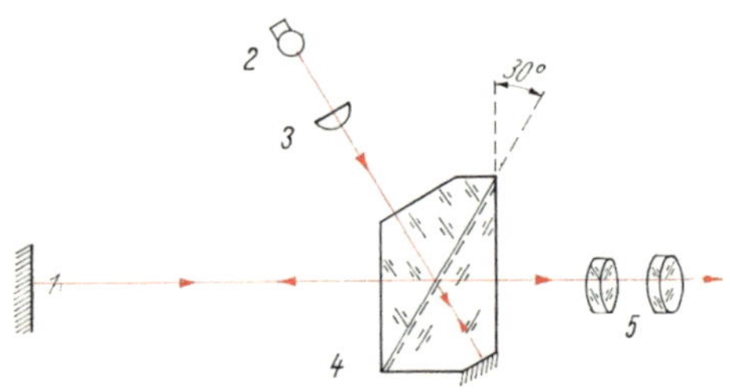

Figure 389 Autocollimation eyepiece by *Monchenko*
1 mirror; *2* light source: *3* condensor; *4* cross hair; *5* eyepiece

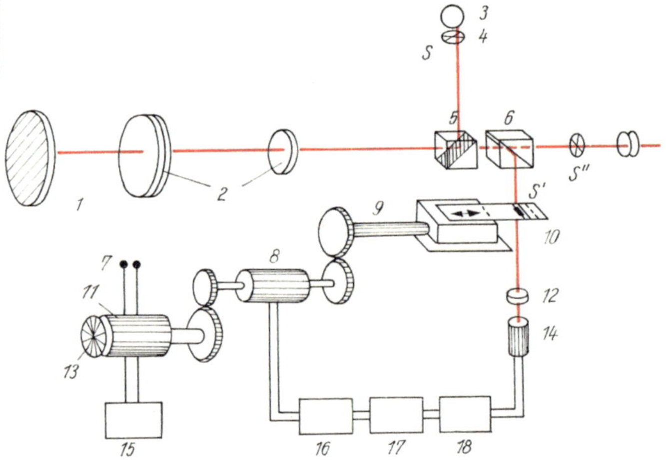

Figure 390 Principle of the electro optical autocollimator TA 58 (Rank Prec. Ind.)
1 plane mirror; *2* objective lens; *3* light source; *4* cross hair; *5* beam splitter cube 1; *6* beam splitter cube 2; *7* voltage supply; *8* servo motor; *9* spindle; *10* swinging slot; *11* potentiometer; *14* photo cell; *15* digital voltage meter; *16* servo amplifier; *17* discriminator; *18* amplifier

a photo cell utilizing modulated or polarized light (figure 390). If the measuring device does not have motor driven parts, the bundle of rays reflected by the mirror is split and falls onto two photo cells. The location of the image in respect to its reference position is then determined by the difference of the two photo-electic currents.

In electro-optical autocollimators by Rank, England (figure 390), the cross hair S is imaged at S' where a slot with the width of the cross hair oscillates at 50 cycles with an amplitude of twice the slot width. The rays pass through it onto a photo cell, whose A/C voltage is demodulated without phase change in a discriminator. The smoothened and amplified D/C current activates a servomotor, who shifts the oscillating slot via a spindle. When slot and cross hair image coincide, the photo cell generates a A/C voltage of 100 cycles, such that the servo motor does not receive any current after demodulation. The potentiometer, which is also activated by the servo motor, releases a voltage proportional to the position of the oscillating slot, resp. to the shift of the cross hair image, which is processed analogue or digitally.

For a few years, gas lasers (He-Ne-Laser, wavelength 632.8 mm, red light) have also been used for the previously mentioned purposes for surveying work in construction industry, mining and mechanical engineering, namely setting out, control- and monitoring surveys, plumbing for high elevations and great depths, aligning of conveyor belts, axes, and communication- and similar lines, with accuracies in the centimetre and millimetre range, occasionally even higher. These beams have a very narrow bundle of rays. A beam of 2 mm diameter leaving the instrument and having a divergency of 2′, generates a light spot of 62 mm at 100 m, which can be reduced with telescopic optics. These lasers have high light density, high efficiency, as well as excellent monochromatic character and directional stability. Gas lasers operate continually without cooling.

The laser beam can be used as:
– straight line for alignments and setting out, e.g. for staking roads and railroads, monitoring crane rails, construction of machine fundaments, alignments of power dams
– plumbline, e.g. for smoke stacks and bridges
– reference plane for measurements of inclination and settlements, area levelling or profiles, e.g. for meliorations or stadia surveys
— reference beam for placing pipelines and monitoring machines in direction and elevation, e.g. by digging in mining and tunneling, placing of supportive walls, and sewerage ditches.
Gas lasers consist basically of an optical resonator, formed by two spherical mirrors with an active amplifying medium (Helium Neon gas-mixture). One of the mirrors is semi permeable, so that part of the energy stored in the resonator can be released and utilized as starting power of the osillator. The He-Ne-Mixture in the laser tube is aggravated by D/C current until emission of the beam occurs. The laser tube is vacuum-tightly closed by plane plates, which, because of their inclination angle (Brewster angle), cause that a light ray parallel to the tube axis is not reflected at their surface, but passes the closed tube without loss.
Laser alignment devices consist of the gas laser, an instrument to generate voltage (12 V accumulator or net) and the optics for bundeling or diverting of the rays (cylindrical lens).
For subjective sensing, screens with holes, ground glass screens etc. are used where the position of the light spot is manually determined, while photo diodes, sensing the laser are used for objective readings. Photo diodes linked with a steering device for the machine, permit a fully automatic monitoring in tunnelling and melioration works with the laser as guide beam.
Laser instruments are either built as separate instruments, or obtained with lasers attached to surveying instruments. There are differences in range based on intensity and beam divergence. Due to low intensity, the human eye is not directly in danger.

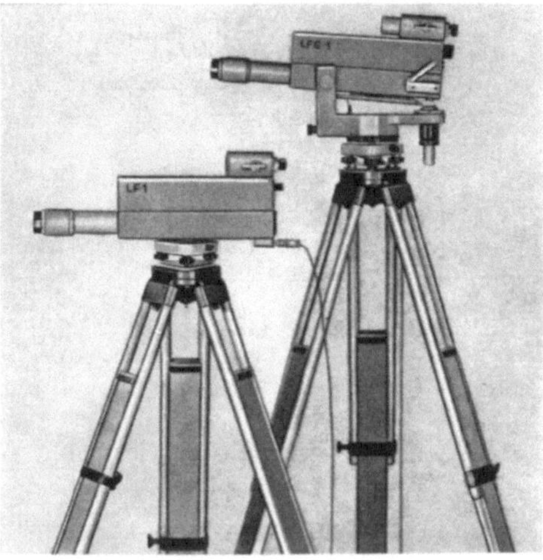

Figure 391 Laser alignment instruments LF 1 and LFG 1 (Zeiss, Jena)

Separate Laser Alignment Devices have a laser tube either in a special support or rigidly connected to the telescope of a theodolite or level. Bull's eye or spirit levels serve for levelling, while movements in small ranges are accomplished with a micrometer. Occasionally, a horizontal- and vertical circle are available.
In the laser alignment devices LF-1 (dimensions: 54 mm × 24 mm, 4.6 kg) and LFG 1 (figure 391) by Zeiss, Jena, gas laser (0.8 mW intensity) electronics, telescope and level bubbles (20″ and 8′ sensitivity), as well as rigidly set collimator (15 × or 30 × magnification) are combined to one unit. A 12 V battery serves as power supply. The LFG 1 (dimensions: 54 mm × 15 mm × 34 mm, 9.5 kg) with horizontal circle (1 gon resp. 1° least count) and fine movement, can be inclined with a measuring screw. Reflective screens (figure 393) are used to make the laser point visible for alignments, while transparent screens (figure 392) are used to find points on the line.
The laser beam can be formed with cylindrical lenses in the horizontal- or vertical plane. The setting accuracy is ±1 mm/100 m with a range of 0.5 km in daylight and 1 km at night.

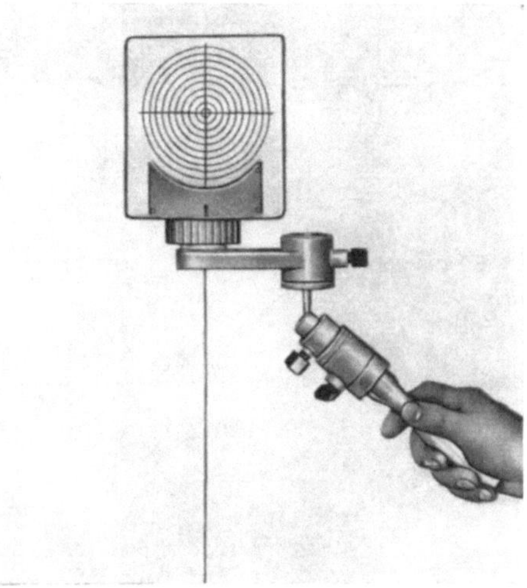

Figure 392 Transparent target screen

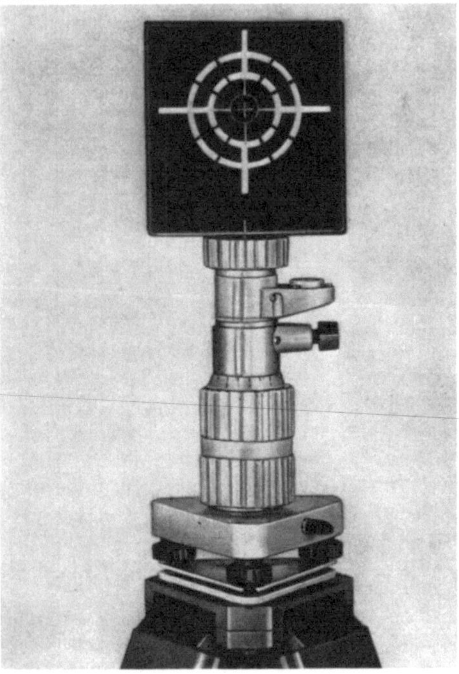

Figure 393 Light reflecting target screen

Figure 394 Construction laser VSE 20 (Laser Light)

Figure 396 Alignment laser with theodolite ST 200 (Rank Prec. Ind.)

Figure 395 Surveying laser LS 132 (GEO-Feinmechanik)

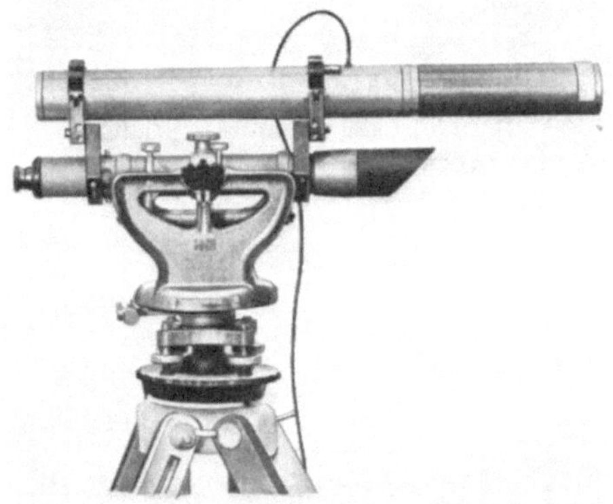

Figure 397 Construction laser (Keuffel & Esser, U.S.A.)

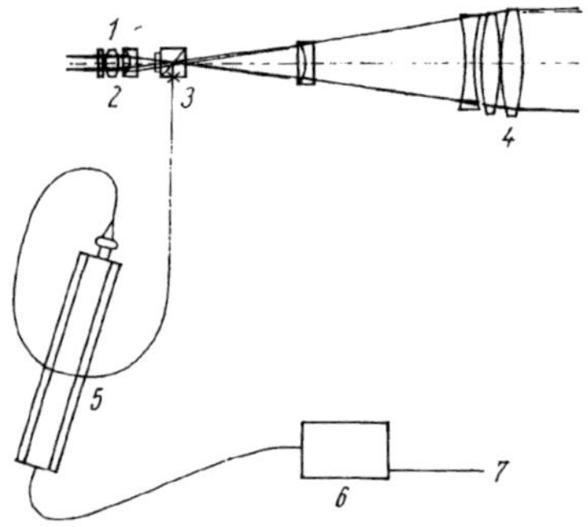

Figure 398 Principle of laser theodolite DKM 2-AL (Kern)
1 eyepiece; *2* filter; *3* beam splitter cube; *4* objective lens; *5* laser; *6* laser supply unit; *7* power source

Separate laser instruments have been developed by combining lasers with theodolites, levels or special tribracks, namely in U.S.S.R. (e.g. laser theodolite LT 56, laser level LN 56, laser alignment device LV 55), in Poland (laser theodolite ULIG-KP 1, laser level ULIG-KP 2, laser theodolite APLO-KP 4), by Spectra-Physics, U.S.A. (laser theodolites LT-2, LT-3 and LT-611, canal laser RK 3), by Rank Precision Industries, England (alignment laser) by Siemens, Munich (laser theodolites LGN 9136, LG 68, GL 681, laser level Gradomat), by Laser Light, Switzerland (construction laser VSE 20, figure 394), by Visomat, Switzerland (multipurpose laser LS 4), by Stolz, Switzerland (High construction laser 035), by Covi, FRG (levelling laser HKT 1), by Geo-Feinmechanik, FRG (tunnelling laser LH 132, and 134 for hanging position, and LS 132 (figure 395) and 134 for standing position, plumbing laser LL 132, underground construction laser TL 7), by SLOM, Paris (alignment laser) and others. The intensities of these lasers are 1 to 5 mW, the diameter of the beam at the instrument 8 to 18 mm, occasionally even less, at 200 m distance about 25 to 50 mm. The laser tube can usually be tilted by $\pm 30°$. Bull's eye bubbles and occasionally tilting and readable tubular bubbles serve for levelling. The instruments are 400 to 600 mm long with a mass of 7 to 10 kg.

Figure 399 Wild T 2 with laser eyepiece GLO 2

Laser Attachments mounted with the aid of special laser holders or lasers adapters onto the telescope of theodolites transform them to laser instruments. The laser beam is then eccentric to the collimation axis of the theodolite. A Spectra Physics Laser can be attached to the theodolite STN by SLOM, Paris via a telescope adapter, while it can be mounted onto the ST 200 by Rank (England) via a special plate (figure 396). Similar to these instruments applies to the alignment laser by Keuffel & Esser, U.S.A. (figure 397).

By bringing the beam of a 5 mW laser 120 T by Spectra Physics into the telescope between eyepiece and cross hair, concentric exit of the laser beam is obtained for the Kern DKM 2-AL (figure 398). The laser beam therefore does not serve for measurements, but for illumination of the cross hair. It is projected onto a target by the objective lens of the telescope. Its range is 300 m by day light, and 700 m at night. The thickness of the measuring cross is 7 mm at 100 m, and 27 mm at 400 m distance. The telescope of the DKM 2-A has a 30 × magnification. The DKM 2-AL equipment has a mass of 28.5 kg.

For the Nikon NT 2 and the Topcon TL 2, the laser light is brought from the laser tube to the attached telescope with a reverting prism.

For the Wild theodolites T 1, T 16 and T 2, a laser eyepiece with dividing cube and fiber optics for transmission of the laser light can be exchanged with the standard eyepiece (figure 399).

Laser and power pack are fastened to a tripod leg. Thus the theodolites become laser theodolites, one can directly sight to a point or direct the beam with the horizontal and vertical angles in to the desired direction and tilt. Simple construction lasers, however, require that the laser beam is set to a reference point.

3. Instruments to Measure Elevations and Elevation Differences

Elevation differences are measured in order to determine the position of points in the vertical plane. These are then used to compute elevations in a unique system referred to mean sea level. In surveying, mainly three methods are used to measure elevation differences, namely

1. geometric levelling
2. trigonometric heighting
3. barometric altimetry.

The given order signifies accuracy and required effort for the methods. The highest accuracy can be obtained with geometric levelling, the effort required, however, is the greatest. When selecting the approach, required accuracy and the terrain are the main criteria.
In geometric levelling, the elevation differences are directly determined with the aid of a horizontal line of sight. To achieve this, the effect of gravity is utilized, under whose influence the free surface of a liquid becomes horizontal, a liquid in connected tubes comes to the same level, a pendulum comes to rest in direction of the gravity, and the bubble centre of a level moves to the highest point of the ground arc.
In trigonometric heighting, the elevation differences are computed using zenith distances (vertical angles), and distances to the target.
In barometric altimetry the elevation differences are determined from differences of the measured air pressure.

3.1. Simple Instruments for Geometric Levelling

3.1.1. Hydrostatic Levels

Hydrostatic levels are based on the theorem, that in connected tubes the calm surface of a liquid forms a horizontal plane. Its line of sight is automatically level.
The simple water level consists of two glass cylinders connected with a tube. It is about half filled with a coloured liquid which gives the necessary horizontal line of sight when sighting across the liquid surfaces. Due to its low aiming accuracy, the instrument is not suitable for surveying purposes.
The hydrostatic level has a long hose instead of the tube, which connects the two glass cylinders, which have millimetre graduation and are encased in metal. Elevation differences can be determined with the aid of special reading devices, but only for points with approximately the same elevation. For the numerical evaluation, additional observations (air pressure, temperature) have to be taken into account.
The hydrostatic level according to *Meisser* (figure 400), has a measuring range in elevation of 100 mm (usable spindle length), 42 mm interior diameter of the cylinder, a 30 to 50 m long hose with 10 mm interior diameter, and a measuring accuracy of ± 0.01 mm (in enclosed spaces). The total mass of the equipment (two containers and back pack) is 31.5 kg. To accelerate the measuring procedure, and to eliminate blunders, an indicator was developed for the instrument, which emits an optical signal when contact is achieved between the liquid surface and the measuring spindle. When setting up on a tripod, bench marks, vertically placed in the ground, can be included in the measurements. A further development ELWAAG by *Thierbach* and *Barth*, F.R.G., has a servo motor with remote control to drive the precision measuring device. Upon contact with the liquid surface, the motor stops. With the hydrostatic level NST-1 (U.S.S.R.), elevation differences of up to ± 200 mm can be measured at 0.1 to 10 m distance to within ± 10 mm.
When measuring, air bubbles are removed by moving the hose, and an index correction is determined by comparing the readings of glass cylinders held beside each other. The glass cylinders are hung onto plugs with known elevation. After the water in the hose has come to rest, the tip of the spindle, which is connected with a measuring drum, is set onto the liquid surface. Several readings of the liquid surface are taken simultaneously on both

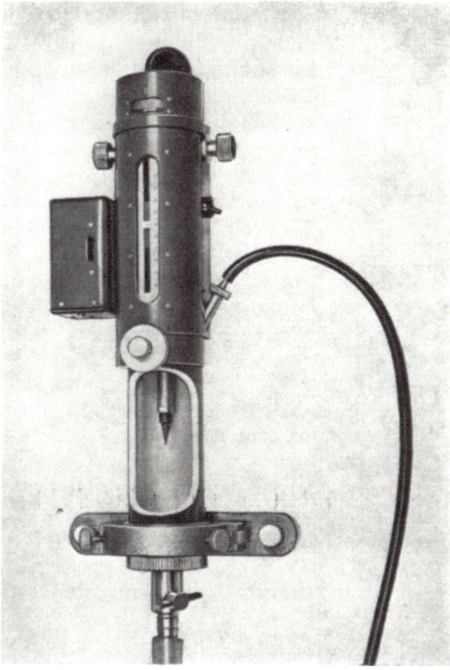

Figure 400 Measuring cylinder of a precision hydrostatic level (Freiberger Präzisionsmechanik)

glass cylinders. The accuracy limit is determined by temperature influences, rather than by the instrumentation. Temperature differences along the measuring path reduce the accuracy.
The hydrostatic level is especially suited for determining and measuring vertical movements on structures, such as bridges, machine foundations, and power dams, i.e. in narrow areas or areas with visual problems. It is also used to prove mining damages and tectonic movements in mountaineous areas.
Automatic hydrostatic levels are designed for the continual control of height points (settlements of the soil).

3.1.2. Pendulum Instruments

In pendulum instruments, the line of sight is determined by a sighting device arranged normal to a rigid plumb and thus automatically levelled. The pendulum level consists of a 1.5 m long rod with a freely swinging rectangular frame with diopter (figure 401). For measuring grades of a slope, the diopter can be shifted along a scale. The standard deviation for the determination of an elevation difference amounts to about ± 0.5 m/km,

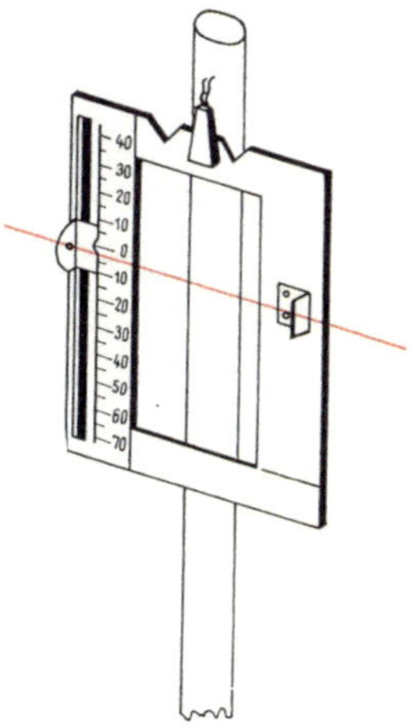

Figure 401 Pendulum level

Figure 402 Cowley level (Rank Prec. Ind.)

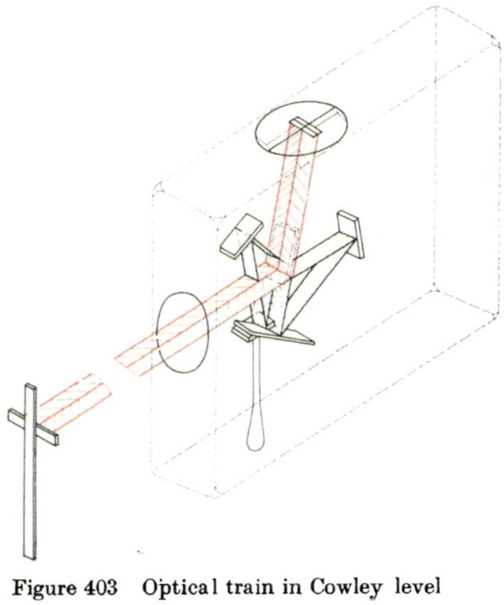

Figure 403 Optical train in Cowley level

provided the sighting distances are not too long. A rectangle prism and a plumb bob can be utilized like a pendulum level. The line of sight is given by the image of the plumb bob in the prism.
Rank Precision Industries, England produces the "Cowley Level" (figure 402) for simple construction measurements. It does not contain lenses or prisms, but two mirror systems. The left one consists of two mirrors under 40° (figure 403), the right of three mirrors, with the central one being connected to a pendulum which is dampened by electric currents. Independent of the tilt of the instrument, horizontal rays leave it under 80° to the horizontal plane. A target is shifted along the graduated vertical staff until it is at the horizontal plane, i.e. until half images coincide when looking into the instrument from the top (figure 403). The standard deviation is ±6 mm/30 m.

3.1.3. Handheld Levels

In anglo saxon countries, hand held levels (figure 404) are used for simple construction measurements. They consist of a tube with a crude tubular bubble mounted to it. Its image is reflected into one half of the field of view, while the object appears in the other half. If the image of the bubble is bisected by the horizontal line on the mirror, the line of sight is horizontal.

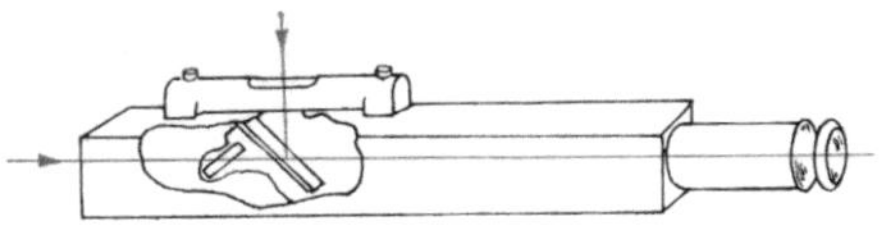

Figure 404 Hand level

3.1.4. Horizontal Straight Edge

The horizontal straight edge consists of the 3 or 4 m long actual straight edge S, equipped with a tubular bubble and of the usually 3 m long measuring rod M, which is graduated in centimetres and has a square cross section. It is set into vertical position with a bull's eye bubble (figure 405).
When measuring, one levels the tubular bubble and reads the elevation difference at the intersection of the vertical measuring rod with the bottom edge of the straight edge. If the points are farther apart than one length of the straight edge, the elevation difference is determined in steps. To obtain correct results, the bottom edge has to be parallel to the bubble axis LL. If this is the case, then the elevation differences Δh_1 and Δh_2, obtained in two positions with the bubble end reversed, are the same. If not, the bubble has to be adjusted using a known elevation difference.

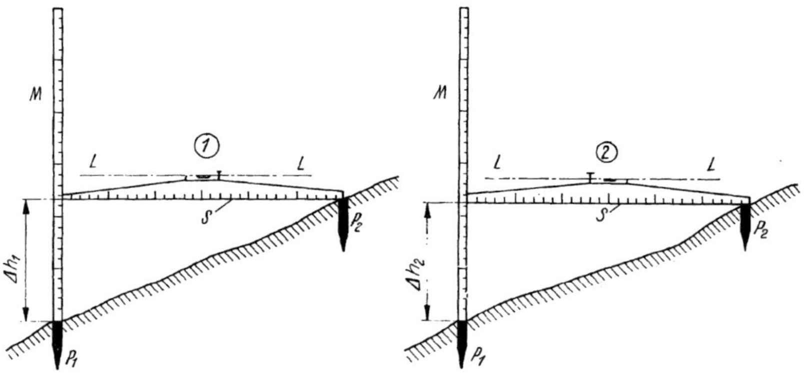

Figure 405 Testing of a horizontal straight edge
1 = 1st position; *2* = 2nd position

The horizontal straight edge is mainly used in very steep and difficult terrain. One can expect a standard deviation of ±0.5 cm per setup; which means a standard deviation for n setups of $\pm 0.5\sqrt{n}$ [cm].

3.2. Levels

Levels are the most important instruments for geometric levelling. Their main elements are a telescope that can be rotated around a vertical axis, and devices to level the line of sight. Some levels are equipped with a horizontal circle and stadia hairs.

3.2.1. Tripod and Fastening Devices for Levels

Tripods for Levels

Levelling tripods are basically the same as for theodolites. However, the possibility of shifting the instrument on the tripod head is missing, because there is no need for centering. Tripods with threaded head, common in England and non-European countries, can be used for both, levels and theodolites. The tripod legs are either rigid or telescopic.
High precision levels are primarily used with tripods with rigid legs.

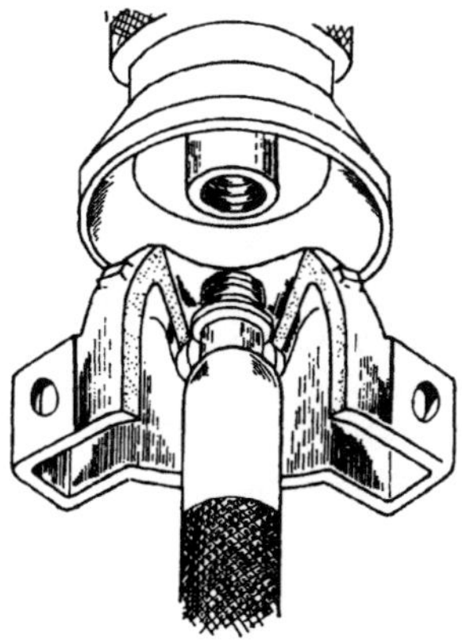

Figure 406 Tripod-instrument connection with sphere head tripod (Rank Prec. Industries)

Connection between Tripod and Level

Like the theodolite, the level is fastened onto the tripod with base plate, spring plate and fastening screw. Occasionally the foot screws are directly placed into the slots of the tripod plate, or the base plate is screwed onto the tripod, which is common for English and non-European instruments. It is becoming more and more common, especially for automatic levels, to use tripods where the bottom part of the instrument is directly tightened with the fastening screw without base- or spring plates. The top of the tripod usually contains a spherical head, on which the bottom part of the level can be shifted and fastened, because of concave (figure 406) or coincal support surfaces (figure 407). Other tripods have tops with plane tripod plate and spherical zones at the sides (figure 408).
When the tripod legs are clamped, the tripod head can be inclined by up to 40 gon. A few types (Sokkisha, Tokyo) have a spherical base plate which is placed onto the tripod with spherical head (see figure 425). Regular plate tripods are used with the Ni 050 (see figure 450) by Zeiss, Jena, and N-10 L (U.S.S.R.) which contain a system of wedge-disks for rough levelling instead of a tribrack with foot screws.

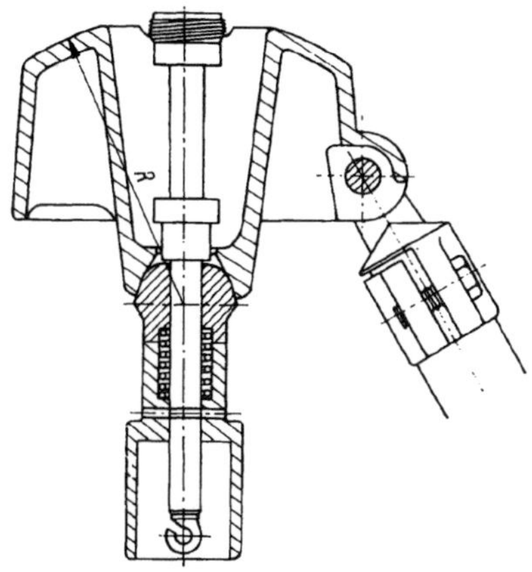

Figure 407 Tripod head with spherical surface based on joint-head principle (Kern)

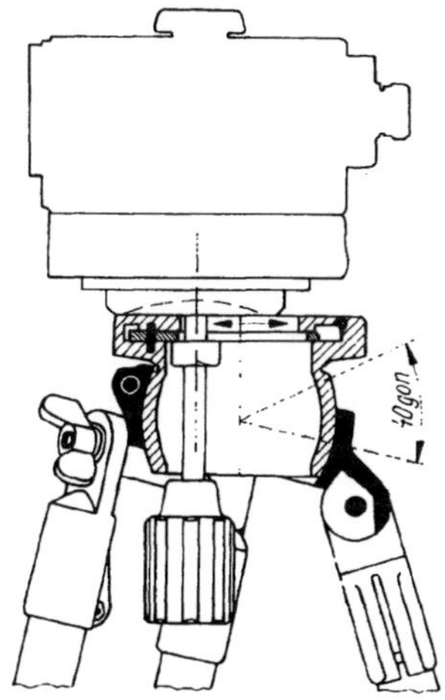

Figure 408 Tripod head of a levelling tripod (Ertel)

Connections of this type between tripod and level are designed to simplify and accellerate the levelling process. After lightly fastening the tightening screw, the bottom part of the instrument is moved along the tripod until the bubble of the bull's eye level is centered. After such rough levelling, the fastening screw is tightened. The instruments usually are not equipped with foot screws, since fine levelling is either accomplished with a tilting screw or else automatically with a compensator.

3.2.2. Structure of the Level

Bottom Part of the Level

The bottom part of a level is rigidly connected with the tripod during measurements. It consists of a tribrack or a cylindrical foot with concave or conical support surfaces, as well as the connection elements to the rotatable top part, which may carry a graduated circle (figure 409).
The tribrack *U* carries the instrument. It is connected to the tripod with a spring plate on a base plate containing part of the thread for the tightening screw. It can be levelled with the aid of three foot screws *F*. Being part of the tribrack, the foot screws rest on the base plate or the plate of the tripod head. They are surrounded by the spring plate which is connected with the base plate by screws and pressed onto the base plate.

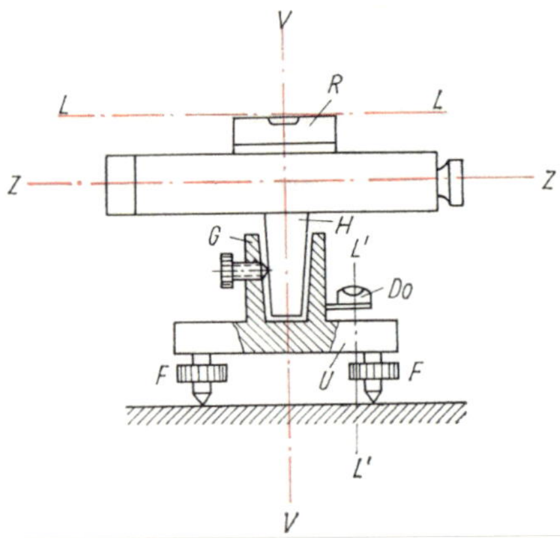

Figure 409 Principle of a level

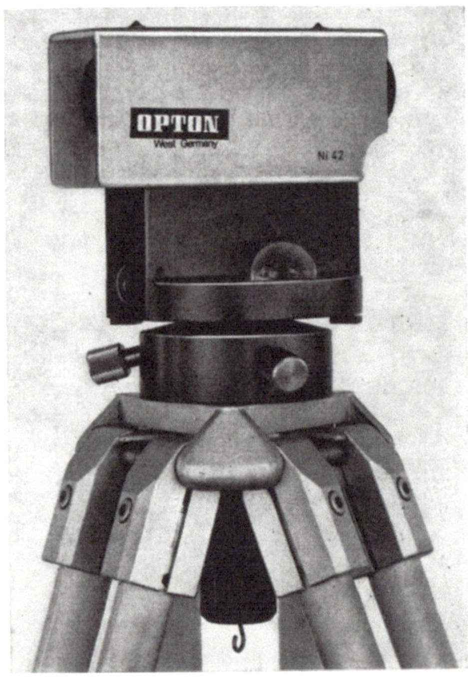

Figure 410 Ni 42 (Zeiss, Oberkochen)

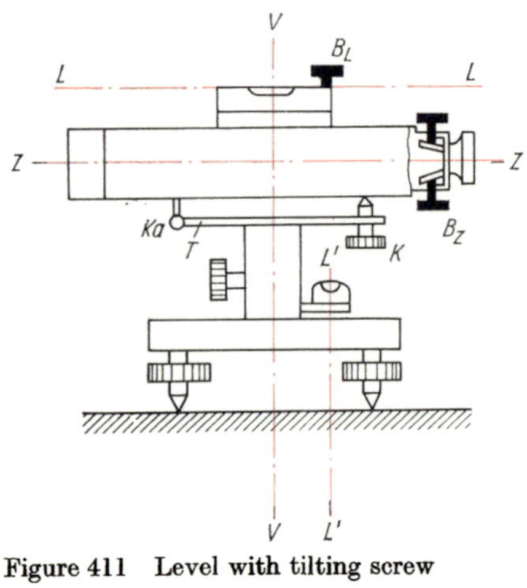

Figure 411 Level with tilting screw

Instruments with cylindrical foot do not have foot screws, spring and base plates, because their foot is directly connected with the tripod head via the tightening screw.
The levels Ni 21 and 42 by Zeiss, Oberkochen are equipped with only two rapidly acting horizontal foot screws for rough levelling. In this case, there is a spherical surface between bottom- and top parts (figure 410). The SNA 1 by SLOM, Paris (see figure 439) has a similar arrangement.
The connecting elements to the rotatable top part are either the socket *G*, which receives the lower end of the vertical axis of the instrument, namely the plug *H* of the top part (figure 409), or else the plug, if the top part is fitted onto it with a socket.

Vertical Axis of Levels

The vertical axis system of a level connects bottom- and top parts. Its primary function is to guarantee the constant position of the axis of the top part after it is set vertical. The vertical axis needs only to be roughly vertical for instruments where the telescope can be tilted, because the telescope is precisely levelled with the aid of the tilting screw.
Therefore the vertical axis systems are of a simple design. Either the axis plug is part of the top part with the socket being at the bottom part or vice versa. The axis plug has a special flange to screw onto the top part, or elese it forms one piece with the telescope body. The axis system is either conical, with the plug resting on a plate- spring for support, or else cylindrical, with the mass of the top part usually supported by an attachment of the plug end. Some instruments have ball bearing axes, similar to theodolite axes (see figure 194).

Top Part of a Spirit Level

The top part of a spirit level is supported by the bottom part such, that it can be rotated. It consists of housing, telescope carrier (for instruments with tilting screw), telescope, and level bubble. Axis plug or socket are either directly attached to the telescope (figure 409), or else connected to the telescope carrier (figure 411), which contains a tilting screw.

Bubble and Telescope of the Level

A bull's eye bubble with about 8 to 15′ sensitivity or a cross bubble at the top part of the instrument is used for rough levelling.
The main element of spirit levels, besides the telescope, is a tubular bubble with 15 to 2″ sensitivity, parallel to the longitudinal direction of the telescope. With it, bubble- and collimation axes can be set parallel, and horizontal, provided the instrument is adjusted and the bubble centred. In reference to the connection between bubble and telescope, as well as telescope and support, levels used to be distinguished as

1. Levels with rigid telescope and fixed bubble (bubble, telescope and support resp. vertical axis are rigidly connected).
2. Levels with rolling telescope and two sided bubble (reversible bubble is rigidly connected with the telescope which can be rolled in its supports).
3. Levels with reversible telescope and reversible bubble (telescope can be reversed in its supports, bubble acts as striding level).

In addition to these, there was a number of other possible constructions with different arrangements and connections between bubble, telescope and support, which required different types of adjustments. The designs with rolling telescope and two sided bubble, resp. reversible telescope and -bubble simplify the instrument adjustment, but require usually an uneconomical effort during production, which cannot be justified.
Therefore, practically all levels are now-a-days built with fixed telescope and bubble connected rigidly to it, often equipped with a tilting screw.
Occasionally, instruments with rolling telescope and two sided bubble are produced. The telescope of the Wild N 2, for example, rests on two V-shaped supports, in which it is held without play by its mass and a pressure spring. Limited by solid stops, it can be rolled into both diametrical positions (figure 412). With this, two rod readings are possible, one with bubble at left, the other with it at the right side. For an adjusted instrument, these are equal while they differ for one which is out of adjustment. Independent of the sighting distance, the arithmetic mean of both readings represents the

Figure 412 Telescope positions for levels with rolling telescope

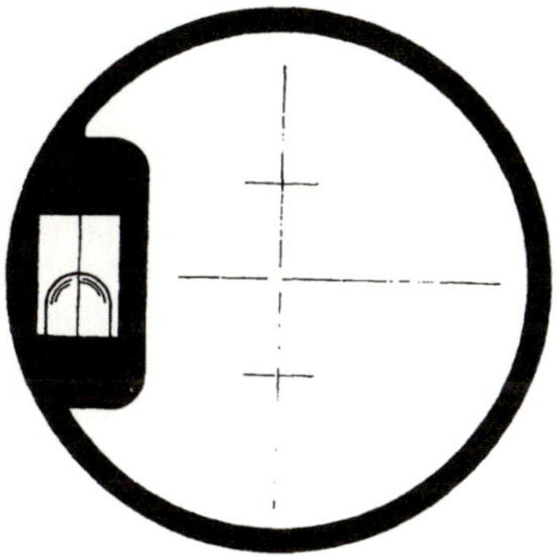

Figure 413 Field of view with image of level bubble (Filotecnica Salmoiraghi 5167)

horizontal sight. Instruments of that kind are especially useful when different fore and back sight distances have to be selected. Levels of that kind are also used for special purposes.
Telescope body and bubble carrier of modern instruments are mostly produced from one piece of cast metal. The bubble is placed such, that it is protected from external influences, especially direct sun rays. The levelling accuracy for setting the vertical axis is increased by refining the bubble readings with the aid of special aids. For low accuracy instruments, the bubble is often reflected by a mirror, and can be observed near the telescope eyepiece. All instruments of medium and high accuracy are equipped with coincidence levels. They permit setting of the bubble to about 1/50th of the sensitivity, which means, at least $\pm 0.5''$ for a $20''$-bubble. For high accuracy instruments, the bubble is usually reflected into the field of view of the telescope (figure 413). This makes the bubble setting more convenient and permits a control of the bubble position when reading the rod.

Devices for Additional Tilting of the Telescope

With the exception of most low accuracy levels, nearly all spirit levels of medium and high accuracy are equipped with a tilting screw *K* (figures 411, 414, 415). With it, the telescope can be slightly tilted around the horizontal axis Ka, which connects the telescope support with the telescope body. Unlike for the theodolite, the vertical axis does not have to be exactly vertical, which can be achieved with the aid of a bull's eye level. For each sighting, the collimation axis is precisely levelled with the tilting screw, prior to reading the rod. Tilting screws have a low pitch (less than 0.5 mm) and operate with lever transmissions. With this, the effect of the screw can be refined at will. This mechanism enables an accurate, convenient and fast levelling of the bubble. Because of the lever transmission, the operating knob can be placed either horizontally or vertically in a position, suitable to the observer. The horizontal axis is placed as near as possible to the vertical axis, so that the instrument height does not change when tilting the telescope.

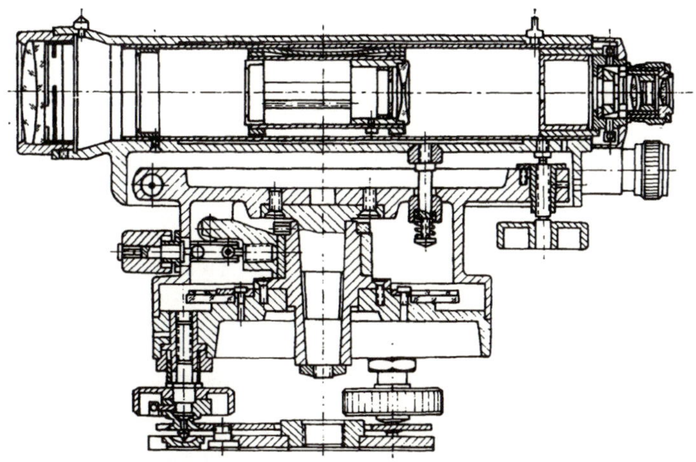

Figure 414 Section through a level with vertically acting tilting screw (Nitac, Fennel)

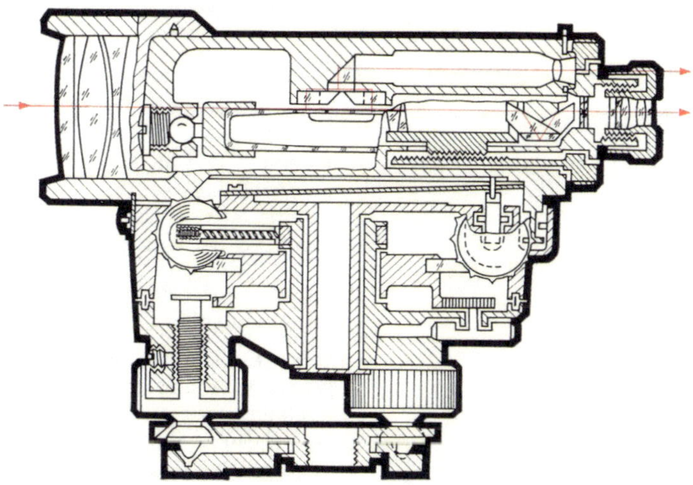

Figure 415 Section through a level with horizontally acting tilting screw (MOM Ni-B 11)

Top Part of Compensator Levels

Spirit levels are more and more displaced by compensator levels, where the bubble is replaced by a tilt compensator which is an automatically acting element. With it, there is no more need for levelling a bubble, so both measuring speed and -accuracy have improved. The instrument needs only to be roughly levelled with a bull's eye level, while the fine levelling of the collimation axis is accomplished automatically by the compensator.

Fundamentals of Compensator Levels

With horizontal collimation axis, a horizontal principal ray, coming from an object point will pass through the horizontal line of the cross hair. If the telescope is inclined by a small angle α against the horizontal plane, the principal ray will hit the reticule above or below the horizontal line S_0 (figure 416) by an amount of

$$a = f \tan \alpha \approx f\alpha \,. \tag{140}$$

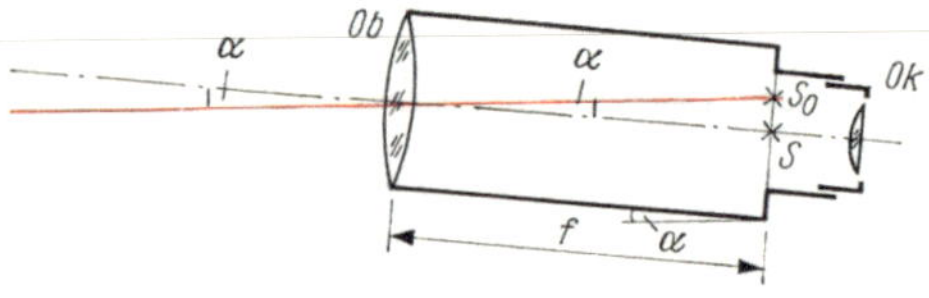

Figure 416 Horizontal principal ray for tilted telescope

In order to achieve, that a horizontal principal ray will intersect the horizontal line of the cross hair in spite of the inclination of the telescope, either the cross hair is monitored such, that it gets from position S to S_0 (figure 417), or else the ray is shifted (figure 418), possibly by parallel shift such, that it will hit the horizontal line of the cross hair. This monitoring is accomplished automatically with the aid of an additional element, called tilt compensator, placed at the bending point K. At that point, the deflection angle β has to be monitored such, that

$$\frac{f}{s} = \frac{\beta}{\alpha} = n \,. \tag{141}$$

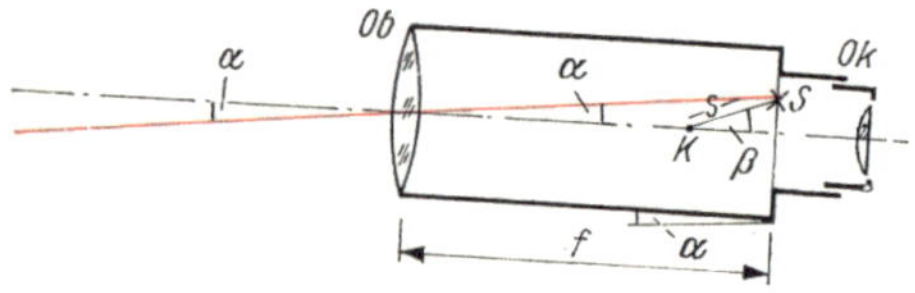

Figure 417 Monitoring of crosshair

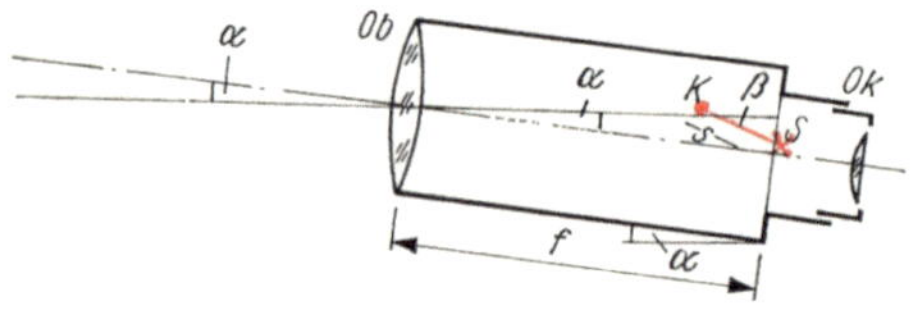

Figure 418 Monitoring of sighting ray by rotation

The value $\beta/\alpha = n$ represents the angular magnification of the compensator, often referred to as compensation coefficient. It depends on the location of the bending point, or by parallel shift of the ray, on the number of reflections. In order to obtain small structural elements, K is usually placed inside the telescope. If $n = 1$, then $f = s$, i.e. the point K coincides with the principal point of the objective lens. For $n = 2$, K is separated from the principal point of the objective lens by $s = f/2$. In order to be able to easily place the focussing lens, and to design the compensator with smaller dimensions and masses, K is often placed closer to the image plane; then $n > 2$.
In spirit levels, the bubble axis always moves under the influence of gravity such, that a tangent touching the interior edge of the bubble tube in the bubble centre is horizontal, resp. normal to the collimation axis.
In automatic levels, elements are used as compensators, which determine the direction of the plumb line, e.g. a pendulum or a liquid surface. They may directly define the collimation axis, or else be additional optical or optical-mechanical design elements. They are combined with other mechanical or optical-mechanical devices for obtaining the respective angular magnification.
A compensator usually has two monitoring elements: an optical element, rigidly built into the telescope (e.g. deflection prism), and a mobile part, affected by gravity (e.g. a pendulum mirror). Occasionally, only the mobile element is considered as being the compensator. A dampening arrangement for the pendulum belongs also to the compensator.
Compensators, which monitor the line of sight by rotation are most common.

The compensator may be
— placed in front of the objective lens
— represented by objective lens and cross hair plate
— used as focussing device
— arranged between the fixed front element and the mobile rear element (focussing lens) of the objective lens
— arranged between rear element of the objective lens and the reticule.

This leads to a variety of solutions. Depending on the angular magnification, one distinguishes:

— compensators with predominantly optical angular magnification (mirrors or prisms with one or more reflections)
— compensators with predominantly mechanical angular magnification (with the aid of jointed quadrilaterals in V or X-shape of wires or ribbons, cross-joints or astatic pendulum devices).

Figure 419 Precision compensator level Ni 002 (Zeiss, Jena); height 310 mm

Compensators with Predominantly Optical Angular Magnification

1. The mobile compensator element is a pendulum mirror hanging on a ribbon at half the objective focal length.

Under the influence of gravity, the mirror moves into the vertical and reflects a ray entering at the height of the crosshair onto it.
In this case n equals 2. The pendulum mirror also serves as focussing element. Therefore, the measurement is not influenced by errors caused by the movement of the focussing lens. Therefore, different sighting distances for back- and fore sights can be tolerated.
This principle is used in the first order level Ni 002 (figure 419) by Zeiss, Jena. The mirror is reversible. By measuring in two compensator positons, the influence of residual errors is eliminated and a quasi-absolute horizon to $\pm 1''$ is obtained for the whole temperature- and tilt range. This avoids, that the line of sight gets out of adjustment. The telescope system (figure 420) consists of the objective lens (*2*) behind the wedge-shaped cover glass (*1*) of the pendulum mirror (*3*) at $f/2$, and the cross hair (*4*), which is fixed onto the objective lens.
The imaging optics, consisting of reversing prisms and the telesystem (*14*), transfers the cross hair (*4*) and the image into the focal plane of the eyepiece (*13*). This, together with the eyepiece, serves for viewing only, and does not affect the line of sight. The pendulum mirror is brought to its second position via the knob (*5*). The fixed micrometer index (*7*) is illuminated via prism (*6*), and imaged on the micrometer scale (*10*) via the imaging optics (*8*), and the pendulum mirror (*9*), which in turn is connected to the pendulum mirror (*3*). This objective micrometer operates according to Abbe's comparator principle, and thus does not generate errors.

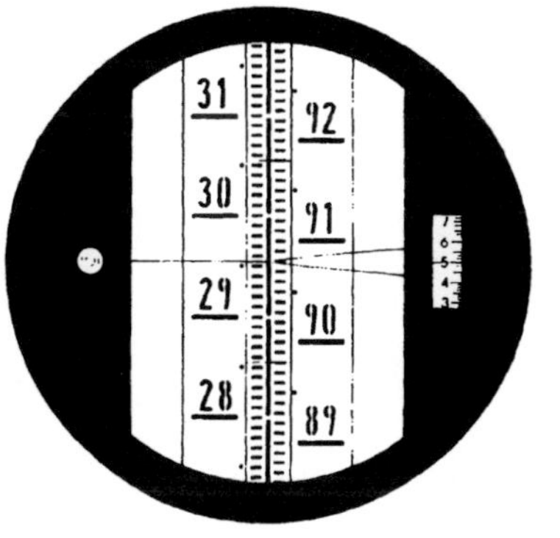

Figure 421 Field of view of the Ni 002

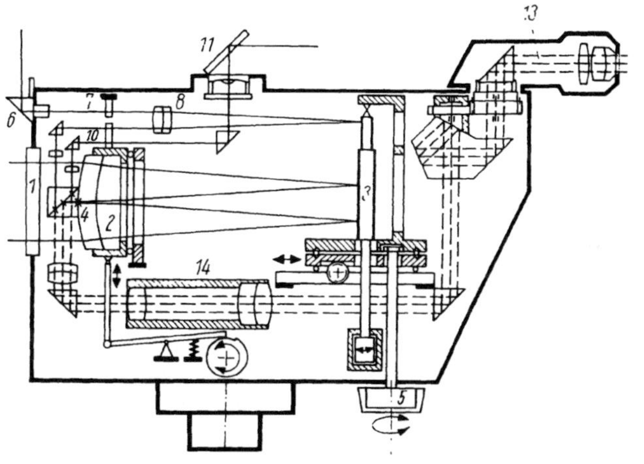

Figure 420 Optical train in the Ni 002

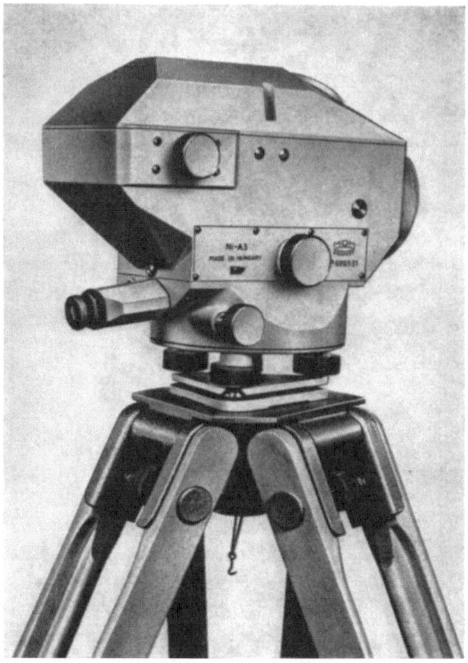

Figure 422 Precision level Ni-A 31 (MOM, Budapest); heigth 175 mm

The scale is imaged in the focal plane of the eyepiece (figure 421), together with the rod, cross hair (*4*) and the transparent bull's eye bubble (*12*), the latter via the mirror (*11*). The swivelling eyepiece, which automatically provides upright images within a 200 gon range, enables measurements from one position of the observer. Therefore, this level is particularly suitable for motorized levelling (precision levelling with accelerated instrument- and rod transportation), as developed by *Peschel* in the GDR.
In the first order level Ni-A 31 (figure 422) by MOM Budapest, the compensator is also placed at $f/2$. In this case, it is a plane mirror sitting on a long physical pendulum. The cross hair plate is cemented to the negative element of the objective lens at the principal plane (figure 423). The rays pass through the objective lens and via a pair of mirrors, arranged at 50 gon to the horizontal, to the pendulum mirror, which serves as focussing element. After another reflection at the pair of mirrors, they pass through a drill hole in the third element of the objective lens onto the cross hair.
Together with the generated image, the cross hair is imaged in the focal plane of the eyepiece. One element of the imaging lens system is attached to the compensator pendulum. With it, one can check whether the tilt is within permissible limits (lines on the left in the field of view, figure 424).

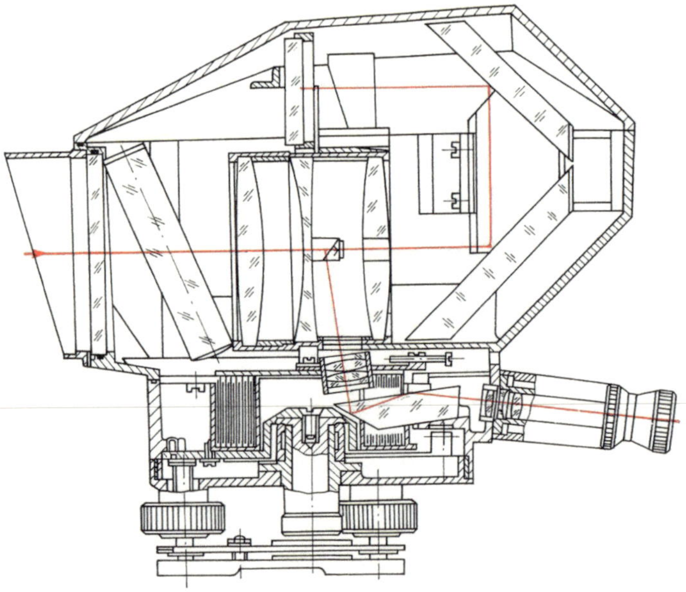

Figure 423 Optical train in the Ni-A 31

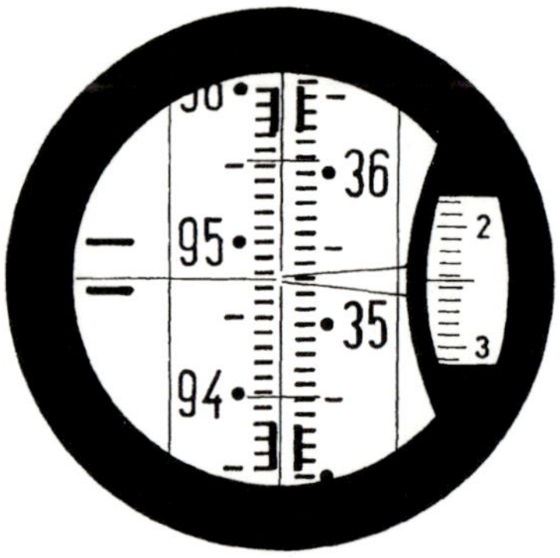

Figure 424 Field of view of the Ni-A 31

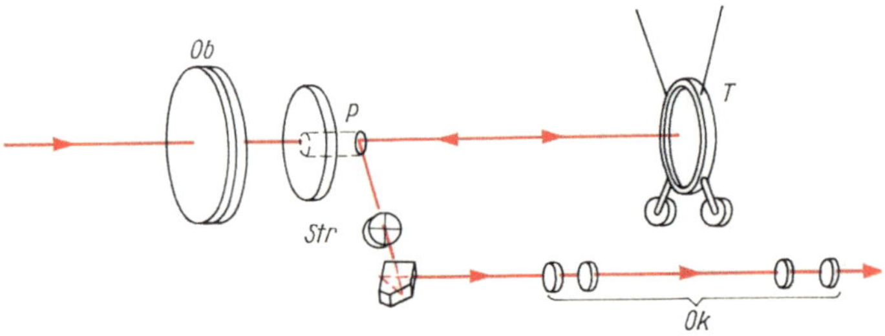

Figure 425 Optical train of AL-2 (Sokkisha), height 130 mm

Figure 426 Engineer's level GK 1-A (Kern); height 135 mm

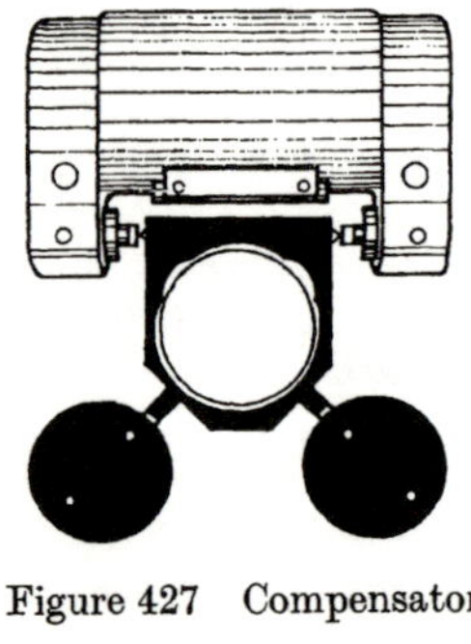

Figure 427 Compensator of the Kern GK 1-A

The levels Al-2 and Al-3 by Sokkisha, Tokyo, have a similar structure. The pendulum mirror, placed at $f/2$ is suspended by two thin wires (figure 425). The horizontal ray reflected by it, is deflected to the cross hair (*Str*) by the central prism (*P*).

2. The mobile compensator element is a roof prism oscillating at the magnetically restrained axis at half the focal length of the objective lens. Again, n equals 2.

In the GK 1-A (figure 426) by Kern, Aarau, the roof prism oscillates in a low friction bearing (figure 427). It consists of a yoke-like permanent magnet, whose magnetic field is closed by the horizontal compensator axis. The rays, reflected from the prism to the objective lens, reach the cross hair plate and the eyepiece via the focussing lens and two prisms (figure 428). Similar is the path of rays in the GK 2-A (figure 429).

3. The mobile compensator element is a mirror surface oscillating in horizontal position at a pendulum.

In the ALG 7 (figure 430) by Officine Galileo, Florence, rays, entering horizontally onto the horizontal pendulum mirror are deflected vertically upwards into the eyepiece of the telescope (figure 431).
At the Ni 42 (figure 410) by Zeiss, Oberkochen, the compensator, which is protected against shock and fire-damp, and which oscillates at a rigid pendulum, is at half the focal length of the objective lens ($n = 2$) supported at an axis, which turns in ball bearings (figure 432). A horizontal ray is diverted by 60° to the compensator by a prism located behind the objective lens. It reaches the cross hair plate and the eyepiece via the focussing lens and a roof prism. The operating range of the compensator ($\pm 1°$) is indicated on the reticule so that its proper functioning can be checked.

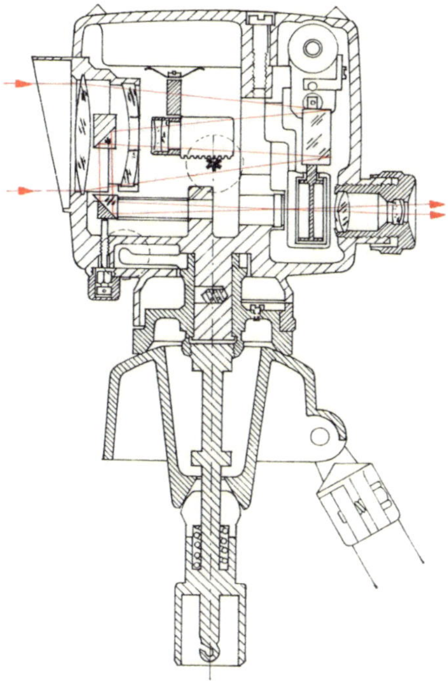

Figure 428 Optical train of the GK 1-A

Figure 429 GK 2-A (Kern); height 175 mm

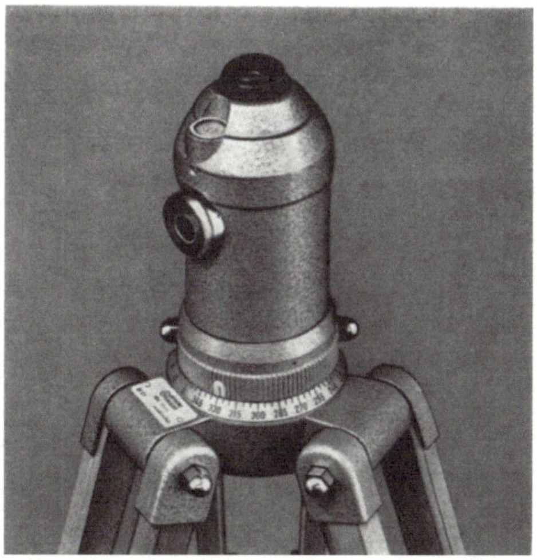

Figure 430 Micro level ALG 7 (Officine Galileo)

At the B-2 and C-3 by Sokkisha, Tokyo, (figure 434) the horizontally oscillating prism is suspended by four wires (figure 434). One side forms a 50 gon angle with the edge of the adjacent prism if the vertical axis is in plumb.
In the Tecomat (figure 435) by Theis, Breidenbach, the rays are totally reflected in the prisms on both sides of the pendulum.
In the NTSK (U.S.S.R.) the horizontally entering rays are deflected by a fixed pentagon prism onto a horizontal oscillating mirror surface, and, after reflection, pass through another fixed pentagon prism into the eyepiece (figure 436).

4. The mobile compensator element is a mirror surface, inclined by 45° to 60° to the horizontal.

If the mirror is rotated by α, a ray falling onto it changes its direction by 2α because of reflection (see figure 47). If the mirror is monitored by a pendulum such, that it always is under the same angle to the horizontal, then it is suitable as a compensator ($n = 2$, figure 437).
This principle is utilized in a simple way in the SNCA by SLOM, Paris (figure 438). In the SNCA, the pendulum mirror, which is inclined by 50 gon to the horizontal, is placed in front of a small vertical telescope. The entry opening for the rays is closed by a cover glass. For simple construction jobs, targets are set with this level. In the SNA 1 (figure 439) by SLOM, Paris, the pendulum mirror is located between the front and rear elements of the objective lens. The rays, reflected upwards by the mirror, are diverted into the eyepiece by a prism.
In the AUTOM and AUCIR (figure 440) by Breithaupt, Kassel the mirrors (prisms) are arranged according to the principle illustrated in figure 441. The axis of the pendulum mirror, which is inclined at 50 gon to the horizontal is hanging on a horizontal prestressed ribbon, perpendicular to the sighting direction. The oscillations of the pendulum are dampened by an inductive brake.
At the VA-1 (figure 442) by Visomat, Switzerland, the pendulum mirror, supported in ball bearings, is placed above a pentagon prism, which diverts the rays through focussing lens and reticule onto the eye piece (figure 443). The operating range for this level is $\pm 4.5°$.
In the Nikon N (figure 444) by Nippon, Tokyo, the pendulum mirror is built differently (figure 445). It hangs on two threads of synthetic rubber. In the NAN 3 (figure 446) by G. Nestle, Dornstetten, the mirror surfaces of pendulum mirror and fixed mirror form a 60° angle with the horizontal (figure 447) The Nikon AP-2 has the same arrangement.

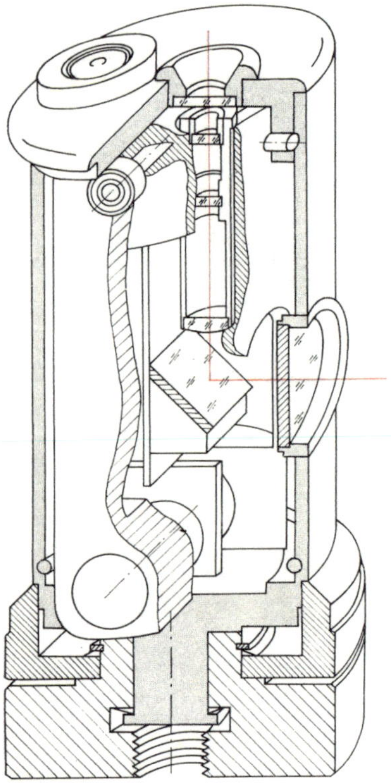

Figure 431 Optical train in the ALG 7

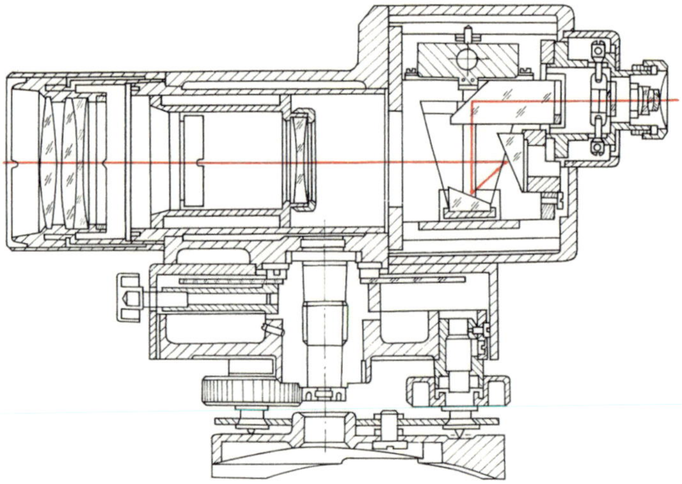

Figure 434 Section through the B-2

Figure 435 Tecomat (Theis); height 125 mm

Figure 432 Optical train in the Ni 42 (Zeiss, Oberkochen)

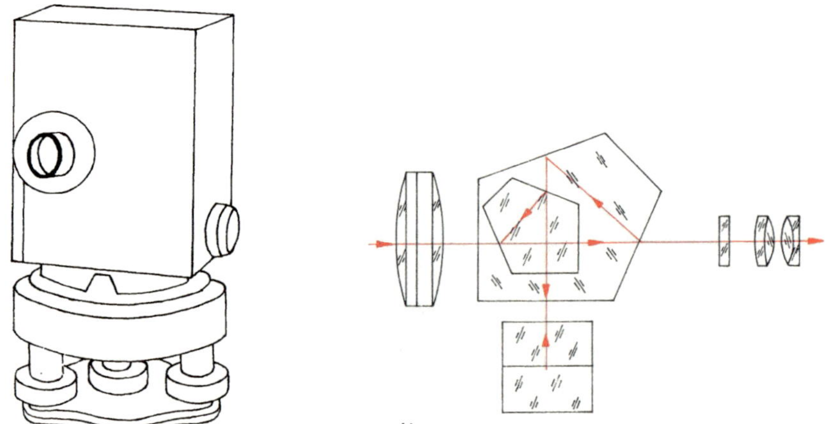

a) b)

Figure 436 NTSK (U.S.S.R.) resp. N-10 KL a) total view; b) optical train

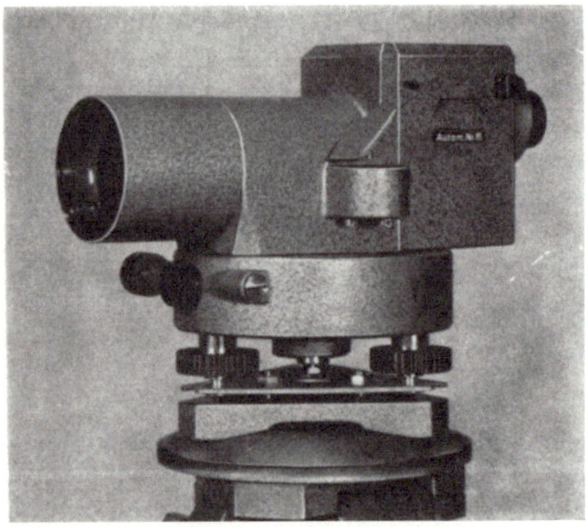

Figure 433 AUTOM (Breithaupt) height 150 mm

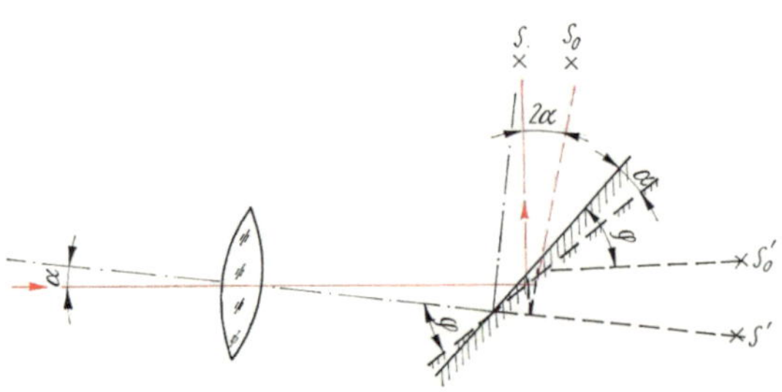

Figure 437 Pendulum mirror

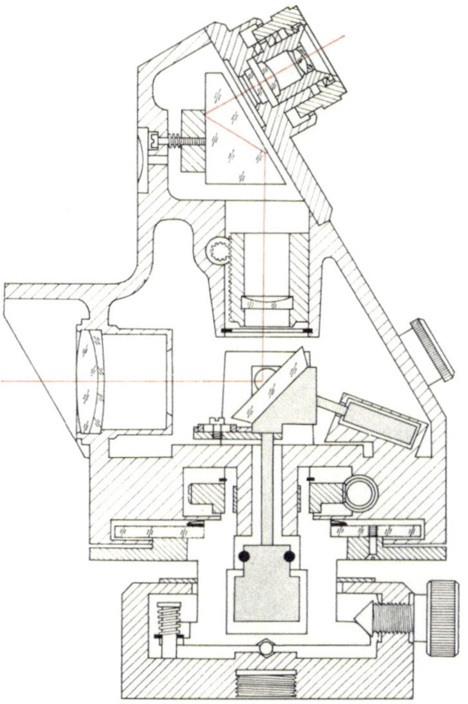

Figure 438 SNCA (SLOM, Paris)

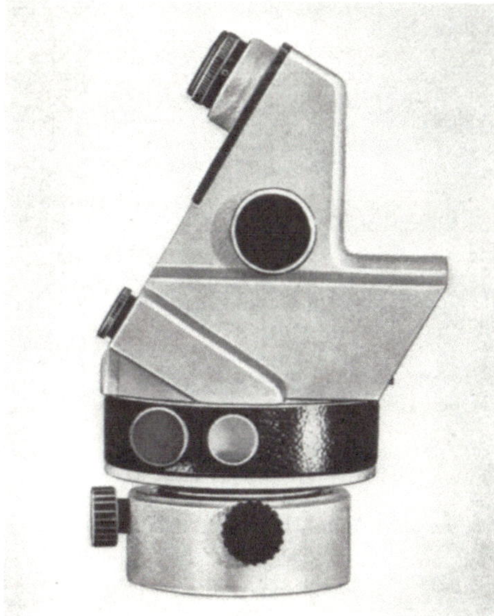

Figure 439 Engineer's level SNA 1 (SLOM, Paris)

1 objective;
2 telescope casing
3 pendulum with mirror;
4 prism;
5 focussing lens;
6 focussing screw;
7 adjusting screw;
8 cross hair;
9 eyepiece;
10 magnetic dampening;
11 slow motion screw;
12 friction clamp;
13 foot screw;
14 spring plate;
15 base plate;

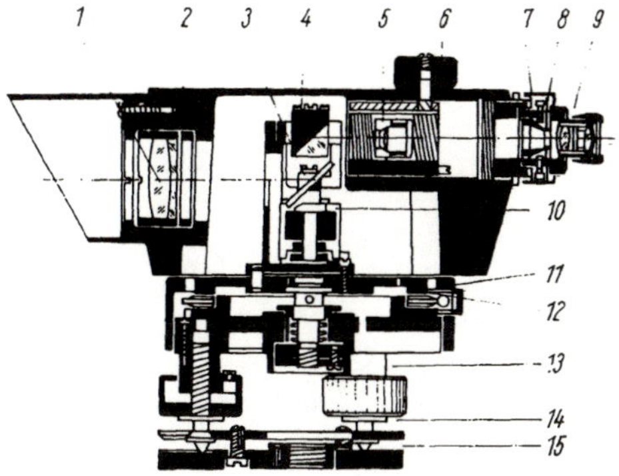

Figure 441 Optical train in the AUTOM

Figure 440 Engineer's level AUCIR (Breithaupt & Son); height 150 mm

Figure 442 Visomat VA-1

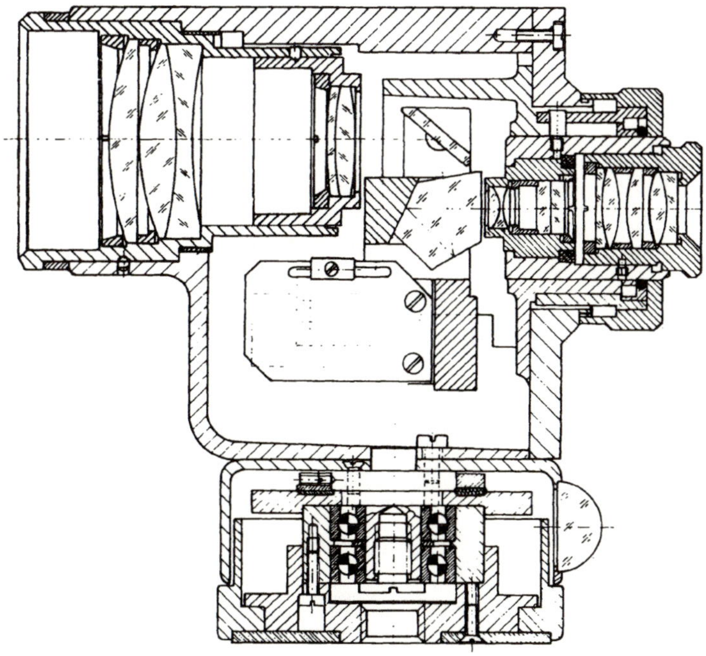

Figure 443 Optical train in the VA-1

Figure 444 Nikon N; height 150 mm

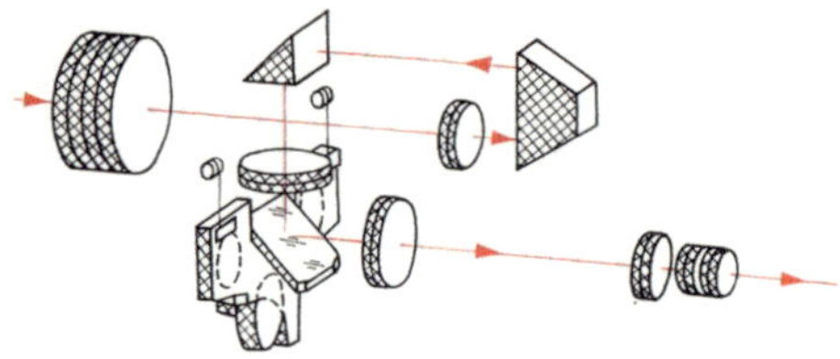

Figure 445 Optical train in the Nikon N

Figure 446 NAN 3 (Nestle)

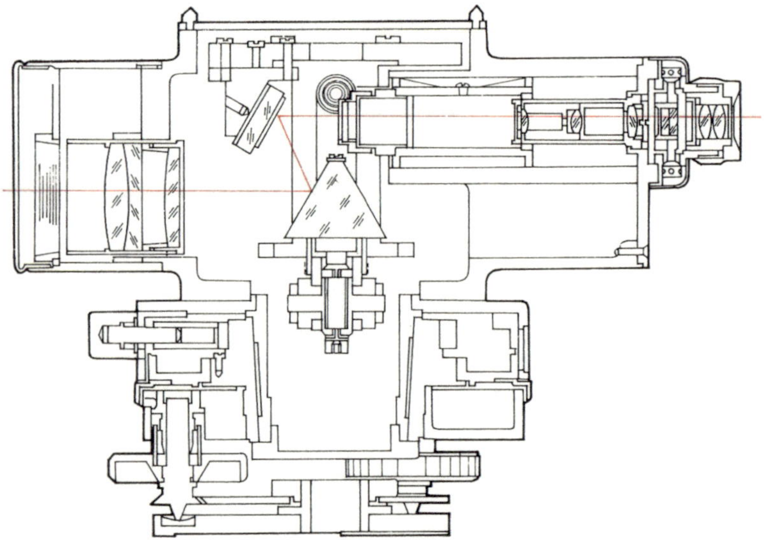

Figure 447 Optical train in the NAN 3

Figure 448 AL-21 (Fuji Surveying Instruments); height 180 mm

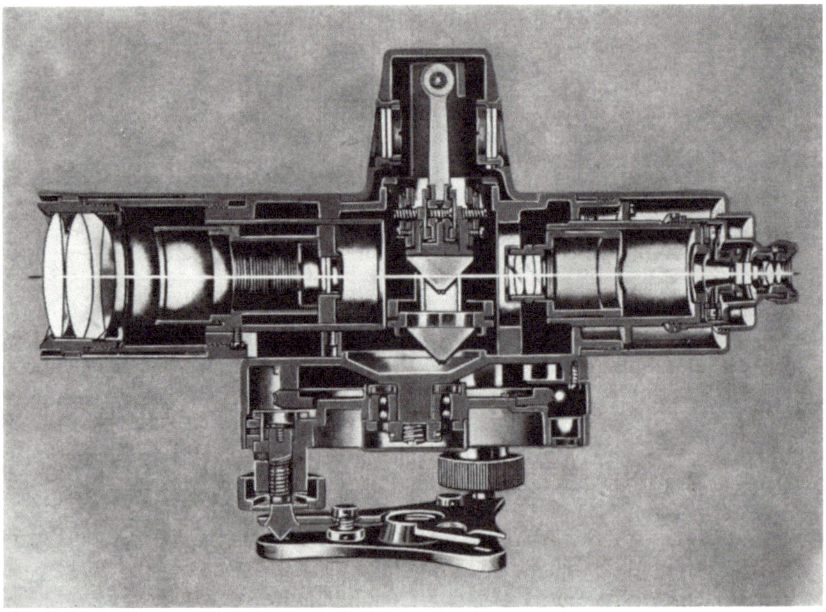

Figure 449 Optical train in the AL-21

Figure 450 Ni 050 (Zeiss, Jena)

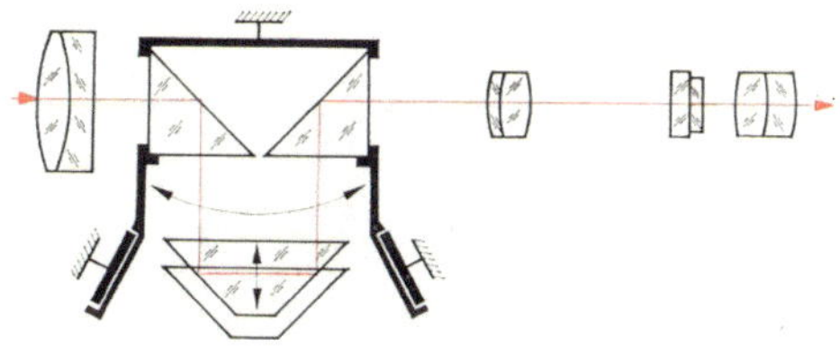

Figure 451 Optical train in the Ni 050

Figure 452 Ni 025 (Zeiss, Jena)

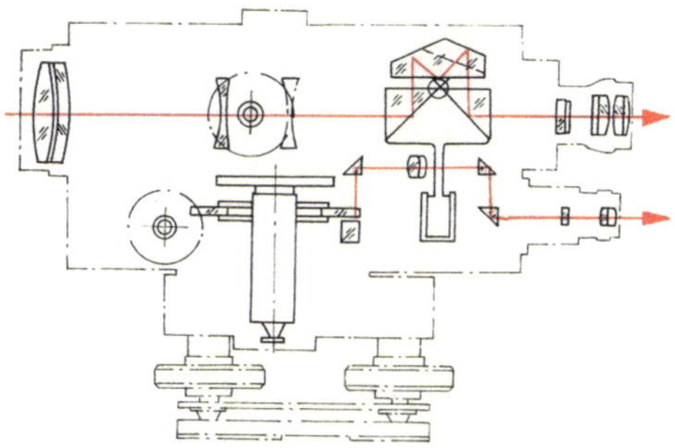

Figure 453 Optical train in the Ni 025

Figure 454 S 27 (Vickers)

5. The mobile compensator element consists of two mirror surfaces, inclined by 50 gon to the horizontal, with a fixed prism in between.

In the AL-21 (figure 448) of Fuji Surveying Instruments, Tokyo, and the S-201 of Path Instruments, New York (figure 449), which is analogous, the surfaces of a rectangular prism with equal sides, arranged as hanging pendulum, are used as compensator. This arrangement is the same for the level Nestor 4 by Nestle, Dornstetten.
The compensator of the Ni 050 (figure 450) by Zeiss, Jena, contains three prisms, with the outside ones hanging in a frame, so they can oscillate. The rays are reflected four times ($n = 2$) (figure 451). The centre prism serves as a means for focussing.
In similar manner, two prisms are fastened to a pendulum with a cross spring joint for the compensator of the Ni 025 (figure 452) by Zeiss, Jena, with the third prism fixed in between (figure 453). Since it is a roof prism, the image is upright and correct. When tilting the level by α, the ray is diverted by 2α because of reflection at the first prism. This angle is doubled by reflection in the second prism, which gives $n = 4$. The compensator is therefore placed at a distance of $f/4$ from the image plane.
In the compensators of the Watts SL 60 to 62 by Rank, England, as well as S 27 (figure 454) and S 77 by Vickers England, the mirror surfaces of the two pendulum prisms B_1 and B_2 are inclined by 30° against the horizontal (figure 455). Between both prisms is a third prism A fixed in a casing, hanging from crossed metal bands.
This prism, being a roof prism, serves to obtain upright images. Horizontal rays entering the instrument, are diverted onto the horizontal line of the cross hair by triple reflection in connection with the suspension.

6. The mobile compensator element consists of two prisms of a Porro System of the second type, oscillating on tapes.

In the Ni-B 3 (figure 456) by MOM, Budapest, as well as in the FN *AK*1 (figure 457) by Fennel, the two exterior prisms of the Porro system, which are rigidly connected, are attached to a pendulum, oscillating on steel tapes (figure 458). The centre prism is fixed in the telescope body. A ray, hitting the prisms, is reflected four times, and diverted by 4α if the level is tilted by α. The parallel shift sideways, which results from the Porro system, is taken into consideration by a sideways shift of the eyepiece for MOM, by a rhombicprism in front of the eyepiece by Fennel. The focussing lens is split into a mobile- and a fixed element.

7. The mobile compensator element is a pendulum prism of a Porro System of the first type.

The pendulum prism of the Porro System of the first type, serves simultaneously for focussing for the Topcon-AT-S by Tokyo Optical Co., as well as for the Topcon U 2 by Tokyo Kogaku Kikai, Tokyo (figure 459).

8. The mobile compensator element is a lens of the telescope, oscillating with a horizontal balance beam.

In the GKO-A (figure 460) by Kern, Aarau, a reversing lens between focussing lens and reticule acts as compensator. It is fastened to a horizontal rigid balance beam whose ball bearing supported axis is located between front element of the objective lens and focussing lens (figure 461). At the end pointing towards the objective lens, there is a compensation piece with the pendulum oscillations are dampened magnetically. Like in a terrestrial telescope, an intermediate image is generated. The image, generated by the objective lens, is imaged upright in the cross hair plane by the reversing lens. The mechanical angle magnification is $n = 1$. The operational range of the compensator is very large, namely ± 27 gon. If the rough levelling is insufficient, a red warning disk appears in the field of view (see figure 480).
When using a line of sight stabilization with a lens compensator based on the principle of the Abat Wedge (1777), a horizontal principal ray can be

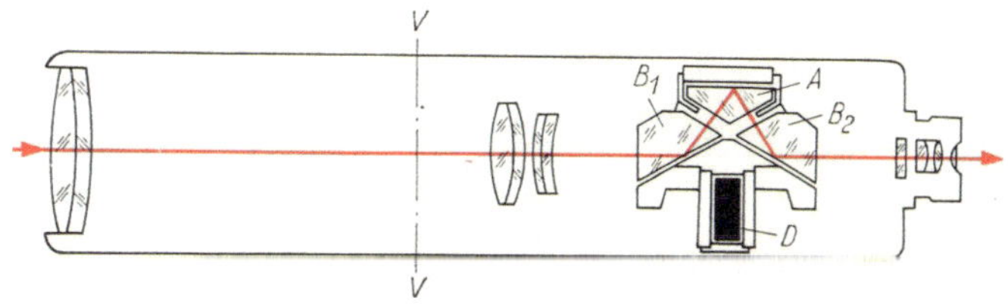

Figure 455 Optical train in the Autoset Level 2 (Rank Precision Ind.) and in the S 77 (Vickers)

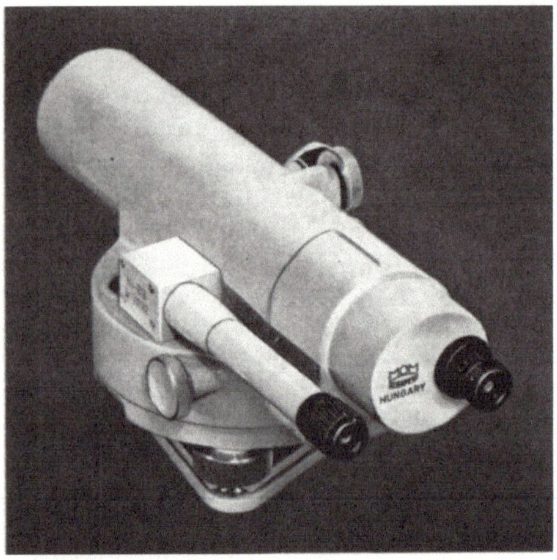

Figure 456 Ni-B 3 (MOM, Budapest); height 130 mm

Figure 457 Construction- and engineer's level FNAK 1 (Fennel); height 150 mm

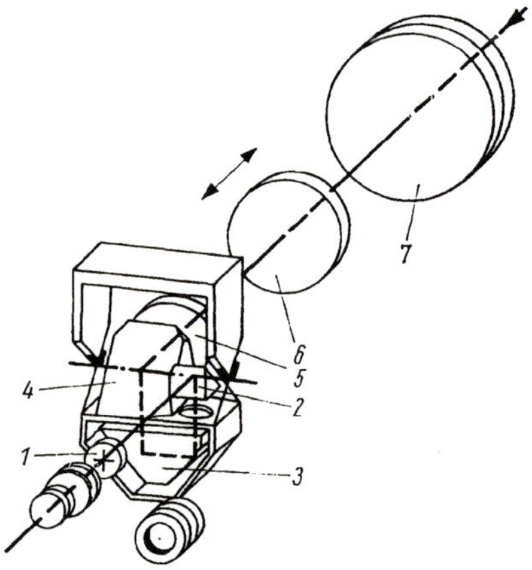

Figure 458 Optical train in the Ni-B 3
1 cross hair; *2* to *4* Porro prism system; *5*, *6* focussing lens; *7* objektive lens

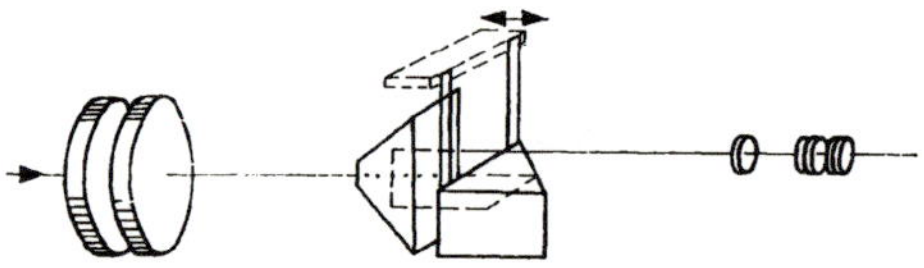

Figure 459 Optical train in the AT-S

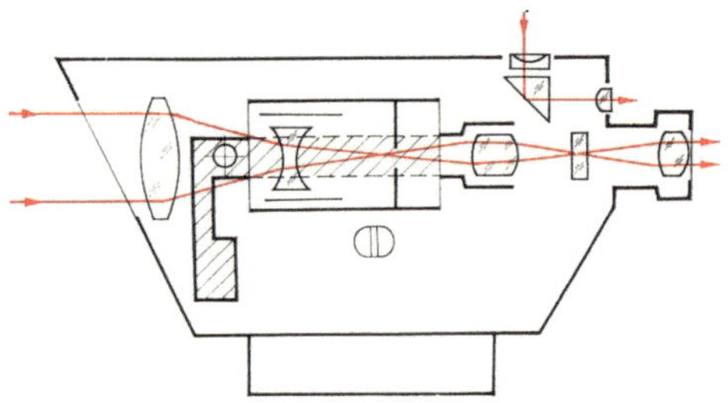

Figure 460 Construction level GKO-A (Kern); height approx. 120 mm

Figure 461 Optical train in the GKO-A

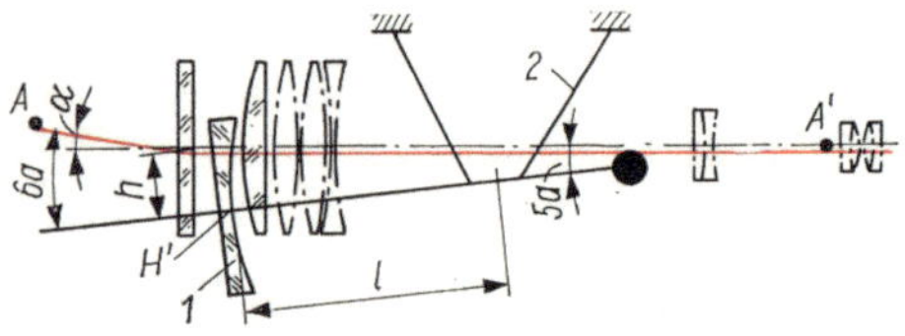

Figure 462 Compensator of the NSM-2 (U.S.S.R.)

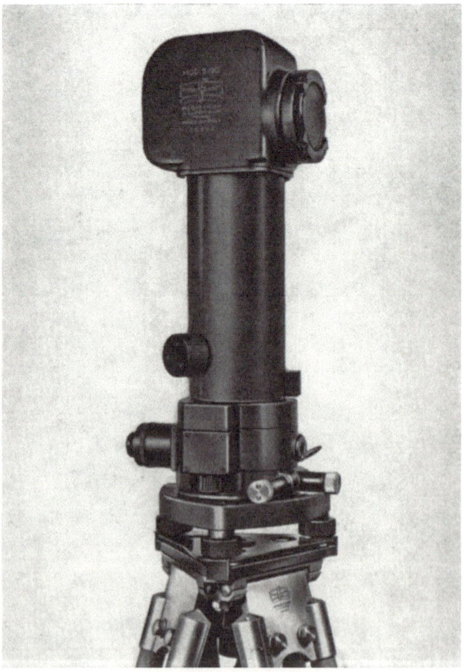

Figure 463 Compensator level of highest accuracy 5190 (Filotecnica Salmoiraghi); height 500 mm

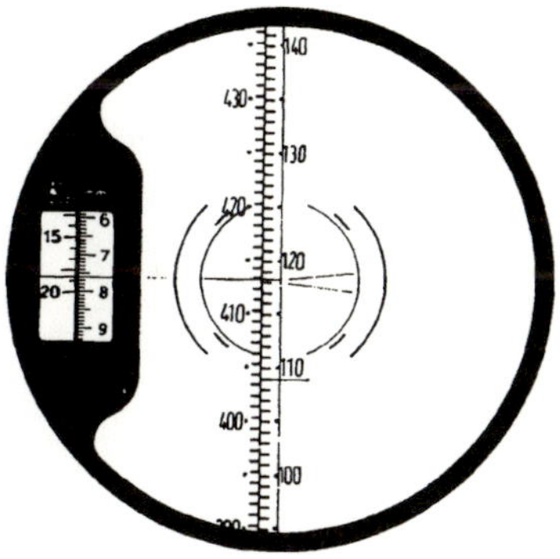

Figure 464 Field of view of the 5190

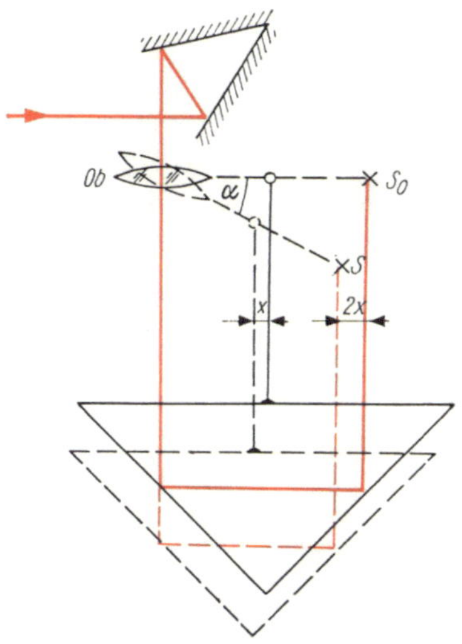

Figure 465 Pendulum prism for optical parallel shift

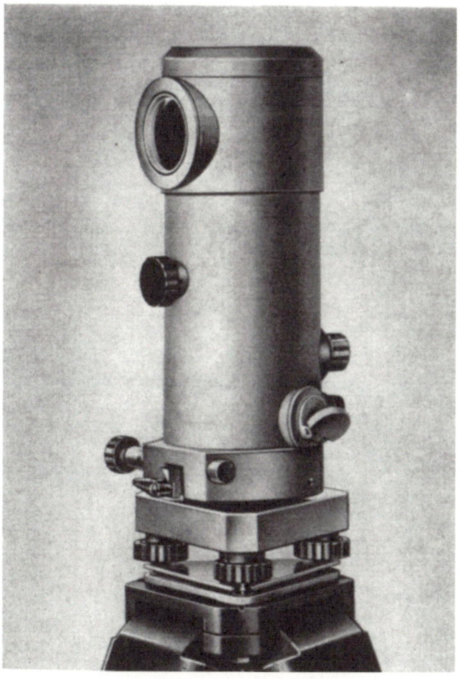
Figure 466 Ni 007 (Zeiss, Jena); height 335 mm

monitored such, that it hits the horizontal line of the cross hair. This is done by shifting of a pendulum-suspended lens normal to the horizontal axis. The compensator of the NSM-2 of the Charkow works for Mining Instruments, consists of a plan-convex positive lens, solidly placed in the objective lens casing, and the negative lens (*1*), oscillating on four wires (*2*) (figure 462). A piece of mass serves as compensation. The wire suspension is so designed, that a tilt α of the telescope results in an inclination of the pendulum system by 6α. The principal point of the negative lens wanders from the optical axis by the amount h, which depends on f and l. The distance l and the focal length f are selected such that the horizontal ray AA' is diverted by α in the opposite direction, i.e. a horizontal ray entering the objective lens, is directed onto the centre of the cross hair. For large tilts, this has the disadvantage of reduced image quality, getting out of adjustment, and others.

9. The mobile compensator element consists of oscillating optical parts in a level with vertical telescope tube.

In the levels of medium accuracy 5173 to 5175 by Filotecnica Salmoiraghi, Milano, the objective lens is suspended in a periscope type telescope by three wires of the length of the focal length. In the first order level 5190 (figure 463) by Filotecnica Salmoiraghi, the cross hair is monitored. The reticule, oscilating on three steel wires of the length f ($f = s$, therefore $n = 1$), always swings to such a position, that optical axis and plumbline coincide. The instrument is focussed by shifting the objective lens below the upper mirrors, which are at an angle of 50 gon to each other. They are rigidly connected to a scale, which is illuminated via another mirror. Together with it, they can be shifted in height by rotating a fine drive, and thus serve as micrometer (figure 464). The principle of stabilizing the line of sight is the reason for the vertical arrangement of the instrument. This has the practical advantage of increasing the distance between terrain and line of sight and thus reducing the influence of refraction near the ground.
In the Ni 007 by Zeiss, Jena, a 100 gon prism is placed between objective lens and eyepiece such, that all entering rays are diverted by 200 gon via double reflection.
If such a prism is suspended on a pendulum with a length of $f/2$, a horizontal ray will always hit the horizontal line of the cross hair (figure 465), because the prism is shifted by the amount x when the instrument is inclined by α, which leads to a parallel shift of $2x$ for the line of sight. In the Ni 007 (figure 466), the pentagon prism (*16*) which is in front of the objective lens (*17*) and the focussing lens (*18*), and protected by the cover glass (*15*) can be tilted normal to the collimation axis and serves as micrometer, which can be arrested (figure 467). The roof prism (*27*), located between compensator (*26, 28, 29*) and eyepiece (*20*), generates an upright image in the cross hair plane (*19*). The optical elements (*21* to *25*) serve as circle reading devices. The optical elements of the AL-31 (figure 468) by Fuji Surveying Instruments, Tokyo are similarly arranged.
In the NT 6 (U.S.S.R.) both, objective lens and reticule are pendulum suspended.

Compensators with Predominantly Mechanical Angular Magnification

1. Jointed quadrilateral to monitor the optical elements of the compensator

In order to arrange the compensator closer to the image plane, n can be magnified mechanically with the aid of jointed quadrilaterals (figure 469). If the base $AB = a$ of the suspension is tilted by α, the reflection base $CD = c$ carrying mirror or prism is inclined by $\beta = n\alpha$.
The angular multiplication is determined by the geometric dimensions as well as the distance between reflection base and centre of gravity. This value is approximately a/c. Therefore for $a = 3c$, β becomes 3α. The jointed quadrilateral acts like a mechanical lever transmission. Combined with the optical compensation, the total amount of angular multiplication is obtained.
The compensator of the Ni 2 (figure 470) by Zeiss, Oberkochen, operates according to this principle. It consists of three prisms, with the two outside ones being rigidly connected to the telescope body, while the central one is suspended on four wires as a pendulum (figure 471). Since the jointed quadrilateral is dimensioned such, that $\beta = 3.7\alpha$, the total angular magnification. amounts to $n = 7.4$, because of the doubling effect of the reflection surface. The prism near the eyepiece is a roof prism, which generates an upright

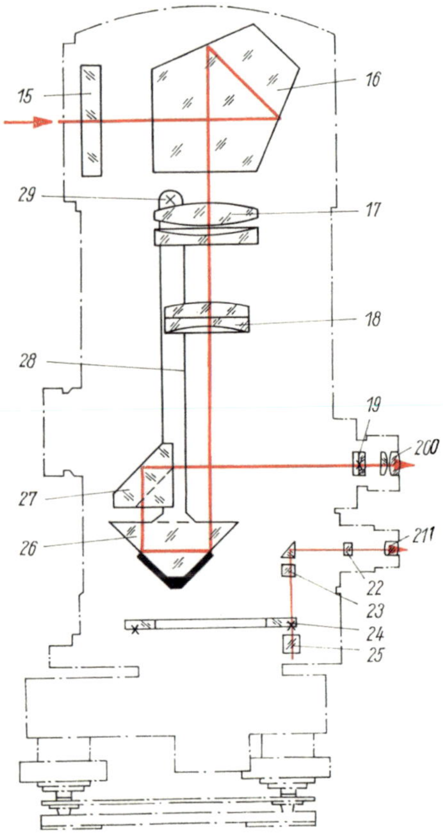

Figure 467 Optical train in the Ni 007 (The type number is an indication of the standard deviation in 0.1 mm/km for levels by Zeiss, Jena; in this case: 0.7 mm/km.)

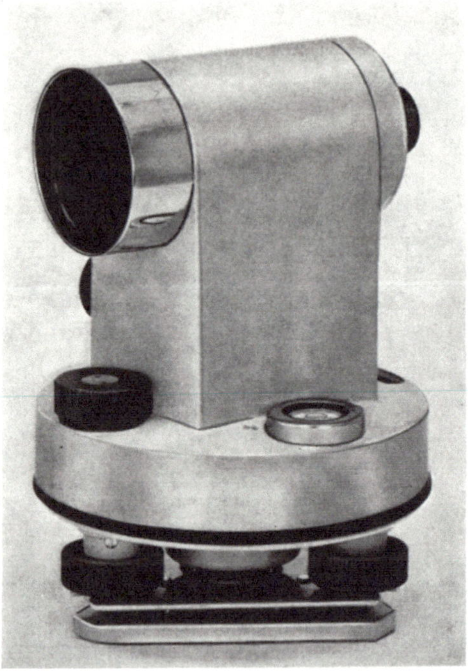

Figure 468 AL-31 (Fuji Surveying Instr.); height 187 mm

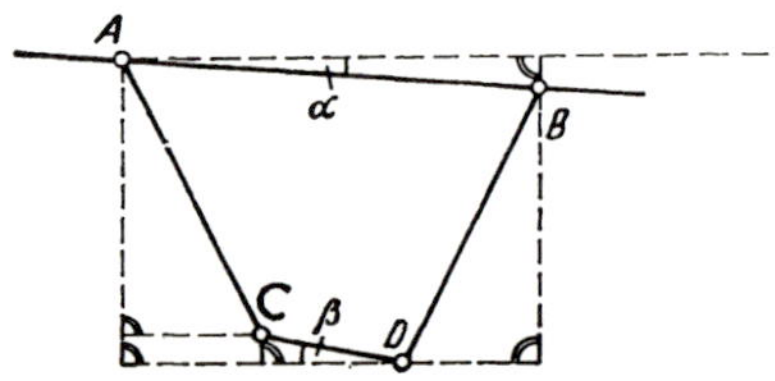

Figure 469 Jointed quadrilateral

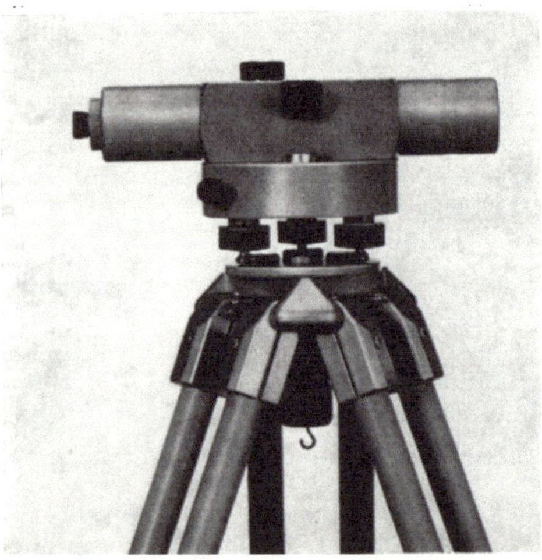

Figure 470 Engineer's and precision level Ni 2 (Zeiss, Oberkochen); height 130 mm

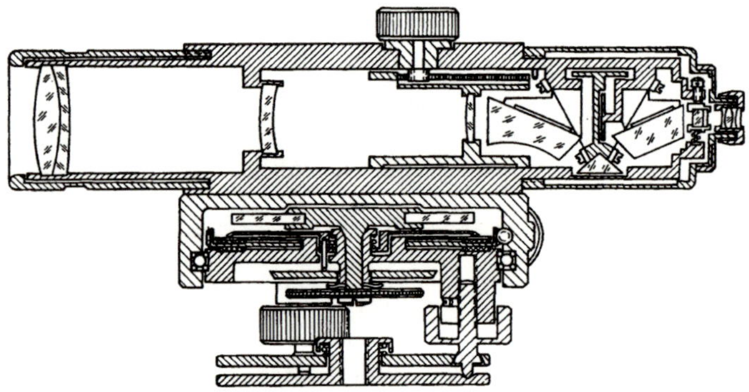

Figure 471 Section through the Ni 2

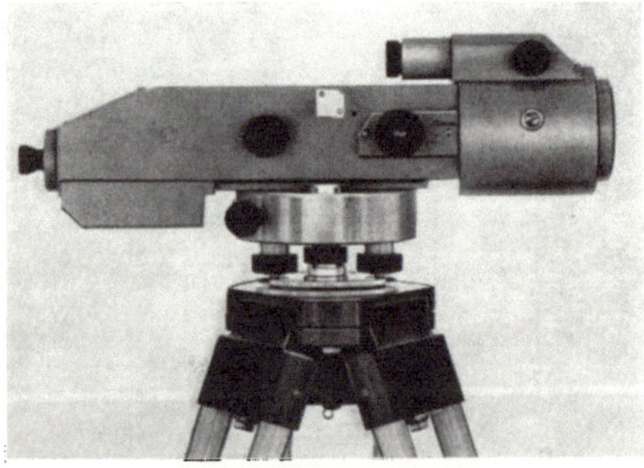

Figure 472 Precision level Ni 1 (Zeiss, Oberkochen); height 180 mm

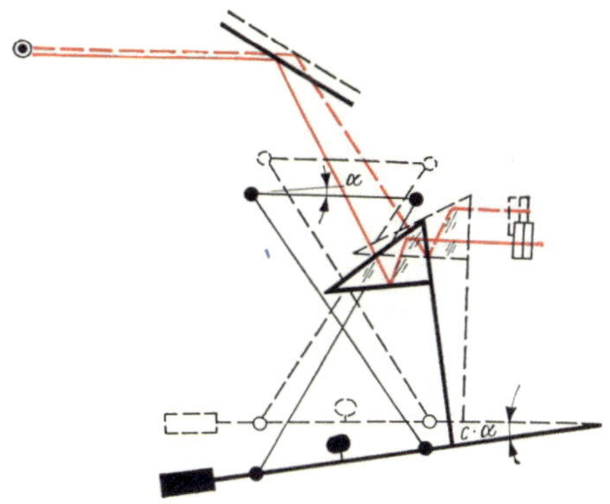

Figure 473 Scheme of the compensator of the Ni 1

image in the field of view. The focussing lens is separated into a mobile and a fixed part, which enlarges the space for the compensator and improves the anallactic effect.
The first order level Ni 1 (figure 472) by Zeiss, Oberkochen, also has a jointed quadrilateral for mechanical angular magnification. Based on the dimensions of the compensator, n equals 16. The internal frequency of the oscillating parts is thus small and is equivalent to a pendulum of 1.12 m length. The compensator is therefore called "long-compensator" with 1.12 m reduced pendulum length (for the Ni 2 it is 0.13 m). External vibrations have virtually no effect on it. With the aid of a double reflecting prism, attached to the jointed quadrilateral-pendulum, the effect of a parallel plate is obtained (figure 473). When tilting the telescope, the ray suffers a parallel shift.
The rotational axis of the prism is in a plane perpendicular to the normal section. Its trace in the normal section bisects the angle between entering and exiting ray. The first fixed prism has a roof edge to obtain mirror reversal, while three times reflection creates the upright image. For all compensators with optical parallel shift of the rays, the position of the rotational axis of the jointed quadrilateral is independent of the focal length.

2. Astatic standing spring pendulum with reversing prism as mobile compensator element

Astatic pendulae are used in the levels KNA, BNA (figure 474) and INA (with tribrack) by Ertel, Munich. They oscillate in a vertically standing elastic spring, which is loaded at the top with a reversing prism as mass element, and connected at the bottom to the base surface with a thin piece of sheet metal (see figure 141). It is characteristic for the spring pendulum, that for a tilt of the telescope by α, it will swing by $n \cdot \alpha$. because of the imbalance caused by the weight of the prism. A horizontal principal ray can be monitored such, that it hits the horizontal line of the cross hair. In the KNA the elastic spring supports a mirror between two prisms.

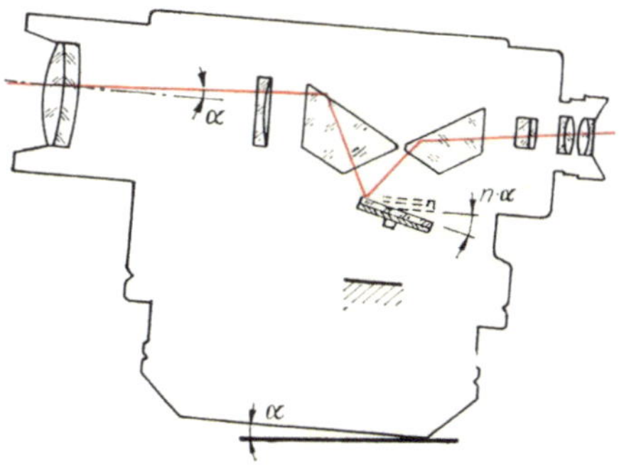

Figure 474 BNA (Ertel, Munich); height 110 mm

3. Cross spring joint with prism as mobile compensator element

The compensator element for the Wild levels NAK 0 (figure 475), NAK 1 (figure 476) and NA 2 resp. NAK 2 (figure 477), consists of a hanging pendulum with prism between focussing lens and cross hair plate. It is rigidly connected to the housing via prestressed suspension bands in a cross type arrangement (figure 478). With this, an effect similar to the one of the astatic pendulum is achieved. The rotational moments of the suspension bands counteract a tilting of the pendulum body due to gravity and remaining tilt of the vertical axis. The mechanically created angular magnification of the rays entering the level amounts to $n = 5$ after passing through the reversing prism. By reflection within the prism it is doubled to $n = 10$. The effective horizontal axis of the compensation element, can thus be placed at a distance of $f/10$ from the reticule. A roof prism is placed in front of the reversing prism, in order to obtain upright and correct images. By depressing a control button (figure 478), the proper functioning of the compensator can be checked. Tipped by a spring, the pendulum body swings from its rest position and returns to it.

Dampening

To enable fast work, the tilt compensator has to come to rest within a short period of time—in fractions of a second. Most commonly, the oscillations of pendulum compensators are dampened by air. To accomplish this for instance, a pressure piston moves in a cylinder. The play between piston and cylinder is kept so small, that air pockets are generated which facilitate the dampening. With magnetic dampening, the oscillations of the metal body of the pendulum, swinging in an inhomogeneous magnetic field, are dampened by inductive currents. In a liquid compensator, the dampening of the oscillations is solely based on the viscosity of the liquid, without additional devices.

Figure 475 Construction level Wild NAK 0; height 110 mm

Specialities of Compensator Levels

By automatic levelling of the line of sight, up to 40% time is saved when levelling with a compensator instrument instead of spirit levels of matching accuracy. The low sensitivity of the compensator against temperature influences is also of advantage, because often one can work without a field umbrella. The measuring range of the compensator is usually $\pm 10'$, but may be $\pm 30'$ and more.

Figure 476 Engineer's level Wild NAK 1; height 133 mm

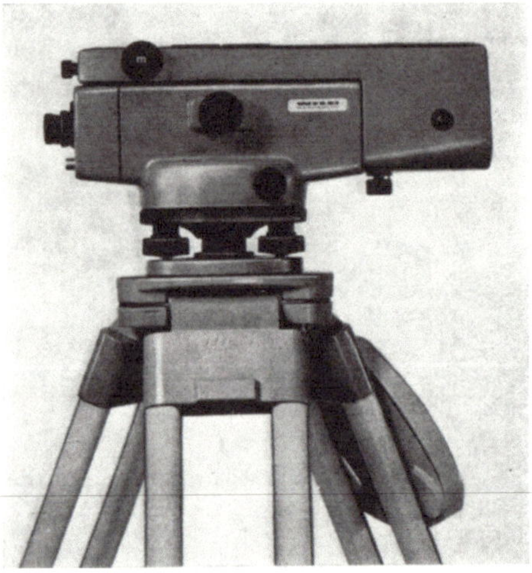

Figure 477 Universal level Wild NA 2 with plane parallel plate; height 150 mm

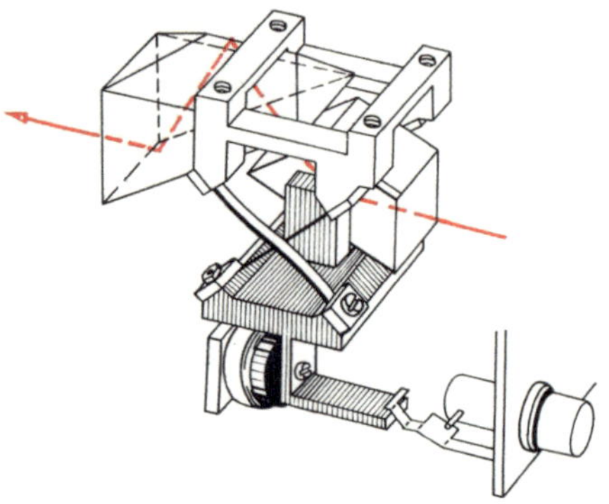

Figure 478 Compensator of the Wild levels

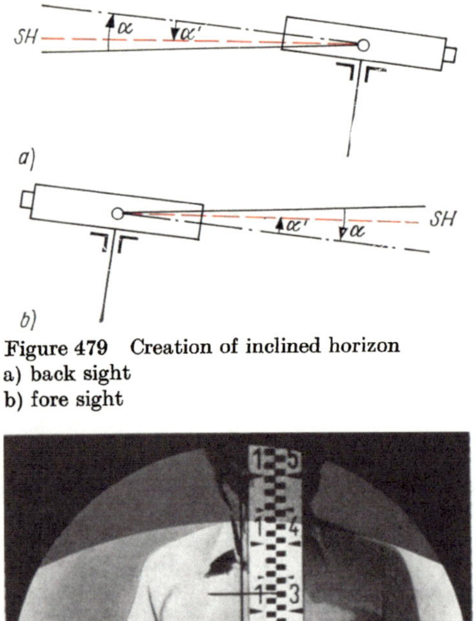

Figure 479 Creation of inclined horizon
a) back sight
b) fore sight

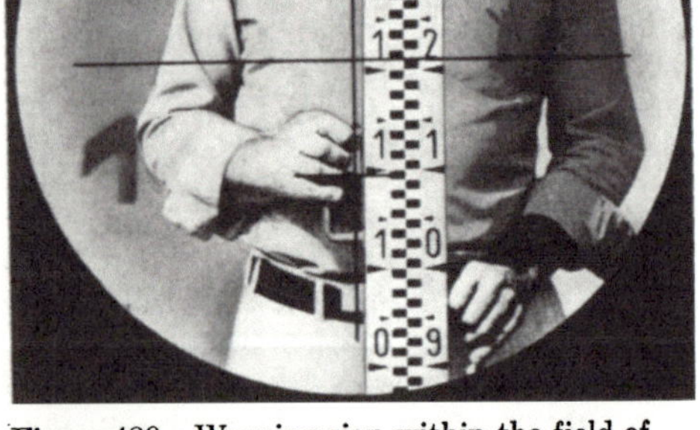

Figure 480 Warning sign within the field of view of a level (Kern GKO-A)

The inclined horizon which has to be specially watched for precision work, is a peculiarity of compensator levels. It is caused by the fact that the compensator can only be levelled mechanically with a limited accuracy. Its magnitude depends on the compensator type, on the tilt, and on the separation of front principal point of the objective lens from the vertical axis. For instance, a residual tilt of 5′ (half the sensitivity of the bull's eye bubble), and a separation of 150 mm between front principal point and vertical axis, causes an error in elevation of 0.4 mm. Either a small undercompensation or overcompensation is possible. If in the back sight a tilt α of the telescope is only compensated to $99\% = \alpha'$ (undercompensation), the line of sight is not horizontal (figure 479a). For the foresight, the same situation arises, but with the opposite sign, because of the vertical axis tilt which remains unchanged (figure 479b). The difference between back sight and fore sight does not eliminate this error. An inclined horizont *SH* is created. If the bull's eye bubble is centered for all set-ups in the same direction (e.g. backsight), the errors due to inclined horizon accumulate. Their influence can cause a difference to 5 mm/km between elevation differences from foreward and return measurements. The mean of both values is free of this error, because in the two directions, the inclination of the horizon occurs in opposite sense.

To largely eliminate inclined horizon, compensator levels have to be well pre-levelled. To check the levelling, the marks of a rigidly mounted reticule in the telescope are imaged in the field of view of some compensator instruments. The oscillating cross hair has to be set within the given range (see figures 421, 424, 463). In other instruments the bubble of the bull's eye level is imaged in the field of view or else a warning sign appears (figure 480). In any case, the carefully adjusted bull's eye bubble of the compensator level is to be well centered. The designer can eliminate inclined horizon to a large extend by positioning the front principal point into the extension of the vertical axis.

For instruments with a periscope type telescope, the objective lens should be located on the vertical axis. Two positions of the compensator of the lovel Ni 002 by Zeiss Jena give an absolute horizon in a way.

Inclined horizon can also occur when changing the focus setting. It is then dependent on the focal length of the objective lens and on the tilt angle. The influence of this error on the levelling error is eliminated, either by using compensators where a change of focus will not generate additional errors, or else by using a compensation which is independent of changes of the focal length. It can also be eliminated by using telescopes where the focal length of the telelens remains constant when changing the focus.

The influence of inclined horizon can also be eliminated by measuring procedures:

1. Centering of the bull's eye bubble for backsight r, reading of rods in back- and fore sights.

2. Centering of the bull's eye bubble for fore sight v, reading of the rods in fore- and back sights.

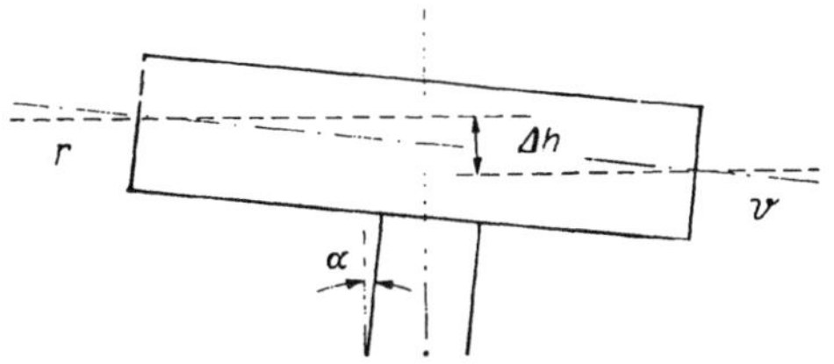

Figure 481 Height shift

For perfectly operating compensators, a shift in height Δh of the objective lens occurs due to a tilt of the vertical axis α (figure 481). It amounts to 0.4 mm for a 150 mm separation of the object principal point from the vertical axis and a 2′ tilt of the vertical axis (compare to sensitivity of bull's eye bubble). It thus has to be considered for precision levelling. By careful adjustment and centering of the bull's eye bubble, this error is also reduced. By taking the mean of both elevation differences, this influence is already eliminated at the station. Alternating methods offer additional possibilities for this.
The actual manufacture of the tilt compensators may lead to further errors, e.g. cross section and material of the wires, angular magnification of the astatic pendulum, degree of dampening. The influence of temperature changes is smaller, if the parts are more evenly affected and the angular magnification is smaller.

Clamps and Fine Motion Drives for Levels

Most levels are equipped with a side clamp (see figures 409, 411, 414, 415) in order to arrest the top part at a specific position of the bottom part. Here, a clamp jaw presses against the plug and thus rigidly connects top and bottom parts for some time. With the horizontally acting fine motion drive, which is connected to the side clamp, the top part can be rotated slightly around the vertical axis in respect to the bottom part after tightening of the side clamp. In this manner, the vertical cross hair line can be set exactly onto the target, which is necessary, for instance with stadia levels.
Low accuracy levels often have neither clamping screws or levers, nor fine drives. The top part is on friction bearing and remains in its position after rotation.
Many modern levels, among others, most compensation levels, contain a friction clamp and an infinite fine drive (figure 482). Top and bottom parts, for instance, are connected via a cog wheel which is placed on the bottom part so that it can rotate. A spiral, placed on the top part, grips into the cog wheel and rigidly connects top part and cog wheel, as long as it is not rotated. For rough rotation, the cog wheel rotates against the bottom part. However, when moving the spiral, the top part is rotated against the cog wheel. The cog wheel is connected to the bottom part by friction, so that top- and bottom parts can be rotated against each other. The sideways movement can be performed with either the right or left hand, because the spiral has appropriate edge rings on both ends.

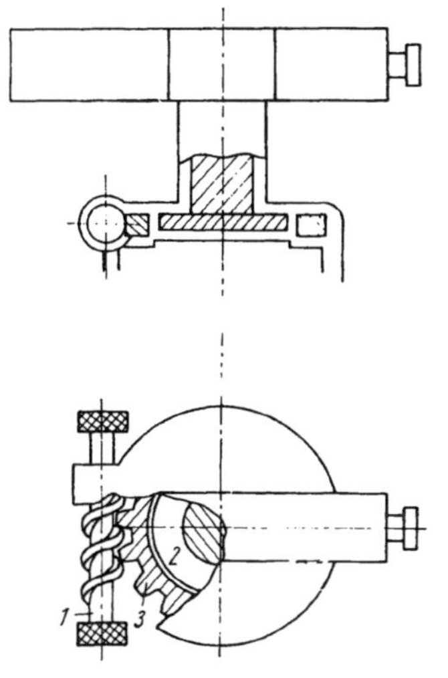

Figure 482 Friction clamp with infinite fine drive
1 snail; *2* central circle; *3* cog circle

Auxilliary Devices for Levels

Auxilliary Devices for High Precision Levels

Modern high precision levels are built with rigid telescope and, if they are spirit levels, with a tilting screw. For precision levels, auxilliary devices are primarily used for the required accurate levelling and for fine rod reading. These are often integral parts of the instruments. The levelling accuracy is increased for spirit levels by using more sensitive bubbles (5 to 10″ sensitivity) and by improving the bubble observation by coincidence reading (split bubble) with the aid of prism systems. The tilting screw is manufactured with such a low pitch, that tilting of the bubble axis and thus the collimation axis is possible within fractions of a second of arc. The tilting screw itself is rarely equipped with a micrometer drum (see figure 506). For spirit levels of high accuracy, occasionally even medium accuracy, the bubble setting is imaged within the field of view of the telescope (figures 413, 483). By the opposite direction of movement of the bubble ends, as obtained with the prism systems, the path of the bubble is doubled. The separation of the two bubble ends represents twice the distance of the bubble movement.
In order to be able to correct the rod reading for tilted collimation axis, based on readings of the bubble, bubble sensitivity p and sighting distance z have to be known. With the readings a_u and a_o (bottom and top, resp. left

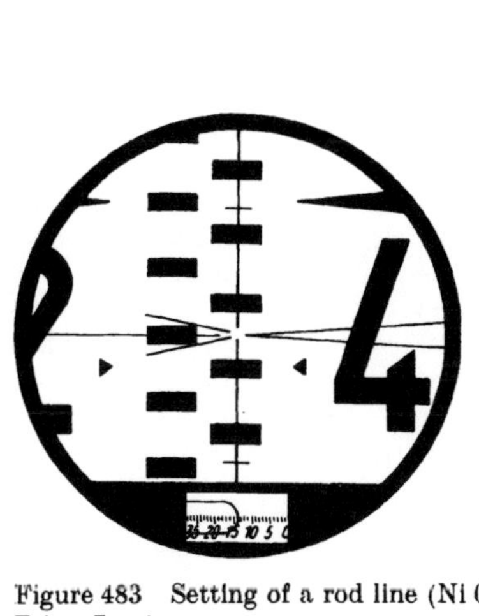

Figure 483 Setting of a rod line (Ni 004, Zeiss, Jena)

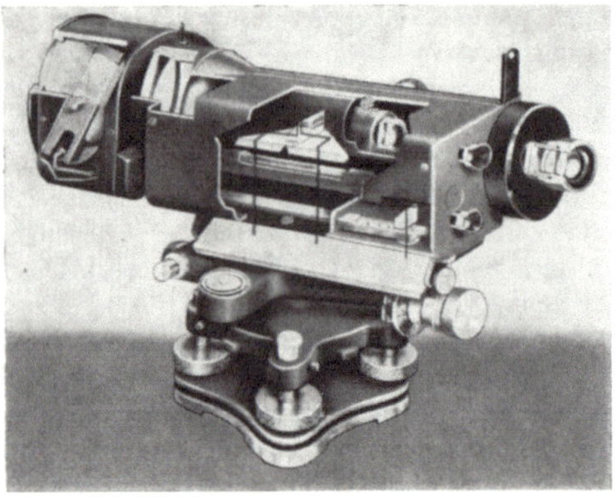

Figure 484 Level with plan-parallel plate micrometer (Wild N 3, old model; height 175 mm)

and right) of the position of the bubble of a coincidence bubble, the correction δh is obtained as

$$\delta_h = \frac{(a_u - a_o)\, pz}{\varrho}. \tag{142}$$

The rod reading is refined by utilizing powerful telescopes and optical micrometers (with rotatable plan- parallel plate) as well as with a reticule with a line wedge. The separation between the nearest graduation line of the rod and the horizontal collimation axis can be measured with the micrometer as function of the rotation of the plan-parallel plate around its horizontal axis normal to the collimation axis. For levels of lower and medium accuracy, this separation is estimated.

The plan-parallel plate is either an integral part of the telescope (figure 484), or attached in front of the objective lens as an auxilliary device. If it is rotated by α from the position, in which it does not change the sighting ray (figure 485), the collimation axis is shifted parallel by the amount $q \approx \approx d \tan\alpha(n-1)/n$, according to equation (11). In this manner the graduation line can be set onto the horizontal line of the cross hair. The amount q can be read on a properly graduated drum, which is connected to the plan-parallel plate. For convenient reading, the drum is placed at the telescope near the vertical axis, and the plan-parallel plate is rotated with a mechanical lever. Since the parallel shift of the collimation axis is very nearly proportional to the tangent of the rotational angle, the drum graduation can have equal intervals if the lever mechanism operates according to the tangent function. For production purposes, this graduation is most economical.

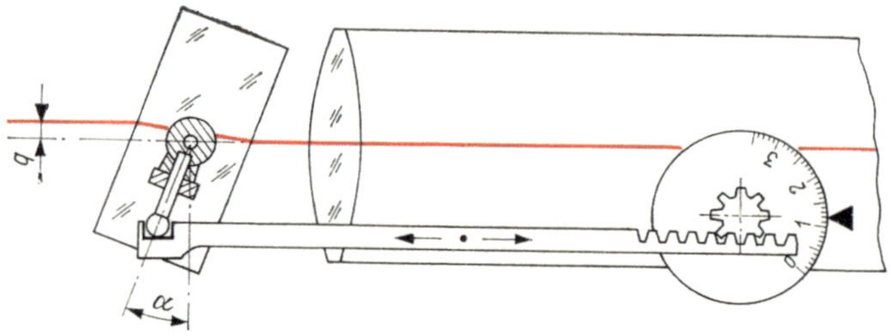

Figure 485 Monitoring of the plan-parallel plate micrometer

Corresponding to the graduation unit of the rods, most parallel plate micrometers are designed for a maximum shift of the collimation axis of 5 mm, a few for 10 mm. The setting is chosen such, that for zero reading of the micrometer, the parallel plate is rotated to one side, while for the maximum reading it is rotated by the same amount to the other side. Thus in the centre position of the micrometer drum, the ray passes through the parallel plate without shift. The rotational angle remains within the limits, where remaining errors due to difference between approximate formula (because of tangent function steering with equal drum graduation) and strict formula, have no practical influence. The drum reading can be transferred into the field of view of the telescope without difficulties. The micrometer drum is usually subdivided into 100 parts, which means, that a micrometer for 5 mm rod graduation has a least count of 0.05 mm. By estimating tenths of the interval, 0.005 mm can be obtained. This, however, is not reasonable because the setting error is beyond this amount.

With the exception of the Soviet precision level N-1 (see figure 507) by *Fefilow* and *Belizyn*, instruments with the parallel plate inside the telescope have not gained practical acceptance because of numerous error sources. The objective lens of the N-1 consists of three components (figure 486), of which the first two (*1*) form a telescopic system similar to Galilei's telescope, while the third one (*2*) serves as focussing lens (which can be shifted by 58 mm). The parallel plate (*3*), which can be rotated by 15° to either side, is placed behind the first two components, which give a near parallel path of the rays. The simple transfer mechanism without lever rods is of advantage.

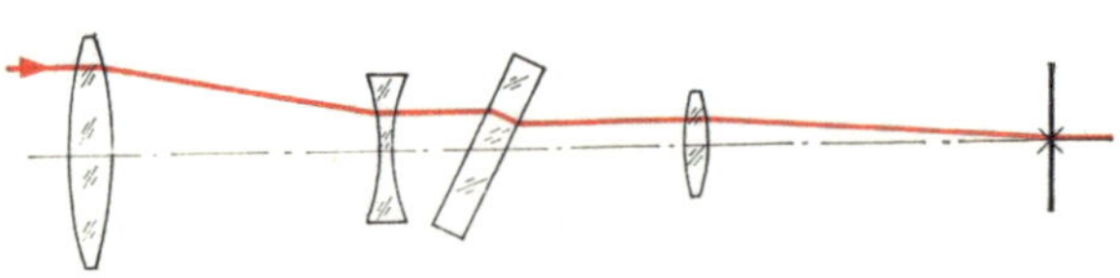

Figure 486 Plan-parallel plate inside the telescope

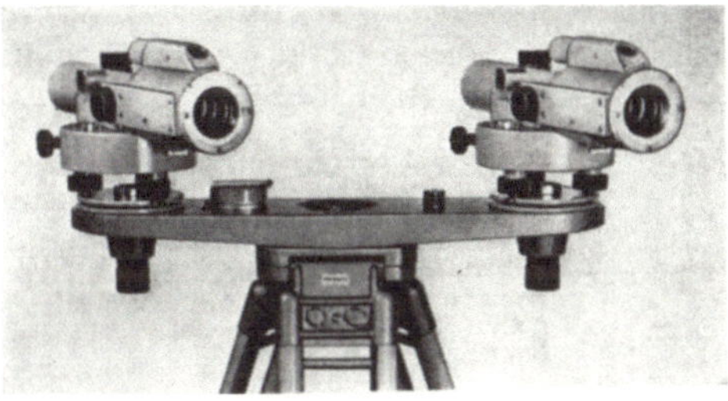

Figure 487 River crossing equipment for the Ni 2 (Zeiss, Oberkochen)

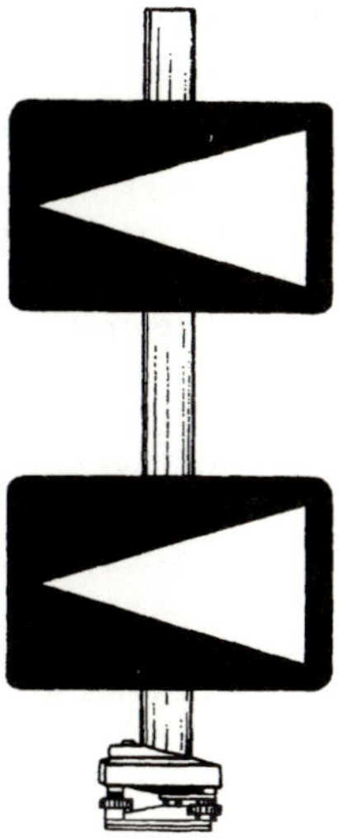

Figure 488 Targets for the river crossing equipment

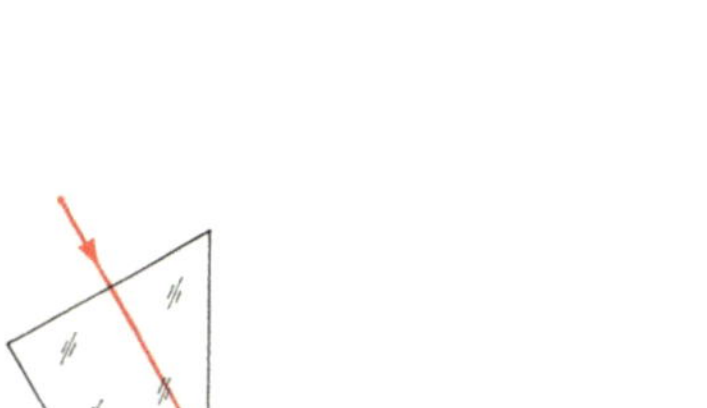

Figure 489 Principle of the prism astrolabe

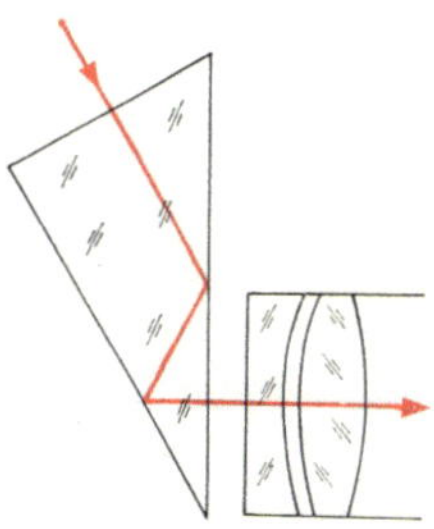

Figure 490 Optical train in the astrolabe for the Zeiss, Oberkochen Ni 2

For precision levelling, rods are used with line graduation on invar tape, which is insensitive to temperature changes. The graduation lines, which appear in varying thickness for different sighting distances, can be set very accurately with the aid of a line wedge on the reticule (see figure 94c). The graduation line of the rod can be centred between the line wedge using lateral fine movement of the telescope (figure 483).

Additional Devices for Levels

The reticules of all levels contain, in addition to the simple cross hair resp. the line wedge, two short horizontal stadia lines placed equidistant above or below the horizontal line for optical distance measurement. Instruments referred to as stadia levels, also have a horizontal circle for rough direction measurements. It is usually subdivided into half or full degrees and can be read with a simple line index (e.g. line microscope) or a vernier.

For river or valley crossing levelling, usually river crossing devices (movable targets at the level rods) are used. Zeiss, Oberkochen, however, produces different river crossing equipment for the Ni 2. On each side two Ni 2's are used, sitting on a common base plate fastened to the tripod (figure 487). Each instrument has a wedge attachment to measurably deviate the line of sight from the horizontal plane. By mutually collimating the instruments of one station prior to commencing with the measurements, one is assured, that the line of sight of one instrument is inclined upwards by the same amount as the one of the other instrument, which is inclined downwards. When sighting to the targets at the other station (figure 488) by rotating the wedges, all instrument adjustment errors are eliminated when using the mean value. Up until now, the instruments had to be interchanged and the measurements repeated in order to eliminate instrument adjustment errors. However, when changing instruments, one cannot be sure that the adjustment remains constant. This river crossing equipment reduces the measuring time and provides better accuracy.

So far, astrolabes have been built either as attachments for theodolites or as separate instruments. Prism astrolabes (figure 489) are used for simultaneous determination of latitude and longitude from stellar observations at equal zenith distances. Zeiss, Oberkochen, produces an astrolabe without mercury horizon as attachment to the Ni 2. The light rays are reflected twice in the prism (figure 490), such that the angle between entering and exit ray is independent of the position of the prism in respect to the entering ray. The collimation axis, deflected by the prism, forms a constant angle with the plumbline. Prior to starting the measurements, the prism is adjusted with the aid of a bubble on its housing until the deflected collimation axis is in the same vertical plane as the telescope axis of the level. The precomputed azimuth of the star is set on the horizontal circle of the level for measurements. The instant of transit of the star through the centre of 10 double lines on the specially inserted reticule is measured with a chronograph or a stop watch. The measurements are suitably evaluated according to the base line method. The company also manufactures a shiftable pentagon prism for the Ni 2 (so does Wild, Heerbrugg, for the NA 2), so that it can be used as optical plummet (with compensator).

Attachable lenses for levels are produced to reduce the minimum sighting distance. Autocollimation devices, eyepiece lamps and zenith eyepieces (sideways or from the top) are manufactured for levels just like for theodolites.

Special laser levels are suitable for area levelling. With the Geoplane 300 (figure 491) by AGA, Sweden, a horizontal reference plane can be obtained on construction sites within a radius of 250 m (corresponding to a working area of 200,000 m^2). The plane is formed by the red laser light (λ = 632.8 nm) of a 1-mW-He-Ne laser, emerging from 2 openings of the instrument head which rotates at 10 revolutions per second. One ray is directed slightly above the reference plane, the other slightly below. The rays have different polarization, and overlap only in the reference plane (figure 492). In it, there is continuous light visible (blinking frequency 20/s), while above and below this plane there is blinking light (frequency 10/s). In this manner, it is easier to find the plane. After rough levelling, the fine levelling is achieved automatically with a compensator (operational range $\pm 12'$). The reading is done by eye or with a photo detector. At 60 m distance, an accuracy of ± 2 mm is reached, while at 250 m it is ± 2 cm. Only one person is required to do the job. Up to 200 measurements can be taken within one hour.

Figure 491 Geoplane 300 (AGA)

The same working principle is applied at, among others, the Laserswinger by Dietzgen, U.S.A., the universal construction laser by LBS Beacon, U.S.A., which is levelled with four bubbles to <0.05%, the laser level meter by Blout & Gorg In., U.S.A. whose laser head rotates with 4 revolutions per second.

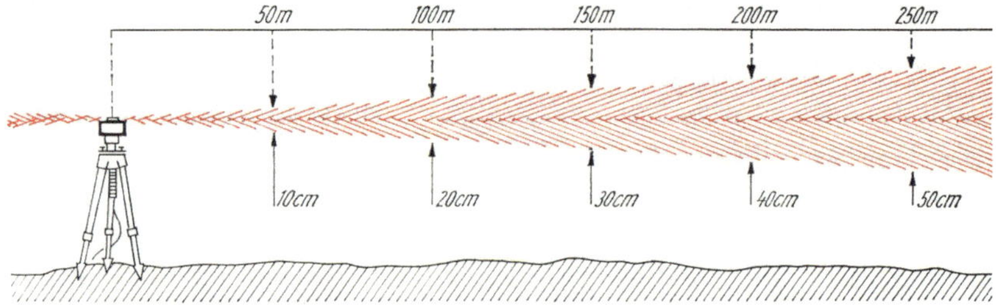

Figure 492 Overlapping of rays for the AGA Geoplane 300

Special Levels

The instrument Ni-C 3 by MOM Budapest for rail settling measurements and levelling, serves specifically for railway work (settlements, staking). It consists of a level and signal disks with diagonal graduation, which can be shifted vertically. Its tripod is clamped onto one rail and braced against the other. Breithaupt & Son, Kassel, produce a similar equipment for rail construction, namely the optical sighting instrument No. 1007 "DRESI". It consists of an aligning telescope which can be turned and slightly tilted, the tripod, and a rod mounted on a support tube.

3.2.3. Overview of Different Types of Levels

Levels are categorized according to special characteristics, such as the element used to level the collimation axis, the connection between level and telescope, the purpose of the instrument, and the measuring accuracy.
According to the element used for levelling the collimation axis, one distinguishes between spirit levels and compensator (or automatic) levels. Spirit levels can be subdivided into the ones with fixed telescope which is levelled via the foot screws or the tripod head (dumpy level), and the ones with a tilting screw, which is used for fine levelling (tilting level).
The categories of spirit levels according to the connection between bubble and telescope, as well as between telescope and standards was given on page 166. For levels with fixed telescope and fixed bubble, bubble, telescope and standards are rigidly connected. For some instruments with reversible bubble, one bubble is rigidly connected with the telescope which can roll in its supports. For instruments with reversible telescope and reversible bubble, the latter is a striding level. Nearly all modern levels are built with fixed telescope. Spirit levels of medium and high accuracy also have a tilting screw.
Categorizing levels according to their purpose is most common. There are construction levels, engineers levels and precision levels, which also corresponds to capacity classification.
The classification according to measuring accuracy is most appropriate. The standard deviation for 1 km levelling direct and return is used, which about matches the instrument purpose

levels of highest accuracy	$\leqq$0.5 mm/km
levels of very high accuracy	$\leqq$1 mm/km
levels of high accuracy	$\leqq$3 mm/km
levels of medium accuracy	$\leqq$10 mm/km
levels of low accuracy	>10 mm/km

Older instruments were awkward in use and hard to transport because of their large dimensions (especially the long telescope) and their considerable mass. The modern instruments however are small, handy and light, but robust. Their elements are protected against dust, humidity and damage by an encased design. Telescope body and bubble carrier are usually produced from one cast metal piece. Telescopes of constant length with interior focussing and comfortable bubble reading are common. Compensator levels gain more and more acceptance. The telescopes usually have stadia hairs.

spirit levels

type	manufacturer	telescope						bubbles			mass			height	diameter of horizontal circle	standard deviation of 1 km levelling	remarks
		Γ_F	diameter of objective lens [mm]	field of view [']	length [mm]	shortest sighting distance [m]	tilting screw	type of reading	bubble sensitiv. level bubble ['']	bubble sensitiv. bull's eye bubble [']	instrument [kg]	container [kg]	tripod [kg]	[mm]	[mm]	[mm]	
1	2	3	4	5	6	7	8	9	10	11	12	13	14	15	16	17	18
NAKUM	F. W. Breithaupt & Son, Kassel	25	30		155	0.9	–	mirror	30	6	1.4	2.2	3.5	130	96	±8	metal circle
NIFIX		20	30	110	127	0.8	vertical	mirror	45	15	1.0	0.5	3.9	88	78	±7	spherical tripod head; figure 493
GKO	Kern & Co., Aarau	18	24	127	127	0.9	vertical	mirror	50	15	0.8	0.5	3.9	80	60	±7	tripod head with joints; friction clamp
SL 10	Rank Prec. Ind. Ltd., England	14	30		170	3.0	vertical	mirror	50	20	1.3	0.7	3.4	100	63	±8	spherical tripod head. SL 18 upright image
8090	Lietz Comp., U.S.A.	18	19		268	3.6	–	–	600	–	1.1	2.4	4.3	95	92	±15	
LD-1	Sokkisha Ltd., Tokyo	18	25	103	150	1.1	–	mirror	50	10	1.7	1.8	4.5	150	(90)	±8	spherical tripod head
Sitemaster	W. F. Stanley & Co., London	14	25		124	1.5	horizontal	mirror	50	–	1.5	1.0	3.8	100	60	±15	spherical tripod head
NT	U.S.S.R.	23	32	90	140	1.5	vertical	coincidence within the field of view	45	10	1.0	0.7	4.5	95	105	±15	spherical tripod head
Ni 52	Opton Feintechnik GmbH, Oberkochen	20	25	124	200	0.9	vertical	mirror	60	30	0.7	1.0	5.5	120	78	±5	spherical tripod head; figure 494

automatic levels

type	manufacturer	telescope						compensator			sensitivity of bull's eye bubble	mass			height	diameter of horizontal circle	standard deviation for 1 km levelling	remarks
		Γ_F	diameter of objective lens [mm]	field of view [']	length [mm]	shortest sighting distance [m]	image	effective optical element	range [']	dampening	[']	instrument [kg]	container [kg]	tripod [kg]	[mm]	[mm]	[mm]	
1	2	3	4	5	6	7	8	9	10	11	12	13	14	15	16	17	18	19
KNA	Ertel-Werk, Munich	19	25	127	160	1.3	upright	standing pendulum	±10	magn.	30	0.9	0.6	4.2	120	78	±8	spherical foot
ALG 7	Officine Galileo, Florence	3.5	11	270	–	1.5	upright	horizontal pendulum mirror		air	30		2.8		160	64	50	spherical foot; figure 430
Cowley Level	Rank Prec. Ind. Ltd., England	–	–		120	1.5	upright	pendulum mirror		air	–	1.2		0.9	140	–	±40	figure 402
SNCA	SLOM, Paris	2	15		–	1.5	upright	pendulum mirror		magn.	15	1.2	2.0		140	70	±20	figure 438
NS 4	U.S.S.R.	30	40	80	188	2.0	reversed	pendulum prism	±15	air	15	2.0	2.8	3.5	122	–	±8	
TN 6		20	29	65	100	0.9	upright	objective lens and reticule as pendulum	±30	magn.	20	0.8	0.7	3.5	150	50	±15	eyepiece viewing from top

Table 24. Levels of medium accuracy (selection)

spirit levels

type	manufacturer	telescope						bubbles			mass			height [mm]	diameter of horizontal circle [mm]	standard deviation of 1 km levelling [mm]	remarks
		Γ_F	diameter of objective lens [mm]	field of view [']	length [mm]	shortest sighting distance [m]	tilting screw	type of reading	bubble sensi. level bubble [″]	bubble sensi. bull's eye bubble [']	instrument [kg]	container [kg]	tripod [kg]				
1	2	3	4	5	6	7	8	9	10	11	12	13	14	15	16	17	18
55	Brunson Instr. Co., U.S.A.	32	37		457	2.8	–	direct	25	–	5.4	5.0		230	–		reversible telescope, Type 45 : 5 kg
NAMAL	F. W. Breithaupt & Son, Kassel	30	40	83	190	1.2	horizontal	mirror	25	6	1.9	2.9	3.5	150	(80)	±5	NAGLA with horizontal circle
5150-A	Filotecnica Salmoiraghi, Milano	22	27		155	1.0	–	mirror	30	–	1.5	0.7	5.6	160	75	±5	container base is part of instrument
Eng. Wye Level	W. E. Gurley & Co., U.S.A.	26 41	43		460	2.4	–	direct	25	–	5.2	3.6	4.7	220	–		can be screwed on; dumpy level with 20″ bubble; 4.6 kg
Light Dumpy		29	35		330	2.0	–	direct	40	–	3.0	2.8	4.7	200	(127)		
N-6370	Dietzgen Corp., U.S.A.	20	32		302	3.0	–	direct	110	–	3.8	2.4			94		4 foot screws; N 6375 with vertical circle; 5,9 kg
Dumpy Level	Keuffel & Esser Co., U.S.A.	32	44		457	2.7	–	direct	20	–	5.0		4.5	210	–		figure 502
110	Lietz Comp., U.S.A.	26	33	60	302	2.0	–	direct	90	3	2.0	2.5		200	78		type 200 with vertical circle (transit level), also 8300 (with 110″ bubble)
TTL-5	Lietz, U.S.A./Sokkisha, Japan	30	40	70	240	2.3	vertical	coincidence bubble	40	6	2.8	2.6	5.0	150	–		spherical tripod head
Eng. Level	Nat. Instr. a. Opht. Glass Ltd., Kalkutta	24	44		300		–	mirror	25	6	3	3.6	7	175	–		dumpy level: direct reading
LG 5b	Officine Galileo, Florence	18	26	90	145	1.7	–	mirror	40	10	1.2	1.0	3.7	125	95	±5	LG 6 with 45″ bubble; ±6 mm/km
B 203	Path Instr. Int., New York	28	45	69		1.9	vertical	coincidence bubble within field of view	15	10	2.6	2.7			–		
NIK 20	NPPGP, Sofia	25	35	82	208	2.5	horizontal	coincidence bubble	20	6	2.1	1.7	4.8	132	80	±5	
SNO	SLOM, Paris	22	25	91	150	1.8	vertical	mirror	25	10	1.2	2.6	3.7	110	(80)	±7	SNOC without foot screws
D 71	W. F. Stanley & Co., London	35	38		310	2.7	–	direct	20		4.8	3.0	5.7	110			
D 20		25	42		190	1.5	horizontal	coincidence bubble	30	45	2.3	2.5	5.7	153	(70)	±5	
D 101		21	25		173	1.5		mirror			1.7		5.7				spherical support

Table 24 (continuation)

N 4	Theis & Co. (FRG)	25	35		160	1.3	vertical	mirror	30	30	1.2		0.8	100	100	±6	figure 503; spherical support Ni 4 with foot screws
N 5									30	15	1.4					±5	figure 497. N 6 with 60″-bubble, without horizontal circle
NL-3	U.S.S.R.	31	40	75	180	2.0	–	coincidence bubble	30	15	2.2	2.8	4.0	160	–	±5	level with inclined collimation axis; figure 514
T-1	Tokyo Sokki, Tokyo	25	38		247	2.1	horizontal	coincidence bubble	120	10	2.8	2.7	4.0	150	–	±5	
S 24	Vickers Ltd., York	19	29	150	130	1.8	vertical	mirror	60	8	1.3	0.7	5.5	110	(60)	±7	friction clamp; spherical support
TS	Topcon, Tokyo	32	47	80	215	1.4	horizontal	coincidence bubble in field of view	40	10	2.0	2.2	4.7	160	–		
NK 01	Wild Heerbrugg AG	19	26	138	160	1.6	–	mirror	60	8	1.5	0.6	3.8	100	90	±5	friction clamp
NK 05		19	25	138	160	0.8	horizontal	mirror	60	8	1.8	1.7	3.8	128		±5	
BNL	Ertel-Werk, Munich	20	25	124	160	0.9	vertical	mirror	60	30	0.7	0.6	4.2	130	100	±6	spherical support

Levels of Low and Medium Accuracy

Levels of low and medium accuracy (tables 23 and 24, figures 493–498), usually called construction levels, are designed for simple technical levelling at construction sites, the setting out of construction heights, short connection loops, simple profiling and area levelling, where the standard deviations more than 10 mm per km, resp. ± 3 to 10 mm per km. The diameter of the objective lens of the telescope ranges from 25 to 30 mm with 10 to 15 times magnification for low accuracy levels, and 15 to 25 times magnification of medium accuracy levels. The bubble sensitivity is 30 to 60″, the operational range of the compensator ± 30 resp. $\pm 20'$, with an error of ± 10.0 resp. $\pm 3.0''$. Usually the bubble is observed with the aid of a trap mirror (figures 493–497). Instruments of this accuracy category normally do not have a tilting screw, and are levelled just with the aid of foot screws or a rough levelling device. The container floor is often used as base plate of the instrument. Clamping screws or -levers are frequently replaced by a friction clamp. Besides spirit levels of low and medium accuracy, more and more automatic levels of this accuracy category are being built (see figures 425–438, 442–454, 457–461, 475). Occasionally the instruments are equipped with a horizontal circle to enable angular measurements in flat terrain.

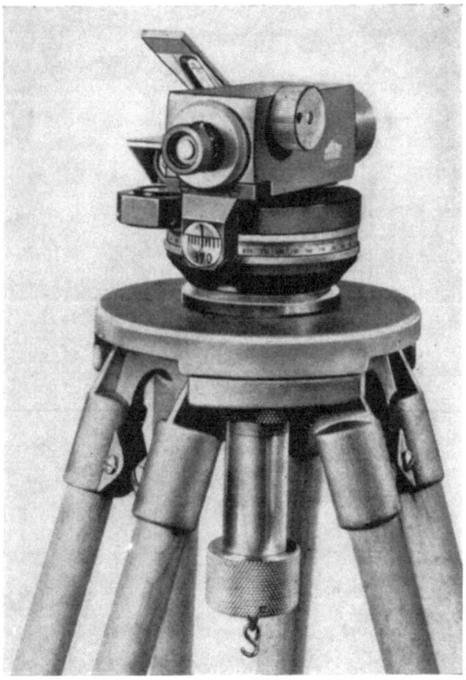

Figure 493 Small construction level No. 4105 "Nifix" (Breithaupt & Son); height 88 mm

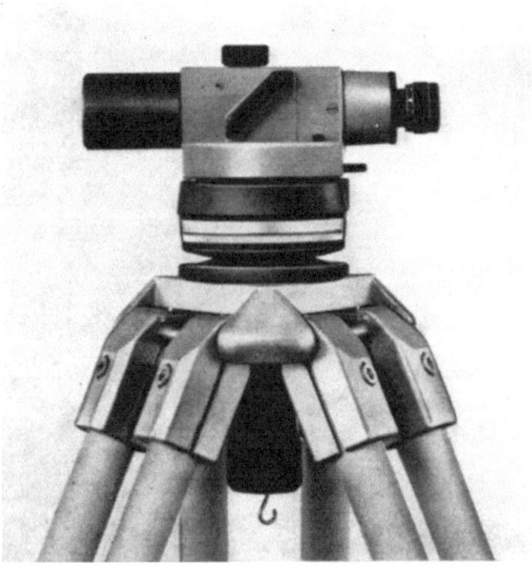

Figure 494 Ni 52 (Zeiss, Oberkochen); heigth 95 mm

Figure 495 LG 4 (Officine Galileo); height 90 mm

Figure 496 N-3 (USSR); height 150 mm

Figure 497 N 5 (Theis); height 80 mm

Table 24 (continuation)

automatic levels

type	manufacturer	telescope						compensator			sensitivity of bull's eye bubble [′]	mass			height [mm]	diameter of horizontal circle [mm]	standard deviation for 1 km levelling [mm]	remarks
		Γ_F	diameter of objective lens [mm]	field of view [′]	length [mm]	shortest sighting distance [m]	image	effective optimal element	range [′]	dampening		instrument [kg]	container [kg]	tripod [kg]				
1	2	3	4	5	6	7	8	9	10	11	12	13	14	15	16	17	18	19
5173	Filotecnica Salmoiraghi, Milano	80	30		(180)	2.0	upright	objective lens is pendulum		air	10	3.4	1.9	5.6	320	(80)	±5	container base is part of instrument; bull's eye bubble within field of view
GKO-A	Kern & Co., Aarau	21	30	103	150	0.8	upright	objective lens is pendulum	±90	air	20	1.9	0.8	3.9	120	60	±5	jointed support; warning disk in field of view; figure 460
6181	Dietzgen Corp., U.S.A.	26	30		216	1.0	upright		±10		10	1.6	2.2			89		type 6182: ±20′ range
6185		26	40		238	0	upright		±15		10	2.3	2.4			89		type 6186: length of telescope: 260 mm
Ni 42	Opton Feintechnik GmbH, Oberkochen	22	30	124	135	1.2	upright	pendulum prism	±60	air	30	1.4	1.0	4.1	150	70	±5	fast levelling with two foot screws; no clamp/fine drive; figure 410
AP-2	Nikon, Japan	20	30	90	235	1.6	upright	pendulum mirror	±5	air	10	2.1	2.4	3.6	130	100		without tribrack: AP
C 3 A	Lietz, U.S.A./Sokkisha, Japan	25	30	85	210	1.6	upright	pendulum mirror	±10	magn.	10	2.2	1.4	4.5	150	80	±5	
U 2	Topcon, Japan	22	30	90	242	1.8	upright	Porro-system of the first type	±10	air	6	2.1	2.2	4.6	145	100		
AT-S		25	40	82	205	2.0	upright	Porro-system of the first type	±10	air	6	2.2	2.5	4.6	140	96		figure 459
S 27	Vickers Ltd., York	21	29	150	130	1.8	upright	2 pendulum mirrors	±15	magn.	15	1.4		4.5	100	70		spherical support; figure 454
Ni 050	VEB Carl Zeiss JENA	18 16	25	144	210	0.9	upright	2 pendulum mirrors	±30	air	25	1.0	1.7	5.0	120	(74)	±5	disk levelling; figure 450

Table 25. Levels of high accuracy (selection)

spirit levels

type	manufacturer	telescope						bubbles			mass			height	diameter of horizontal circle	standard deviation of 1 km levelling	remarks
		Γ_F	diameter of objective lens [mm]	field of view [']	length [mm]	shortest sighting distance [m]	tilting screw	type of reading	bubble sensitiv. level bubble ['']	bubble sensitiv. bull's eye bubble [']	instrument [kg]	container [kg]	tripod [kg]	[mm]	[mm]	[mm]	
1	2	3	4	5	6	7	8	9	10	11	12	13	14	15	16	17	18
NAGUK	F. W. Breithaupt & Son, Kassel	30	40	83	190	1.2	horizontal	coincidence bubble within field of view	20	6	2.1	2.9	3.5	160	–	±2	NAKRE with horizontal circle (80 mm)
N I	Chasselon, Cachan	24	35	90	175	2.2	horizontal	mirror	15	8	2.2	2.0	4.0	100	(65)	±4	NC with horizontal circle
ND 1	China	39	45		180	3.0	horizontal	coincidence bubble	15	10	1.7	–	4	120	–	±2	
5153-B	Filotecnica Salmoiraghi, Milano	22	27	90	155	2.2	–	mirror	30	10	1.7	0.7	5.6	150	57	±4	container base is part of instrument
5168		25	35	90	160	3.0	vertical	coincidence bubble within field of view	30	10	3.3	3.1	7.2	230	(80)	±2	figure 499
L-21	Fuji Surv. Instr., Tokyo	32	45	80	253	1.9	vertical	coincidence bubble within field of view	15	10	2.7	1.1		140	–		
GK 1	Kern & Co., Aarau	22	30	103	118	0.9	horizontal	coincidence bubble	45	12	0.9	0.6	4.7	80	(60)	±4	without foot screws; friction clamp
Ni-B 11	MOM, Budapest	28	40	82	160	1.2	horizontal	coincidence bubble	20	6	2.2	1.7	4.5	137	(76)	±3	Ni-B 21 without horizontal circle; figure 498
E-5	Nikon, Japan	25	40	72	245	2.0	vertical	coincidence bubble within field of view	30	10	2.2	2.1	3.9	150	–	±2	E 6 with horizontal circle 99 mm
S 2		30	45	80	210	2.0	vertical		20	10	2.3	2.5	3.9	140	–	±2	optical micrometer
LG 3	Officine Galileo, Florence	22	30	90	154	1.8	vertical	coincidence bubble	30	10	1.6	1.5	4.7	140	56	±2	
Ni 42	PZO, Warsaw	25	40	80	208	2.5	vertical	coincidence bubble	20	10	2.3	3.8	4.0	145	68	±3	Ni 71 without horizontal circle; Ni 41: 125 mm high
SL 120	Rank Prec. Ind. Ltd., England	25	38	80	200	1.8	vertical	coincidence bubble	50	10	1.9	2.4	5.4	132	57	±2	SL 122 without horizontal circle
SL 130									25	12				120			SL 132 without horizontal circle

SN 1	SLOM, Paris	22	25	91	150	1.8	vertical	coincidence bubble	40	10	1.3	2.8	3.7	110	(80)	±3	SN 1 C without foot screws
TTL 6	Sokkisha Co. Ltd., Tokyo	25	40	75	240	1.8	vertical	coincidence bubble	40	10	2.2	2.5	4.5	140	80	±2	
N-3	U.S.S.R.	30	40	80	175	2.0	horizontal	coincidence bubble within field of view	15	5	1.8	2.0	3.5	152	–	±4	figure 496
S 33	Vickers Ltd., York	28	38	110	194	2.0	vertical	coincidence bubble	50	8	2.0	2.5	4.8	118	70	±4	
S 440		32	45	78	268	3.0	vertical		25	8	2.7	3.8	6.9	118	70	±2	
VBK-1	Visomat AG Switzerland	20	25	138	160	1.2	horizontal	mirror	30	8				110	87	±3	
VN-3		18	25	138	160	1.3	–	mirror	20	–				200	(70)	±3	tilting telescope (VNK) ±5°; 40″ levelling bubble
N 10	Wild Heerbrugg AG	23	32	130	155	1.4	horizontal	coincidence bubble with pointer	60	8	1.7	0.6	3.8	120	(55)	±2.5	NK 10 with horizontal circle; friction clamp
N 2		30	40	97	196	1.6	horizontal	coincidence bubble	30	8	2.2	1.3	6.0	157	(70)	±2	reversible level; friction clamp; figure 512 NK 2 with horizontal circle; 2.8 kg
Ni 030	VEB Carl Zeiss JENA	25	35	90	195	1.8	horizontal	coincidence bubble	30	8	1.8	1.5	4.8	120	75	±3	attachment: parallel plate micrometer ±0.8 mm/km; figure 501

automatic levels

type	manufacturer	telescope						compensator			sensitivity of bull's eye bubble [′]	mass			height [mm]	diameter of horizontal circle [mm]	standard deviation for 1 km levelling [mm]	remarks
		Γ_F	diameter of objective lens [mm]	field of view [′]	length [mm]	shortest sighting distance [m]	image	effective optical element	range [′]	dampening		instrument [kg]	container [kg]	tripod [kg]				
1	2	3	4	5	6	7	8	9	10	11	12	13	14	15	16	17	18	19
Autom	F. W. Breithaupt & Son, Kassel	27	33	66	235	1.5	upright or inverse	pendulum mirror	±45	magn.	8	2.3	2.9	5.0	150	–	±2	friction clamp; AUCIR with metal circle (105 mm); figure 440
BNA	Ertel-Werk, Munich	24	30	103	150	1.3	upright or inverse	prism on standing pendulum	±50	magn.	15	1.4	1.0	4.2	100	80	±4	friction clamp; fast focussing; function curve drive with or without tribrack; figure 474
INA		32	40	76	185	1.7	upright or inverse	prism on standing pendulum	±25	magn.	8	2.0	2.4	3.9	140	67	±2	friction clamp; function curve drive attachment: parallel plate micrometer (±0.4 mm/km)

Table 25 (continuation)

type	manufacturer	telescope						compensator			sensitivity of bull's eye bubble	mass			height	diameter of horizontal circle	standard deviation for 1 km levelling	remarke
		Γ_F	diameter of objective lens [mm]	field of view [']	length [mm]	shortest sighting distance [m]	image	effective optical element	range [']	dampening	[']	instrument [kg]	container [kg]	tripod [kg]	[mm]	[mm]	[mm]	
1	2	3	4	5	6	7	8	9	10	11	12	13	14	15	16	17	18	19
FNAK 1	Fennel (FRG)	25	30	97	200	1.9	upright	Porro-system of the 2nd type	±25	air	15	1.9	1.3	5.2	150	100	±2	friction clamp; figure 457
AL-21	Fuji Surv. Instr., Tokyo	30	42	65	273	1.8	upright	pendulum prism		air	15	2.8	1.5		180	100	±3	figure 448
AL-31		20	25		–	2.0	upright	pendulum prism with parallel shift		air	15	2.2	1.1		187	105	±3	figure 468
GK 1-A	Kern & Co., Aarau	25	45	86	125	2.3	upright	pendulum mirror	±10	air	12	1.6	1.1	3.9	135	60	±2.5	jointed support; figure 426
Ni-B 3	MOM, Budapest	28 32	45	80	272	3.0	upright	Porro-system of the 2nd type		air	8	2.3	2.3	5.1	130	(82)	±2	figure 456; Ni-B 5 and 6 with image of bull's eye bubble within field of view; Ni-B 4 and 5 without horizontal circle
Nestor 4	Gottl. Nestle (FRG)	31	42	69	270	2.0	upright	pendulum mirror		air	10	2.5	2.2	4.4	150	109	±2	
NAN 3		22	30		210	1.5	upright	pendulum mirror		air	10	2.5	1.9	4.4	150	109	±3	figure 446
N	Nikon, Tokyo	28	40	80	221	2.1	upright	pendulum mirror		air	20	2.6	2.0	3.6	150	–	±3	figure 444
AP-3	Nikon, Tokyo	30	30	90	235	1.0	upright	pendulum mirror	±10	air	10	1.9	2.0	3.6	133	100	±2	
AE-3		32	40	80	275	1.6	upright	pendulum mirror	±6	air	6	2.4	2.3	3.6	120	113	±2	
AL-3 EC	Pentax, Hamburg	22	30	90	243	0.2	upright	pendulum mirror	±10	air	10	2.4	1.6		144	100	±3	
S-302	Path Instr. Int. Corp. (U.S.A.)	20	34	90	250	1.5	upright	pendulum prism		air	15	2.2	1.9		150	100	±3	friction clamp
S-201		30	42	65	273	1.8	upright	pendulum prism		air	15	2.8	2.7		180	(100)	±3	friction clamp

SL 70	Rank Prec. Ind. Ltd. (England)	25	36	90	204	2.6	upright	2 prisms	±15	air	8	1.6	2.0	3.4	110	57	±2	spherical support
SL 60		32	40	80	267	1.8	upright	2 prisms	±20	air	8	3.5	2.0	5.5	134	–	±1.2	attachment: parallel plate micrometer; SL 62 with horizontal circle (58 mm)
B-2	Sokkisha Ltd., Tokyo/ Lietz Comp. (U.S.A.)	30	40	65	210	1.8	upright	pendulum mirror between 2 mirrors	±10	magn.	10	2.8	1.4	4.5	150	80	±3	c = +15 cm
SNA 1	SLOM, Paris	23	25		–	1.6	upright	45°-pendulum mirror	±20	air	30	1.6	1.4	5.3	180		±3	without tribrack; figure 439
Toko	Tokyo Optical Co., Tokyo	25	40		208	2.0	upright	Porro-system of the first type		air	6	2.1	1.7	3.3	130		±2	
OL-2	Tokyo Sokki Co., Tokyo	27	40		192	2.5	upright	pendulum mirror		air	20	2.5	2.0	3.5	140	–	±3	
Tecomat	Theis & Co. (FRG)	25 32	35 40	100	220	2.2	upright	pendulum prism		magn.	8	2.4	2.4	5.0	125	(120)	±3	figure 435; friction clamp; infinite rough- and fine drives
NS 3	U.S.S.R.	30	34	80	140	3.0	inverse	lens compensator	±15	air	7	2.0	2.8	1.9	138	–	±4	2′-cross level; spherical support
S 77	Vickers Ltd., York	32	39	80	267	1.8	upright	2 pendulum prisms	±20	air	15	2.7	5.0	5.4	133	58	±1.5	friction clamp; attachment: parallel plate micrometer
VA-1	Visomat AG (Switzerland)	22	40	206	155	1.0	upright	pendulum mirror	±270	magn.	8				145	(70)	±4	centering head; figure 442
NAK 1	Wild Heerbrugg AG	24	36	110	201	1.0	upright	pendulum mirror	±15	air	8	2.2	1.5	3.8	133	92	±1.5	figure 476; button for functional control; with parallel plate: ±1 mm/km
NAK 0		20	30	124	188	0.9	upright	pendulum prism	±15	air	8	1.8	1.5	3.8	120	100	±2.5	figure 475
Ni 025	VEB Carl Zeiss JENA	20	30	100	195	1.5	upright	2 pendulum prisms	±10	air	8	1.7	1.7	3.4	140	(50)	±3	figure 452

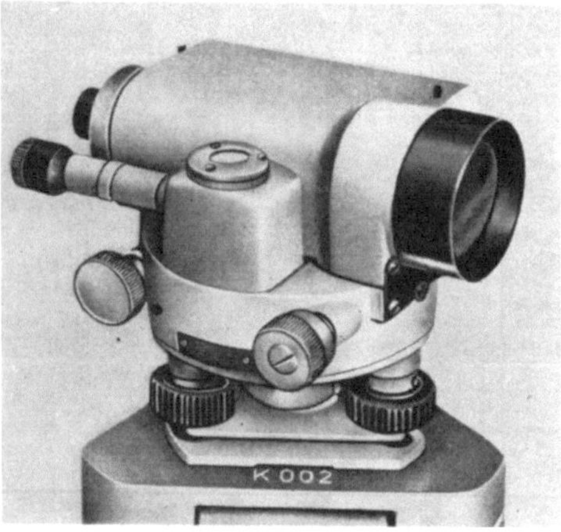

Figure 498 Ni-B 11 (MOM, Budapest)

Figure 499 Engineer's level 5168 (Filotecnica Salmoiraghi); height 230 mm

Figure 500 Engineer's level Wild N 10; height 130 mm

Figure 501 Little engineer's level with tilting screw, without horizontal circle (Fennel) height 155 mm

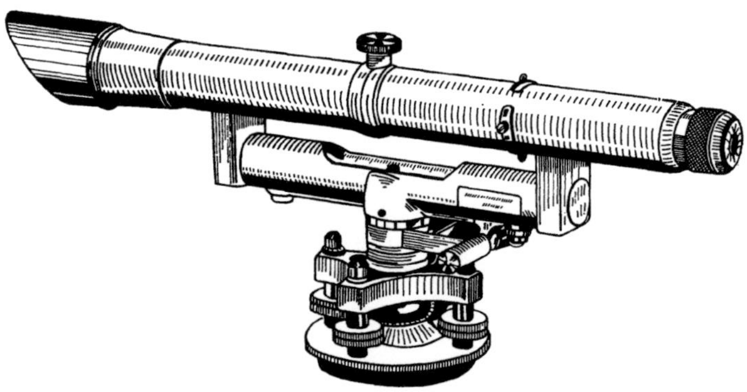

Figure 502 Paragon dumpy level (Keuffel & Esser); height approx. 220 mm

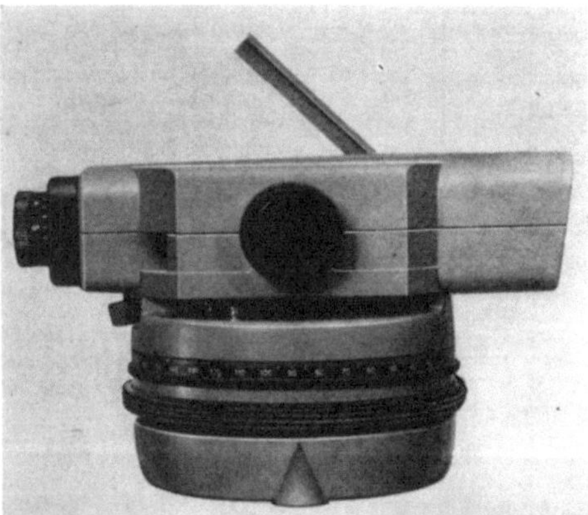

Figure 503 N 44 (Theis & Co.); height 80 mm

High Accuracy Levels

High accuracy levels (table 25, figures 499 – 504), usually called engineer's levels, are used for 3rd order levelling, area levelling and profiling for earth work computations, stadia surveys and staking in flat terrain, and levelling for engineering purposes, where a standard deviation of ± 1 to 3 mm per km is required. The diameter of the objective lens of the telescope is generally 30 to 40 mm with a 25 to 30 times magnification. The bubble sensitivity is 20 to 30″. Usually, split bubbles are used, whose image is often reflected into the field of view of the telescope. Nearly all spirit levels have fixed telescope with fixed bubble and tilting screw. Occasionally, levels with reversible bubble and telescope that can roll in its supports, are built (figure 512). The operational range of the compensator for automatic levels (see figures 455 – 458, 468, 476 and 477) is $\pm 10'$, with an error of $\pm 1''$.

Table 26. Levels of very high accuracy (selection)
spirit levels

type	manufacturer	telescope						bubbles			mass			height	diameter of horizontal circle	standard deviation of 1 km levelling	remarks
		Γ_F	diameter of objective lens [mm]	field of view [']	length [mm]	shortest sighting distance [m]	tilting screw	type of reading	bubble sensitiv. level bubble ['']	bubble sensitiv. bull's eye bubble [']	instrument [kg]	container [kg]	tripod [kg]	[mm]	[mm]	[mm]	
1	2	3	4	5	6	7	8	9	10	11	12	13	14	15	16	17	18
5169	Filotecnica Salmoiraghi, Milano	30	45		160	3.5	vertical	coincidence bubble within field of view	20	10	3.4	3.6	7.2	270	–	±1.3	with parallel plate micrometer: ±0.3 mm/km
GK 23	Kern & Co., Aarau	30	45	86	170	1.8	horizontal	coincidence bubble within field of view	18	6	1.5	1.9	4.6	110	62	±2	figure 504; rough-fine focussing; sliding clutch; with parallel plate micrometer: ±0.5 mm/km
LG 2	Officine Galileo, Florence	29	40	90	177	1.8	vertical	coincidence bubble within field of view	20	10	2.0	2.0	4.7	150	56	±1	
SN 28	SLOM, Paris	28	40	103	150	1.7	vertical	coincidence bubble	20	15	1.8	4.1	4.7	146	80	±2	with parallel plate micrometer: ±1 mm/km
N 2	U.S.S.R.	40	52	55	390	2.0	horizontal	coincidence bubble within field of view	10	5	6.0	4.8	6.0	195	–	±1	micrometer scale within the field of view of the telescope

Table 26 (continuation)

automatic levels

tpye	manufacturer	telescope						compensator			sensitivity of bull's eye bubble [']	mass			height [mm]	diameter of horizontal circle [mm]	standard deviation for 1 km levelling [mm]	remarks
		Γ_F	diameter of objective lens [mm]	field of view [']	length [mm]	shortest sighting distance [m]	image	effective optical element	range [']	dampening		instrument [kg]	container [kg]	tripod [kg]				
1	2	3	4	5	6	7	8	9	10	11	12	13	14	15	16	17	18	19
AL-2 E	Pentax, Hamburg	32	45	72	265	1.5	upright	pendulum prism	±10	air	10	2.6	1.9		139	(98)	±1	
GK 2-A	Kern & Co., Aarau	32.5	45	80	200	2.2	upright	pendulum prism	±20	magn.	10	3.5	3.1	4.7	190	62	±0.7	figure 429; with parallel plate micrometer: ±0.3 mm/km
Ni 2	Opton Feintechnik GmbH, Oberkochen	32	40	79	270	3.3	upright	prism on jointed quadrilateral	±15	air	10	2.1	3.3	4.5	130	(75)	±0.7	figure 470; with parallel plate micrometer: ±0.3 mm/km Ni 21: horizontal circle; fast levelling with 2 foot screws
B-1	Sokkisha (Japan)/Lietz (U.S.A.)	32	45	80	265	2.4	upright	mirror on four wires	±10	magn.	10	3.0	1.5	4.5	180	80	±1	rough- and fine focussing; with parallel plate micrometer: ±0.5 mm/km
NA 2	Wild Heerbrugg AG	32 40	45	83	246	1.6	upright	pendulum prism	±15	air	8	2.4	1.8	5.7	138	(70)	±0.7	figure 477; with parallel plate micrometer: ±0.3 mm/km. NAK 2 with horizontal circle 2.9 kg
Ni 007	VEB Carl Zeiss JENA	31.5	40	79	–	2.2	upright	pendulum prism	±10	air	8	3.9	2.8	5.6	335	(63)	±0.5	figure 466; with arrested micrometer: ±2 mm/km

Figure 505 Wild N 3; height 175 mm

Often high accuracy levels have a horizontal circle for angular measurements in flat terrain. Many instruments have friction clamps with infinite fine movement. Instruments of this category are most widely used. To increase the range of application, many manufacturers produce parallel plate micrometers as attachments, which increases the accuracy.

Figure 504 Engineer's level GK 23 (Kern); height 110 mm

Figure 506 Geodetic level S 550 (Vickers); height 150 mm

Levels of Very High and Highest Accuracy

Levels of very high and highest accuracy (table 26 and 27, figures 505 – 507) usually called precision levels, are used for bench mark levelling of highest accuracy (1st and 2nd order), monitoring of bridges and power dams, precision levelling in industry, settlement determination in mining areas and other high precision levelling, where a standard deviation of ± 0.1 to 0.5 mm or less than 0.5 mm per km has to be reached. The diameter of the objective lens of the telescope is 45 to 60 mm with 30 to 50 times magnification. The bubbles are all split bubbles with a sensitivity of 5 to 20''. Often the image of the bubble, together with its graduation is reflected into the field of view to facilitate observation of the bubble during rod reading. For automatic levels (figures 419 – 424, 463, 466, 470 and 472), compensators with an operational range of 5 to 10'' with an error of 0.1 to 0.5'' are used. There are no horizontal circles. For better setting and reading of the rod graduation, the reticule has a line wedge, and the instrument is equipped with a parallel plate micrometer. Otherwise the structure (e.g. fixed telescope, precise tilting screw) is similar to the high accuracy levels, however the dimensions of the instruments are larger, the screws finer, and the tripods heavier.

Figure 507 Precision level N-1 (U.S.S.R.); height 200 mm

3.2.4. Testing and Adjusting of Levels

Axes and Adjustment Devices of Levels

Like any other surveying instrument, the level has errors which adversely affect the observation results. Even instruments of the same series are different, because mounting and adjustment lead to different accuracies. Furthermore, the adjustment may change during measurements. Therefore, the instruments have to be tested periodically (e.g. once a week) and adjusted if necessary.
To avoid the influence of small residual errors on the observation results, the measuring procedure is chosen such that they are eliminated.

A level has the following axes (see figure 411):

a) the vertical axis VV
b) the collimation axis ZZ
c) the axis $L'L'$ of the bull's eye bubble, and for spirit levels
d) the bubble axis LL of the tubular bubble, and for levels with reversible or rolling telescope
e) the telescope axis FF, which is also called mechanical axis.

Table 27. Levels of Highest Accuracy (selection)

spirit levels

type	manufacturer	telescope						bubbles			mass			height	diameter of horizontal circle	standard deviation of 1 km levelling	remarks
		Γ_F	diameter of objective lens [mm]	field of view [']	length [mm]	shortest sighting distance [m]	tilting screw	type of reading	bubble sensitivity: level bubble ['']	bubble sensitivity: bull's eye bubble [']	instrument [kg]	container [kg]	tripod [kg]	[mm]	[mm]	[mm]	
1	2	3	4	5	6	7	8	9	10	11	12	13	14	15	16	17	18
NABON	F. W. Breithaupt & Son, Kassel	42	50	86	335	2.0			8	–	6.6	4.1	7.5	220	–	±0.2	1′ cross level
L-11	Fuji Surv. Instr., Tokyo	28	40	66	266	1.4			10	8	3.0	3.0		155	–	±0.5	
55/PL-5	Sokkisha (Japan)/Lietz (U.S.A.)	42	50	65	355	2.5	horizontal	coincidence bubble within the field of view of the telescope	10	3.5	5.1	3.8	4.0	175	–	±0.4	1′ cross level
N 1	U.S.S.R.	49	60	50	420	4.2			10	2	6.9	4.2	6.5	233	–	±0.5	figure 507; 0.8 mm bubble graduation
S 550	Vickers Ltd., York	36	51		300	4.6	vertical		15	20	5.4	3.7	6.7	150	–	±0.3	figure 506
N 3	Wild Heerbrugg AG	46	52	60	297	0.5	horizontal	coincidence bubble	10	4	3.5	2.5	7.6	175	–	±0.2	figure 505; panfokale optic

Table 27 (continuation)

automatic levels

type	manufacturer	telescope						compensator			sensitivity of bull's eye bubble [′]	mass			height [mm]	diameter of horizontal circle [mm]	standard deviation for 1 km levelling [mm]	remarks
		Γ_F	diameter of objective lens [mm]	field of view [′]	length [mm]	shortest sighting distance [m]	image	effective optical element	range [′]	dampening		instrument [kg]	container [kg]	tripod [kg]				
1	2	3	4	5	6	7	8	9	10	11	12	13	14	15	16	17	18	19
5190	Filotecnica Salmoiraghi, Milano	39	45	80	–	3.9	upright	reticule as pendulum	±5	air	10	6.9	6.5	7.2	500	–	±0.3	figure 463
Ni-A 31	MOM, Budapest	30 40 50	67	60	(210)	2.5	upright	pendulum mirror	±13	air	10	3.0	3.6	4.5	175	–	±0.2	figure 422; c = 370 mm
Ni 1	Opton, Feintechnik GmbH, Oberkochen	40 (50, 30)	50	63	407	1.4	upright	prism on jointed quadrilateral; x-long compensator	±8	air	5	5.2	4.6	7.5	180	–	±0.2	figure 472
Ni 002	VEB Carl Zeiss JENA	40	55	77	(370)	1.5	upright	pendulum mirror at $f/2$	±10	air	8	6.5	5.1	6.2	310	–	±0.2	figure 419: quasi-absolute horizon; swivelling eyepiece

Since elevation differences are determined with the aid of a horizontal straight line, the primary condition for each spirit level is: The collimation axis has to be horizontal for a levelled spirit bubble, i.e. for horizontal bubble axis:

Collimation axis ZZ has to be parallel to the bubble axis LL ($ZZ \parallel LL$). For fast and convenient operation of spirit levels, the following minor conditions are requested:

$LL \perp VV$, $L'L' \parallel VV$

horizontal line should be horizontal for levelled bubble, and there should be no cross divergence of the bubble. However, it is possible to obtain correct measuring values if these minor conditions are not fulfilled. Spirit levels may have the following adjustment devices (see figure 411):

B_L at the bubble
B_Z at the reticule.

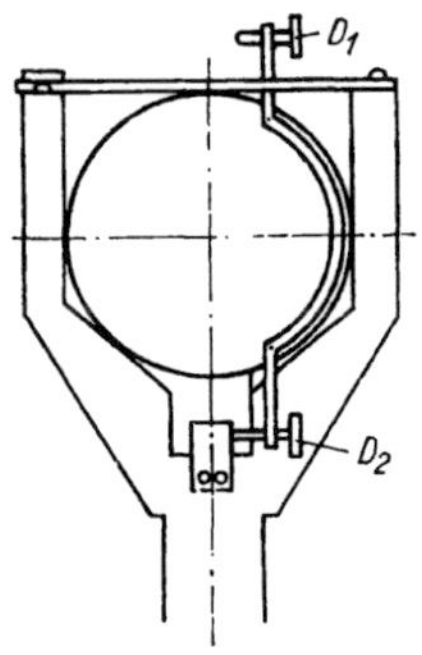

Figure 508 Bearing of a rolling telescope

For levels with reversible or rolling telescope and reversible bubble, an additional pair of adjustment screws D_1 and D_2 (figure 508) are necessary to make the horizontal line horizontal.

The primary condition for compensator levels is:
Within the operational range of the compensator, the collimation axis has to be horizontal.

The screws B_Z at the reticule are used for adjustment. Some types achieve adjustment with a slightly wedge shaped glass in front of the objective lens or with an adjustment screw at the compensator. Each level equipped with a bull's eye bubble, contains the necessary adjustment devices for it. On modern instruments, the adjustment screws are usually covered and accessible only after removal of a protective cap or cover. Due to the rigid connection of bubble, telescope and vertical axis, the rules for the testing and adjustment of spirit levels are simple. Besides testing and adjustment, the instruments have to be checked for the characteristics and operation of their main elements.

Testing and Adjustment of a Level with Fixed Telescope, Fixed Bubble and without Tilting Screw

Since bubble, telescope and vertical axis are rigidly connected, the vertical axis has to be correctly vertical for levelling of the collimation axis. Testing and adjustment are to be done in the following sequence, so that a subsequent adjustment does not nullify a previous one.

1st minor condition: Bubble axis LL perpendicular to vertical axis VV ($LL \perp VV$). Testing and adjustment as in section 1.2.1.

2nd minor condition: Axis $L'L'$ of the bull's eye bubble parallel to the vertical axis ($L'L' \parallel VV$).

If the bubble of the bull's eye level is not centered after the 1st minor condition is fulfilled, it is centered with the aid of adjustment screws.

3rd minor condition: Horizontal line is horizontal for centered bubble.

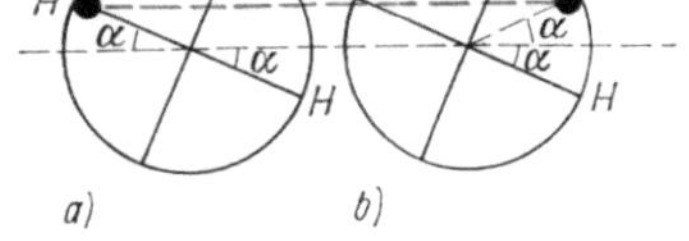

Figure 509 Testing of cross hair position

After levelling of the instrument, one sets a point P at the edge of the field of view onto the horizontal line HH of the cross hair (figure 509). If P does not remain on the horizontal line when the telescope is turned, the cross hair has to be rotated by α loosening the clamping screw and turning of the support ring. This is to be repeated for checking purposes.

4th minor condition: Vertical planes through bubble axis LL and collimation axis ZZ are parallel (bubble is free of cross divergence).

A cross divergence can be noticed by sideway tilting of the level. One reads the value "a" at an approximately 20 m distant rod with the instrument being horizontal with centered bubble, and the telescope placed above one foot screw. Then one turns one of the two back foot screws by two or three revolutions and again the cross hair is set to "a" by using the other back foot screw. If the bubble moves, a cross divergency is present, which is to be eliminated completely by using the horizontally acting adjustment screws of the bubble. If this is not possible, the cross hair plate has to be shifted horizontally until — when repeating this procedure — the bubble remains centered.

Main condition: Collimation axis ZZ parallel to bubble axis LL ($ZZ \parallel LL$).

The levelling rod is placed at points A and B in equal distance from the instrument S_1 (30 to 50 m) which with centered bubble, yields the readings a_1 resp. b_1 (figure 510). The correct elevation difference $\Delta h = a_1 - b_1$ is independent of a tilt error, i.e. the angle between collimation- and bubble axes. Afterwards the instrument is placed as close as possible to B at S_2. After centering the bubble, one reads b_2 on the rod at B.

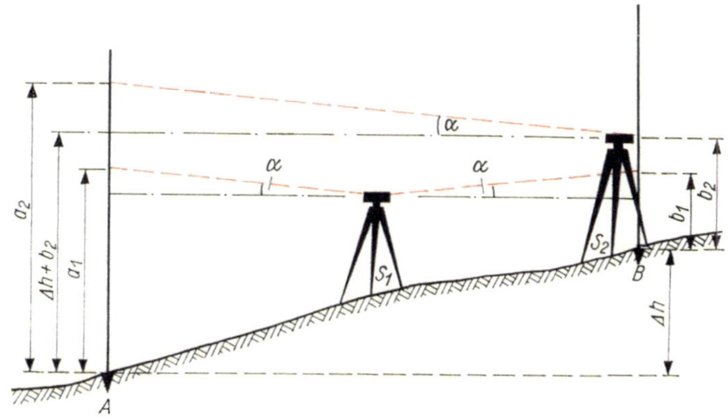

Figure 510 Testing for main requirement $ZZ \parallel LL$ from the middle (Peg test)

The value b_2 can be obtained in various ways: — one places the instrument far enough from B, so that a normal reading can be taken with the shortest possible sighting distance, or — one looks through the objective lens into the telescope to direct the tip of a pencil or a needle to the proper reading at the rod, or — one places the instrument such, that the objective lens touches the rod, reads the top and bottom of the lens housing and uses the mean as b_2.
If there is no collimation axis error α, the reading at the rod A is $(\Delta h + b_2)$. If the reading a_2 is not equal to $\Delta h + b_2$, the reticule is to be shifted using the adjustment screws B_Z until the desired reading $(\Delta h + b_2)$ is obtained.

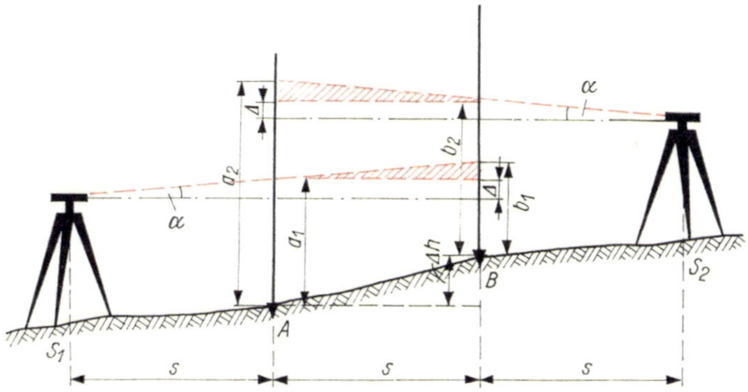

Figure 511 Testing the main requirement $ZZ \parallel LL$

The adjustment screws have to be consecutively loosened and tightened. They cannot be overtightened, because otherwise damaging tensions can occur. For checking purposes, the procedure is to be repeated.

Example

	Station *1*	Station *2*
Reading at A	2.132 m	2.449 m
Reading at B	1.281 m	1.580 m
Δh	+0.851 m	+0.869 m

should be: $\Delta h + b_2 = +0.851\,\text{m} + 1.580\,\text{m} = 2.431$ m

Check after adjustment:

	Station *1*	Station *2*
Reading at A	2.184 m	2.413 m
Reading at B	1.332 m	1.562 m
Δh	+0.852 m	+0.851 m

Judging from the arrangement of this peg test (level in the centre), it is apparent that small remaining errors are without influence onto the result if equal distances for fore- and back sights are maintained.
Due to the different sighting distances from the stations S_1 and S_2, this method requires large focussing changes which can cause errors because of the path of the focussing lens.
In the following procedure, this disadvantage does not occur to the same extent.
A 45 to 60 m long line is divided into three equal sections s (figure 511). One places one rod each onto the points A and B of the central section and uses the other points S_1 and S_2 consecutively as instrument stations. With centered bubble, the readings a_1 and b_1 respectively a_2 and b_2 are taken. Since the hatched triangles are equal, the correct elevation difference between A and B is

$$\Delta h = (a_1 - \Delta) - (b_1 - 2\Delta) = (a_2 - 2\Delta) - (b_2 - \Delta)$$

with this, one obtains for a_2

$$a_2 = a_1 - b_1 + b_2 + 2\Delta. \tag{143}$$

If there is no tilt error, Δ equals zero. When adjusting the collimation axis, the cross hair is to be shifted until the desired value

$$a_2 = a_1 - b_1 + b_2 \tag{143a}$$

is reached.
When adjusting precision levels, the uneven influence of earth curvature and refraction has to be taken into account for different sighting distances. According to *Kukkamäki* it amounts to

$$p = -0.0168 \cdot 10^{-8}\, s^2. \tag{144}$$

For the most common sighting distances it is

s =	10	20	30	40	50	60	m
p =	−0.02	−0.07	−0.15	−0.27	−0.42	−0.60	mm

Example ($s = 20$ m): $p_{20} = -0.07$ mm, $p_{40} = -0.27$ mm

Rod	Station S_1 Reading [mm]	S_1 Corrected Reading [mm]	S_2 Reading [mm]	S_2 Corrected Reading [mm]	Theoretical Value [mm]
A	a_1 = 1721.5	1721.43	a_2 = 2133.5	2133.23	2125.83
B	b_1 = 1408.2	1407.93	b_2 = 1812.4	1812.33	
Δh_A^B		313.5		320.9	$d = -7.4$

From S_2 the cross hair has to be set to $a_2 = 2125.83$ mm.

For medium and low accuracy levelling p does not have to be considered. To adjust a level as accurately as possible, the sighting distances should be in the same range as for the anticipated job. If these range between 20 and 40 m, the adjustment should also be carried out between 20 and 40 m.
A collimator in a laboratory can also be used for setting collimation- and bubble axes parallel. If the objective lens of the instrument to be tested is aimed at ∞ focus onto a collimator, which is a few decimetres away, the parallel bundle of rays entering when the bubble is centered, is united to a point on the horizontal line. The cross hair of the collimator is also imaged there, if the instrument is adjusted (see figure 119). A separation between the collimator cross hair and the cross hair of the level has to be completely eliminated with the aid of B_z.

Testing and Adjustment of a Level with Fixed Telescope, Fixed Bubble and Tilting Screw

The 1st minor condition "bubble axis normal to vertical axis" is cancelled because this orthogonality is immediately lost when tilting the telescope. Therefore, one adjustment device would be sufficient.
The instrument is levelled with a bull's eye bubble, and the bubble of the tubular level is centered via the tilting screw before each reading.

Testing and adjusting for the 2nd (axis $L'L'$ of bull's eye bubble parallel to vertical axis VV), 3rd (horizontal line horizontal for levelled instrument) and 4th minor condition (bubble free of cross convergency) are as for levels without tilting screw.
The primary condition (collimation axis parallel to bubble axis) is to be checked as for levels without tilting screw. The adjustment however, can be performed with either B_L (adjusting the bubble after the desired reading was set with the aid of the tilting screw) or with B_Z (after the bubble was centered with the tilting screw).

Testing and Adjusting of a Level with Rolling Telescope, Fixed Reversible Bubble and Tilting Screw

Testing and adjustment for the 1st and 2nd minor conditions is accomplished as for levels with fixed telescope and tilting screw, while for the 3rd minor condition (horizontal line horizontal for levelled instrument) it follows the procedure for instruments without tilting screw.
The adjustment is not performed by rotating the reticule, but by turning of the telescope and changing of the stopping screws D_1 and D_2 (see figure 508). The 4th minor condition (cross divergency) is controlled by turning the telescope.

The primary condition (collimation axis parallel to bubble axis) is fulfilled, if

a) the telescope axis FF and the parallel bubble axes L_1L_1 and L_2L_2 are parallel
b) the collimation axis coincides with the telescope axis.

It is still possible to obtain correct measured values, even if requirement b) is not fulfilled.
To test for the primary condition, one reads the rod at the horizontal line in both telescope positions after the bubble had been centered. The mean of the two readings corresponds to the horizontal sight, if the difference exceeds 2 mm at 30 m sighting distance when using an arrangement as in figure 511, one can also check whether the position of the coincidence prisms of the reversible bubble remains unchanged. The position of the prism can only be corrected in a workshop.
Levels with rolling telescope and reversible bubble (figure 512) can be tested from one station with two readings. A level with fixed telescope and fixed bubble however, can only be tested from two stations with four readings, but it requires less adjustment devices.

Testing and Adjustment of a Compensator Level

Since compensator levels do not contain tubular spirit levels, only the following conditions exist:
Minor conditions: Axis $L'L'$ of the bull's eye bubble parallel to vertical axis VV ($L'L' \parallel VV$).
Testing and adjustment according to section 1.2.2.
Horizontal cross hair line to be horizontal within the operational range of the compensator. Testing and adjustment is the same as for spirit levels.

Primary condition: The collimation axis should be horizontal within the operational range of the compensator.

Testing and adjustment are like for spirit levels with fixed telescope. Some compensator levels of various manufacturers permit adjustment of the collimation axis by turning a slightly wedge shaped cover glass in front of the objective lens, instead of using B_Z. For the Wild NA 2, it can be adjusted with the adjustment screw shown in figure 478 at the lower left, which activates an additional tilting moment for the prism pendulum via a spring, and thus tilts the collimation axis.

Figure 512 Wild N 2 heigth 157 mm

Testing and Adjusting of a Spirit Level of Older Design

Because of the multitude of bubble arrangements and connections between bubble and telescope, resp. standards in older instruments, it is not suitable to discuss the adjustment methods for these instruments in detail. Basically, the methods for levels with and without tilting screws can be applied to all other instruments, provided one considers them as instruments with fixed telescope and disregards other possibilities.

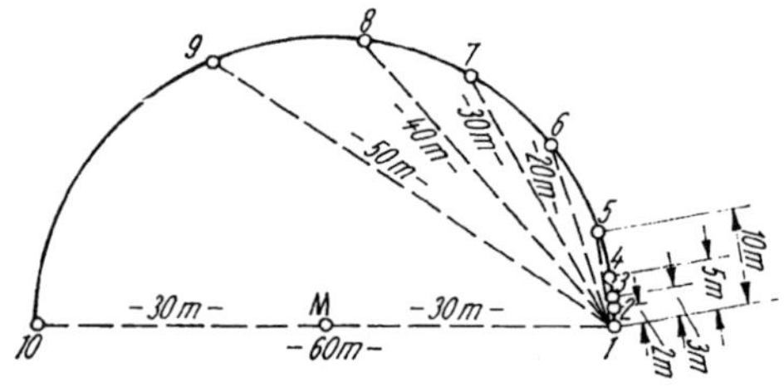

Figure 513 Measuring arrangement to study the movement of the focussing lens

Testing of the Telescope and the Path of the Focussing Lens

Characteristics and efficiency of the telescope are tested according to the methods described in section 1.1.2.
Knowledge on the path of the focussing lens, which changes the position of the collimation axis and affects the parallelity of collimation and bubble axes, respectively disturbs the horizontal position of the collimation axis, is of interest. Therefore one determines the elevations of points on a semi circle around *M* (figure 513) by levelling from the middle, and a second time from point *1*. While there is no focussing change when levelling from *M* because of equal sighting distances, the focussing has to change when levelling from point *1*. The differences between the elevation differences from both level readings are an indication of the elevation errors caused by the path of the focussing lens. For high precision instruments the influence of earth curvature and refraction has to be considered.

Testing of Bubbles and Compensators

The bubbles are tested according to the procedures described in section 1.2.
The proper functioning of a compensator is tested as follows:
The tribrack of a level is oriented towards an approximately 40 m distant rod in such a way, that one foot screw is under the telescope when sighting to the rod. After levelling of the instrument, the foot screw is slowly rotated while the rod reading is continuously monitored. When the reading changes (e.g. after 1/8 of a revolution), the compensator has reached the limit of its working range. The bubble of the bull's eye level cannot be properly centered. The foot screw is then turned back to the other limit of the compensator. Between these two extremes, the rod reading should change corresponding to the elevation change of the level when turning the foot screw by not more than about 0.2 mm. Any necessary correction of the compensator should be left to a service mechanic.
The mean error m_k of a compensator can be obtained for constant temperature during measurements, from the standard deviation m_n of the tilt, and the standard deviation m_b of the setting of the horizontal line on the rod as

$$m_k = \pm \sqrt{m_n^2 - m_b^2}\,. \tag{145}$$

The standard deviation of the tilt is determined from the residuals v from the mean of a series of, e.g. $n = 10$ readings on a fixed horizontal target, as

$$m_n = \pm \sqrt{\frac{[vv]}{n-1}}\,. \tag{146}$$

Before each reading, the compensator is aggravated by turning of the telescope. The standard deviation of setting is determined for equal sighting distance (e.g. 40 m) with the compensator being at a complete rest, using a series of readings of the target as for equation (132). After each reading, the rod graduation is centered again using the parallel plate micrometer. If the instrument is not equipped with one, the target has to be shifted and directed back for each reading.

Testing of Tilting Screw and Parallel-Plate Micrometer

To test the tilting screw, the level is placed onto a level trier such, that the axes of both instruments are parallel. The tilting screw is brought to its extreme position, and the bubble is centered first roughly with the foot screws, then with the measuring screw of the level trier. Then the tilting screw is rotated in steps of one full revolution each, and the bubble is centred with the measuring screw of the level trier, whose position is read. This is continued until the bubble is at the other end. From the back- and fore measurements, plus an additional series (with telescope turned by 200 gon) the pitch of the tilting screw can be determined. Also one can determine whether the pitch is even, especially in the central areas, and whether there is a back lash in the tilting screw.
In order to obtain the least count of the parallel plate micrometer, one sights for instance to a 20 cm long 0.8 mm graduation (e.g. a thermometer in 10 m distance) and, after centering of the bubble, sets several graduation lines in fore and back sequence, each time reading the drum. From the differences of the drum readings, i.e. for six series, and the intervals of the graduation to be compared, the least count is computed. It is compared to the theoretical values from equation (11) for the shift of the sighting ray due to different

tilt angles of the parallel plate. This comparison provides an overview of type and magnitude of graduation changes and of the quality of the micrometer.

Determination of the Standard Deviaton of a Level for 1 km Double Levelling

The standard deviaton of double levelling is determined from back- and fore levelling of 10 nearly equally long sections of a loop of about 5 km length. The slope should not exceed 3%. The measurement should be carried out on different days, although each section should be measured on the same day in both directions. The differences $d = \Delta h_1 - \Delta h_2$ are formed from the measured elevation differences of the sections fore- (Δh_1) and back (Δh_2). The standard deviation m_h of the mean of back- and fore measurements for 1 km is obtained as

$$m_h = \pm \frac{1}{2}\sqrt{\frac{1}{n}\left[\frac{dd}{l}\right]} \tag{147}$$

n denotes the number of sections, and l the length of the section in km. If the differences d contain a systematic component (e.g. inclined horizon for compensator levels), equation (147) becomes

$$m_h = \pm \frac{1}{2}\sqrt{\frac{1}{n-1}\left[\frac{d'd'}{l}\right]} \tag{148}$$

where

$$\left[\frac{d'd'}{l}\right] = \left[\frac{dd}{l}\right] - \frac{[d]^2}{L}. \tag{149}$$

3.2.5. Level with Inclined Line of Sight

For technical purposes, a level with inclined line of sight is manufactured in U.S.S.R. The telescope of this instrument contains a device (figure 514) which transfers the image of the horizontal line onto the left half of the field of view (horizontal line (*4*) onto mirror coated surface (*3*) in figure 515). This line is engraved on the glass plate (*2*) and illuminated through the protection glass (*1*). The transfer utilizes the mirror (*3*), connected rigidly with the spirit level, the prism (*5*) and the mirror coated surface of prism (*6*). In the right half of the field of view, the fixed cross hair (*1*) and the image of the rod graduation (*2*) is visible, against which the horizontal line changes its position due to tilting of the telescope.

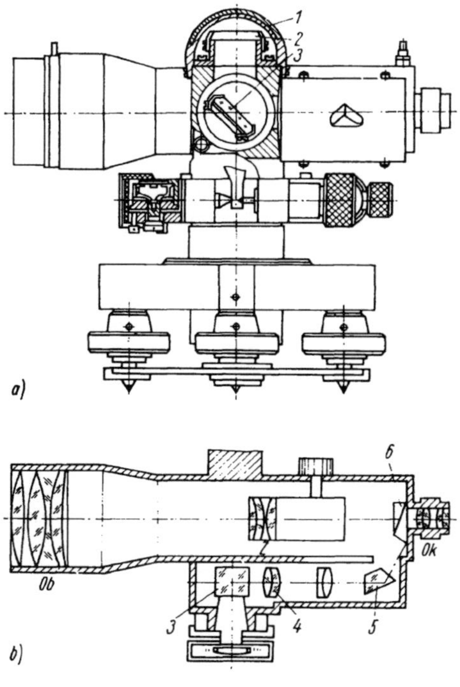

Figure 514 Sections through the NL-3
a) vertical
b) horizontal

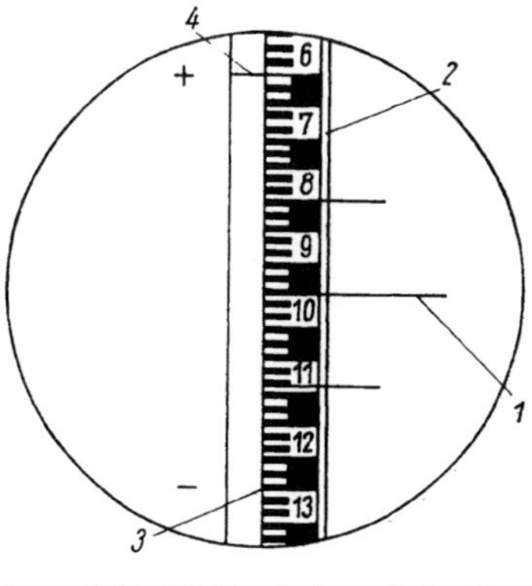

Figure 515 Field of view of the NL-3

The elevation difference between instrument and rod station is obtained from the rod readings h_m resp. h_h at the centre cross hair and the horizontal line. With k as coefficient of the level (5 for the NL-3), i as instrument height, and z as target height, the elevation difference is

$$\Delta h = k(h_m - h_h) + i - z. \tag{150}$$

By levelling from between the rods and reading of the same line in back- and fore sight, the elevation difference between the rod stations becomes

$$\Delta h = k(h_{h_r} - h_{h_v}) \tag{150a}$$

The tilt range is $\pm 3°$, the standard deviation about ± 2 mm.

3.2.6. Levelling Rods

In addition to the level, a scale—the levelling rod—is needed for differential levelling. With it, the vertical distance is read between the points and the horizontal line of sight of the telescope.

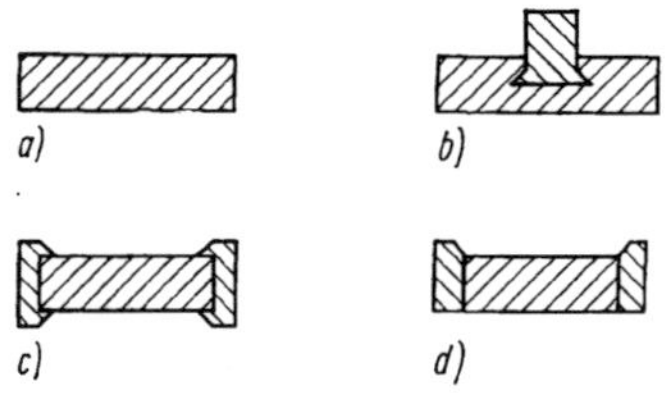

Figure 516 Cross sections of levelling rods

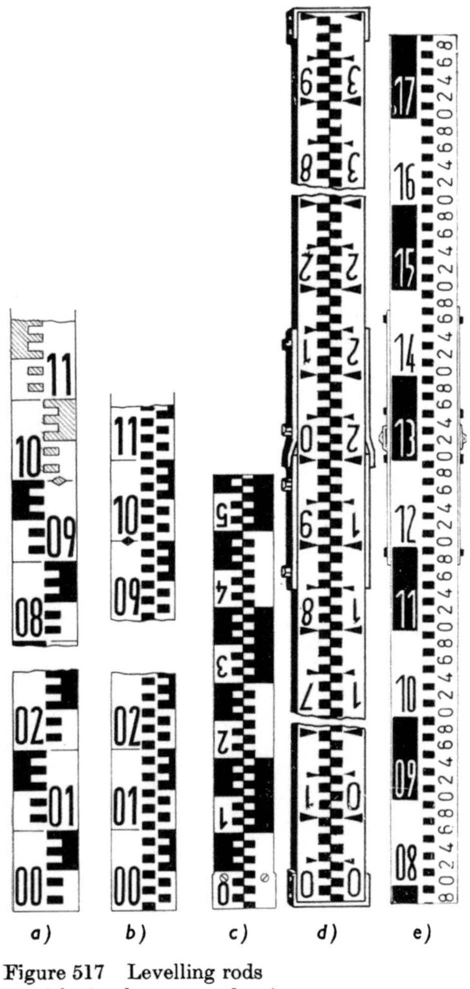

Figure 517 Levelling rods
a) with simple area graduation
b) with checker board graduation
c) same (Wild, Heerbrugg)
d) same (Kern)
e) same with partially digitized rod image (Zeiss, Oberkochen)

Simple Levelling Rods

Levelling rods are usually made of dry, straight grained pine- or fir wood. They are 3 or 4 m long and 8 to 10 cm wide.
Simple levelling rods have a rectangular cross-section (figure 516a) and are usually protected against bending by a brace along the centre (figure 516b) or with two along the sides (figure 516c), which leads to *T*- or *I* shaped cross section. The two edge braces also serve as protection of the graduation against damage. On other rods, the graduation is located in a slight depression (figure 516d). At its bottom end, the levelling rod has a fitting of hardened steel with a plane surface exactly normal to the rod axis, which is located at the origin of the graduation. Deviations of it (zero-errors) have no influence on the levelling, if the same rod is placed on top of all points. Since this error occurs equally often when reading back sights and fore sights, it is eliminated during the calculation of an elevation difference. If two rods with different zero errors are used, each one has to be read equally often in back and fore sight to eliminate the influence of the zero error. The number of instrument set-ups should be even between benchmarks, so that the first and last readings are taken at the same rod.
On both sides of the rod, there are handles at about 1.25 m above the origin of the graduation. Hanging rods are produced for special measurements, e.g. in mining. They have a hook at their upper end, and are only a bit longer than 1 m. The separation between the hangerpoint and the origin of the graduation should be known.
According to their type, one distinguishes rigid rods, folding rods, extension rods and reversible rods. Rigid rods are made of one piece of wood and are normally 3 m long. For the 3 to 4 m long folding rods, the side with the graduation is so hinged, that it is protected by the side braces. This rod is more convenient for transportation. These types of rods are most often used for technical levelling, even more than the extension rods, which are 1.5 to 4 m long and even more convenient for transportation. Reversible rods are only produced as rigid 3 m rods. They are primarily used for precision levelling and serve as a safeguard against blunders and one sided estimation errors. For checking purposes, the graduations on front and back are shifted by an uneven amount (3.335 m), which means that the difference between the readings of both sides has to equal this amount. For low order measurements, a 2 m folding scale can be used consisting of 20 cm long sections (pocket rod), which can be mounted onto a wooden rod of equal length for easier handling on the job.
The levelling rods have several layers of oil paint as protection against humidity and as base for graduation. The graduation is added onto the white front surface (light yellow, for Zeiss, Oberkochen) with a stencil or a graduation machine. The standard deviation amounts to about ± 0.15 mm per 1 m with a maximum error of a single rod position of ± 0.15 mm. Rods with area graduation, used for low and medium accuracy are subdivided into m, dm, half dm, and cm. The decimetre areas are separated by a 1 mm wide line. The area graduation (figure 517a) is alternatively 1 m black, and 1 m red, on white background, starting with black. Often a rhombic mark (20 mm wide, 10 mm high) in the colour of the next interval is placed at every full metre for stadia readings and at every half metre a circular mark (10 mm diameter) is placed having the colour of the metre-interval.
The checkerboard graduation (figure 517b to d) shall prevent estimation errors caused by optical delusion, when reading millimetres by comparing black and white areas (a white field on black back ground appears to be bigger than a black one of equal dimensions on white background: "irradiation"). Numbering is usually at every dm. The numerals are large, clear and of equal thickness, for easy recognition. They are placed above the corresponding line. The black arabic numerals of modern rods are mostly upright, and 40 to 60 mm tall (figure 517e). Up to now, the majority of rods had numerals upside down, because the astronomic telescopes of levels generated reversed images. Nearly all the telescopes of modern levels, however, generate upright images. The backside of the rods, except for reversible ones, contains red and white caution hatching (200 mm wide) for measurements in traffic. Figure 517f shows the graduation of a rod with lines in 1/100th of a foot.

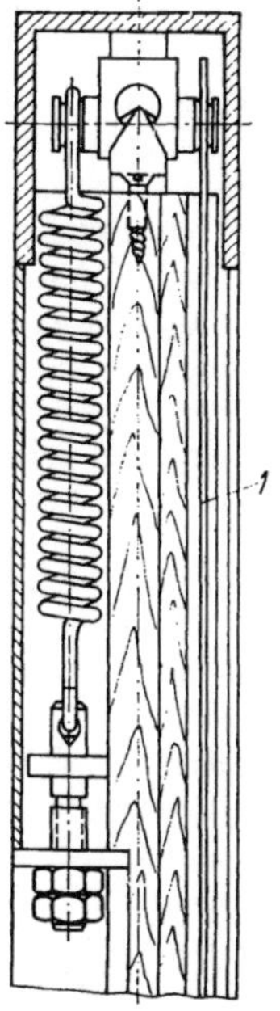

Figure 518 Device for pulling an invar tape
1 invar tape

Special Rods for Precision Levelling

To avoid irregular changes in length, which are possible for wooden rods, invar rods are used for precision levelling. They consist of a rigid, usually 3 m long staff of wood or light metal, into which a 25 mm wide invar tape is placed under tension. While 3 m long wooden rods can change by 0.12 to 0.24 mm for 10 K temperature difference, an invar tape of equal length can maximally change by 0.04 mm. Therefore, the rod metre of an invar rod remains practically constant within a large temperature range. Furthermore, wooden rods can change their length by as much as several tenths of a millimetre per metre due to penetrating humidity, while invar rods are not influenced.

The invar tape rests on six supports within the body of the rod. It is fastened to the bottom end and pulled with a spring on the upper end via a lever with 200 N pull (figure 518). The graduation is added to the invar tape with the aid of a 3 m long invar stencil (figure 519). To achieve this, the invar tape, which is painted white (light yellow for Zeiss, Oberkochen), is positioned under tension under a support with a rectangular window. After the stencil is placed on top, black paint is sprayed for marking the graduation. All graduations are line graduations, because they can be read more accurately with the wedgeline of the reticule, than the boundary between two areas with a horizontal line. The maximum error in placing the graduation lines

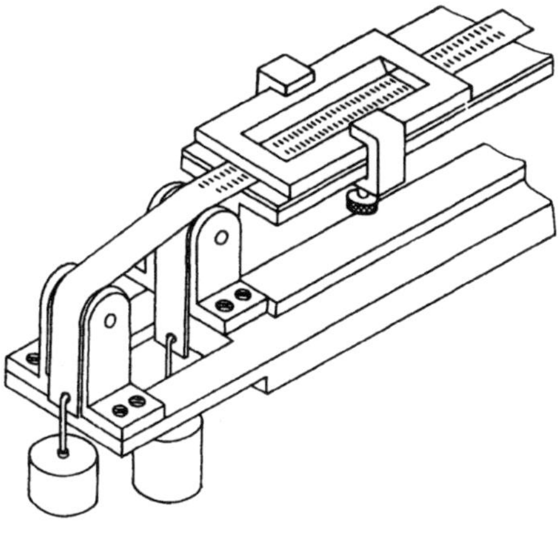

Figure 519 Stencil for engraving the graduation onto an invar tape

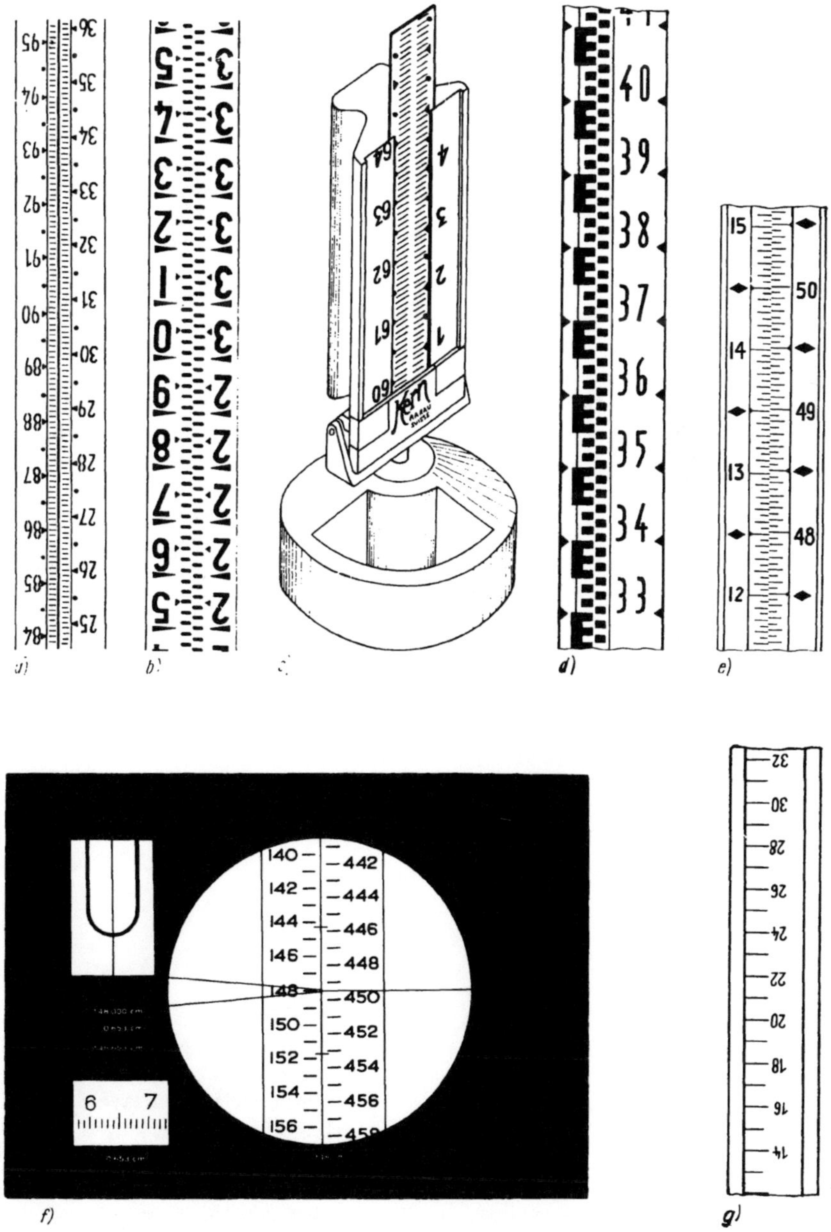

Figure 520 Invar tape rods (selected)
a) with two graduations, shifted against each other (Zeiss, Jena)
b) with 1/2 cm graduation (Zeiss, Jena)
c) made of light metal (Kern)
d) with reflex film (Zeiss, Oberkochen)
e) with two graduations shifted against each other (Rank Prec. Ind.)
f) with two graduations shifted against each other, having 1 cm line separation (Wild, Heerbrugg)
g) Industry rods (Wild, Heerbrugg)

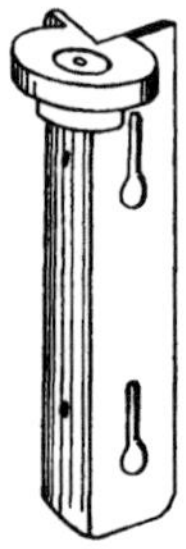

Figure 521 Rod level

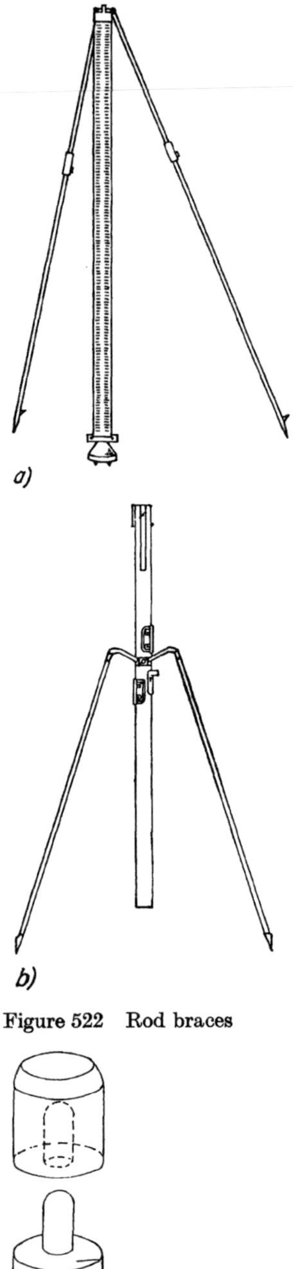

Figure 522 Rod braces

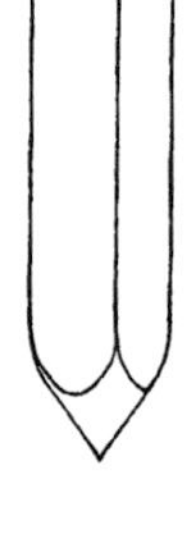

Figure 524 Ground peg

according to the above method is ±0.1 mm per 1 m. With special procedures, standard errors of the rod metre of ±0.01 mm are reached. ±0.03 mm are considered to be tolerable for any rod metre. On the white staff, there are black Arabic numerals on both sides of the invar tape. Like simple rods, invar rods have two flapping handles, and a fitting of hardened steel with plane surface at the bottom end. The back of the rod again has hatched caution lines. The graduation of the rods shown in figure 520a has 5 mm separation. The left and right graduations are shifted by 25 mm against each other, which serves as a check by providing elevation differences for back- and fore sights, and also improves the accuracy. Since for larger distances, the thin graduation lines cannot be clearly recognized, invar rods with 1/2 cm graduation are produced (two shifted cm rods), which have 3 mm thick lines (figure 520b), but do not permit a check.
The invar rod by Kern, Aarau is placed in a light metal frame (figure 520c). The one by Zeiss, Oberkochen, is produced with a reflective coating (figure 520d), like the simple rod produced by this company. Such reflective rods are very advantageous for work in tunnels, at night, etc.
The invar rod by Rank Precision Industries has two graduations, shifted by 83.730 mm ± 0.3 mm, with the second one starting with decimetre 32 (figure 520e). The rod by Wild, Heerbrugg, has two graduations, shifted by about 55 mm, and a line separation of 1 cm (figure 520f). The parallel plate micrometer of the N 3 permits a parallel shift of the collimation axis of 10 mm. For industrial measurements, Wild produces light invar rods of 90 and 180 cm length with three different, interchangeable legs (figure 520g).

Rod Accessories

A bull's eye bubble with 5′ to 20′ sensitivity is used to set the levelling rod vertical. It is fastened to the side or back of the rod at about 1 m height, or else is mounted onto a metal guard which can be rested against or snapped onto the rod (figure 521). After transportation, the bubble has to be checked. During measurement it is advisable to support the rod with two braces, which are held together with the handles (figure 522). With them, the rod can be held steady and vertical above a point. Changeable braces with screw threads, attached to the upper end of the rod, are produced for invar rods. The rod can stand alone and can be rotated without changing the braces.
Base plates are used to solidly set the rod at the turning points. These are heavy cast iron plates with a curved top or a nipple (figure 523). Base plates with two nipples provide a check reading if the rod is placed first on top of one and then the other nipple. The base plates have three short steel tips to be pressed into the ground. Occasionally 20 to 30 cm long iron bolts, which are about 3 cm thick, are pounded into the ground and removed after use (figure 524). Their upper end, which is a semi sphere, is protected by a cap when they are driven into the ground. When using base plates, a metal rod shoe can be attached to the bottom of the rod. This rod shoe has a depression fitting the nipple which enables a secure setting of the rod and prohibits sliding.

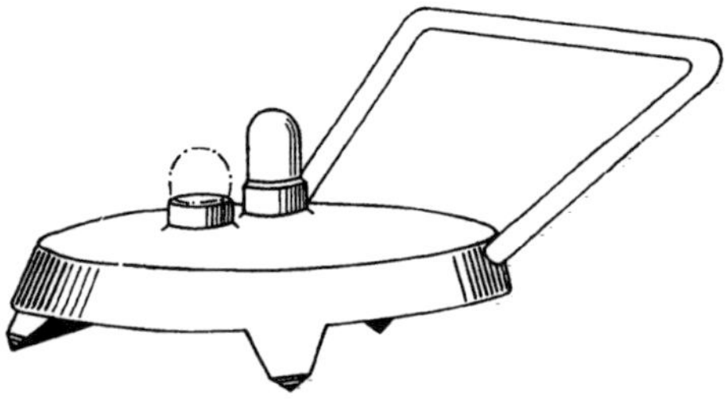

Figure 523 Base plates

Testing of Levelling Rods

Prior to using a levelling rod, and at regular intervals, it is to be checked for its suitability for levelling of the desired accuracy. Errors in graduation, plane setting surface (setting errors), and differences in the zero position have to be determined, and the bull's eye bubble has to be adjusted. With folding and extension rods, care has to be taken, that the parts are perfectly connected and that they do not deviate from the vertical direction when erect.

Adjustment of the Bull's Eye Bubble

To test the bull's eye bubble, the rod is made vertical with the aid of a free swinging plumb bob, hung from about 3 cm above. If the bubble is not centered, it is to be adjusted with the aid of adjustment screws.

Determination of the Zero Error

To determine the zero error of a pair of levelling rods, several baseplates are placed at different terrain heights or on some steps within the range of one instrument set-up. With a levelled instrument 20 to 30 m distant, both rods are read subsequently on each point. The mean of the differences between corresponding readings on both rods is the zero error. This error has no effect when an even number of instrument stations is used.

Determination of the Error of the Bottom Surface

The levelling rod is placed onto a ground bolt at, for instance six points evenly distributed over the bottom surface in addition to the central one (figure 525). If differences in the reading are apparent between these settings and the central one (e.g. $\geqq 0.15$ mm for precision levelling), even though the instrument remained level, the bottom surface is either not plane or not normal to the rod graduation. These irregularities should not exceed the observation accuracy. The rod should therefore always be placed centrally onto a bolt or nipple of a base plate for the measurements.

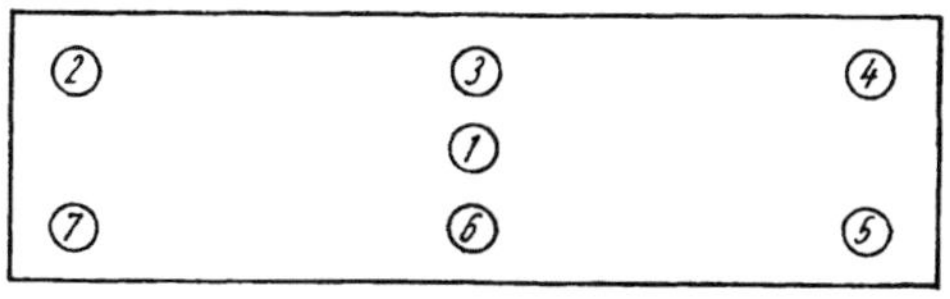

Figure 525 Position of the support points for the levelling rod when testing the rod support area

Testing of the Rod Graduation

By testing the rod graduation, the interior graduation accuracy as well as the length of the rod metre are determined. A precise ruler is used to check for irregular graduation errors of medium accuracy rods, while for precision rods accurate normal metres (calibration bars) or comparators are needed. For good graduations, the irregular errors are below 0.1 mm. The mean length of a rod metre is determined with a normal metre. Its beginning and end marks covering 1 m are lines with additional graduation lines at 0.2 mm intervals. A magnifying glass is used to take the readings. The temperature t is determined with the thermometer embedded in it. The usuable length of the normal metre is obtained by inserting the proper value into equation (176).
For testing, the normal metre is placed directly on top of the graduation, and the position of the boundaries is read on both scales. This is repeated at different places, e.g. every dm. The temperature and scale corrections of the normal metre have to be added to the differences obtained. The mean of the single readings represents the average rod metre.
Example: Determination of the mean rod metre of a wooden levelling rod with area graduation.

Rod Section	Reading at normal metre left mm	right mm	Difference mm	Mean mm	Mean of measure-ments mm
1st metre section	+0.6	−0.2	+0.4		
	−0.4	+0.9	+0.5	+0.45	
2nd metre section	+0.5	−0.1	+0.4		
	−0.4	+0.8	+0.4	+0.40	+0.38
3th metre section	+0.6	−0.3	+0.3		
	−0.2	+0.6	+0.4	+0.35	
4th metre section	+0.6	−0.3	+0.3		
	−0.1	+0.4	+0.3	+0.30	

Equation of normal metre: $l = 1\text{ m} + 0.020\text{ mm} + 0.0115(t - 20)\text{ mm}$

Temperature of the normal metre
prior to test +10.8 °C
after the test +10.6 °C
mean +10.7 °C

reduced length of the normal metre
at +10.7 °C 0.99991 m
mean of measurements +0.00038 m
mean rod metre 1.00029 m

Figure 526 1-m-double image compensator PROEM (Breithaupt & Son)

For simple levelling jobs, deviations of up to 0.2 mm per metre can be tolerated if no large elevation differences are encountered.
When testing with a normal metre directly on top of the graduation, there is the danger of damaging it. The body temperature may cause an irregular temperature distribution not shown on the thermometer of the normal metre. Two persons are required to take the readings. These disadvantages are avoided with a rod compensator operating according to the coincidence principle, e.g. the 1 or 2 m double image compensators (figure 526) by Breithaupt & Son, Kassel. The readings are reflected via mirrors into a common field of view, while its lines are coincided with a parallel plate micrometer (figure 527).

Table 28. Determination of the mean rod metre

equation of test ruler (normal metre) $1 = 1000\text{ mm} + 0.04\text{ mm} + 0.0055\,(t_N - 20)\text{ mm/K}$

length of comparator	before rod comparison	after rod comparison
comparator temperature t_K	19.6°C	20.6°C
test ruler temperature t_N	19.8°C	20.4°C
rod temperature t_L	21.0°C	20.0°C
length of test ruler at $t_N : 1\,m + x$	+2.9 μm	+6.2 μm
micrometer readings M_N when comparing with the test ruler μm	154	150
	149	144
	155	145
	150	145
	151	142
	148	148
	153	150
	157	146
	151	142
	152	146
M_N (mean) μm	152.0	145.8
$K = N_t - M_N$ mm	999.8509	999.8604
length of comparator K (mean)	999.8556 mm	
ΔK	+9.5 μm	

length of rod metre

rod metre	t_K	micrometer readings M_L when measuring the rod metre			ΔK at T_K	M_L	v
		coincidence 1	coincidence 2	mean			
[m/2]	[°C]	[μm]	[μm]	[μm]	[μm]	[μm]	[μm]
0.6 – 2.6	20.2	186	189	187.5	+5.7	193.2	– 7
1.0 – 3.0	20.2	183	180	181.5	+5.7	187.2	– 1
1.4 – 3.4	20.2	201	203	202.0	+5.7	207.7	–21
1.8 – 3.8	20.2	213	214	213.5	+5.7	219.2	–33
2.2 – 4.2	20.4	173	173	173.0	+7.6	180.6	+ 6
2.6 – 4.6	20.4	175	170	172.5	+7.6	180.1	+ 6
3.0 – 5.0	20.4	150	148	149.0	+7.6	156.6	+30
3.4 – 5.4	20.6	155	157	156.0	+9.5	165.5	+21
						$\sum = 1490.1$ M_L (mean) = 186.3 μm	+ 1 [vv] = 2993

length of the mean rod metre at 20.5°C $L = K + M_L = 999.8509\text{ mm} + 0.1863\text{ mm} = 1000.037\text{ mm}$
standard deviation of the mean rod metre $m_L = \pm\sqrt{\dfrac{[vv]}{n(n-1)}} = \pm\sqrt{\dfrac{0.0030}{8 \cdot 7}} = \pm 0.007\text{ mm}$
$L = 1000.037\text{ mm} \pm 0.007\text{ mm}$

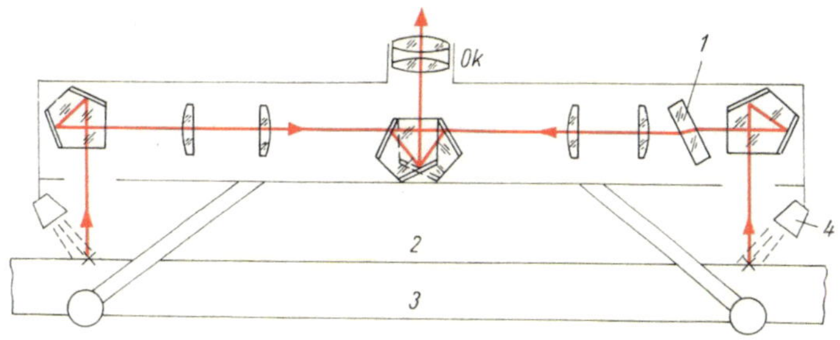

Figure 527 Scheme of a levelling rod comparator (Coincidence principle)
1 parallel plate micrometer; *2* invar tape; *3* rod; *4* reflector

First the comparator length (figure 528a)

$$K = N_t - M_N \tag{151}$$

is determined by measuring the normal metre

$$N_t = 1\,\text{m} + a + \alpha(t + t_0)\,\text{m} = 1\,\text{m} + \Delta N \tag{152}$$

with the aid of the micrometer reading M_N. Then, after positioning the rod to be tested, the rod metre is determined from dm to dm (figure 528b) with the micrometer reading M_L as

$$L = K + M_L = 1\,\text{m} + \Delta N - M_N + M_L \tag{153}$$

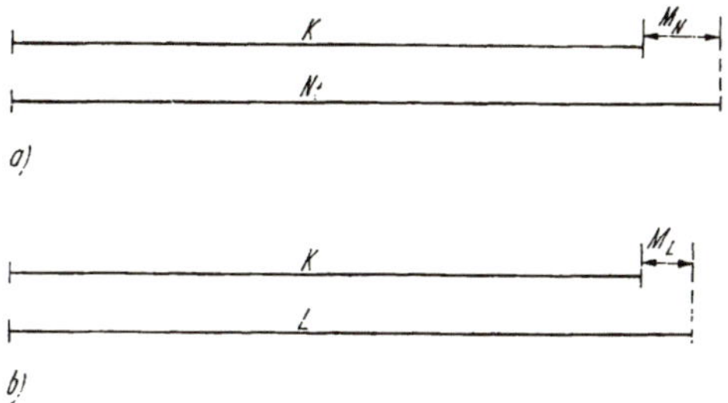

Figure 528 Principle of the determination of the rod metre

K is defined by the origin of the micrometer scale. At the end, the length of the normal metre is determined again. The mean of first and last measurement is used in the calculations. Occasionally, the comparator housing contains a metre for comparison.

The check from dm to dm is recommended for invar rods after longer transportation, otherwise the determination of the rod metre at three positions of the rod every 3 months is sufficient. The standard deviation of the rod graduation should not exceed ± 0.02 mm.

Example: Determination of the average rod metre of the invar rod no. 107 by Nestler with the levelling rod travel comparator of the Geodetic Institute of the Technical University Dresden (table 28).

For levelling rods with shifted graduations, the length of the comparator is determined before, in between, and after the rod comparison.

Rod tests at central locations are faster and more accurate as they utilize laser interferometers. The rod to be checked is continually shifted in horizontal position (6 cm/min) under a stationary photo-electric microscope. The shift is measured via the interferometer by counting the interference strips with the aid of an electrical counter. The microscope releases an impulse when the edge of a graduation line passes through, which in turn results in a marking of the measuring scale, which is formed by the laser ray. The measuring data are transferred onto punched tape. About 1.7 hours are required to test a rod with 1/2 cm graduation (1200 measuring values), and 5 hours with two repetition measurements. In this manner, the measurements are performed automatically, and not only the mean rod metre is determined, but also the errors of all graduation lines.

3.3. Instruments to Measure Vertical Angles

3.3.1. Simple Instruments to Measure Vertical Angles

For simple height measurements on construction sites, approximate profiling etc., especially for reducing slope distances measured with staves or tapes, small instruments (clinometers) without tripod or with a pole support are used.

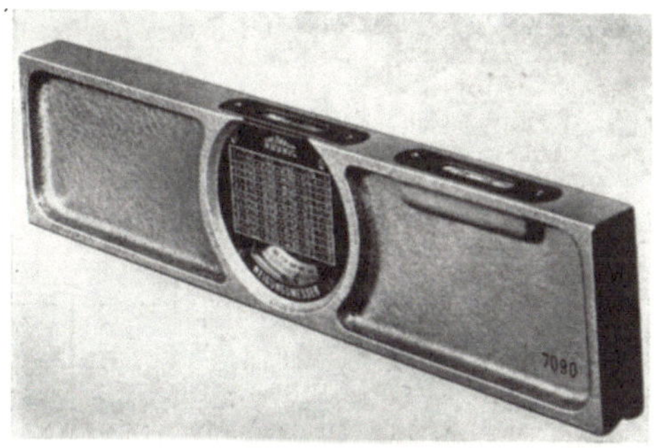

Figure 529 Clinometer NEMES (Breithaupt & Son)

Clinometers to Place on Top

These instruments are used to measure the inclination of plane bases.

The clinometer arc consists of a triangular wooded frame with two legs for setting onto base, e.g. a measuring staff. A graduation starts at the centre of the arc, which is read with the aid of an index or pendulum attached to the frame.

At the bubble clinometer by *Wimmer*, the index is set plumb with the aid of a tubular bubble. In addition to the degree readings, usually the reduction of slope distances for a certain staff-length can be read on the graduation.

The clinometer NEMES (figure 529) by Breithaupt & Son, Kassel, consists of a cast piece of light metal with an *I* shaped cross-section (300 mm × 74 mm × 23 mm). A graduated circle with 56 mm diameter and 1° least count, which is supported by ball bearings and which can rotate around a horizontal axis serves as measuring element. It has a one sided ballast and therefore always sets itself according to gravity. The instrument also has a 2′ spirit level.

Clinometers to Hang

The inclination of strings or wires in mining is determined with the aid of a hanging clinometer. This instrument has a semi-circle of light metal which is graduated to both sides from 0 to 100 gon (resp. 0 to 90°) in steps of 1 gon/4 resp. $^1/_4$° or $^1/_5$°. It is hung onto the string or wire with its hooks which are opened in different directions. A plumb bob, which hangs on a lady's hair (or perlon thread) from an opening in the centre of the circle, is used for reading the tilt angle.

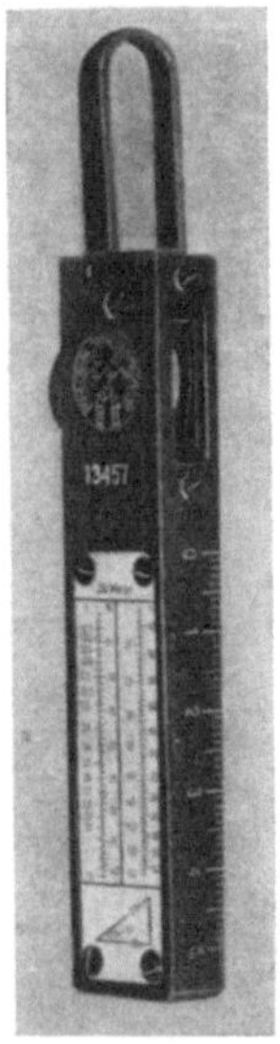

Figure 530 Pendulum clinometer

Clinometers with Sighting Device

Pendulum clinometers have a graduation which is visible in the sighting magnifying glass. The instrument shown in figure 530 for instance, has a 1° (1 gon) graduation (from $\pm 0°$ to $\pm 92°$) and a 5% graduation (from $\pm 0°$ to $\pm 100\%$). With the instrument in the zero position of the pendulum, the zero line is horizontal. The clinometer is held between thumb and index finger, so that the pendulum can swing properly. In order to reduce the swinging time, the hand is held against a range pole or the forehead. One looks with one eye into the sighting magnifying glass and with the other to the target and reads its height on the degree- or the percentage graduation. The reduction of slope distances measured with a 20 m tape is done with the reduction table attached to the side of the instrument. For rough levelling, the zero mark of the graduation is used as index. The pocket clinometer by Möller, Optical Works, Wedel (Holstein) has a box with graduation which, together with its hanging device, can be snapped into the plumb body, which forms the frame (110 mm × 25 mm × 1 mm). Pendulum clinometers which can be folded together (height, when open, is 190 mm), by SLOM, Paris permit measurements of tilts up to ± 40 resp. 100%, of angles up to ± 25 resp. 50 gon, and distances for 1.70 and 2.00 m bases. Fennel's clinometer (figure 531) consists of a perforated circular sector of 190 mm radius onto which a telescope is fastened rigidly. This telescope has a 22 mm objective lens opening, is 152 mm long, and has a 5 × magnification. With the index on the socket for the staff, tilts of up to 40% can be read to within $^1/_2$%.

Within the field of view of the telescope, the spirit level has to be observed with a small mirror. The target staff has the same height as the clinometer, which means that the line of sight is parallel to the tape (the mass of the instrument is 1.5 kg, while the mass of the wooden container is 2.9 kg).

In the clinometer according the *Brandis*, often named after Wolz (figure 532), a wheel shaped disk swings within a cylindrical capsule. The degree- or

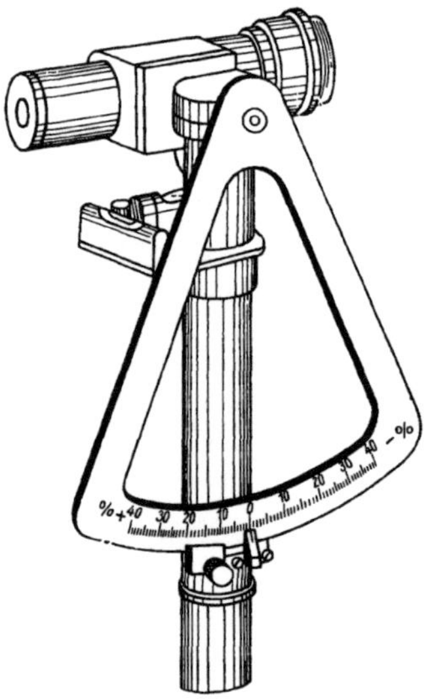

Figure 531 *Fennel's* clinometer

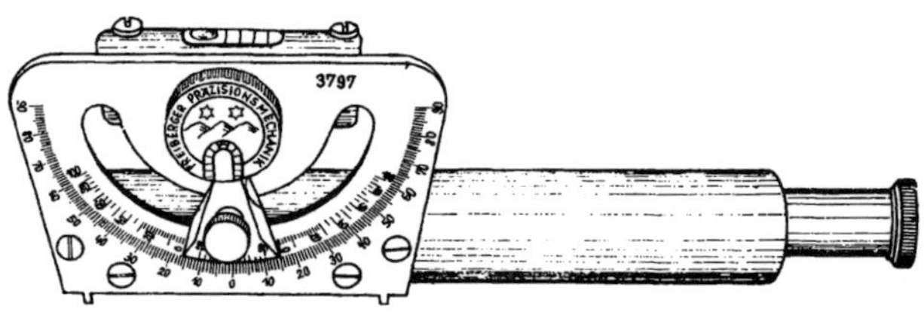

Figure 533 Clinometer according to *Abney* (Abney level)

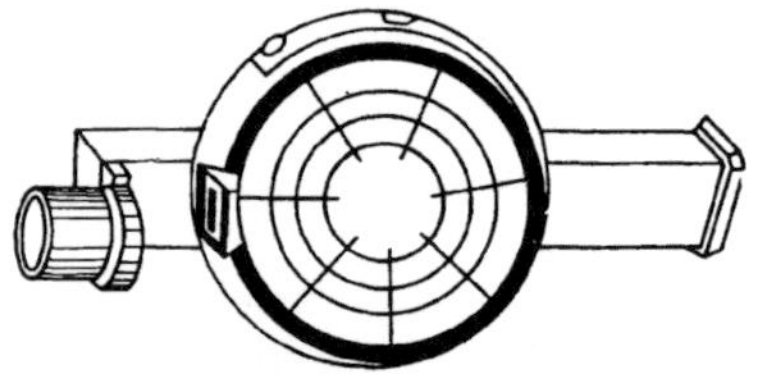

Figure 532 Clinometer according to *Brandis*

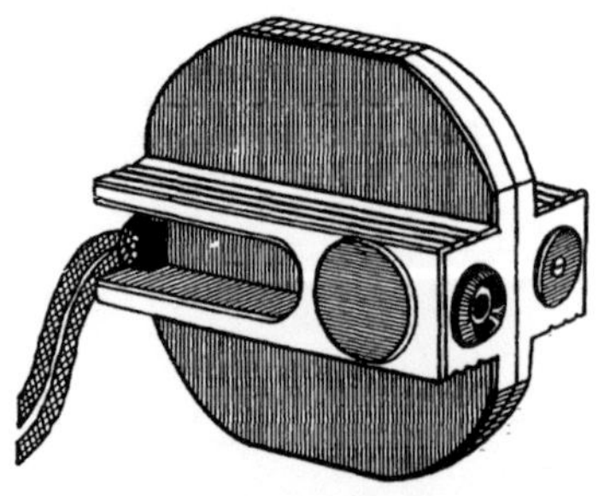

Figure 534 Freehand clinometer NECLI with liquid dampening (Breithaupt & Son); mass: 300 g

percentage graduation is read on the cylinder surface with a magnifying glass through a glass covered section of the capsule. A diopter is fastened to the capsule.

The *Abney* level (figure 533) which is widely spread in the Anglo-saxian world, has a diopter tube rigidly connected to the clinometer arc. An index, rigidly connected to a spirit level, can be rotated around the centre of the degree- and percent graduation. The bubble image is reflected into the field of view by a mirror in the left half of the diopter tube, and bisected by the horizontal hair, directed to the target.

The hand held clinometer by *Breithaupt* (figure 534) consists of a broken telescope, with a transparent graduated circle suspended in its image plane. This circle with liquid dampening can swing freely and is oriented to the vertical with a pendulum. Together with the target, one can see a gon- and a percent graduation when looking into the eyepiece at left, and a degree- and reduction graduation on the right (figure 535). A standard deviation of 0.4 gon is being claimed.

Figure 535 Field of view of the NECLI

Testing of Clinometers

Setting clinometers give correct results, when the plumb line from the support of the pendulum passes through the zero point of the graduation for a horizontal base. This is the case, when the two readings of the clinometer in direct and reverse positions are the same. If the setting clinometer is out of adjustment, it has to be corrected by the manufacturer, unless the index error can be eliminated by the user by shifting the scales. The setting clinometer gives correct results, if the mean of the readings in both positions is used. Hanging clinometers are tested in a similar manner. Instruments with sighting devices are tested by measuring the angle from both ends of the slope distance. Both values have to be the same but with a different sign. If there is an error, it can be adjusted with the adjustment device.

3.3.2. The Use of the Theodolite for Vertical Angle Measurements

The most important instrument for measurements of vertical angles is the theodolite, equipped with vertical circle and proper reading device.

Vertical Circle and Reading Device

Most theodolites have a vertical circle for measuring vertical angles. This graduated circle is mounted such, that its centre is located in the horizontal axis of the telescope, and that its plane is vertical as soon as the vertical axis of the theodolite is plumb. The diameter of the vertical circle usually is a bit smaller than for the horizontal circle. Usually, the vertical circle is rigidly connected to the telescope and thus follows all its tilting motions. With the height clamp, it can be clamped together with the telescope onto the standard, while a slow motion screw is used for the fine setting. Instruments designed for astronomical purposes, occasionally have vertical circles that can be changed.

The reading devices (vernier with magnifying glasses or reading microscope) are rigidly fastened to the telescope standard. In modern instruments, the reading positions of the vertical circle, made of glass, are reflected into the field of view of the reading eyepiece directly beside the telescope (figures 213, 219, 223, 226, 235, 239, 243). For theodolites of medium and low accuracy, readings are taken at usually only one position (see figures 213, 219).

In the field of view of the reading microscope, the reading positions of the vertical circle appear together with the ones for the horizontal circle, or also sequentially after inserting a prism into the light path coming from the horizontal circle (figures 214, 220, 222, 224, 241, 242, 247).

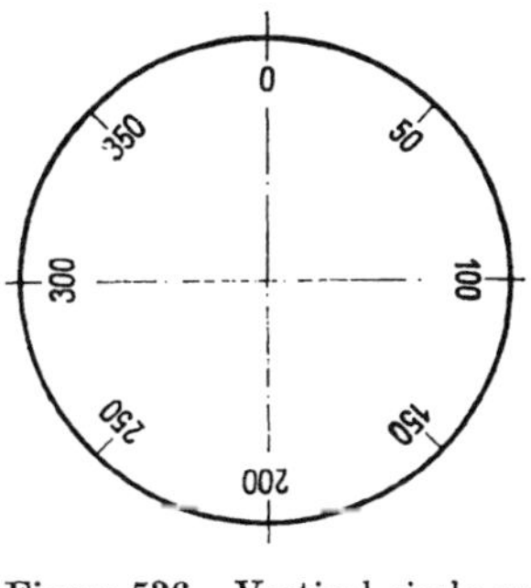

Figure 536 Vertical circle graduation

The graduation is engraved either onto the basic surface (figure 536) of the cylindrical circle, or else — for some simple instruments — onto the cylindrical surface. Nearly all vertical circles are graduated from 0 to 400 gon (resp. 0 to 360°) with the zero reading at the zenith, increasing readings for direct—and decreasing readings for reverse telescope position. The telescope positions are marked on the instrument, with the direct one generally having the vertical circle to the left. Zenith distances are read, and the sum of the readings in both positions amounts to 400 gon. The vertical circles of older instruments are graduated with semi circles (twice from 0 via 90 to 180°) or quadrants (four times from 0 to 90°) and attached such that vertical angles can be read.

Vertical angles can be computed from zenith distances:

$$\alpha = 100\,\text{gon} - \zeta \qquad (154)$$

Reading devices for the vertical circle at one or two circle positions totally correspond to the ones for horizontal circles and are placed at the level of the horizontal axis. To secure a constant position of the reading device, which is needed for the measuring of vertical angles, the platen level, needed to set the vertical axis plumb, is not sufficient. The theodolites are equipped either with a tubular level, which is parallel to the plane of the vertical circle, or with an automatic height index to eliminate the influence of a remaining tilt of the vertical axis on the measurement of vertical angles.

Height Index and Index-Bubble

At older instruments, the index used for reading the vertical circle, and the tubular bubble are rigidly fastened to the telescope standard. The bubble of this tubular level can only be centered with the foot screws, which has to be done before each sighting.

The vertical angle can also be measured in connection with a telescope- or levelling bubble, which is rigidly connected to the telescope and which is usually built as double- or reversible level. The telescope has to be levelled with the aid of this bubble to read the vertical circle. This reading is then subtracted from the reading obtained when the telescope was pointed to the target.

An index bubble, which is rigidly connected to the reading device (figure 537) is most suitable for the measurements. Height index and index bubble are mounted on a plate, which is directly in front of the vertical circle and supported by the horizontal axis of the instrument. With the height index screw, it can be tilted independently of telescope- and vertical axis position within a limited range to permit centering of the index bubble. It is essential, that this is done before each reading, if proper measurements are to be obtained.

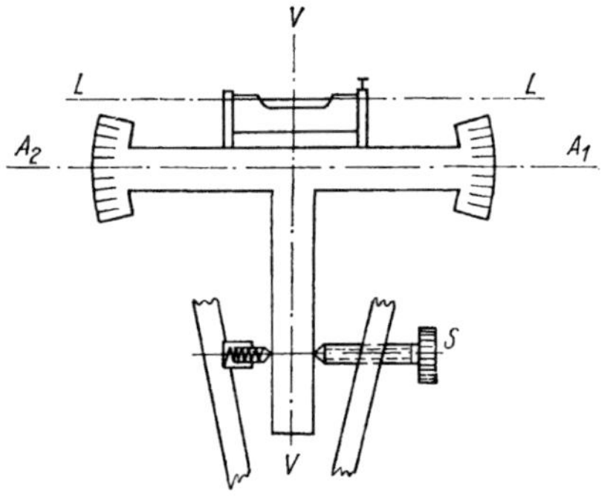

Figure 537 Height index bubble

Automatic Height Index

The component of a remaining vertical axis tilt in direction of the target plane is compensated by the automatic height index, and the vertical circle reading is automatically referred to the plumbline, which eliminates the need of centering a level bubble. The principle of the compensator is the same as for automatic levels. It works as follows (figure 538): For an exactly plumb vertical axis VV and exact levelling of the collimation axis ZZ of the telescope, the index I is opposite to the reading position A at the vertical circle. A zenith distance of 100 gon is read. Due to a vertical axis error α, $V'V'$ is not vertical, and the index is now in position I'. The reading is still correct, because the vertical circle underwent the same movement, and the reading position A moves to A'. However, the collimation axis ZZ was also rotated by α to the position $Z'Z'$. When resighting the original target, the vertical circle, which is rigidly connected to the telescope, is rotated by α, and the correct reading position A' (in this case 100 gon) opposite to I' is rotated back to A. Therefore a value which includes the error α is obtained at the index I', unless the reading position A is projected onto the index of the reticule of the reading microscope.

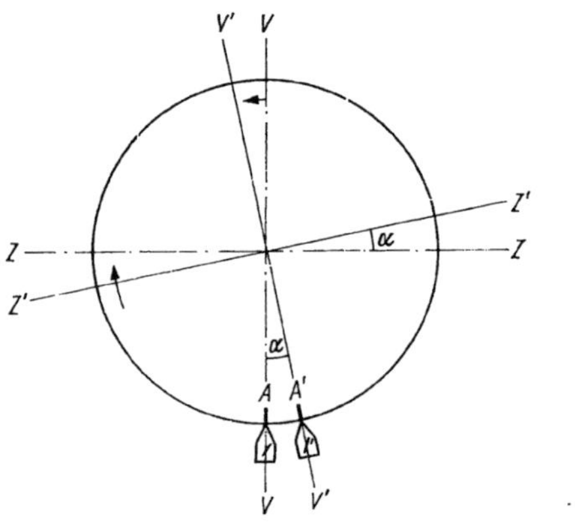

Figure 538 Operational principle of an automatic height index

As in levels, one part of the compensator is fixed in the instrument, while the other one moves according to gravity. Automatic height indices use primarily

- a free swinging pendulum
- a pendulum with forced oscillations
- liquid compensators.

The operating range is generally 4 to 20 mgon. The height index of high accuracy theodolites usually has the accuracy of a 1″ bubble. It can therefore be used to improve the levelling for steep sights.

Automatic Height Index with Free Swinging Pendulum

In the theodolite by Dietzgen (see figures 306, 320), the mobile compensator element of the automatic height index is a pendulum prism, which always swings into the plumb line. Together with the reading objective, -prism

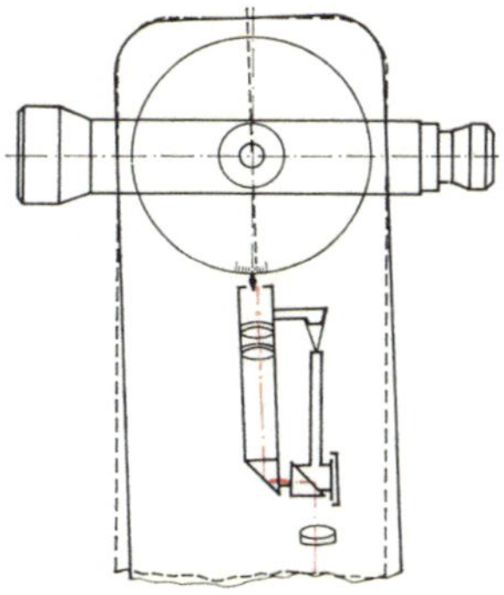

Figure 539 Automatic height index of the 6020-A1 and A 2 (Dietzgen)

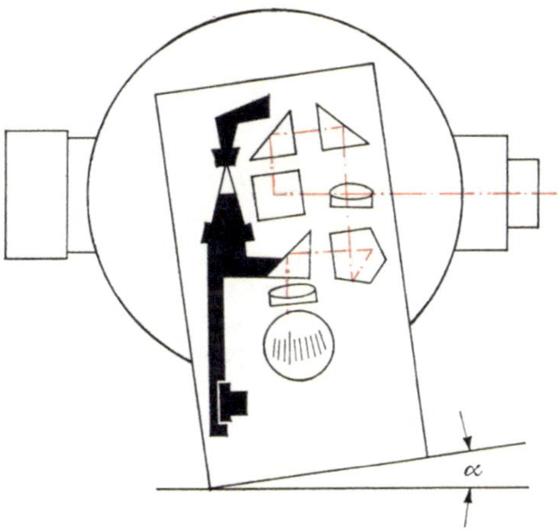

Figure 540 Automatic height index of the FT 1 A (Fennel)

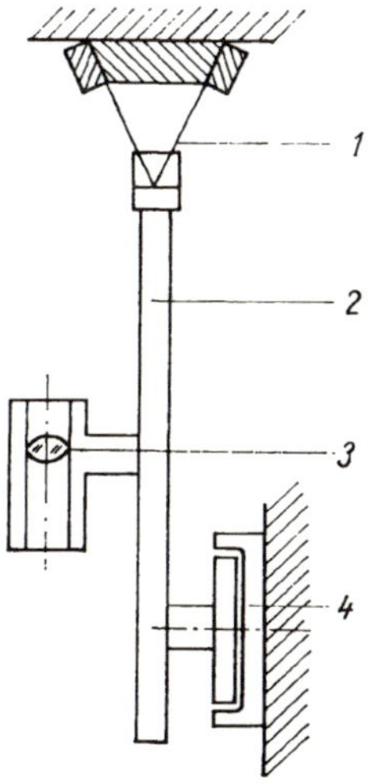

Figure 541 Automatic height index of the TEAUT (Breithaupt & Son)

1 spring joint;
2 pendulum;
3 objective;
4 air dampening

and -eyepiece, it influences the optical train such, that the reading position of the vertical circle, which is located in the vertical plane through the horizontal axis, is imaged onto the index of the reticule (figure 539).

The FTlA and FTRA by Fennel are equipped with a similar automatic height index (figure 540). The reading position, located in the vertical plane of the horizontal axis, is imaged onto the reticule index with a pendulum prism and additional prisms and lenses, which are fixed. Swinging plan parallel plates serve as compensators in the Wild theodolites T16 (figure 305) and T2 (figure 225). The rays, emerging from the reading positions, located in the vertical- resp. horizontal plane of the horizontal axis, pass through the imaging system without deflection.

A warning sign indicates for the Wild T-16 when the operational range of the compensator is exceeded and a better levelling of the instrument is required. The proper functioning of the automatic height index of the Wild T-2 can be checked with the aid of a control knob. The theodolite Tpr has two connected prisms, which are pendulum suspended by two pairs of crossed band springs.

The automatic height index of several theodolites is the pendulum supported objective lens, which—as part of the reading optics—always swings into the plumbline. In the TEAUT by Breithaupt & Son, Kassel, the mobile compensator element is a system with air dampening, suspended from a spring joint (figure 541).

In the theodolite 4149A by Filotecnica Salmoiraghi, Milano, the objective lens is pendulum supported on three threads. The theodolite is equipped with an optical bubble to check the proper functioning of the automatic height index. A circle, resp. point, engraved on a glass plate located behind the light entrance, is imaged together with the reading position through the imaging system onto the reticule plane. It appears to have the shape of the bubble of a bull's eye level (see figure 339). When measuring, this circle should always be within a ring engraved on the reticule.

The optical bubble, coupled with the automatic height index, permits not only the elimination of the tilt component of the vertical axis in direction to the target but also the semi automatic correction of the component normal to it (with the foot screw). With this, the vertical axis can be set plumb to within 5″ in direction to the target.

In the Soviet theodolite T5, a fixed element is placed in front of the pendulum component, which is suspended by four wires with 42 mm length (figure 542). The reading position A is therefore always projected onto the index I, which for a tilted vertical axis is rotated by the same amount within the vertical plane of the horizontal axis. The setting accuracy of the height index is $\pm 6''$. The pendulum compensators for the Kern K-1A (see figure 303) and the Kern K1-RA (figure 729) are similarly constructed. The compensation of the remaining tilt of the vertical axis within the target plane is accomplished by the swinging component (*3*) of the double objective (*3*, *4*) (figure 543). The length of the pendulum is equal to the graduation radius r of the vertical circle (*1*). It is supported with little friction at position (*5*) within precision ball bearings. Due to vertical axis tilt, the circle position

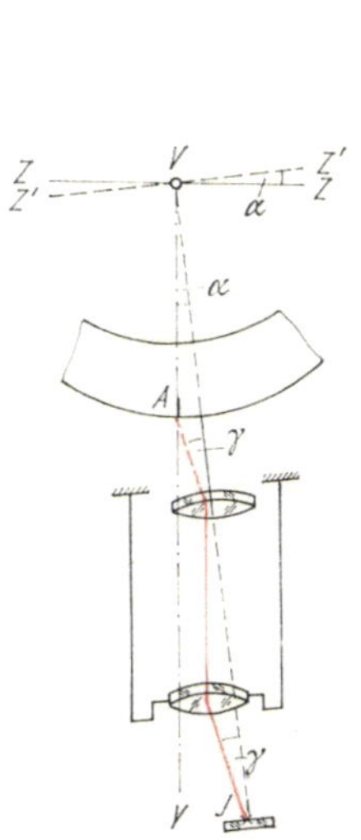

Figure 542 Automatic height index of the T 5 (U.S.S.R.)

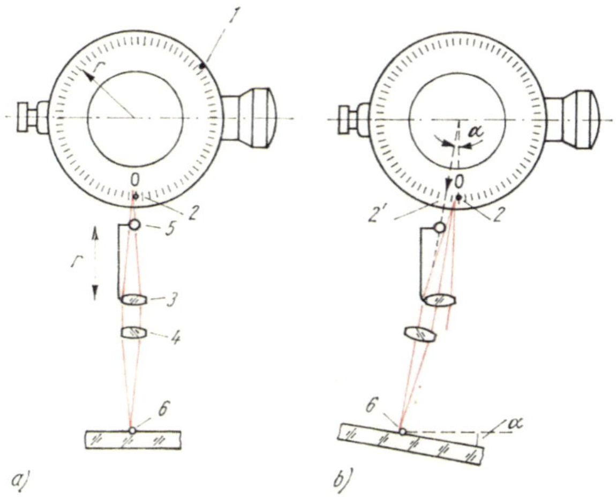

Figure 543 Operation principle of the pendulum compensator of the Kern K 1 A and K 1-RA

(*2'*) instead of (*2*) would be imaged on the index (*6*), i.e. the vertical angle would be incorrectly read by the angle α. Since the pendulum objective component swings with its centre vertically below the pendulum support (*5*), and since the path of rays is parallel between the elements (*3*) and (*4*), the circle position (*2*) is projected onto the index (*6*), and thus the vertical axis tilt is eliminated. In the Kern K1-S the half of the double objective, which is farther away from the circle, is pendulum suspended.
The automatic height index in theodolites of the D-series by MOM, Budapest, as well as in theodolites by Lietz/Sokkisha, U.S.A./Japan contains a swinging objective lens as mobile comparator element.

Automatic Height Index with Pendulum and Forced Oscillations

The automatic height index of the THEO 020A, THEO 010A and DAHLTA 010A by Zeiss, Jena contains a spring pendulum with prism as mobile comparator element. The objective lenses (*3*) and (*4*) and two reversing prisms (*2*) and (*5*) are located on the pendulum body (*10*). This pendulum is suspended from the telescope standard by a joint spring (*11*). Its oscillations are dampened by the air dampening (*12*) (figure 544). The reading position of the vertical circle is imaged onto the reticule (*6*) via the prisms (*2*) and (*5*) and the objective lens (*3, 4*). For a tilted vertical axis, the point of suspension of the pendulum is outside the plumb line, and the pendulum body (*10*) rotates around the instant rotational centre *M*. By proper selection of the spring, one can achieve that the optical centre of the objective lens arrives in the vertical plane of the horizontal axis (i.e. the pendulum does not swing by α but rather by α'), and the reading position *A*, which in this plane is imaged onto the reading index. The image, created in the plane of the reticule (*6*) is observed via the pentagon prism (*7*), and the objective lens (*8*) and the eyepiece (*9*) of the reading microscope. The automatic height index of the Th 2 by Zeiss, Oberkochen contains the X-long compensator with 12.3 m reduced pendulum length used in the Ni 1 (see figures 472, 473). Its jointed quadrilateral carries two plan-parallel plates at its lower end (figure 545).

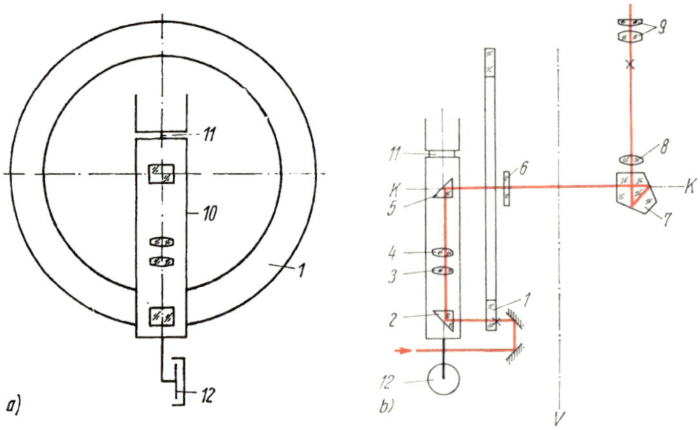

Figure 544 Automatic height index of the THEO 020 A and THEO 010 A (Zeiss, Jena)

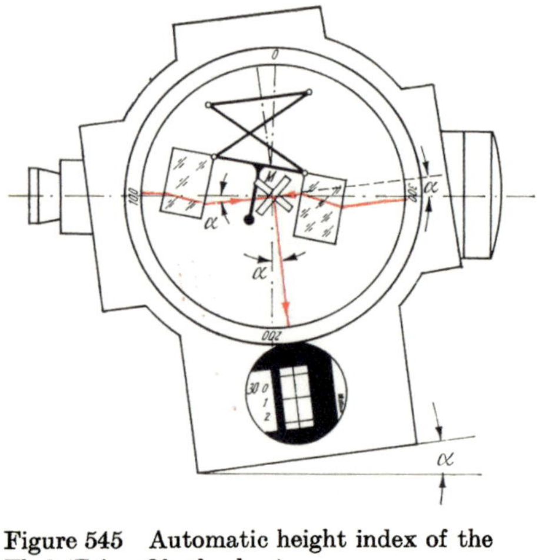

Figure 545 Automatic height index of the Th 2 (Zeiss, Oberkochen)

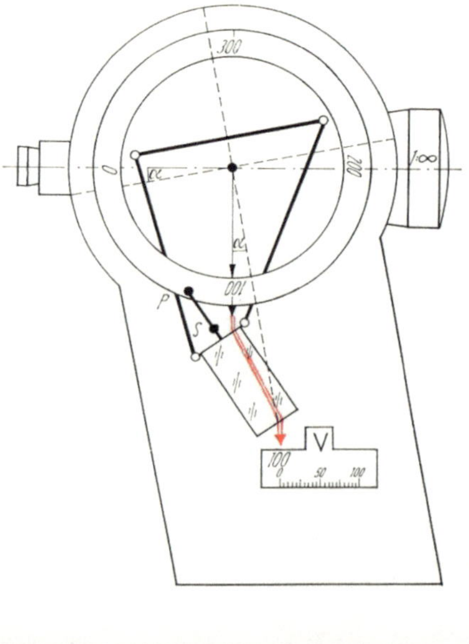

Figure 546 Automatic height index of the Th 42, Th 43, RTa 4 (Zeiss, Oberkochen)

For horizontal sight, the rays emerging from the circle positions 100 and 300 gon reach the point M in the horizontal axis where they are combined on equally long paths without shift by the parallel plates, provided the vertical axis is exactly plumb. A tilt in the vertical axis is multiplied by the jointed quadrilateral by a certain factor, so that the rays are shifted by the plates according to the parallel plate equation.
The automatic height index of the Th 42, Th 43, and RTa 4 by Zeiss, Oberkochen, is based on the jointed quadrilateral of the Ni 2 (see figure 469). Through reinforcement of the suspending wires, it forces a different tilt to the base of a swinging plan-parallel plate, than that given by gravity (figure 546). The centre of gravity of the pendulum was raised from S to P with the aid of a piece of mass on top of the parallel plate. With this, the reduced pendulum length is increased, and swinging behaviour becomes more favourable.

Automatic Height Index with Liquid Compensator

The parameters of liquid compensators (see figure 146b) can be computed such, that the reading positions located in the horizontal or vertical plane of the horizontal axis are projected onto the index of the reticule. This is achieved with the imaging system and the liquid wedge generated by a tilt in the vertical axis. In the Wild T 1 (see figure 219a), the rays from the vertical circle pass throught a container with silicone oil before hitting the reading index. If the vertical axis is plumb, the bottom of the container and the surface of the silicone oil form a plan-parallel plate which is penetrated by the rays without deflection. If the vertical axis is tilted by α, the liquid forms a wedge with the angle α. Based on the refractive index of silicone oil, the rays are deflected by $(n - 1)\,\alpha$, which compensates the influence of the tilt of the vertical axis on the reading of the vertical circle (figure 547).
In the DKM 2-A by Kern, Aarau, the rays are totally reflected on a liquid surface between the two circle reading positions, and deflected according to the vertical axis tilt. When the vertical axis VV is plumb, rays coming from the circle position A_1, are imaged via prism P_1, liquid surface and prism P_2 in the plane of the diametrically opposite circle position together with it (figure 548). For a tilt α of the vertical axis, the prisms P_1 and P_2 as well as the housing of the liquid compensator deviate by α from the horizontal plane. If a fixed mirror would be in place of the liquid, the wrong zenith distance $\zeta' = \zeta - \alpha$ would be obtained. Because the liquid surface remains horizontal, A_1 is totally reflected into A_1' in the plane of the diametrically opposite circle position by 100 gon + 2α. The mean of the values A_1' and A_2 corresponds to the actual zenith distance ζ.

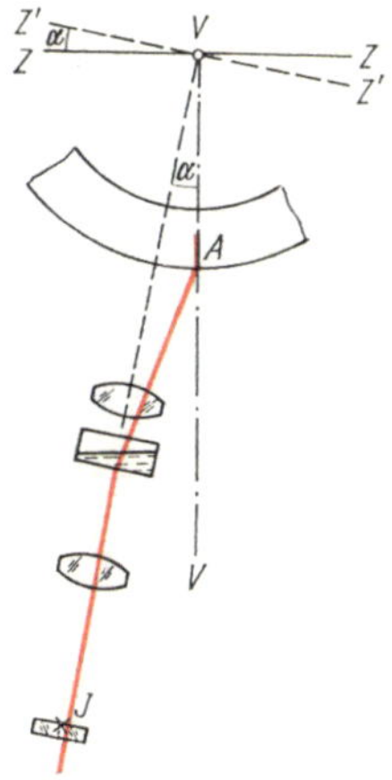

Figure 547 Automatic height index with liquid compensator (principle)

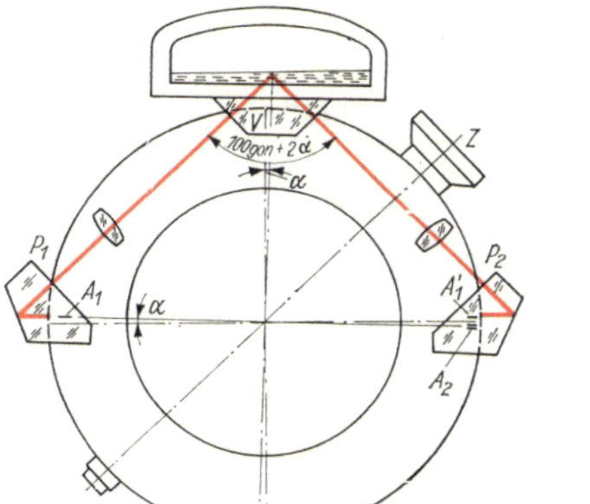

Figure 548 Operation principle of the liquid compensator of the DKM 2-A (Kern)

Testing and Adjustment of the Vertical Circle and its Reading Device

Index Correction

When measuring horizontal angles, the horizontal circle remains fixed, while the reading index follows the movements of the telescope. When measuring vertical angles, however, the height index remains fixed, while the vertical circle follows the movements of the telescope within its vertical plane. A horizontal angle is obtained by setting and reading of the angular values in two directions. To measure a vertical angle, however, requires only the setting of one direction and its angular reading, because the second side is determined by the position of the reading device, which is set with the aid of a spirit level or an automatic index. The vertical angles are free of error, if

a) the projection of the collimation axis of the telescope into the plane of the graduated circle coincides with the diameter of the graduated circle at the positions 100 and 300 gon (resp. the normal to the radius through 0 gon).

b) the axis of the index level or of the automatic height index is parallel (resp. normal) to the straight line connecting the two diametrically opposite reading positons (resp. for only one reading position, it refers to the radius of the vertical circle passing throught this index).

The deviation v_{ζ_1} and v_{ζ_2} of this introduce an error into the vertical angle which has to be corrected as index correction (figure 549):

$$v_\zeta = v_{\zeta_1} + v_{\zeta_2}. \tag{155}$$

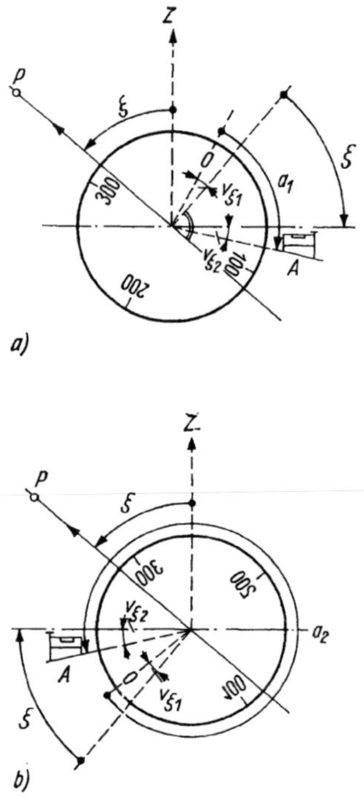

a)

b)

Figure 549 Determination of the index correction

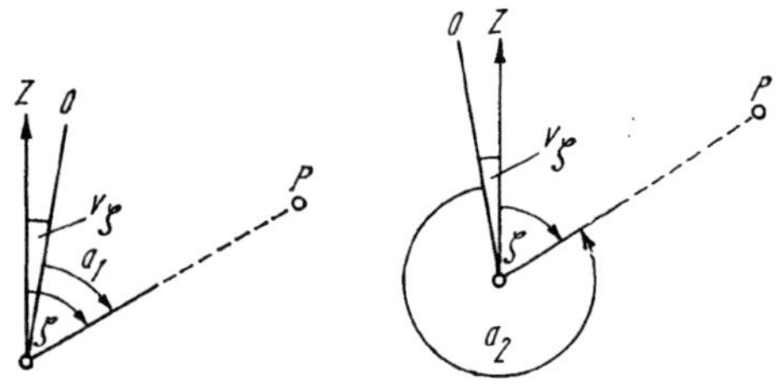

Figure 550 To the determination of the index correction

While v_{ζ_1} is caused by erroneous adjustment of the cross hair, v_{ζ_2} is generated by erroneous adjustment of the index level. If both errors have the same amount, but opposite sign ($v_{\zeta_1} = -v_{\zeta_2}$), zenith distances free of error are obtained. It is sufficient to consider v_ζ. v_ζ is basically a zero error, which appears in similar form in instruments with automatic height index.
When sighting to a point P in both telescope positions (figure 550), one obtains the reading

$$a_1 = \zeta - v_\zeta \tag{156a}$$

in direct position (circle on the left), and

$$a_2 = 400 \text{ gon} - \zeta - v_\zeta \tag{156b}$$

in the reverse position (circle on the right).
A subtraction gives $a_1 - a_2 = 2\zeta - 400$ gon. Therefore the zenith distance becomes

$$\zeta = \frac{(a_1 + 400 \text{ gon}) - a_2}{2}. \tag{157}$$

This value is free of the index error. It is a prerequisite that v_ζ is equal in both telescope positions. Therefore the index bubble has to be centered before each reading.

Determination of the Index Correction and Adjustment of the Height Index

It is easier to measure and calculate zenith distances if the index error is small or non existing. Since measurements in one telescope position require the application of the index correction, it has to be either known or eliminated. In order to determine v_ζ, the sum of the readings in both telescope positions is found with equations (156a) and (156b) as $a_1 + a_2 = 400$ gon $- 2v_\zeta$. This gives

$$v_\zeta = \frac{400 \text{ gon} - (a_1 + a_2)}{2}. \tag{158}$$

When the level bubble is carefully centered before each reading, v_ζ is constant at the same instrument station, and the sum $a_1 + a_2$ is also constant. This provides a good check for the measurements.

Example:

Station	Target	Telescope Position	Reading [gon]	Index Correction [mgon]	Corrected Zenith distance [gon]
S	*P*	*I*	54.1633	+10.1	54.1734
		II	345.8165	+10.1	345.8266
		Σ	399.9798	+20.2 $= 2v_\zeta$	400.0000

For the adjustment, the correct zenith distance is set with the index screw, and the bubble movement of the index level is eliminated with the adjustment screws of the bubble. At theodolites with coincidence micrometer, the correct reading is set at the micrometer drum, then the graduation lines of the vertical circle are coincided using the index screw, and the bubble is adjusted.
At theodolites with automatic height index, the correct reading is set. Then the cross hair is moved with the aid of its adjustment screws such, that the target appears in the centre of the cross hair. For Zeiss, Oberkochen and Wild instruments the target is set and adjusted to the correct reading, while for the Dietzgen theodolites the correct reading is set on the micrometer and the main graduation is coincided during the adjustment procedure.
For theodolites with telescope bubble, first the parallelity of bubble and collimation axes has to be obtained by following the same procedure as for a level with double bubble. If the bubble of the corrected spirit level is centered, the reading of the vertical circle in this (horizontal) position of the telescope directly represents the index correction, provided vertical angles are read, while for zenith distances it would be the difference to

100 gon. The index error is eliminated by shifting the index with its adjustment device such, that the reading 100 gon (resp. 0 gon) appears for a centered level bubble.
At theodolites with bubble on the telescope standard, first the correct reading is determined, and then the index error is eliminated. To achieve this, first the correct reading is set at the vertical circle with centered bubble, then the cross hair is brought to the target with the aid of the adjustment screws, or else the cross hair is set to the target, and the correct reading is obtained by shifting the height index.

Testing for Graduation Errors of the Vertical Circle

The graduation errors of the vertical circle are determined according to the same principle as for the horizontal circle.
The reflection method, where the target is first reflected and then directly sighted, is most suitable. For this, a surface mirror is placed in front of half of the objective lens with a ring-clamp. By rotating the mirror, any test angle can be set. If a prism is placed instead of the mirror, the test angle remains constant for a given prism. Since the vertical angle generally cannot be moved, a number of targets is arranged in a circle around the horizontal axis of the instrument at 5 gon or 4.5° intervals. These intervals need not be very exact, as the desired reading can be obtained with the vertical fine movement after sighting to the target. The distances between targets and horizontal axis however, should not differ by more than 0.5 m.
First the reflection of the lowest target is sighted at, and the index is set to the correct reading with the vertical slow motion screw. Then the target is directly sighted at. In the first series, one measures from the bottom to the top, in the 2nd from top to the bottom, so that two complete sets are obtained. In this manner the graduation corrections for the test sector of e.g. 100 gon or 99° are determined according to the distribution of the targets.

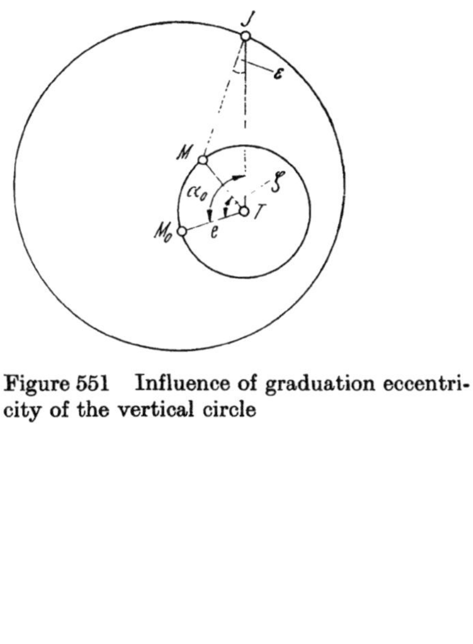

Figure 551 Influence of graduation eccentricity of the vertical circle

Testing for Graduation Eccentricity

The graduation eccentricity of the vertical circle is tested in the same way as for the horizontal circle. Since the position of the reading device cannot be changed in respect to the movable vertical circle, the circle position remains the same, and thus the amount of the eccentricity is constant. The influence of this eccentricity is eliminated in the mean of the readings of two diametrically opposite circle positions.
For theodolites with only one reading position, the situation illustrated in figure 551 exists. M_0 designates the position of the centre of the graduation for the zenith distance 0 gon, while T represents the centre of rotation of the vertical circle on the horizontal axis, and I indicates the position of the reading index. When tilting the telescope by the zenith distance ζ, M_0 moves to M. According to equation (97), the influence of the eccentricity on the circle readings in direct and reverse telescope positions becomes, provided that the index error is neglected:

$$\varepsilon_1 = \varepsilon_{\max} \sin(\alpha_0 - \zeta);\ \varepsilon_2 = \varepsilon_{\max} \sin(\alpha_0 - 400\text{ gon} + \zeta) = \varepsilon_{\max} \sin(\alpha_0 + \zeta).$$

In the mean of the readings in both telescope positions according to equation (157), the following influence of the eccentricity remains:

$$\varepsilon_{1,2} = (\varepsilon_{\max}/2)[\sin(\alpha_0 - \zeta) - \sin(\alpha_0 + \zeta)]$$

or

$$\varepsilon_{1,2} = -\varepsilon_{\max} \cos \alpha_0 \sin \zeta. \tag{159}$$

The influence of graduation eccentricity at theodolites with only one circle reading is not eliminated when measuring in direct and reverse.
Since the eccentricity also influences the index error, a second eccentricity component is obtained according to equation (159) as

$$\varepsilon_{v\zeta} = \varepsilon_{\max} \sin \alpha_0 \cos \zeta. \tag{160}$$

While the first component has the largest influence for flat sights ($\zeta \approx 100$ gon) according to equation (159), the second one is negligibly small for these sights, and is eliminated like the index error when reading in both telescope positions. As evident from equation (160), the index error should be determined with near horizontal sights.
For $e = 2$ μm and a circle diameter of 70 mm, $\varepsilon_{\max}$ equals 3.6 mgon. Studies of different theodolites resulted in an average value of $\varepsilon_{\max} = 2$ mgon. It

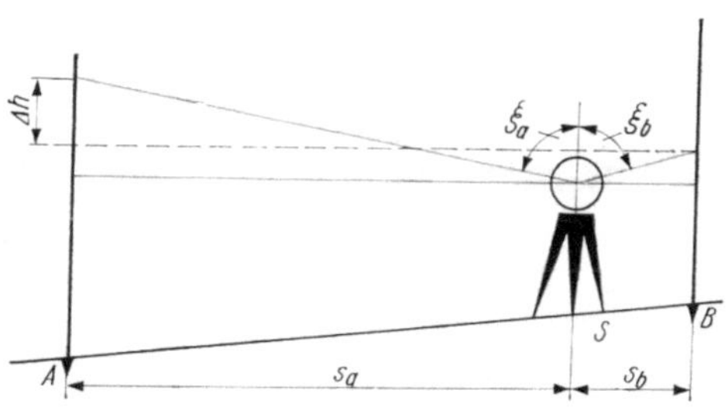

Figure 552 Determination of the graduation eccentricity of the vertical circle for single index theodolites

may therefore be necessary to consider the influence of graduation eccentricity when measuring vertical angles with theodolites with one reading position. The first component $\varepsilon_{1,2}$ can be determined as follows (figure 552): In flat terrain, the elevation difference Δh between the points A and B is determined by differential levelling. Then the theodolite is set up at S, located at about the same elevation. It is suitable to select the sighting distances s_a as 40 to 50 m, and s_b as 5 to 10 m. At both rods, equal graduation lines near the horizontal are sighted, and the angles ζ_a and ζ_b are read at the vertical circle. With these zenith distances and the distances s_a and s_b, the elevation difference $\Delta h'$, which is influenced by the graduation eccentricity is computed.
The component $\varepsilon_{1,2}$ is obtained from the comparison of the determined elevation differences as

$$\varepsilon_{1,2} = \frac{(\Delta h - \Delta h')\,\varrho}{s_b - s_a}. \tag{161}$$

The eccentricity e can be obtained by multiple determinations of different zenith distances and by solving a system of equations with consideration of equations (159) and (160).

Other Tests of the Vertical Circle

Other tests are performed similarly to the ones for the horizontal circle, i.e. determination of reading errors, determination of the sensitivity of the index bubble.
The standard deviation m_e of the automatic height index can be determined by separating the coincidence error m_k from the total error m_g:

$$m_e = \pm\sqrt{m_g^2 - m_k^2}. \tag{162}$$

The value m_g can be determined by alternately turning the micrometer and the fine movement screw clockwise and counter clockwise and setting the index of the reading microscope onto a fixed point of the graduation and reading it. In order to determine m_k, the vertical axis is tilted until the compensator is beyond functioning. This is repeated several times.
The standard deviations m_e can also be determined by multiple pointings to a collimator with the compensator being vibrated with the aid of the vertical slow motion screw. The reading error m_a, which is determined before and after the measuring series with arrested compensator has to be separated from the total error m_g. This reading error contains the coincidence error m_k and the pointing error m_z.

$$m_e = \pm\sqrt{m_g^2 - m_a^2}. \tag{163}$$

The standard deviation m_ζ of a zenith distance measured in both telescope positions can be determined from measurements of four zenith distances each, to four targets located in different directions. For each target, the index error is to be computed according to equation (158) using the observations in both telescope positions, and averaged for the four readings.

$$(v_\zeta) = \frac{[v_\zeta]}{4}. \tag{164}$$

With the residuals to all observations

$$v_{v_\zeta} = (v_\zeta) - v_\zeta \tag{165}$$

the standard deviation becomes

$$m_\zeta = m_{v_\zeta} = \pm\sqrt{\frac{[v_{v_\zeta} v_{v_\zeta}]}{n-4}} \tag{166}$$

where $n = 16$ is the number of observed zenith distances.

Automatic Recording of Vertical Angle Measurements

Like horizontal angles, vertical angles can be automatically recorded according to the principles described on page 105. With incremental scanning they can only be obtained if the telescope is directed into the zero direction — i.e. exactly horizontal or vertical — prior to commencement of the interval counting. The vertical reference is realized with the aid of a pendulum.
It is also possible to measure vertical angles via a liquid sensor by resistance measurements. This is accomplished with a cylindrical ceramic container,

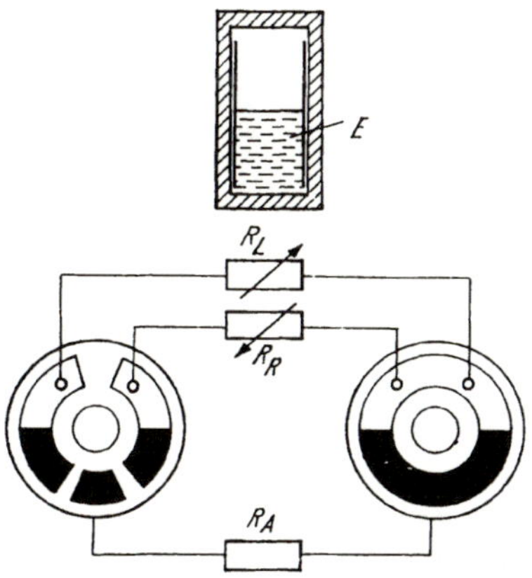

Figure 553 Principle of the angle sensor

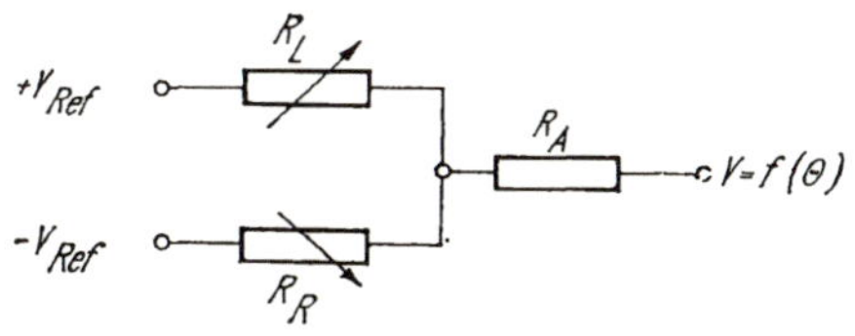

Figure 554 Equivalent circuit of the angle sensor

which is half filled with a conducting liquid. The side walls of this container have ring shaped electrodes. This ring is divided into 3 sectors (figure 553). When the left and the right electrode are covered equally, the container is level ($\alpha = 0$ gon). When tilting the measuring part, the electrodes are submerged at different depth in the liquid. The resistance, measured across the electrodes, is reduced on the deeper immersed electrode and increases on the opposite side.
The central electrode always remains submerged, within the operational limit of ± 33 gon. The resistance measured between this electrode and the ring opposite to it serves as a comparison for the compensation of resistance changes for different temperatures of the liquid. The resistances R_L and R_R obtained with the liquid, dependent on the tilt α, lie together with the compensator resistance in a junction of circuits at which point the reference voltage $\pm V_{\text{Ref}}$ is applied (figure 554). The output is a voltage proportional to the tilt. Thus the measurement of vertical angles is reduced to voltage measurement:

$$V = R_A \left(\frac{1}{R_R} - \frac{1}{R_L} \right) V_{\text{Ref}}$$

The voltage is measured with a digital volt meter. The vertical angles are displayed digitally. For $\alpha = \pm 22$ gon, a measuring accuracy of ± 15 mgon is to be expected.
This principle is utilized in the electronic tacheometer HP 3810A (see page 302). Another possibility is used in the HP 3820A (figure 766).

3.3.3. Sextant

With a sextant, space angles can be measured, and thus horizontal and vertical angles. It is used in marine navigation for measuring vertical angles to stars, occasionally, also for measuring angles between terrestrial targets. This free hand instrument is of little importance in surveying.
The sextant is an angle mirror device, where the angle between the mirrors can be varied. It consists of a rastered frame made of light metal in the shape of a sector of a circle of about 20 cm diameter ($^1/_6$th of a circle, therefore the name "sextant") (figure 555). The arc of the sector has a silver coated metal strip which serves as graduation. On the graduation side of the frame (figure 556) there are

a) the alidade A, with a plane mirror S_1 fastened normal to the graduation surface in its centre of rotation. It is equipped with a clamp, fine motion screw and 1′ vernier.
b) a telescope F with low magnification (e.g. 3.5 times), parallel to the graduation surface
c) the plane mirror S_2 which covers half of the field of view of the telescope and which is rigidly fastened on the frame normal to the graduation surface
d) two sets of colour filters of different transparency for solar observations.

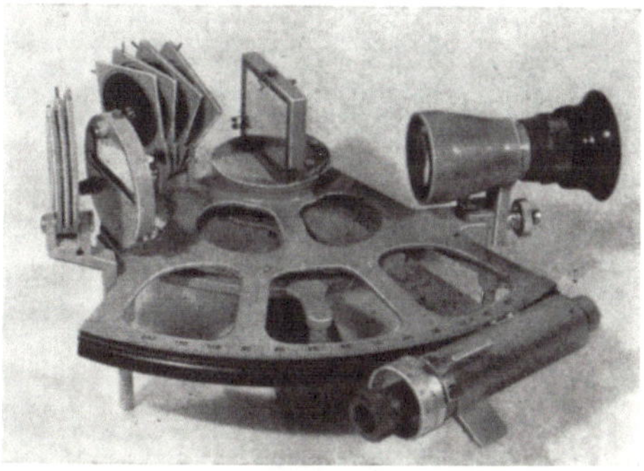

Figure 555 Drum sextant (Freiberger Präzisionsmechanik)

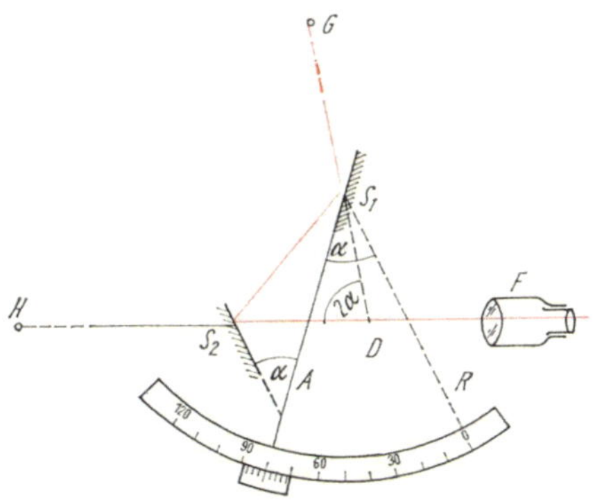

Figure 556 Operational principle of the sextant

The structure of the sextant is based on the theorem that a ray reflected by two mirrors is deflected by twice the angle enclosed by the mirrors (see figure 50). Two images appear in the field of view of the telescope. One image is generated by rays coming directly into the telescope from direction H, while the other one is generated by rays coming from direction G and being reflected by the mirrors S_1 and S_2. The images from G and H — the reflected and direct one — are combined, if the angle between the mirrors S_1 and S_2 is half of the angle at D where the principal rays GD and HD intersect. α can be read with the reading device at the graduation (figure 556). The graduated circle is numbered such that the full value of the tilt angle is directly read.
In addition to vernier sextants and drum sextants (with simple reading without magnifying glass), scale sextants are produced, where the measured angle is read with a scale microscope at the side of the telescope.
The sextant is held with its handle such, that the graduation surface is roughly vertical and directed towards the star. The telescope is directed towards the horizon. By rotating the alidade with the mirror S_1, the image of the star is coincided with the image of the horizon.

3.4. Instruments for Barometric Levelling

The air pressure decreases with increasing elevation at a rate of 1 mbar for 7,5 m (at sea level) to 11 m (at about 3500 m elevation). It furthermore depends on air temperature, and, to a lesser degree, on humidity and geographic latitude. Since it can also change due to meteorological conditions, air pressure measurements do not yield absolute heights, but only elevation differences.
The air pressure is determined by comparison with opposing pressure. This can be obtained with the pressure of a mercury column (mercury barometer), with the gravity of a spring acting on a metal box under vacuum (aneroid barometer), with the air volume under constant tension in a closed container (gas barometer), or with the surface tension of steam for boiling water (hypsometer).
Compared to differential- or trigonometric levelling, barometric heights are less accurate. For elevation differences of up to 200 m within small areas, a height error of ± 1 to ± 2 m can be expected. With precision barometers accuracies of ± 0.3 to ± 0.8 m can be reached. A standard deviation of a single elevation of ± 3 to ± 4 dm can be obtained when having overcast sky, stable meteorological conditions and no air movement.

3.4.1. Mercury Barometer (Hg-Barometer)

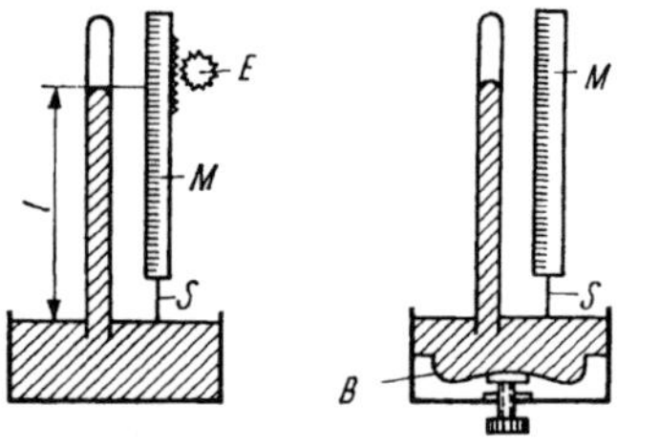

a) b)
Figure 557 Container barometer

Structure and Types of Hg-Barometer

According to design, there are container barometers, siphon barometers, and container siphon barometers.
The container barometer (figure 557) consists of a container filled with mercury into which a vertical pipe is immersed with its open bottom end. The top end is molten closed, and there is a vacuum within the pipe. The air pressure forces the mercury upwards into this pipe. The length of this mercury column is measured with a scale M that can be set onto the Hg surface. This scale can be either shifted against the pipe (figure 557a) or fixed to the container (figure 557b, where the Hg-surface is brought to the tip S by raising or lowering of the bottom B of the container).
At the siphon barometer, the lower, open end of the pipe is bent to a U-shape. Usually the pipe is widened within the reading range. A "Bunte" air trap catches air that rises in the corners Z and prevents it from entering into the vacuum. The scale is either fixed (figure 558, where l is the difference between readings on both Hg-surfaces) or can be shifted (its zero line is set to the lower Hg-surface, while the reading is taken on the upper one).
In the container-siphon barometer (figure 559) the siphon is replaced by a container with movable bottom B so that both mercury surfaces can be raised or lowered by the same amount. By measuring on different positions of the scale, an accurate value for l can be obtained.
According to their use, one distinguishes between normal- and station barometers. Normal barometers are used for observations at important stations and for calibrating other barometers with container-siphon barometers being most suitable. Station barometers are used as stationary barometers in the centre of the measuring region, with container barometers being most suitable.

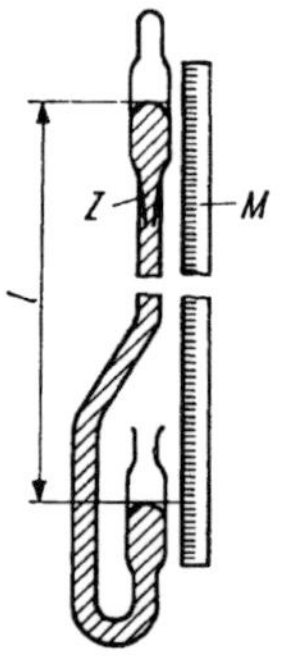

Figure 558 Siphon barometer

Corrections of Hg-Barometers

Due to the external influences mentioned earlier, the readings on Hg-barometers have to be corrected.

Temperature Correction V_t. Since both mercury and scale are temperature dependent, the barometer reading B is reduced to B_0 at the temperature 0°C:

$$B_0 = B(1 - \gamma t + \alpha t). \tag{167}$$

With $\gamma = 180 \times 10^{-6}$ as cubic coefficient of expansion for mercury, and α as linear expansion coefficient for the scale (e.g. brass: $\alpha = 19 \times 10^{-6}$), and the temperature t, the temperature correction for a brass scale becomes

$$v_t = -\frac{B}{614}\frac{t}{10}. \tag{168}$$

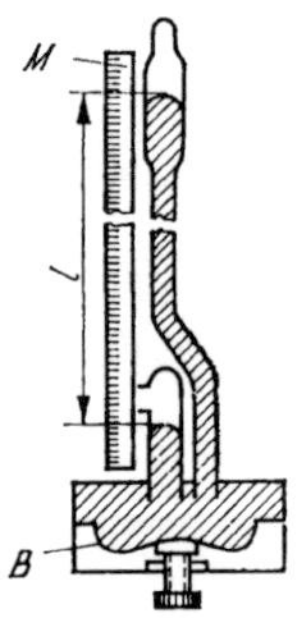

Figure 559 Container-siphon barometer

Capillar Depression v_k. Because of capillar depression, the Hg surface is slightly held back and curved, so that the barometer reading is too small by v_k (figure 560). The capillar depression depends on the inside diameter of the pipe and on the height of the meniscus k. Its amount can be obtained

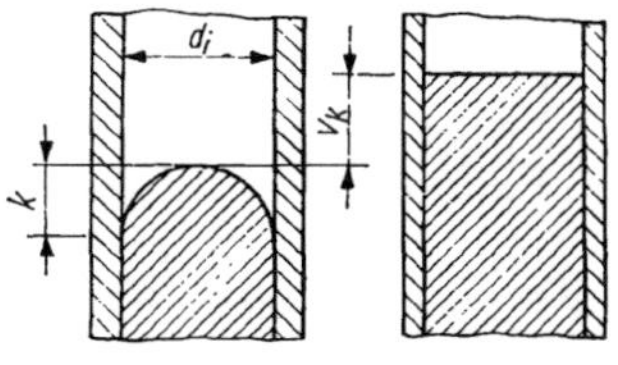

Figure 560 Capillar depression

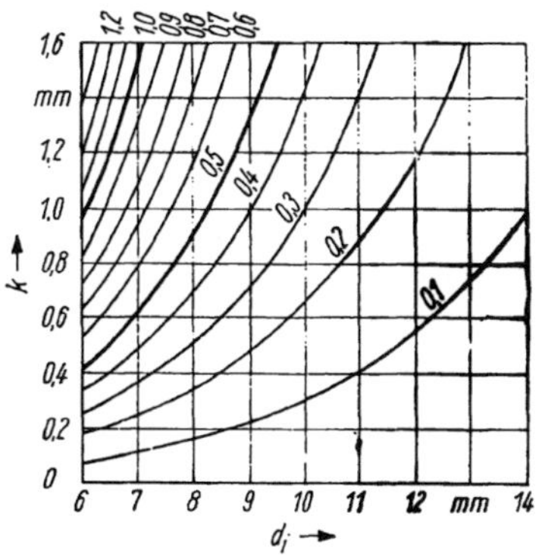

Figure 561 Diagram for determining capillar depression

from tables or diagrams (figure 561). For pipes with a diameter of more than 20 mm it does not appear. It is cancelled for siphon- and container-siphon barometers if the diameter of the pipe is the same on top and bottom.

Gravity Reduction v_φ and v_H. Since the weight of the mercury column is the product of its mass m and the gravity acceleration g, and since the latter changes with latitude φ and sea level elevation H, the measurements are reduced to sea level at 45° latitude with

$$v_\varphi + v_H = B_0^{45} - B = -B\beta \cos 2\varphi - \frac{2BH}{r} \tag{169}$$

where β is a gravity dependent coefficient. The following formulae are more convenient for practical computations:

$$v_\varphi = -264 \cdot 10^{-5} \cos 2\varphi B;\ v_H = -3 \cdot 10^{-7}\, HB. \tag{170}$$

Station Correction v_{St}. The station correction v_{St} of a barometer is the difference between the readings of a barometer corrected with v_t and v_k and the readings at a normal barometer at the same location. It is determined as the mean of several comparisons.

Reduced Air Pressure. The barometer reading reduced to $t = 0$ °C, $\varphi = 45°$ and sea level is obtained as

$$B_0 = B + v_t + v_k + v_\varphi + v_H + v_{St}\,. \tag{171}$$

The standard deviation of the measurement of air pressure amounts to about ± 0.2 to 0.25 mbar.

Aneroid barometers are best suited for field work.

3.4.2. Aneroid Barometers

Barometer with Membrane Box

The main element of these barometers is a flat air tight box with near vacuum with a rippled membrane soldered onto it. It is called Vidi box for its inventor. This box is protected against collapse by a strong spring which counteracts the external air pressure. The membrane rises or falls according to air pressure changes by about 0.004 to 0.08 mm per mbar. Since the air pressure is to be determined to 0.13 mbar, changes on the box have to be obtained with an accuracy of 0.5 to 1 µm. To read such small amounts, the box movements are either mechanically magnified with a lever system, or measured microscopically, or magnified optically or a combination of these is used. In the *Naudet* barometer (figure 562) the box D is encased in a metal housing G. Its membrane carries a column S. The spring F which is only fastened on one point, is connected to it. The small vertical movements of the membrane are mechanically magnified by the lever system H and transferred onto the index Z via the small link chain K, and transformed into a rotation. With the index, the pressure can be read on a scale (least count 1 mbar) and estimated to 0.1 mbar.

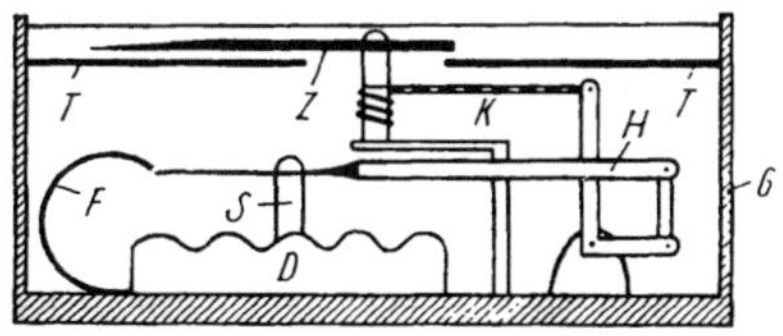

Figure 562 Naudet barometer

The barometer by the Swed *Paulin* (figure 563), which is manufactured in the U.S.A, also works with mechanical transmission. The movements of two membrane boxes are transferred with virtually no friction loss. They generate linear movements on a bridge B fastened to a flat spring F and a spiral spring S_1. A torsion spring joint T_g magnifies these significally and transforms them into rotations of the index I. These rotations are only used for zero display. The pressure from the membranes acting on the bridge in zero positon is a measure for the air pressure. For its determination, a measuring screw with index M is activated. This causes the eccentric disk E to push the sphere K towards the outside. A swivel joint connected with the sphere, transfers the force onto the compensation spring S_2 and onto the bridge B. The position of the index M indicates this force, i.e. the air pressure. It can be read on a scale (line interval 0.5 mm) to 0.2 mbar (resp. 2 m elevation), and estimated to 0.02 mbar. The elevation range is e.g. 1620 m.

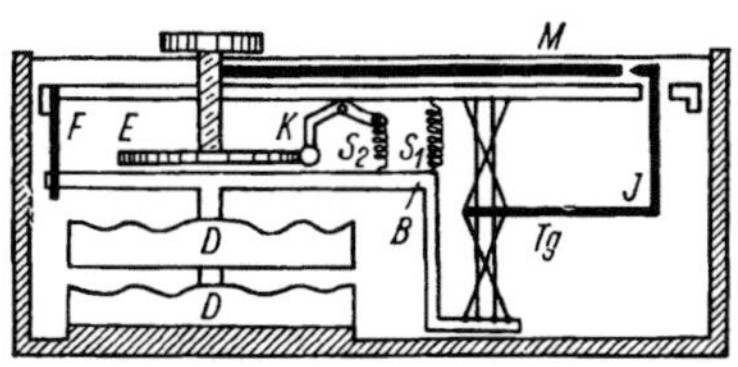

Figure 563 Paulin barometer

In the precision barometer FA-181 by Wallace & Tiernan, Günzburg, and Belleville, U.S.A., the movements of a completely stabilized vacuum capsule spring, made of copper-beryllium, are transferred nearly without friction via spring joints and a cog wheel segment with pin onto an index.

The circular scale, which is 76 cm long, is subdivided for two complete revolutions with a least count of 2 m (measuring range -300 to $+900$ m), or 2.5 m (range -300 to $+1800$ m) or 5 m (range -300 to $+4000$ m). The instrument has a diameter of 18 cm and a mass of 2.3 kg.

Figure 564 Terrain altimeter 3 B 4 (Revue Thommen, Waldenburg, Switzerland)

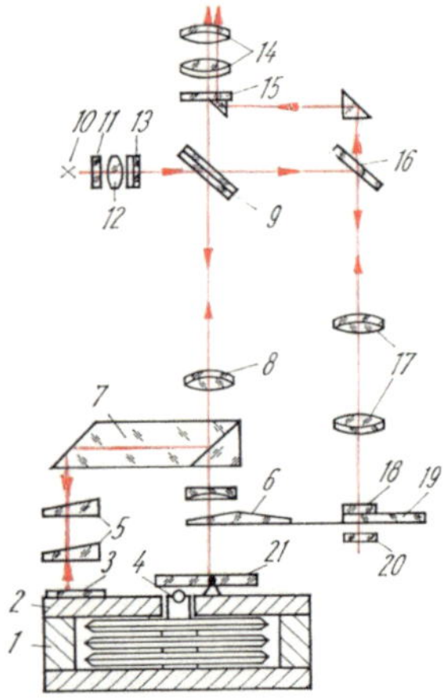

Figure 565 Principle of the optical micro barometer OMB-1

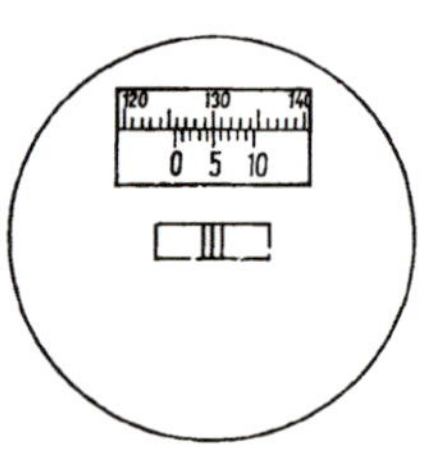

Figure 566 Field of view of the OMB-1

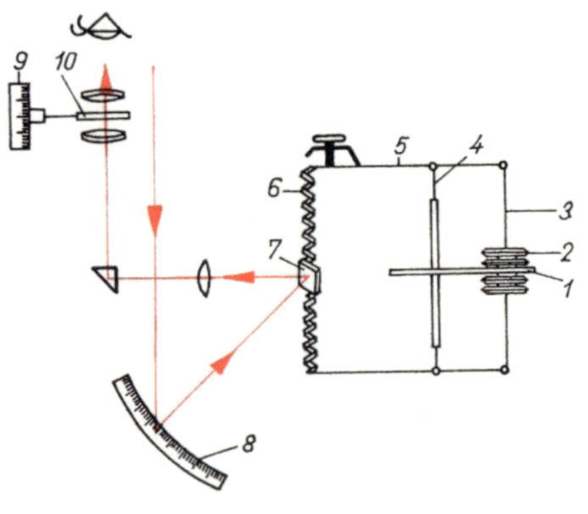

Figure 567 Principle of the spring micro barometer MB-63

The Barolux precision aneroid barometer by Fueß, R., Berlin has a mechanical-optical magnification of the measuring movements of three boxes. The motions of the membranes are transferred with levers onto a glass scale, which is magnified and imaged after a number of reflections with the aid of a light bulb (least count: 0.2 mbar).
The terrain height meters of the 3B Type by Revue Thommen, Waldenburg, Switzerland (figure 564) also utilize three coupled boxes (made of copper-beryllium). The box movement is transferred via cogwheel- and lever transmissions onto an index such, that the elevation can be directly read on a circular scale with linear 2 m graduation (0.8 mm line interval). The measuring range amounts to -500 to $+2500$, 3000, 4500, or 5000 m (500 m per revolution) with a reading accuracy of ± 3 m. This instrument has a mass of 1.25 kg, including a steel container, and dimensions of 90 mm $\times$ 100 mm $\times$ 128 mm. The type 3B1 has a larger measuring range (3000 resp. 10,000 m) is, however, less accurate (± 5 resp. 20 m).
The Soviet "optical micro barometer OMB-1" contains a set of three membrane boxes with an expansion of 4.8 mm at 930 mbar. This expansion is determined by double auto-collimation to ± 3 µm, which yields an accuracy of ± 0.6 mbar for the air pressure. The set of boxes is fastened to the floor of the container (figure 565). When the air pressure changes, the mirror (*21*) at the lid (*2*) is tilted around a horizontal axis by the free end of the set of boxes, while mirror (*3*) is at the wall. In reference to the fixed mirror, the tilt angle is measured with the aid of auto-collimation system, optical micrometer and reading microscope. The light, passing from (*10*) through filter (*11*), condensor (*12*) and glass plate (*13*) (with reference line), is diverted onto the prism block (*7*) by the semi permeant mirror (*9*) and the objective lens (*8*), and from there onto the mirrors (*3*) and (*21*). After reflection, it reaches the diaphragm (*15*) in the image plane. The lines of the glass plate (*13*) are observed with the eyepiece (*14*), while the wedges (*5*) are used for adjustment purposes. By pressure changes, the image of the single line generated in the field of view by the tilted mirror, is shifted in respect to the image of the double line. The amount of the shift is determined with the compensation lens (*6*) at scale (*19*) and vernier (*18*), which are projected into the field of view by the objective lens (*17*) (figure 566). Illumination occurs via the mirrors (*16*) and (*20*). 400 scale intervals correspond to a tilt of 3° for mirror (*21*). The least count is 0.2 mbar, the vernier reads to 0.02 mbar, the instrumental error is ± 0.04 mbar and the measuring range is 100 mbar. The instrument has a mass of 3.8 kg. The standard deviation of elevations is about ± 0.8 m.
In the Soviet microbarometer MB-63, the movements of two sets of two membran boxes (*2*) opposite to each other on a common support (*1*) are transferred into a torsion of a double cylindrical sping (*6*) via the beams (*3*) and the levers (*5*) with intermediate beam (*4*) (figure 567). This leads to a rotation of mirror (*7*) which is measured with the scale (*8*) (0.5 Torr least count) and the micrometer eyepiece (*9, 10*). The mirror turns by 5′ for a box movement of 1 µm (0.13 mbar pressure change). The measuring range of 100 mbar can be changed. The instrument has a mass of 2.8 kg.

Barometer with Tube-Spring

A circular spring tube with elliptic cross section and vacuum, is the pressure sensitive element, named after its inventor *Bourdon*. When the air pressure increases, the separation of the ends of the tube decreases, because the outside of the tube is loaded more than the inside because of its larger surface. This movement is read as pressure change on a scale with the aid of an index.
The microbarometer Gb 5, developed by *Graf* and manufactured by Bodenseewerk Geosystem, Überlingen, contains a tubular spiral spring with nine coils of copper-beryllium (figure 568). Its lower end supports a mirror with four index lines, whose movements can be read on a fixed scale with the aid of an autocollimation system. The spring can be rotated around its longitudinal axis which increases its measuring range of 50 mbar 15 times to 800 mbar.
The optical magnification of the movement is completely without friction. The least count is 0.13 mbar, the internal measuring accuracy is 0.013 mbar, and the mass is 3 kg.

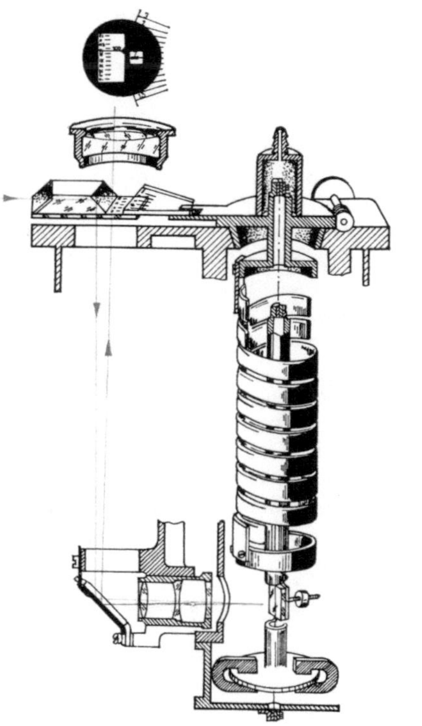

Figure 568 Measuring device in the micro barometer Gb 5

Barometer with Gas Spring

In barometers with gas spring, a gas volume under known tensions serves as elastic reacting force against the external air pressure. A barometer of this kind was built according to instructions by *Haalck* (height measuring error ± 30 cm for short distances between stations). The development by *Stribling*, *Broding* and *Wilhelm* (U.S.A.) is more accurate. Here the gas chamber is connected to the outside air with a capillar tube carrying a graduation.
A drop of oil within it moves when the pressure changes. The measuring accuracy is about ± 0.04 mbar.
A gas spring is free of friction and does not have any elastic after effects. It is, however very dependent on the temperature (the gas has to be kept constant within a temperature of 0.01 K, because a temperature change of 0.0037 K represents a pressure change of 0.13 mbar). Because of large apparatus, caused by this, these instruments are not suitable for field work.

Corrections for Aneroid Barometers

The direct readings of aneroid barometers have to be refined with the temperature correction v_t (especially because of the influence on the elasticity of the springs), the graduation correction (because of too narrow or too wide intervals due to insufficient matching of transfer mechanisms and graduation) and the graduation correction v_{St} (or for the zero error of the graduation line 1013). Because the values v_t and v_{St} can be considered as linear, one needs to determine only the temperature expansion coefficient a and the graduation coefficient b, while the station correction is constant. With this and the reading F on the aneroid barometer the reduced value B_0 for $t = 0$ °C becomes

$$B_0 = F + at + b(1013 - F) + c. \tag{172}$$

To determine a, b, and c, the direct reading F is compared to the reduced readings B_0 for $t = 0$ °C of a calibrated mercury barometer at various pressures and temperatures. Because of elastic after effects and aging, which cannot yet be effectively counteracted, these values have to be determined yearly.
The temperature coefficient a is determined from readings of the aneroid barometer and a calibrated Hg-barometer at different temperatures (outside and inside the laboratory resp. inside a climatized chamber). The differences between the Hg-barometer readings reduced to 0 °C and the readings on the aneroid barometer are plotted in a graph. The slope of the best fitting straight line represents the temperature coefficient a (figure 569). It shall not exceed 0.15. The graduation coefficient b and the zero correction c are obtained by comparing readings reduced to 0 °C for the aneroid and mercury barometers at various air pressures (e.g. when climbing a mountain, by observing over a long period of time, or in a pressure chamber). If the differences of the reduced readings (ordinate) are plotted against the readings (abscissa) in a rectangular coordinate system, the slope of the best fitting straight line represents the graduation coefficient b, and the ordinate for 1013 mbar the zero correction c (figure 570).

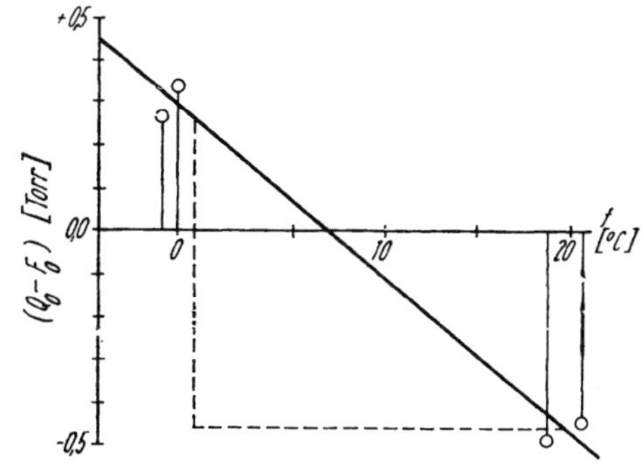

Figure 569 Determination of the temperature coefficient

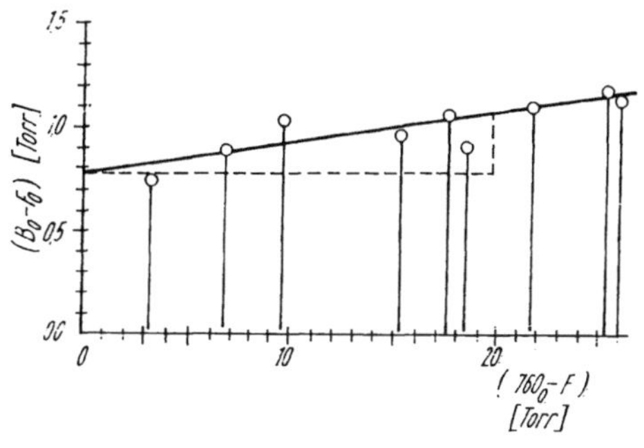

Figure 570 Determination of graduation coefficient and zero correction

3.4.3. Hypsometers

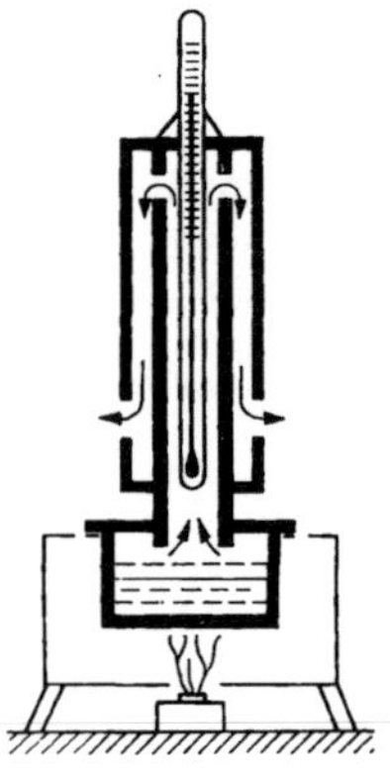

Figure 571 Hypsometer

The boiling point of water depends on the air pressure. It falls for decreasing pressure. Since a temperature difference of 0.1 K corresponds to a pressure difference of 3,6 mbar, the boiling temperature has to be determined to 0.01 K if the barometric pressure is to be safely obtained to within ±0.4 mbar. This is achieved with hypsometers (figure 571). A thermometer with a fine capillar tube of borosilicate glass and a scale graduated to 2 mbar hangs in a cover tube which extends into a water container and which is sealed on top with the rubber ring. The thermometer, whose Hg-reading is taken at the upper end, shall not extend by more than 5 mm from the tube, so that it is not cooled by the surrounding air. The flame of alcohol burner used to heat the chemically pure water in a container, is protected against draft by a screen. A wire net prevents water drops to splash onto the thermometer ball. A water vapor surrounds the thermometer on all sides and escapes through a vent.
The reading is best taken a few minutes after the water starts boiling. After each boiling, the thermometer has to be shaken violently and abruptly to clean the capillar tube of overdistilled mercury. Zero reading and graduation unit are determined by comparison with a mercury barometer. There are no other corrections to be applied. The hypsometer is especially used on expeditions to monitor aneroid barometers. Its accuracy is approximately ±0.3 mbar.

3.5. Fully Automatic Elevation Measuring Devices

Fully automatic elevation measuring devices ("level automats") automatically record elevation differences. These instruments, attached to a bicycle or automobile are based on a pendulum to measure the slope angle α and a distance counter, which records the distance section Δs covered by one wheel (figure 572). The elevation difference is:

$$H_A^B = \sum \Delta h_i = \sum \Delta s \sin \alpha \tag{173a}$$

resp. for continuous recording

$$H_A^B = \int_A^B ds \sin \alpha . \tag{173b}$$

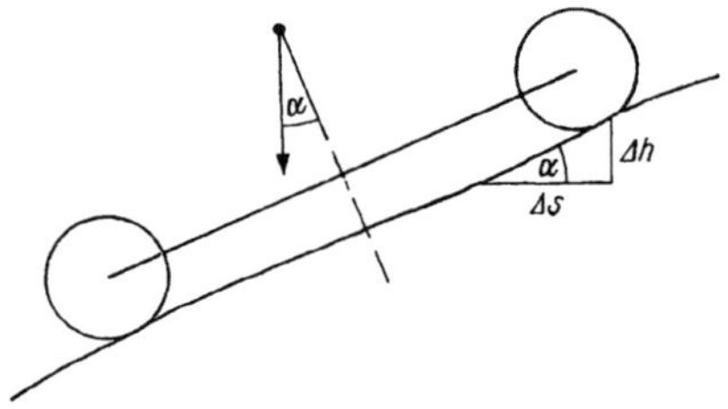

Figure 572 Principle of automatic altimeter

The Soviet instrument WA-1 computes automatically the elevation difference when driving over the terrain. It also records the profile of the path covered, and permits readings of elevations and elevation differences on counters. The unit elevation differences for small segments of the path are added (discrete principle). If the wire at the end of the pendulum (*1*) touches the contact arm (*4*) of the globoid drum (*2*) which is driven by the electromotor (*3*), then four contacts are established (figure 573). The pair of contacts (*II*) is connected to a second pair (*I*). By rotating the contact drum, the pairs are sequentially closed which monitors a thyratron relay (Thyratron is a switch or monitor for large loads; namely a triode filled with mercury vapor). Depending on the slope angle α and the form of the contact comb, the separation of the impulses given to the electric circuit, change. When $\alpha = 0$, they are equal, while for an inclinded system their difference is proportional to $\sin \alpha$. The oscillations from the relay reach the electromagnets (*6*) and (*7*), which are activated by a reduction (*8*), connected to the wheel (*9*) of the vehicle, and which move the height counter (*11*). The armature (*10*), attracted by the electro magnet, rotates by an angle proportional to the length of the path covered. When the second impulse arrives,

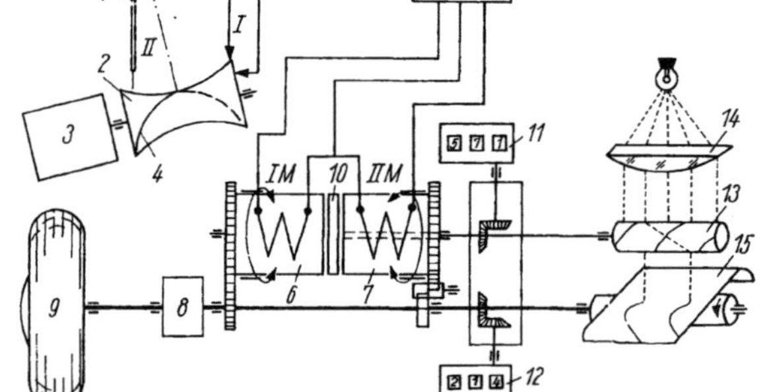

Figure 573 Scheme of an automatic altimeter (WA-1 M)

the armature is attracted by the second electromagnet and rotates in opposite direction. By this recording in two directions, a difference is formed in the reductor, proportional to the elevation difference.
While the counter (*11*) indicates the elevations based on the revolutions of the armature, the counter (*12*) displays the path covered by the wheel (*9*) of the vehicle.
The elevation measuring automats are equipped with a device for recording the profile of the path covered onto photographic paper, which is unrolled proportionally to the distance covered. The terrain profile is generated with the aid of a "light roller", which is a cylinder with drill holes and which follows the movement of the armature (*10*). Rays from a wide parallel bundle pass through the drill holes of the cylinder and generate an uninterrupted line on the photographic paper. The instrument is installed in an automobile. During a seven hour working day, up to 90 km path, a profile can be obtained with the WA-56 to an accuracy of ± 4 to 10 cm per km double levelling. For route design and photo control determination 60 to 65 km are covered in a day. Discrete elevations of points are determined with a standard deviation of about ± 25 cm. The maximum inclinations, which can be covered, are $\pm 15°$. On the photographic paper, the elevations are recorded to 0.10 m, and the distances to 1.0 m in 1 : 4000. The cassette contains 100 m photopaper which is 100 mm wide. For the instrument AWA (1960), based on impulse counting, the measuring reliability is increased while dimensions and mass are reduced.
Similar instruments are known in the U.S.A., which, like the "Johnson Ground Elevation Meter" by Sperry-Sun Rand, are mounted on an automobile or a three wheeled trailer. During measurements, a pendulum measures the tilt of the frame of the car and generates a current proportional to the sine of the tilt. Distance and speed are determined with the aid of the rear measuring wheel, which generates electrical impulses or signals in a frequency proportional to the speed. Both measurements are evaluated in an electronic integration unit, and the results are read on scales and counters.

4. Instruments for Distance Measurement

In surveying the following methods for distance measurements are used:
1. mechanical (direct) distance measurements
2. optical (indirect) distance measurements
3. electronic distance measurements.

For mechanical distance measurement, the distance is determined by sequentially placing the distance measuring tool. The errors which occur often cannot be avoided.
Optical distance measurement provides significant savings in time compared to mechanical measurements, because the terrain between the terminal points of the distance to be measured, has to be entered in exceptional cases only. However, a good organization of the work and a well functioning field crew are necessary. The instruments are suitable for tacheometric and topographic jobs, as well as for traversing and detail surveys. Depending on the instrument type, the range is about 200 to 1000 m.
The instruments for electronic distance measurement distinguish themselves by high measuring speed, considerable range (up to 50 km and more), high accuracy and independence of local conditions.
The accuracy of distance measurements depends on the instruments and methods used (table 29).
The result of a distance measurement indicates how often the measuring unit is contained in the measured distance. The basic unit for length is the metre. It was originally defined as the separation of the central lines of line groups on the International Metre Prototype (figure 574). This standard, kept at the International Office for Measures in Breteuil, Sevres near Paris, consists of platinum-iridium and is defined for an equilibrium temperature of ice and pure, aired water at a pressure of one physical atmosphere.

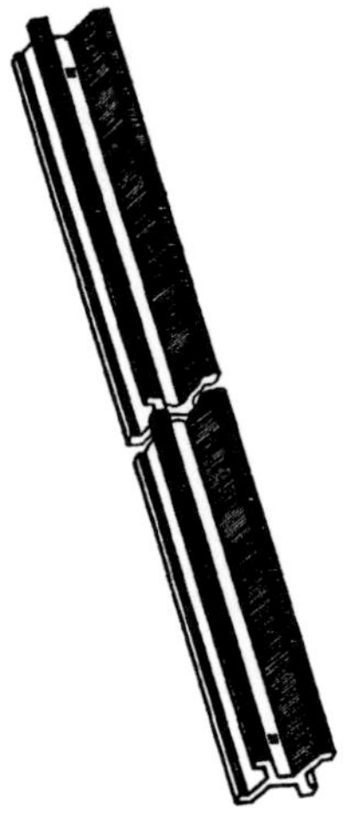

Figure 574 Prototype of the metre

Table 29. Accuracy of Geodetic Distance Measuring Instruments

instrument or auxiliary device	accuracy
pacing	1 : 50 up to 1 : 100
measuring wheel, odometer	1 : 100 up to 1 : 500
field dividers	1 : 1000
tape with alignment poles	1 : 1000 up to 1 : 3000
rolled measuring tape	1 : 3000
steel tape, fully supported with temperature and slope corrections applied	1 : 10.000
measuring rods (horizontal)	1 : 10.000
stadia (up to 250 m)	1 : 300
stadia with wedges (up to 200 m)	1 : 5000
double image tacheometer with rotating wedges (up to 200 m)	1 : 5000
double image tacheometer with half lenses (up to 1000 m)	1 : 1000
subtense bar and theodolite (up to 150 m without auxiliary base)	1 : 5000
24 m invar tapes with tension device	1 : 50.000
microwave distance meters (up to 50 km)	1 : 300.000
electro-optical distance meters (up to 50 km)	1 : 1.000.000
electro-optical distance meters for short range (up to 3 km)	1 : 300.000
electro-optical precision distance meters (up to 3 km)	1 : 1.500.000
24 m invar wires (up to 2 km)	1 : 1.000.000
interference comparator (864 m)	1 : 15.000.000

Since the accuracy of this International Metre is not sufficient anymore for modern needs, this definition of 1889 was revoked at the XIth General Conference for Measurements and Weights on October 14, 1960, and replaced by the optical metre, based on the wavelength of the crypton isotope 86. This is a natural and indestructable standard, internationally defined as "The metre is the 1,650,763.73 times multiple of the wavelength of radiation emitted by ^{86}Kr when changing from state $5d_5$ to state $2p_{10}$ and propagated in vacuum." With other words: the metre represents the length of 1,650,763.73 wavelengths of the orange red radiation emitted by the inert gas Crypton 86. In October 1962, the International Committee for Metre Definition accepted three secondary sources of radiation for the metre definition. These are lamps without electrodes who, when stimulated in a high frequency field, emit Crypton 86, Mercury 198 resp. Cadmium 114.

4.1. Instruments for Direct (Mechanical) Distance Measurements

4.1.1. Simple Means for Distance Measurements

The distance of a path can be determined from the marching time and -speed, which is about 5 to 6 km per hour for adults. Pacing can be used as an aid. According to Jordan, the length of a pace of a man in relaxed walk usually equals half his eye height. To ease the job, pace counters (pedometers) are used.

With a measuring wheel (figure 575), the distance is computed from the number of revolutions and the circumference of the wheel. Occasionally, this is used to mark kilometres on roads.

Field dividers having wooden legs braced with a crosspiece and usually a 2 m opening, are used in agriculture for dividing fields or staking drainages.

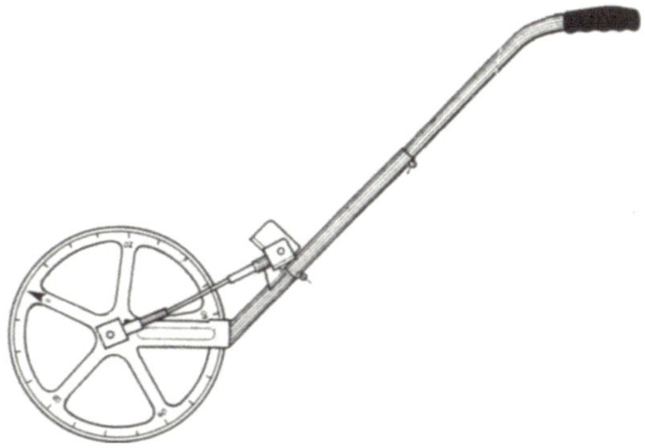

Figure 575 Measuring wheel

4.1.2. Auxiliary Devices for Distance Measuring Instruments

Range poles are used for alignment. They consist of coniferous wood or light metal, with either round or triangular cross section (6 travelling range poles with 26 mm wide sides can be combined to a set with circular cross section), are usually 2 m long and divided into 0.50 m long sections by alternatively red and white paint. They sit in shoes with the same cross section (figure 576), with the triangular ones being narrower at that part. Range poles with fluorescent foil simplify working in fog, woods or at dusk. Recently, range poles are manufactured which have a wooden core near the

Figure 576 Range pole

tip which is surrounded by an aluminium pipe which carries a 1 mm thick cold resistant plastic coating. Occasionally, range poles consist of two parts, which are connected with a socket (in U.S.A. also by a screw connection).

Three legged range pole tripods (figure 577) provide support in a terrain where the range poles cannot stand by themselves.

The range poles are set vertical with a plumb bob or a rod level (figure 521). A plumb bob is a cone shaped or pear shaped metal piece, hanging on a piece of string. It may also have a cross section as shown in figure 578, and has a mass of 150 to 300 g. The tip *Sp* has to be in line with the string *S*.

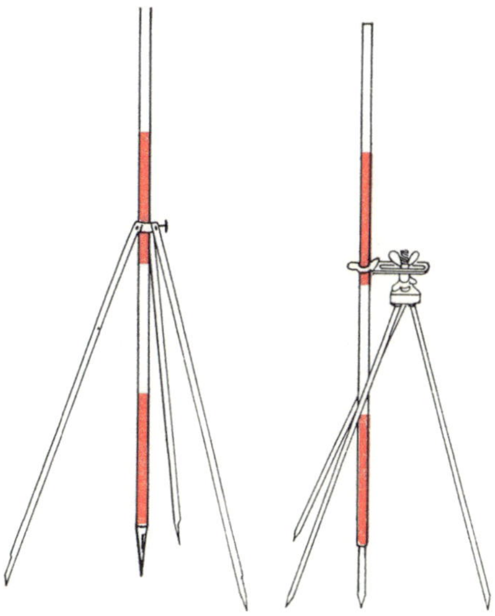

Figure 577 Range pole tripod

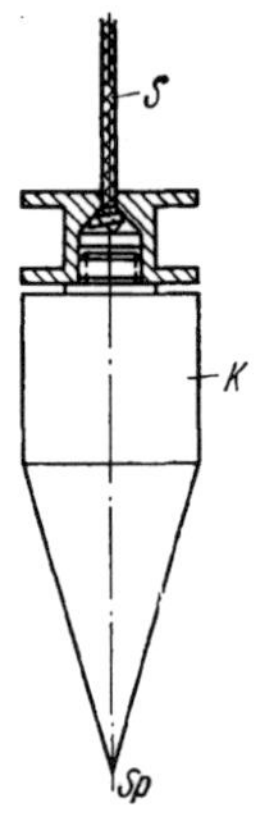

Figure 578 Plumb bob

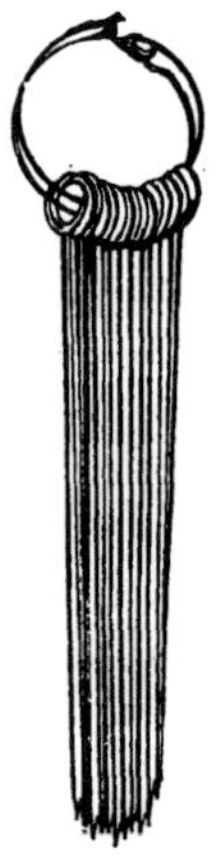

Figure 579 Taping pins

In order to check this, the plumb bob is hung up. Then, after the body K has come to rest, it is rotated by turning the string between the fingers. The tip has to remain in one spot and shall not move along a circle. The plumb rod, a rigid plumb, is rarely used. It consists of a round rod with a few centimetres thickness and a heavy iron tip on the bottom end. If the bull's eye level on the upper end is centered, the rod axis is vertical. This can be checked with a plumb bob.
A striding level (see figure 529), placed in the centre, is used to make the measuring rod horizontal. It is also possible to measure along the slope, determine the slope angle of the rod and compute the reduction to horizontal. These devices often have graduations directly indicating the reduction values. The value d, which has to be subtracted from the rod length l, is

$$d = l(1 - \cos\alpha). \qquad (174)$$

It is the difference between l and the projection $l' = l\cos\alpha$. For a 5 m rod and 10 gon slope it amounts to

$$d = 5(1 - \cos 10\,\text{gon})\,\text{m} = 5(1 - 0.9877)\,\text{m} = 0.0615\,\text{m} \approx 62\,\text{mm}.$$

Taping pins of 30 to 40 cm length with 3 mm diameter made of galvanized iron wire (figure 579) are used to mark the end points of tape lengths and to count these. They are pointed on one end and formed to a ring at the other, and usually are arranged in sets of 10 on a wire ring. Climometers (see section 3.3.1.) are used to determine the inclination for slope taping.
Tape clamps can be used to immediately connect broken tapes or to support a tension handle after being clamped at any place on the tape. To read millimetres on a tape with decimetre graduation, a special ruler of about 12 cm length can be used. A pin in its zero point can be inserted into the hole graduation of the tape. At both ends are clamping devices with springs.
The air temperature which for precise measurements has to be considered because of its influence on the length, is usually determined with a whirling thermometer. The whirling thermometer has a glass tube with a white scale M (least count 1 K, measuring range -30 to $+40$ °C) within a protective pipe S and screwcap K on the bottom. The reading slot is protected by a telescopic brass casing F during transportation (figure 580). The air temperature is read on the scale after the thermometer had been swung around on a string until the reading remained unchanged. Since the temperature of the tape usually does not coincide with the air temperature, thermometers clamped onto the tape (e.g. sticking- or resistance thermometers) are used for high precision tape measurements, especially in the sun.
Tension handles or dynamometers (figure 581) are used to obtain uniform tension of the measuring tape. They consist of spring balances with carabine hook and handle. Generally tension values of 50, 100 and 150 N are given. After the tension handle is hooked onto the tape, it is pulled by hand until the desired tension for the measurement is indicated. For all precision measurements, temperature, and tension have to be considered. For simple measurements, the errors caused by them are smaller than errors due to inaccurate holding, sag etc.

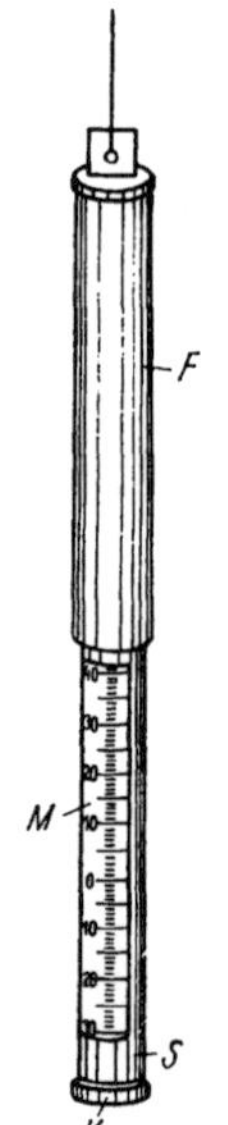

Figure 580 Whirling thermometer

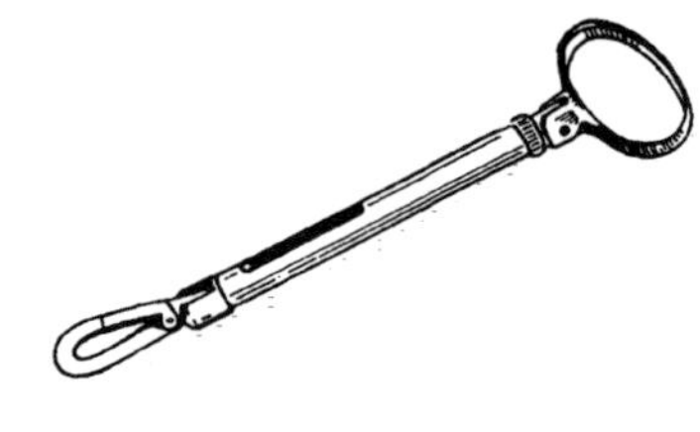

Figure 581 Tension handle

4.1.3. Measuring Rods

Types of Measuring Rods

Measuring rods are 3 or 5 m long and made of well aged knot free pine wood. They have an oval cross section, whose major axis decreases from about 5 cm in the centre to about 3 cm at the ends. The ends are fitted with either blunt or wedged steel shoes of at least 4 mm thickness, with the latter being oriented at 100 gon to each other (figure 582). As a protection against humidity and for easier recognition of the metre sections, the rods are painted metre by metre with either black and white, or white and red oil paint.
Measuring rods are always used in pairs. When measuring, one starts with a "black" rod, which means that all readings within a white section on both rods belong to an uneven metre count. The metres are denoted by large

square brass marks, half metres by small squares, and decimetres by round ones; the centimetres are estimated.

Temperature changes have little influence on the length of the rods, because the temperature expansion coefficient of wood is small (3 to 9×10^{-6}). On the other hand, wood changes with humidity, which means that rod measurements on wet days may be erroneous. The permissible change in length for one rod is +1.8 to −0,9 mm for 3 m rods, and +3.0 to −1.5 mm for 5 m rods.

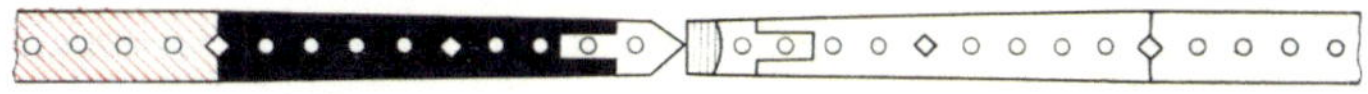

Figure 582 Measuring rods with edge shaped ends

Calibration of Measuring Rods

Excellent measuring results are obtained, if the measuring tools are regularly checked against a norm. This is done with the aid of a rod comparator. Such a comparator (figure 583) consists of a solid horizontal base T (e.g. a smooth plank or a T-iron beam), onto which two steel edges K_1 and K_2 are mounted, separated by a distance a, e.g. 5.01 m.

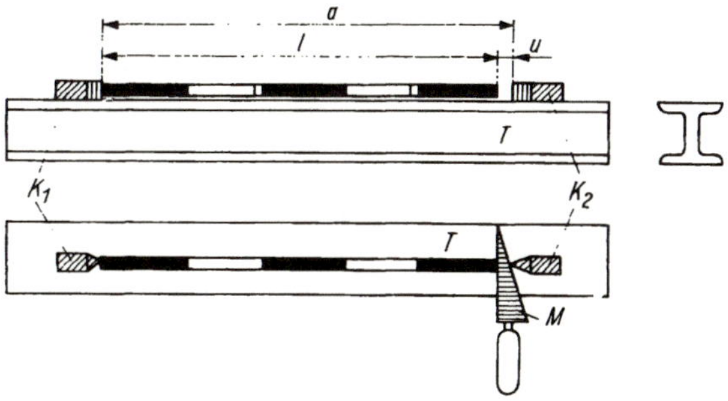

Figure 583 Rod comparator

The line connecting K_1 and K_2 is marked carefully. The distance a between these edges is measured with calibrated test rulers with wedge- ends and with a measuring wedge (see page 233). Once the comparator length is known, the rod length can be determined. The rod to be calibrated is placed between the edges, and the difference between the edge separation and the rod length is determined with a measuring wedge. For precision work, the rod lengths are determined four times each, once each lying on front and back for touching one and then the other measuring edge. Afterwards the comparator length is remeasured, this time starting with the second test ruler.

4.1.4. Measuring Tape

It is faster and easier to measure with tapes than with rods. However, due to the danger of breakage, tapes require good care and also are more sensitive to temperature changes than rods.

Types of Tapes

Tapes for land surveyors usually are 20 m long, 12 or 20 mm wide, 0.4 mm thick and reinforced at the ends with little plates. The even metre values are denoted on both sides by square brass disks, the odd metres by larger round disks, half metres by smaller round disks, and the decimetres by holes or brass rivets (figure 584). At 5 m and 15 m, rectangular number

Figure 584 Marking on the tape

plates of brass with the numeral 5 are fastened to both sides, with copper plates at 10 m carrying the number 10. Attached to both ends are doubly rotatable brass rings (figure 585), which can be slid over alignment poles. Origin and terminal points are located within the end rings on which their position is marked by an engraved line. The tapes are made of steel with a temperature expansion coefficient of 11.5×10^{-6} at 20 °C. The difference to the nominal length has to be within +3.0 to −1.5 mm for a fully support

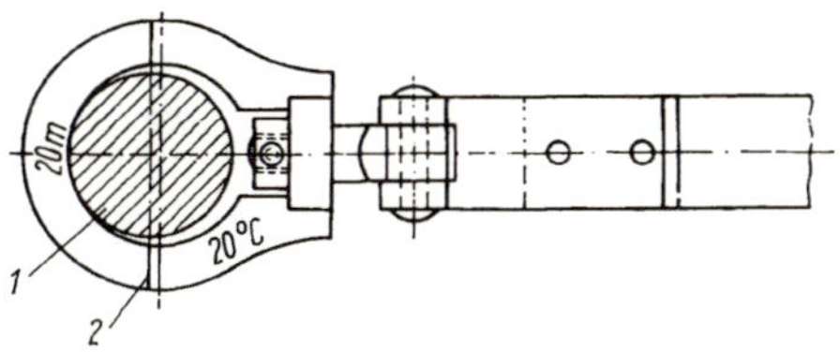

Figure 585 End of tape
1 alignement; *2* zero line pole

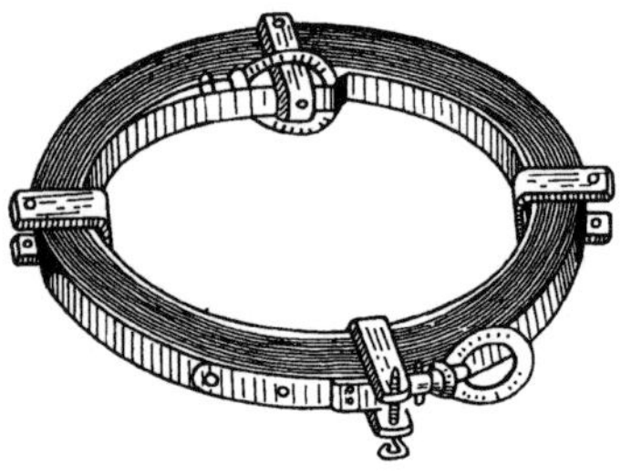

Figure 586 Rolled up tape

a)

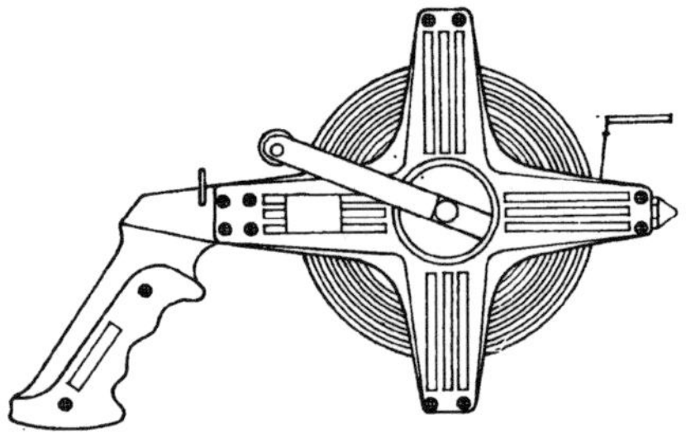

b)

Figure 587 Tape on reel

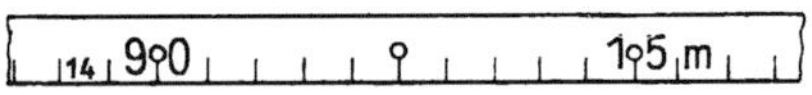

a)

b)

c)

Figure 588 Graduation of reel-tape

ed tape with 150 N pull at +20 °C. For transportation, the tape is rolled onto a metal ring (figure 586). The alignment poles have a diameter of 30 mm, are about 1.35 m long and have a pointed iron shoe with a cross pin for stepping it into the ground. This pin also prevents sliding of the tape rings when moving along.
Rolled tapes are 10, 20, 25, 30 or 50 m long with a cross sections of 13 mm × 0.2 mm. Their length has to be within ±1 mm per 10 m at 20 °C with 100 N pull under fully supported conditions.
These tapes consist of steel or plastic. Metal tapes have a continuous centimetre graduation, with the first decimetre being subdivided into millimetres (figure 587). For each decimetre number, the corresponding metre value is also given, which means that metre, decimetre and centimetre can be read with one look. This graduation is engraved, which is only possible on one side of the tape. Lines and figures are blackened (figure 588a). It is of advantage, that the graduation cannot rub off. To reduce the danger of breaking, occasionally protective edges are engraved on both sides. Other tapes have graduation lines and numbers raised (figure 588b), and have a polished appearance. The metre numbers are placed on polished spots and thus can easily be recognized. If the tape edges remain raised for protection (figure 588c), the tape around graduation numbers and protective edges is edged three times as deep as normal, so that they remain clearly readable even after long usage. These tapes are also more resistant against breakage. Since tapes made of carbon steel or non rusting steel can rust resp. become shiny; they are tempered. Fibre glass tapes are completely insensitive to weather influences. Presently there is a trend towards using plastic coated steel tapes. The tape is fastened to a reel with handle (figure 587), whose main parts are made of brass. By flipping the crank handle, the tape can be arrested in any position. Tension is applied manually. The graduation begins either on the edging, or for precise tapes on the tape itself (figure 589).
Precision tapes, especially for mining surveying, are made of special steel and are 20, 30, or 50 m long. Shaft tapes for vertical measurements are up to 1200 m long with a cross section of 11 × 0.4 mm² and a temperature expansion coefficient of 11.4×10^{-6}. They are rolled onto special block frames. Decimetres and half metre values are denoted by beads, while full metres have brass signs. The ends of the tape and its core are similar to the land surveyor's tapes. Presently, 100 m steel precision tapes with 4 mm × 0.3 mm cross section are used for accurate direct distance measurements (e.g. in engineering surveys) with tension handle and ends support. Even metres are numbered, and full metres are indicated by small rivets. For high precision work, invar tapes are used.

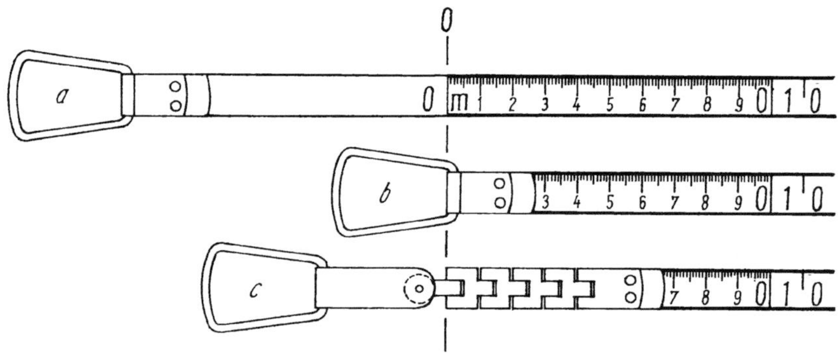

Figure 589 End of the reel tape
a) with graduation origin on the tape
b) with graduation origin on the edging
c) with null point joint of brass

Tape Standardization

The actual length of tapes is determined with a comparator, who matches the tape length (figure 590). On the surface of the measuring rail there are two (or else at every 5 m) metal plates M_1 and M_2 with millimetre graduation, whose zero lines are separated by approximately the tape length. The distance a between the zero lines—the comparator length—is determined with calibrated test rulers and a measuring wedge.
Test rulers are supplied in wooden crates together with a calibration certificate (in the GDR by the Bureau of Standardization, Measurements and Control of Goods). A set of test rulers (figure 591) consists of two metre rulers, designated with A and B. They have a length of (1000 ± 0,02) mm with a square cross section (10 mm × 10 mm) and are made of non hardened

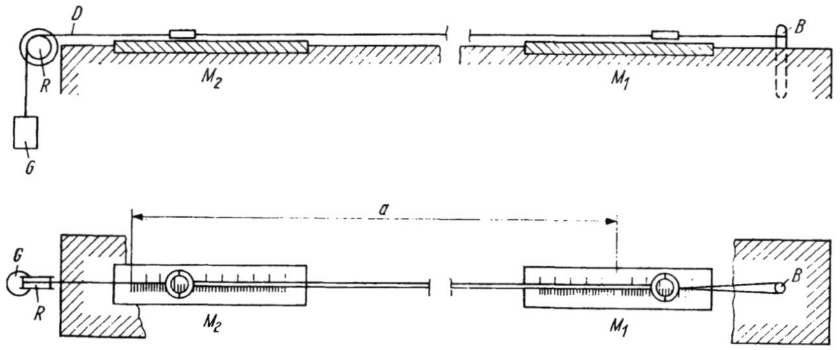

Figure 590 Tape comparator

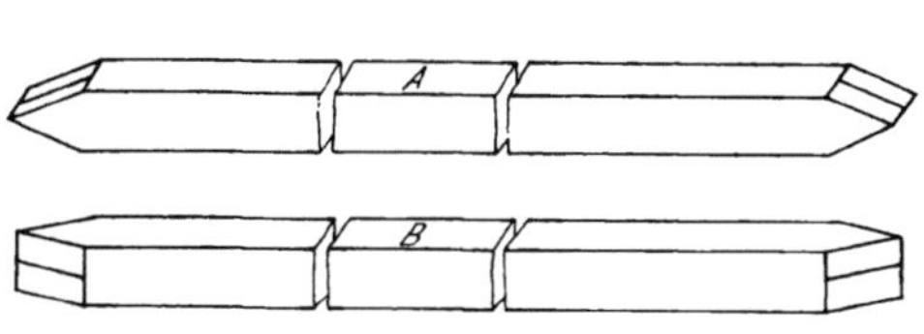

Figure 591 Set of test rulers

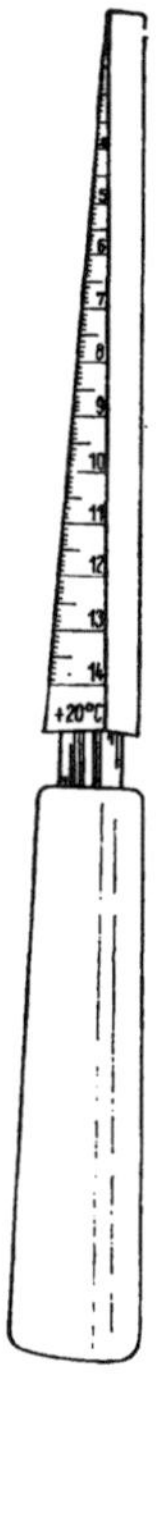

Figure 592 Measuring wedge

steel with a temperature expansion coefficient of $(11.5 \pm 1.5) \times 10^{-6}$ at +20 °C. (Normal metre rulers are also produced of other materials with small temperature expansion coefficient). Based on its calibration, the equation of the normal metre is given for each ruler as

$$L_t = 1\text{ m} + a\text{ mm} + \alpha(t - 20)\text{ mm}. \quad (175)$$

The first term represents the standard length of the test ruler at +20 °C The second term contains the deviation from the standard length at 20 °C (permissible up to ±0.02 mm) caused by its manufacture. The third term corresponds to the change in length of the normal metre due to differences between the working temperature t in the comparator room and the reference temperature.
The measuring wedge (figure 592) consists of steel with an expansion coefficient of $(11.5 \pm 1.5) \times 10^{-6}$ at +20 °C and is equipped with a wooden handle. The wedge angle is 5°42′18″, so that one millimetre on the wedge graduation corresponds to one tenth of a millimetre of the width to be measured. The least count is 0.1 mm. The measuring wedge is always placed between the edges such, that its long side fits along a horizontal edge.
The length of the comparator is determined as follows: The test ruler A is pushed against one edge of the comparator. The ruler B is laid against it such that both rulers lie exactly on the straight line connecting the two edges of the comparator. While ruler B is held, ruler A is carefully moved backwards and then placed on the other side of ruler B. This procedure is repeated until the other comparator edge is reached. The small section between ruler A and the edge of the comparator is measured twice with the measuring wedge.
Afterwards the tape is fastened to the bolt B. On the other end the mass piece G, which produces the tension is attached via a tension wire D (or a tension tape). This tension wire is laid over the guide roll R. The tape length is determined simultaneously on both end marks of the comparator. The temperature is read both at beginning and end of the measurements. Its mean value is used in the computations. Finally, the comparator length is measured again, this time starting with test ruler B.
The true tape length for the temperature during the calibration is obtained from the difference between comparator- and tape lengths. This length has to be reduced to the reference temperature of +20 °C. The change in length Δl_t of a tape of the length l, having an expansion coefficient α between the temperature t and the reference temperature +20 °C amounts to

$$\Delta l_t = l\alpha(t - 20). \quad (176)$$

Example: Test of a 20 m tape.

1. Determination of the comparator length (table 30) (length of test rulers is reduced to given temperature)
2. Determination of the tape length with 100 N tension (table 31).

After each reading, the position of the tape in respect to the comparator marks is shifted, to obtain different readings.
If there is no comparator nearby—e.g. in remote areas—and if the requirements are not too high, an inexpensive comparison device can be obtained, using for instance a bowling alley as solid and plane base. Two pieces of paper with a well defined line on it are taped onto the base at a distance corresponding to the tape length. The comparator length is measured along a chalk line with test rulers, or else with a recently calibrated steel tape, or with compass and ruler. The tape length is then compared to this distance.

Table 30. Determination of the Length of the Comparator

test ruler	1st measurement		2nd measurement	
	fore [m]	back [m]	fore [m]	back [m]
A	1.000.022	1.000.034	1.000.039	1.000.040
B	31	56	62	65
A	23	34	39	40
B	32	56	62	65
A	23	35	39	40
B	32	56	62	65
A	23	35	39	40
B	33	56	63	64
A	25	35	39	40
B	33	58	63	64
A	26	36	40	40
B	34	58	64	65
A	27	36	40	40
B	35	58	64	64
A	27	36	39	41
B	36	58	63	64
A	28	37	40	41
B	38	59	63	64
A	28	37	40	42
B	1.000.038	1.000.059	1.000.064	1.000.065
wedge	0.008.880	0.008.505	0.008.335	0.008.280
Σ	20.009.474	20.009.434	20.009.359	20.009.329
mean	20.009.454 m		20.009.344 m	

length of comparator (mean of 1st and 2nd measurements)
K_{20} = 20.009.40 m ± 0.000.03 m

Table 31

front of tape					back of tape				
t	readings at scale		difference u		t	readings at scale		difference u	
	left	right	$r - l$	mean		left	right	$r - l$	mean
[°C]	[mm]	[mm]	[mm]	[mm]	[°C]	[mm]	[mm]	[mm]	[mm]
18.0	80.7	92.0	+11.3	+11.4	18.5	79.1	87.9	+8.8	+8.7
	83.5	95.0	+11.5			86.1	94.8	+8.7	
	84.6	96.1	+11.5			88.4	97.0	+8.6	
	87.0	98.3	+11.3			92.0	100.7	+8.7	

The following equation for the measuring tape can be obtained as result of this comparison:

$$L = L_0 + a + l_\alpha(t - 20\,°\mathrm{C})\,\frac{\mathrm{mm}}{\mathrm{K}} + b(P - P_0)\,\frac{\mathrm{mm}}{\mathrm{N}}. \tag{177}$$

The first term represents the standard length, e.g. 20 m. The second term indicates the amount by which the length of the tape differs from the standard at the reference temperature and -tension. The third term corresponds to the change in length of the tape caused by the difference between the working temperature and the reference temperature according to equation (176). A 20 m tape for instance, expands by 2.3 mm when the temperature increases by 10 K, i.e. a distance of 100 m is measured too short by 11.5 mm. If a measuring error of ± 0.5 mm is to be reached per tape length, the temperature has to be determined to ± 2 K for a 20 m steel tape. Due to differences between air- and tape temperature resp. temperature differences at various locations on the tape, temperature errors of ± 5 K are possible, which cause an uncertaincy in a distance of 100 m of 6 mm. The fourth term corresponds to elastic expansion due to different tension. Based on Hooke's law, the change in length becomes

$$\Delta l_P = \frac{Pl}{PE} \tag{178}$$

where P equals the tension, l the length and F the cross section of the tape, and E the module of elasticity of the material (i.e. $E = 20$ GPa for steel). Therefore a tension of 20,000,000 N per square centimetre is necessary to double the length of a steel body. If the difference between working tension and reference tension is introduced into equation (177) instead of P, the change in length due to a deviation from the reference tension becomes

$$\Delta l_P = \frac{(P - P_0)l}{FE}. \qquad (179)$$

If, for instance a tension of 100 N is used instead of 150 N for a 20 m tape with 20 mm × 0.4 mm cross section, the change in length amounts to

$$\Delta l_P = \frac{(100 - 150) + 20 \text{ Nm}}{20 \times 0.4 \times 20 \text{ mm}^2 \text{ GPa}} = -0.6 \text{ mm}.$$

If a standard deviation of ± 0.5 mm shall be obtained per tape length, the tension has to be within ± 50 N for steel tapes.
Example for the use of equation (177): A steel tape with the equation

$$L = 20 \text{ m} + 0.8 \text{ mm} + 0.23(t - 20\,°\text{C}) \text{ mm/K} + 0.10(P - 15 \text{ N}) \text{ mm/N}$$

has a field length at $t = +28$ °C with an average tension $P = 80$ N of

$$L = 20 \text{ m} + 0.8 \text{ mm} + 1.84 \text{ mm} - 0.70 \text{ mm} = 20.0019 \text{ m}.$$

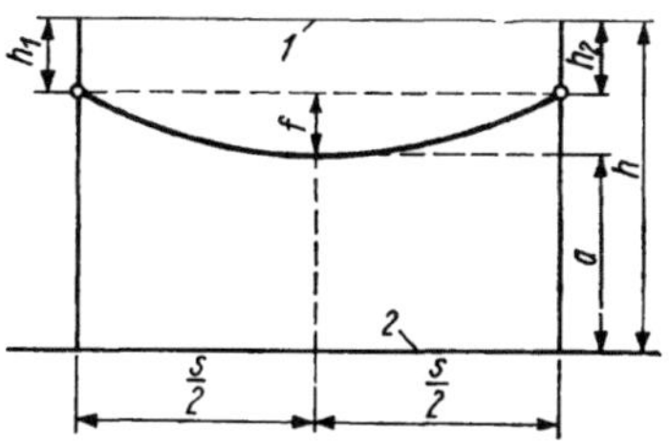

Figure 593 Sag
1 horizon of level; *2* floor

Longer tapes (i.e. 100 m tapes) are standardized like invar wires and -tapes (see section 4.1.4.).
The difference in length Δl_f between a fully supported tape and end support only, can be computed using the sag f or the formula

$$\Delta l_f = \frac{g^2 l^3}{24 P^2 n^3} \qquad (180)$$

where g equals the product of mass of the tape per metre length and gravity acceleration in 10^{-2} N/m, l equals the distance in m, P the tension in 10^{-2} N, and n the number of sections by support of the tape. Δl_f is obtained in metres. The sag f (figure 593) can be measured with a level and scale. Δl_f becomes

$$\Delta l_f = -\frac{8}{3}\frac{f^2}{s} \text{ with } f = h - a - \frac{h_1 + h_2}{2}. \qquad (181)$$

If a 50 m tape has a sag of 1.00 m, the distance is 49.95 m instead of 50 m.
At central agencies, laser interferometers instead of mechanical reference rulers may serve for tape standardization purposes. The He-Ne-Laser in turn is directly related to the primary standard, namely the radiation of a Kr_{86} lamp. The measuring tapes are checked by visual comparison of the graduation lines with the reference marks in microscopes. Their position, respectively separation can be changed, is however, always recorded by interferometric means. An accuracy of 10^{-6} of the length can be obtained with such comparators.

4.1.5. Baseline Measuring Apparatus with Invar Wires or -Tapes

Invar wires or -tapes are used for distance measurements of highest precision, e.g. for measuring base lines for national triangulation nets and sides in local triangulation nets for engineering surveys. The *Jäderin* method is used in this case.
Invar is an alloy consisting of approximately 36% nickel and 64% steel, whose expansion coefficient is extremely small (2 to 5 × 10^{-7}). The wires are usually 24 m long and have a circular cross section with 1.65 mm diameter. They are equipped with 8 cm long millimetre graduations at both ends, which lie in the wire axis and have numbers running in the same direction (figure 594), so that distances of 24 m ± 5 cm can comfortably be measured. The wires are rolled onto drums with about 50 cm diameter for transportation. This has to be done under extreme care in order to avoid sudden changes in the wire length due to bends, tension and shock. For the measuring period, the wires are stored in a tent, and are supported at the ends only.
For the measurements, the wires are pulled with mass pieces of 10 kg ± 1 g at their end supports. Their wire bows are fastened onto a straining string with straining rope via a snap-hook. The string lies on the roll of a pulling device, onto which the mass pieces are connected. As a rule, Finnish (Witram's) tension tripods are used as pulling device. They consist of a approximately 1.6 m high metal frame with a point at its lower end for stepping it

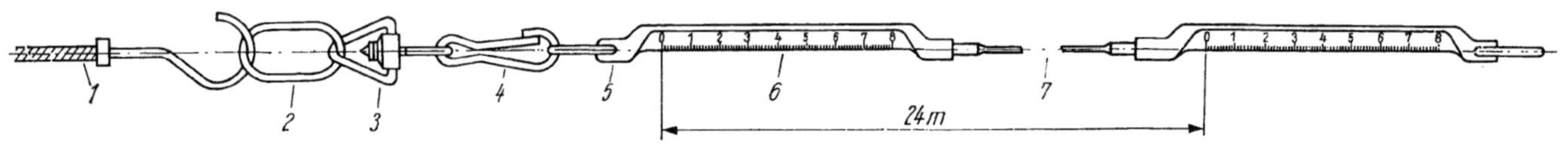

Figure 594 End of measuring wire with suspension
1 pull string; *2* ring; *3* joint; *4* snap hook; *5* bow; *6* scale; *7* measuring wire

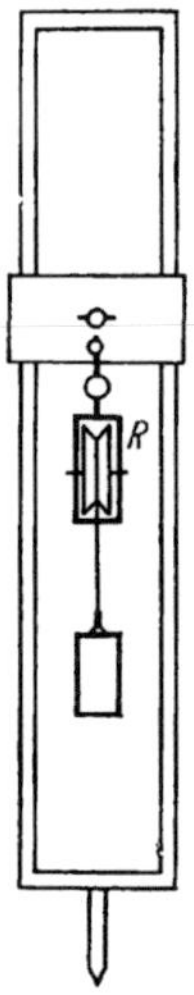

Figure 595 Finnish (Witram's) tension tripod

into the ground and with a movable ring (figure 595). The roll R is hanging on the ring with a gimble joint so that it automatically orients itself into the measuring direction due to the pull.
The base is subdivided into sections of approximately 24 m length. At the end points, measuring tripods with movable tripod heads are set up, or else 3 m long wooden posts with 10 cm diameter are rammed into the ground until they extend 80 cm above the ground. Vertical marks with curved surface, carrying an engraved cross are used as reading index (Jäderin-plugs). They are used either in form of plugs fitting into the measuring tripods, or else equipped with wooden screws (for the posts).
For the measurement, the wires, supported at the ends only, are pulled with the tension tripods and mass pieces, and then placed section by section onto the measuring marks of posts and tripods. On verbal command, two observers read the millimetre graduations simultaneously on both ends and estimate tenths of millimetres. Several readings are taken on each scale (e.g. four times). After each reading, the wire is slightly moved along its longitudinal axis to obtain different readings. Each measurement is carried out with several wires (at least three), to be safe of random changes of the wire length. For each wire length, the air temperature is measured with a whirling thermometer to 1/10 K for subsequent reductions. The speed of a good wire measurement is about 40 min per km. Before and after each job, the measuring wires are calibrated on a check base (in the G.D.R. a 960 m long one near Potsdam) or on an interference comparator.
The elevation differences between the measuring marks, which are required for the reduction of the section length to horizontal distances, are determined by differential levelling. In order to obtain the horizontal distance reduced to the reference surface sufficiently accurate, additional reductions have to be applied, such as for deviation from the straight line, plumbing of the end points, wire extension, and differences of the mass pieces from their theoretical mass. Invar tapes are not as sensitive to bending as invar wires

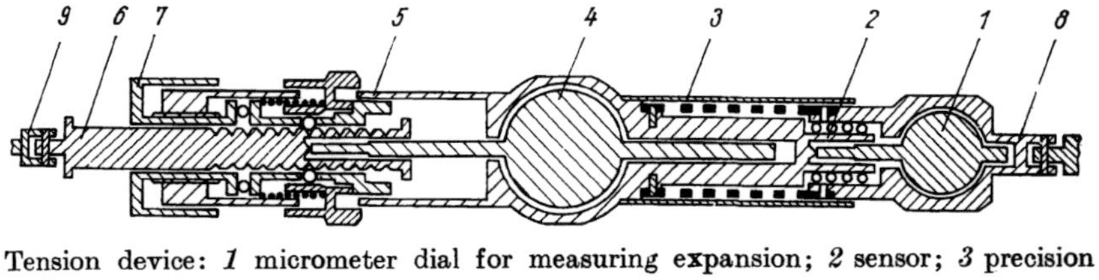

Figure 596 Distometer ISETH (Kern & Co.)

Tension device: *1* micrometer dial for measuring expansion; *2* sensor; *3* precision spring;
Measuring device: *4* micrometer dial for measuring lengths; *5* sensor; *6* rough setting; *7* fine setting; *8*, *9* joint

and can be rolled easier. They usually are also longer, which increases the measuring speed at times by nearly 50%. Invar wires, however, are less influenced by side winds as the invar tapes.
The distometer ISETH (figure 596) by Kern, Aarau, a mechanical precision instrument for measuring distances with the aid of invar wires, is primarily used to determine distance changes during deformation measurements, but also for the monitoring of pillar movements, calibration bases, etc. With this instrument and an invar wire of 1 to 50 m length, changes in distance up to 100 mm can be determined to ± 0.02 mm by wire lengths up to 20 m, and 1×10^{-6} s for longer wires. The equipment consists of the invar wire and the instrument which combines the tension- and measuring devices. The tension device — a precision spring whose expansion represents the pull acting on the wire - keeps the wire under the required tension during measurements. With the aid of a micrometer dial (figure 597), the measuring device provides the separation of distometer and the end of the invar wire fastened to it. The other end is fastened with a gimble joint to the end point of the section. The invar wire is cut to the required length for the sections to be measured and has a connection piece attached to both ends. A calibration ruler serves to check, calibrate and adjust the micrometer dials.

Figure 597 Distometer, reading: 74.57 mm

4.2. Instruments and Devices for Optical Distance Measurements

In optical distance measurements, the distance is determined geometrically within the parallactic triangle (figure 598) with base b and the parallactic angle γ as

$$s = \frac{b}{2} \cot \frac{\gamma}{2}. \tag{182}$$

The permissible errors of b and γ can be estimated for the most unfavourable case for a desired distance accuracy by differentiating equation (182). With insignificant neglections, this gives

$$\frac{ds}{s} = \frac{db}{b} + \frac{d\gamma}{\gamma}. \tag{183}$$

If, for instance, a distance of 100 m is to be determined with an accuracy of 20 cm, i.e. a relative error of 1 : 500, each term on the right hand side of

Figure 598 Parallactic triangle

equation (183) has to have an accuracy of 1 : 1000, assuming equal contribution to the error. This leads to $db \leqq 1$ mm and $d\gamma \leqq 0.64$ mgon for $b = 1$ m. If, however, the same distance is to be determined to within 2 cm, i.e. a precision of 1 : 5000, then we require for a 2 m base $db = 0.2$ mm and $d\gamma = 0.13$ mgon.
To satisfy these conditions and to keep the measuring effort within tolerable limits, the base b, and particularly the parallactic angle have to be determined very accurately. Slope distances have to be reduced to horizontal either by computation or else automatically with reduction devices.

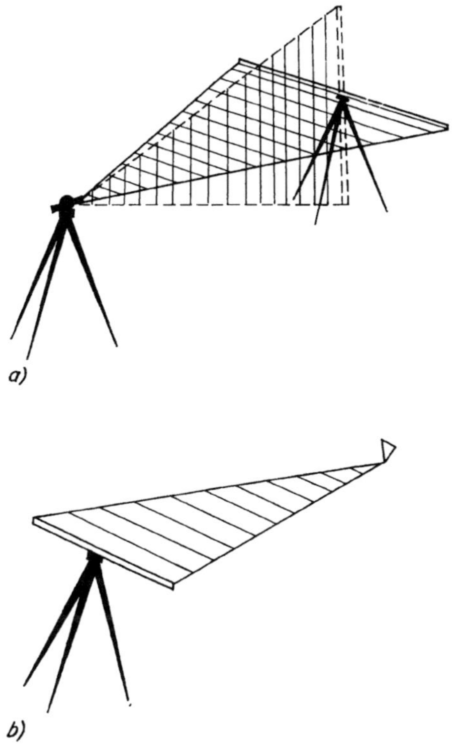

Figure 599 Position of base for optical distance meters

Depending on the position of the base, the optical distance meters are categorized as:

1. Distance meters with base at the target (the base ends are sighted from the vertex of the parallactic angle) (figure 599a).
2. Distance meters with base at the station (the vertex of the parallactic angle is sighted from the ends of the base) (figure 599b).

Depending on the measuring principle, optical distance meters are categorized as (figure 600):

1. Distance meters with fixed (constant) base and variable parallactic angle. The distance error increases proportionally to the square of the distance, which is rather unfavourable for longer distances.
2. Distance meters with variable base and fixed parallactic angle. The distance error increases proportionally with the distance.
3. Distance meters with variable base and variable parallactic angle. In this case the parallactic angle is monitored such, that it becomes a function of the vertical angle (see section 5.2.).

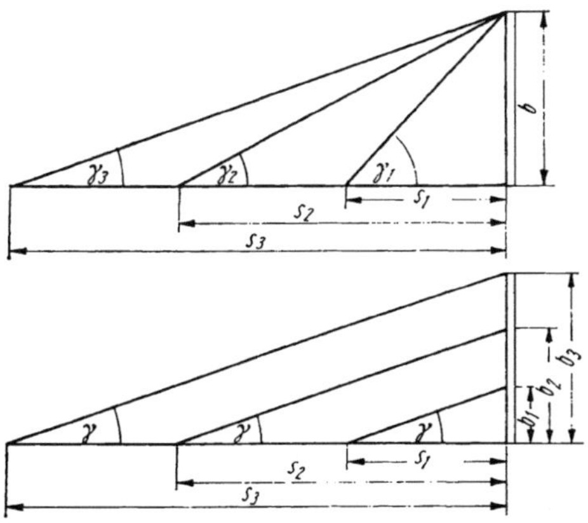

Figure 600 Base and parallactic angle for optical distance meters

The sighting to or from the base ends can be done sequentially, i.e., in two telescope positions, or simultaneously, i.e. with double image or stereoscopically.
The base can be arranged horizontally or vertically, or even tilted (but normal to the sighting direction). Measurements with horizontal base are more accurate than with vertical one. When sighting to a vertical base especially near the ground, the sighting rays pass through layers of air with different density and thus different refraction characteristics. The measuring result is therefore noticably influenced by refraction near the ground. Measurements to horizontal bases however, are virtually free of refraction influence. But the horizontal rod has the disadvantage, that — especially in areas with varied vegetation — it is rather bulky, and that it always needs to be set up exactly horizontal and normal to the sighting direction. With a vertical rod it is only necessary to have it plumb. This is much easier to accomplish and thus permits a faster working speed. Therefore, the horizontal base is used for jobs with high accuracy requirements, while the vertical base is used for stadia- and topographic surveys.
The different possibilities when combining the given characteristics, has naturally led to a great number of designs of optical distance meters.

4.2.1. Distance Meters with Base at the Target

Stadia Instruments

The reticules of nearly all telescopes of instruments of low, medium and high accuracy carry two stadia lines placed at a distance $p/2$ parallel to the centre cross hair. These generate a constant parallactic angle (figure 601), which is the basis of all stadia instruments.
For the measurements, a rod (levelling rod) is held vertically at the target Z, while the instrument with stadia hairs is set up at the station S. Using the parallactic triangle, the desired distance s can be determined from rod section l, read on both stadia hairs, and the given separation p of these hairs. From

$$\frac{s - (a + f)}{f} = \frac{l}{p}$$

one obtains

$$s = a + f + \frac{p}{f} l. \tag{184}$$

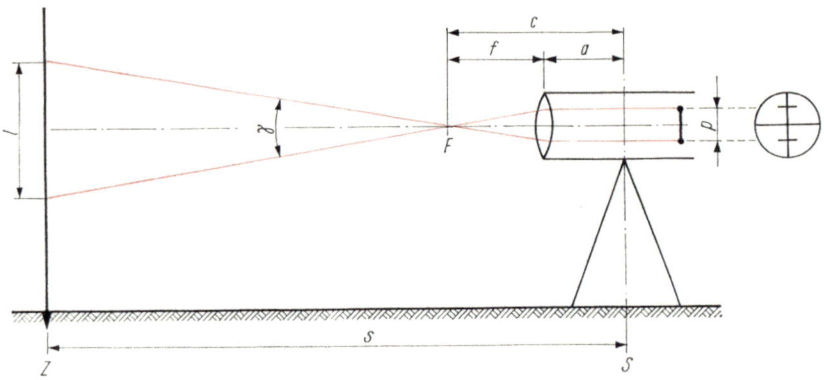

Figure 601 Determination of a distance s using a telescope with stadia hairs

The previously common telescopes with external focussing have a constant focal length f. The separation $a + f = c$ of the vertex of the angle from the vertical axis of the instrument is denoted as addition constant, while the fraction $f/p = k$ is referred to as multiplication constant. For horizontal sights, the horizontal distance is then

$$s = c + kl. \tag{185}$$

The addition constant amounts to 0 to 0.5 m for these telescopes. By chosing $p = 0.01\, f$, the multiplication constant is designed to become 100.
For $k = 100$, the parallactic angle becomes 636.6 mgon (34′23″), because of $f : p/2 = (\cot \gamma/2)$. Its vertex is fixed and is referred to as anallactic point (anallactic means not moving).
Telescopes with internal focussing have an objective lens with two components. By moving the focussing lens, the equivalent focal length of the system changes (and therefore also the multiplication constant k because of $k = f/p$) and also the position of the focal point in object space (and therefore the addition constant c). In this manner the strictly linear correspondence between s and γ according to equation (185) is lost. The anallactic point moves, and c and k become functions of the distance s. *Roelofs* proved that the functional relationship is satisfied by a hyperbola. If one imagines that the lower stadia hair is imaged throught the telelens together with the vertical line in the object space, i.e. on the rod at a distance s, and if the focussing lens is shifted, the image point describes a curve, which is the line of sight. It is the branch of a hyperbola, which can be approximated by a straight line (figure 602).

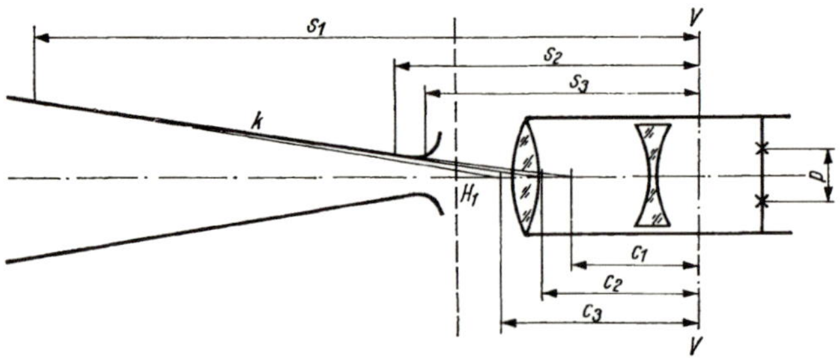

Figure 602 Hyperbola $1/p = f(s)$ for distance measurement for a telescope with interior focussing for the assumption of constant $k = f'_\infty / p$

Depending on the measuring method, a chord, tangent or asymptote to the hyperbola is used to obtain its tilt $\varkappa$ (which determines the multiplication constant $k = l/\varkappa p$) and its separation from the axis, which provides the addition constant. Straight lines parallel to the asymptote of the hyperbola branch are used if k is kept constant at $k = f_\infty/p$.
The image, created in this manner (figure 602) can be considered such, that the value c is reduced for shorter sights and constant k, and that the anallactic point moves. Therefore, the anallactic point is better denoted as "quasi anallactic point" for telescopes with internal focussing.
By proper selections of the focal lengths of the front element of the objective lens and of the focussing lens, the designer of the telescope can achieve that the value c remains within limits which are suitable for geodetic measurements. Furthermore, the quasi anallactic point is placed into the vertical axis for a rather large sighting range, e.g. its separation from the vertical axis is kept smaller than 0.001 s in order to permit computations with $c = 0$. These instruments are therefore called anallactic instruments. The designer sets the value k to a certain round figure, usually 100, when tuning the equivalent focal length f' to the amount kp by changing the telescope length and the lens separation.
The distance s for anallactic instruments is then

$$s = kl. \tag{186}$$

The following errors in the distance s, due to differences of c from 0 and of k from 100 (errors of 0,1% are quite possible) are obtained for a Wild instrument with the objective focal length $f_2 = -101.00$ mm and a telescope length of approximately 156 mm:

s	5	10	20	50	100	∞	[m]
Δs	+16	+7	+4	+1	+1	0	[mm]

For a Zeiss (Oberkochen) Ni 2, the following corrections s were obtained:

s	3.3	4	5	6	8	10	12	14	16	>16	[m]
Δs	+83	+70	+54	+43	+28	+20	+13	+9	+4	0	[mm]

It is therefore advisable to consider a correction Δs, which is a function of the distance, for sightings of less than 15 m, according to

$$s = 100\,l + \Delta s. \tag{187}$$

Often this is provided by the manufacturer in tabular form.

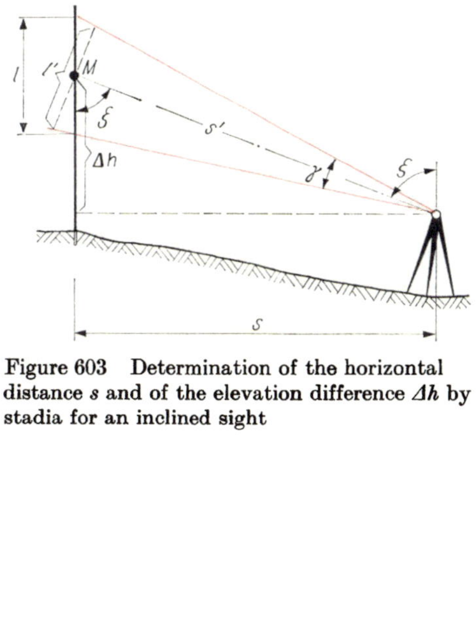

Figure 603 Determination of the horizontal distance s and of the elevation difference Δh by stadia for an inclined sight

Utilization of Stadia for Inclined Sights. If the collimation axis of a stadia instrument when sighting to a vertical rod has a zenith distance of $\zeta \neq 100$ gon, the rod section l is not normal to the collimation axis, but forms the angle ζ with it (figure 603). With the component l' normal to the collimation axis, the slope distance s' becomes for anallactic instruments, according to equation (186):

$$s' = kl'. \tag{188}$$

For $\zeta = 60$ gon and $l = 100$ cm, one can use $l' = l \sin \zeta$, with the neglected value being less than 0.01 mm. With this, equation (188) becomes

$$s' = kl \sin \zeta \tag{189}$$

with $s = s' \sin \zeta$, the horizontal distance becomes

$$s = kl \sin^2 \zeta. \tag{190}$$

The elevation difference Δh between the horizontal axis of the instrument and the point M on the rod, determined by the centre of the cross hair, is obtained as

$$\Delta h = s \cot \zeta. \tag{191}$$

It can also be computed directly from the rod section l using $\Delta h = s' \cos \zeta$ and equation (189) as

$$\Delta h = kl \sin \zeta \cos \zeta = \frac{1}{2} kl \sin 2\zeta. \tag{192}$$

If vertical angles α are read, equations (190) and (192) become

$$s = kl \cos^2 \alpha \tag{190a}$$

$$\Delta h = \frac{1}{2} kl \sin 2\alpha. \tag{192a}$$

Equations (190) and (192) are transformed with equation (187) and some simplifications into

$$s = (kl + \Delta s) \sin^2 \zeta \tag{190b}$$

$$\Delta h = \frac{1}{2}(kl + \Delta s) \sin 2\alpha . \tag{192b}$$

These simplifications generate an error in the distance s of 29 mm for 100 m with $\Delta s = 0.30$ m and $\zeta = 60$ gon, and of 15 mm in the elevation difference Δh.
For the determination of s and Δh with equations (190) and (192) resp. (190a) and (192a), stadia tables, stadia slide rules or nomograms can be constructed and used.

Determination of the Stadia Constants. The stadia constants can be determined with the aid of test distances or -angles. The multiplication constant k of a telescope with internal focussing can be determined with the aid of the parallactic angle γ, whose sides are the collimation axes defined by the stadia hairs when sighting to a distant target at the horizon. If $k = 100$ is to be determined to ± 0.1, the angle γ has to be measured to ± 0.6 mgon. If the angular measuring accuracy of the instrument is sufficient, one sights to the target several times with one stadia hair, then with the other, and reads the vertical circle each time. Since the target is practically at infinity, the separation between actual anallactic point and the vertex in the vertical axis of the instrument can be neglected. With the mean value for the angle γ, k is obtained as

$$k = \frac{1}{2} \cot \frac{\gamma}{2} . \tag{193}$$

Since angular measurements of such high an accuracy are not possible with stadia instruments, the telescope has to be placed such, that the vertical line is horizontal. With a high precision theodolite the angle γ can then be directly measured by focussing to infinity, provided that its vertical axis is plumb and works properly.
k and c can be determined together by comparing optically measured distances with mechanically measured ones, e.g. with calibrated tapes. Due to the change of the constants for telescopes with internal focussing, this method is not theoretically correct, provides however, suitable results if distances of less than 10 to 15 m are not used.
Since for short distances the influence of k is minimal, and for long ones the influence of c, one choses two distances, e.g. 15 and 100 m. According to equation (185) the equations $s_1 = c + kl_1$ and $s_2 = c + kl_2$ can be used for this and give

$$k = \frac{s_2 - s_1}{l_2 - l_1} \tag{194}$$

$$c = s_1 - \frac{s_2 - s_1}{l_2 - l_1} l_1 . \tag{195}$$

The standard deviations for k and c are computed from the standard deviations m_i for l_i as

$$m_k = \pm \frac{k}{(l_2 - l_1)} \sqrt{(m_1^2 + m_2^2)} \tag{196}$$

$$m_c = \pm \frac{k}{(l_2 - l_1)} \sqrt{(l_2^2 m_1^2 + l_1^2 m_2^2)} . \tag{197}$$

Example: The distances $s_1 = 15.010$ m and $s_2 = 100.10$ m were measured mechanically. The mean values of the rod sections as read with the stadia instrument were $l_1 = 0.1502$ m $\pm$ 0.210 mm; $l_2 = 1.002$ m $\pm$ 1.14 mm. According to equations (194) and (197):

$k = 99.89 \pm 0.14$; $c = +1$ cm $\pm$ 3 cm.

The constants are obtained more accurately, if more than two distances and a least squares adjustment are used. On a horizontal test line, starting at about 10 m from the instrument station, points are monumented every 20 m up to the maximum required range e.g. 120 m. The distances are measured with a calibrated tape (with application of temperature correction) and also several times (e.g. five times with a slight tilt of the telescope for each) optically. The mean of these measurements is computed for each

Table 32. Determination of the Constants k and c of the Theodolite THEO 020 A (VEB Carl Zeiss, JENA) No. 600479

s_i [m]	l_i [m]	$100\, l_{i\,(\text{mean})}$ [m]	Δs_i [cm]
1	2	3	4
9.99	0.100 0.099 0.099	9.93	+6
20.08	0.201 0.201 0.201	20.10	−2
40.40	0.403 0.403 0.404	40.33	+7
60.10	0.601 0.600 0.600	60.03	+7
79.90	0.798 0.799 0.799	79.87	+3
100.11	1.001 1.001 1.001	100.10	+1
120.40	1.204 1.203 1.205	120.40	±0

optically measured distance. By graphical adjustment of these values k and c are obtained as follows:
From $k \approx 100$ we get $k = 100 + \Delta k$, and with equation (185) $s = 100l + \Delta kl + c$, or after rearrangement $s - 100l = \Delta kl + c$. This equation can be considered as the equation of a straight line $y = mx + b$, where $m = \tan\alpha$ is the directional factor, and $y = b$ for $x = 0$. If one plots (like in figure 604) the values $100l_i$ and $\Delta s_i = s_i - 100l_i$ in a rectangular coordinate system, and places an adjusting straight line by visual estimation through these points, then the section on the Δs axis represents an approximation to c, and the directional factor the value k.

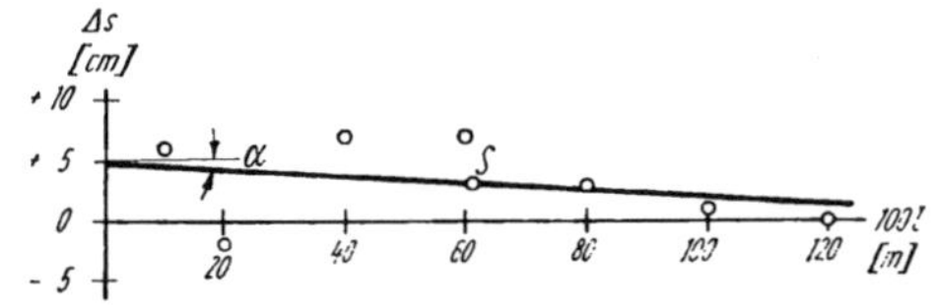

Figure 604 Determination of c and k by graphical adjustment

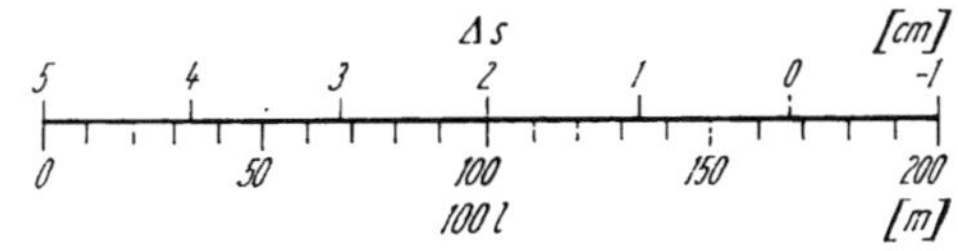

Figure 605 Diagram for determining the values Δs based on the distances s

Example: Determine the constants k and c of the THEO 020 A Nr. 600479 (Table 32).
According to figure 604:
$c = +0.05$ m, $\Delta k = -0.03$, $k = 100 + \Delta k = 99.97$.

Since the accuracy of k and c depends primarily on the accuracy of the determination of the rod section l the rods have to be set up exactly vertical (possibly with braces), and millimetres have to be read for short sights with a mm ruler. Multiple measurements of the rod sections also increase the accuracy. One can also use the following procedure: The points are staked exactly at 10 m, 20 m, 40 m, 60 m etc. from the instrument station. The rod is placed on six to ten points placed symmetrically at a few decimetres from the central point. For each rod section $\Delta s_i = s_i - 100l_i$ is computed. The values Δs_i for each of the monumented central points are obtained as the mean of the individual differences. For example, the following values were obtained for the group of points for $s = 20$ m:

s_i	19.52	19.66	19.84	20.16	20.34	20.48	[m]
l_i	0.194	0.196	0.198	0.201	0.203	0.205	[m]
Δs_i	+0.12	+0.06	+0.04	+0.06	+0.04	−0.02	[m]

which lead to a mean value of $\Delta s_{20} = +0.05 \pm 1.4$ cm. The standard deviation m_i of the mean values Δs_i shall not exceed ± 0.05 to 0.10% of the distance.
The adjusted values Δs_i can be obtained from figure 604. The Δs value for any distance can be obtained from a graph derived from it (figure 605).
k can be determined faster and more accurate with a collimator (see page 53) with 0.1 mm scale. The separation p of the stadia hairs is directly measured for ∞ setting. With the focal length f_k of the collimator, k becomes $f_k/p_k = f/p = k$. The value has an error of about $\pm 0.05\%$.

Accuracy of Stadia Instruments. The accuracy of distances measured by stadia depends primarily on the sighting accuracy, and also on the quality of the telescope and the accuracy of the rod reading. Furthermore, it is influenced by ground refraction.
Instruments with 20 to 30 × magnification, used for stadia measurements, permit millimetre estimation on centimetre rods up to about 100 m distance, and centimetre readings up to about 300 m. Because of $k \approx 100$, generally 3 m rods are used, whose graduations permit the determination of the rod section to about 0.1%. The stadia range is about 300 m for horizontal sights, and 150 m for sights inclined by 50 gon. For longer distances, the upper (resp. lower) stadia hair and the centre cross hair can be used to read the rod section. The distance is then obtained by multiplying the rod section by $2k$ instead of by k. Based on experience, the setting error on the rod is smaller than the reading error. Therefore, and also for easier calculation of the rod section, one stadia hair is usually set onto an even graduation value, and the rod is read on the second one.
If the rod is not plumb, a distance error is introduced, which is quite significant for inclined sights and which increases with the tilt angle. If the upper end of a rod deviates by only 5 cm from the vertical, the horizontal distance for 100 m and 50 gon has an error of 1.7 m. To avoid errors because of inclined rods, it is advisable to set the rod plumb with the aid of a level with well adjusted bull's eye bubble, and to brace it with two range poles.
The influence of ground refraction has to be considered as unknown external error influence, which cannot be corrected numerically. It can be up to five times as large as the reading error. Since the refraction has a maximum immediately above the ground, it is advisable not to use the lower part of the rod up to 1 m or 1.5 m, especially during sunny periods. Heat waves, indicating turbulent air movements caused by heating of the ground, lead to irregular reading errors.

The constants can be kept in all cases within $\pm 0.10\%$. The distance errors caused by this are in the same magnitude as inavoidable observation errors. In general, one can assume an uncertaincy of 0.2% for distances of 20 to 130 m with moderate inclination, and of 0.4% for up to 300 m. It may reach 1% of the distance under unfavourable conditions.

Double Image Tacheometer (*Wedge Attachment*)

If an achromatic glass wedge is arranged in front of the objective lens of the telescope such that one part is covered, the bundle of ray coming from the image plane is split in two: One will pass through the uncovered part of the objective lens and continue straight, while the other is deflected by the angle γ from its direction (figure 606).

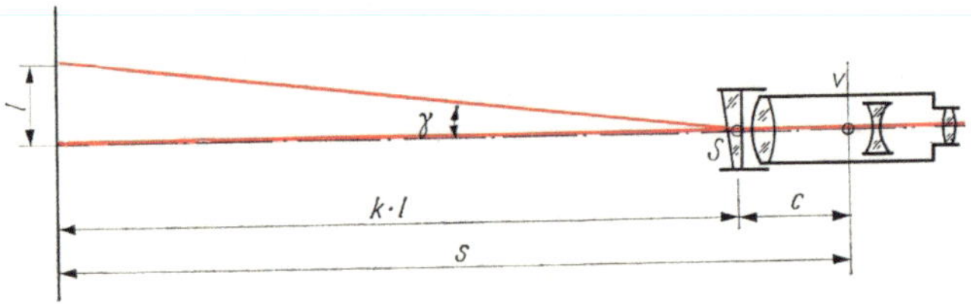

Figure 606 Optical train for the double image distance meter with wedge attachment

When sighting to a rod at the target, two images, shifted by the rod section l, are generated in the field of view of the telescope (double image). The index contained in one image directly shows the rod section l in the other one. This is a significant advantage over stadia, where the reading resp. setting of the rod with the stadia hairs cannot be accomplished simultaneously.
The parallactic angle γ is constant, while the rod section l—the base—changes depending on the sighting distance. Knowing the value of γ, the distance s can be determined from the rod section.
The vertex S of the parallactic angle γ, which is the anallactic point of the double image tacheometer, is usually located in the wedge. Therefore an addition constant c has to be considered when determining s.
From figure 606 it follows that

$$s = c + l \cot \gamma \tag{198}$$

or with k for the constant value $\cot \gamma$, like for stadia

$$s = c + kl. \tag{185}$$

By proper selection of refractive index n and of the prism angle α one can achieve that k becomes a round figure, because of $k = \cot \gamma$, and $\gamma = \alpha(n-1)$ according to equation (17). For $k = 100$, $\gamma = 636.62$ mgon (34′22.6″). With $n \approx 1.5$, α becomes 1.27 gon.
In order to obtain a distance accuracy of 1 : 5000, i.e. an error $m_s = \pm 2$ cm for $s = 100$ m, the error m_γ of the deflection angle γ cannot exceed

$$m_\gamma = m_s \frac{\gamma}{s} \approx \pm 0.13 \text{ mgon} (0.4''). $$

Since such an accuracy cannot easily be reached during manufacture, the wedge is rotated in its casing around its refractive edge until the theoretical reading is obtained. One can also rotate the two components of the achromatic wedge, made of flint- and crown glass, against each other until the theoretical reading is obtained, or else correct for the residual error of the wedge by turning a slightly wedge shaped cover glass. The troublesome addition constant $c = c' + c''$ (c' is the separation of the anallactic point S

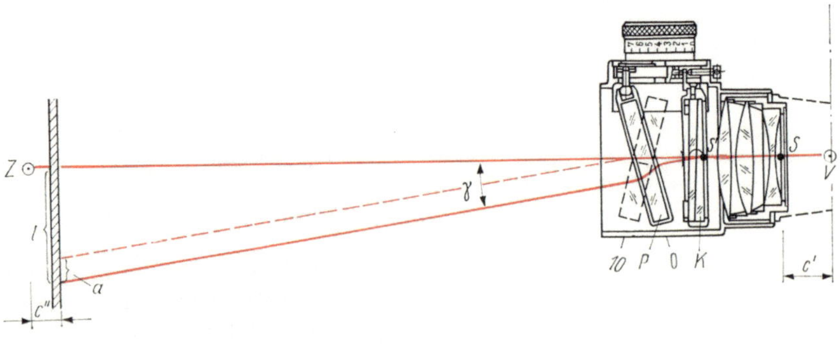

Figure 607 Optical train for the double image distance meter with wedge attachment and plane-parallel plate (Wild DM 1)

from the vertical axis V, c'' the one of the rod graduation from the target Z, figure 607) is eliminated by shifting the rod index by c/k in direction of the graduation. The distance s is then

$$s = kl. \tag{186}$$

To eliminate the influence of ground refraction, a horizontal rod is used. Therefore, the normal section of the wedge has to be also horizontal. The wedge is furthermore cut such, that it only covers the central strip of the objective lens (figures 609, 615). If it were only placed in front of the lower half of the objective lens, then the rays through it pass through the upper half of the exit pupil and into the pupil of the eye which is placed in this plane for observations. The rays passing through the free upper half of the objective lens, pass through the lower half of the exit pupil and the eye pupil. Since the eye pupil is rarely curved simultaneously, a small shift of the two images occurs on the retina. In this manner, personal errors occur, which can amount to up to 20 cm for 100 m distance. These can be largely eliminated, if either the two semi circles of the exit pupil are superimposed by special arrangements (see figures 736 to 738) or else when one pupil symmetrically surrounds the other one. The latter is the case if the distance wedge only covers the central strip of the objective lens.

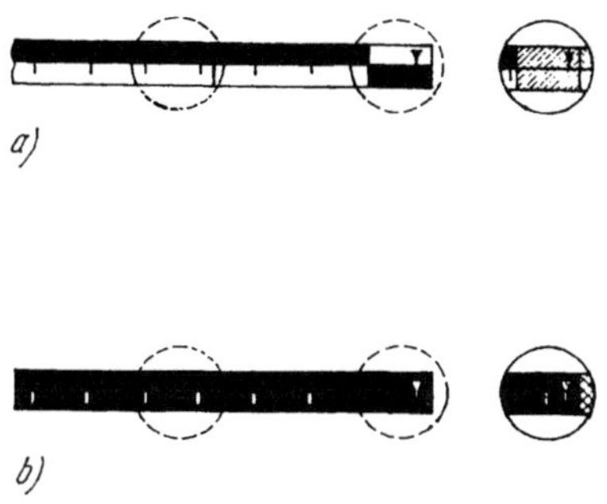

Figure 608 Painting of rod for reading of mixed images

Since both images are superimposed, a mixed image is obtained which makes it nearly impossible to read the rod section. This is improved by special painting of the rod (figure 608). The rod is divided into halves along its longitudinal axis. One half contains the main graduation, the other one contains the index. The areas opposite to graduation and index are black. If black graduation lines are painted onto white surfaces, a grey mixed image is generated where the deeply black graduation lines stand out. If white graduation lines are painted onto black surfaces, gray graduation lines appear on black background on the double image. It is impossible to measure short distances with this rod, because the area opposite the index has to be black. They are determined either directly with a measuring tape or else with the aid of an auxiliary rod fastened to the main one.

Because of the uneven value of s, the rod index usually does not coincide with a line of the rod graduation. Therefore a vernier is used. To further improve the reading accuracy, the remaining separations between vernier- and rod graduations for a given s, are measured with a parallel plate micrometer P, arranged in front of the wedge K with the same position and dimension (figure 607). By rotating the parallel plate, a line of the rod graduation can be coincided with a vernier line, and the amount of the shift is read on the measuring drum of the micrometer. The parallel plate shifts the vertex S of the parallactic angle, the anallactic point, from inside the wedge back to S'. This amount changes the addition constant, which is considered by shifting of the rod index.

Since the parameters of double image tacheometers with constant parallactic angle are independent of the telescope parameters, wedge attachments are produced to be placed onto the objective end of the telescope of a theodolite or alidade. The range is determined by the graduation length of the rod and the separation of the zero lines of vernier and graduation. Since the remaining rod section, which is to be measured with the micrometer, shall not exceed 5 mm because of the thickness of the parallel plate and its tilt angle, the vernier is matched to this value. The larger the least count on the rod, the wider its lines, so that they can be better seen over long distances. For a large rod interval however, the vernier becomes very long. To obtain ideal parameters, the range is chosen as to be within 200 m.

The telescope with attached wedge is brought to equilibrium by a counter mass, clamped to the eyepiece end of the telescope. For the distance measurement, double image and vernier are sharply set near the centre of the field of view. One vernier line is coincided with one line of the rod graduation using the micrometer screw. Metre and decimetre readings are obtained on the rod, centimetres on the micrometer drum. Double coincidence is used to increase the reading accuracy. The distances obtained represent the slope distances s' between instrument- and rod stations. The horizontal distance s and the elevation difference Δh are obtained with the zenith distance ζ to

$$s = s' \sin \zeta \tag{199}$$

$$\Delta h = s' \cos \zeta. \tag{200}$$

To measure the angles with attached wedge, usually a metal strip can be clapped in front of the wedge. The measuring accuracy of attached wedges with micrometer is about 1 : 5000 for 100 m slope distance.

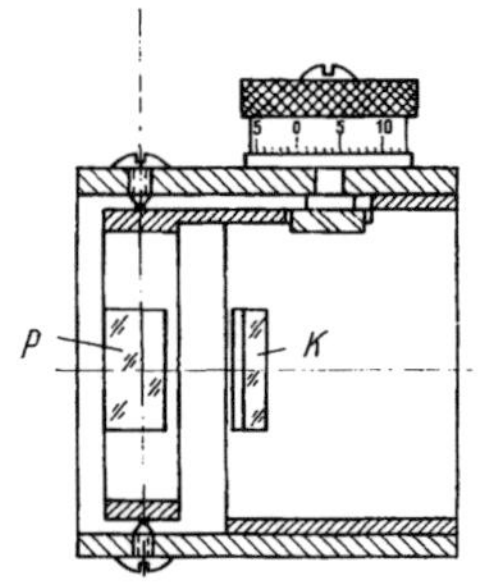

Figure 609 Section through Dimess Wedge (Zeiss, Jena)
P parallel plate; *K* glass wedge

Types of Attached Wedge Instruments. The Dimesskeil 002 by Zeiss, Jena is a wedge attachment with parallel plate micrometer (figure 609). The rod equipment for it (figure 610) resembles the one for the Redta 002, except for the rods. The wooden rods have a black 2 cm graduation on a white thin steel sheet. Distances up to 21 m have to be measured with an auxiliary rod, screwed onto the main one. The distances between vertical axis and anallactic point (9 cm), and rod plane and tripod pipe (4 cm) are considered in the vernier shift. The rod section can be measured with the vernier to an accuracy of 2 mm. The sighting ray can be shifted with the aid of a parallel plate micrometer (drum with 20 intervals) up to 2 mm, so that the remaining interval can be read to the centimetre of the distance. Distances between 21 and 122 m are determined with the interior vernier (figure 611), distances between 72 and 172 m with the exterior vernier (+5, i.e. +50 m). Distances between 72 and 122 m can be measured with both verniers. For distances beyond 172 m, the Dimess rod can be used as a subtense bar. The wedge type endmarks indicate a 2 m base (rod graduation 4.4 cm in front of the rod station! temperature correction!).
The white square, standing on its edge, serves as target for vertical angle measurements.

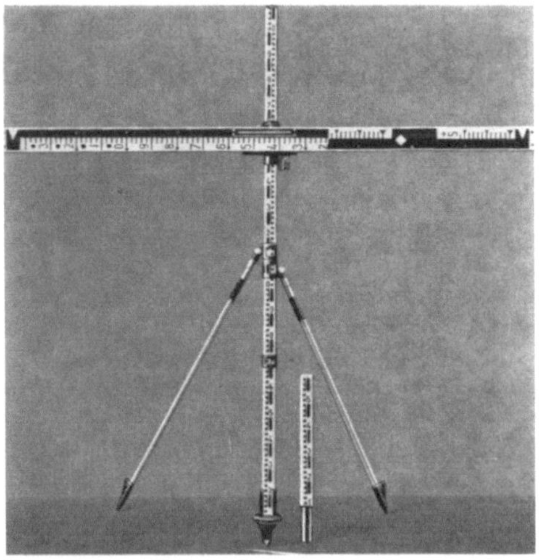

Figure 610 Dimess rod on stand (Zeiss, Jena): length 2.09 m

Figure 612 Rod image for reading with distance wedge DM-M
Reading at rod: 117.40 m; at drum: 0.12 m; together: 117.52 m

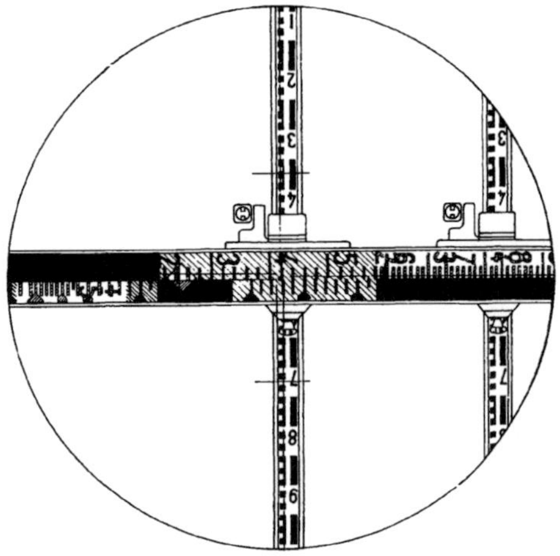

Figure 611 Field of view with Dimess wedge 002. Reading at main graduation: 36 m; at the vernier: +0.80 m; together: 36.80 m + centimetres at drum

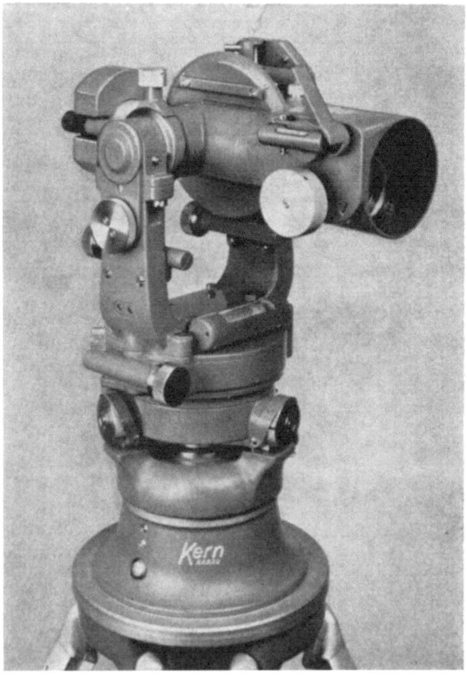

Figure 613 DKM 2 with reducing distance wedge DR (Kern)

The non reducing distance wedge DM-M with optical micrometer by Kern, Aarau, can be used with Kern theodolites of medium accuracy. The 1.7 m long light metal rod contains a 2 cm graduation on a steel tape, which is white on a black base. It also contains two 19 cm verniers with 10 intervals, with which the distance can be read to 20 cm. Distances are read between 20 and 50 m on the interier vernier, between 50 and 100 m on the interior or exterior vernier (+30), and between 100 and 150 m on the exterior vernier (figure 612). A 25 cm auxiliary rod can be attached to the main rod at the tripod pipe for distances below 20 m. The reducing distance wedge DR (figure 613) by Kern contains a pair of rotating wedges. It is monitored by the observer with the aid of a spirit level. If the level bubble is centered manually by turning a knob, the wedges are rotated via a cog wheel transmission, corresponding to the inclination of the telescope. The deflection of the sighting ray is reduced proportionally to the cosine of the vertical angle, so that the rod reading provides directly the horizontal distance (see page 293). The 1.50 m long rod has a 1 cm graduation, and two verniers with 20 intervals, separated by 30 cm, which enables a reading of 1/20 of the least count of the rod, i.e. 5 cm in distance (figure 614). The measuring range is 20 to 130 m.

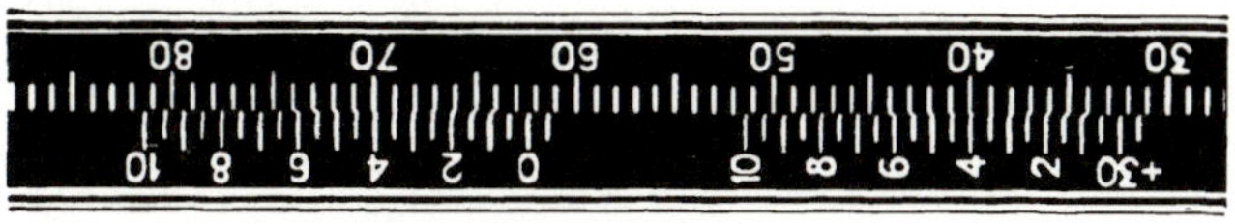

Figure 614 Rod image for reading with distance wedge DR. Reading: 62.35 m

The distance wedge Wild DM 1 (figure 615) with micromete rhas a 1.6 m long wooden rod with a 1 cm graduation, starting at 10 cm, and with two 10 part verniers on white background (figure 616). The interior vernier is to be used for distances up to 100 m, the exterior one for 60 to 150 m. The rod equipment consists of a vertical tripod pipe with centimetre graduation to set the rod to instrument height, two braces with movable clamps, of the rod carrier with cross type guide rail and lock of a vertically movable diopter and of a target for horizontal angle measurements (figure 617).

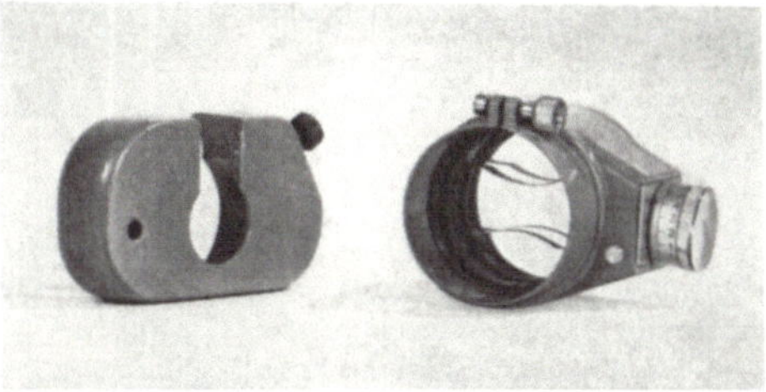

Figure 615 Distance wedge Wild DM 1 with counter weight

Rank Precision Industries, Leicester, manufacture wedge attachments with micrometer and corresponding 1.6 m long rods of mahogoni wood, whose 10 part vernier is placed on the first 20 cm of the rod. MOM. Budapest produce wedge attachments and 1.7 m long distance rods with two 10 part verniers for distance measurements from 20 to 140 m.

Figure 616 Rod image for reading with Wild DM 1. Reading: 58.44 m

In the distance meters built in U.S.S.R., the attached wedges also have the function of an optical micrometer. One half of the objective lens is covered by a system of two glass wedges, while the other is covered by two plan-parallel plates, serving as protection glass. The wedges are arranged such, that the system produces the magnification $\Gamma_k = 90/89$ and the parallactic angle 0.6366 gon ($k = 100$). The magnification of an optical system consisting of two specifically arranged glass wedges is based on the fact that a parallel bundle of rays hitting it, will exit again as parallel bundle, however, the dimension of the bundle has changed in the principal section of the system: $\Gamma_k = \xi_1/\xi_4$ (figure 618).

If a measuring rod is read with such a distance meter, attached to the telescope, then according to equation (185) $s = c + k$. The addition constant c is considered by shifting the zero line of the vernier on the rod. If the telescope is tilted by $\Delta\gamma$, the rod section l is shifted by

$$\Delta l = \Gamma_k \Delta_1 - \Delta_2 = \Delta(\Gamma_k - 1) \tag{201}$$

because $\Delta_1 \approx \Delta_2$ (figure 619). An 89 cm long rod with 1 cm graduation and a 10 part vernier is used with this distance meter. Because of $\Gamma_k = 90/89$ the vernier is not only imaged by the wedge on the main graduation, but also enlarged by 90/89, i.e. to 89 mm × 90/89 = 90 mm (9 rod intervals). The same rod image as with the usual wedge attachments is seen in the field of view. If the telescope is tilted so far, that its cross hair covers the whole vernier, the image of the vernier is shifted by 1 mm according to equation (201). The rod section is obtained from the rod reading at the zero line of the vernier, the coinciding vernier line, the vernier line located in front of the horizontal line of the cross hair, and by estimation of the separation between these two lines (figure 620). One can also use a horizontal rod instead of a vertical one, but will obtain the slope distance.

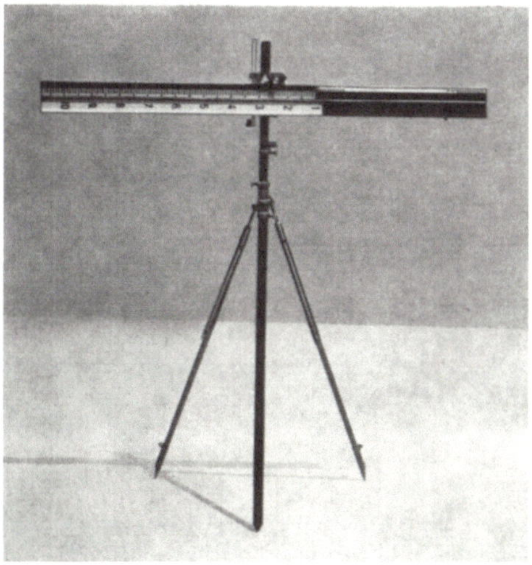

Figure 617 Rod stand with horizontal rod for Wild DM 1

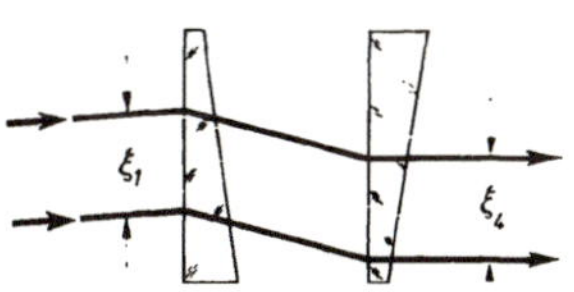

Figure 618 Bundle of rays for the differential distance meter

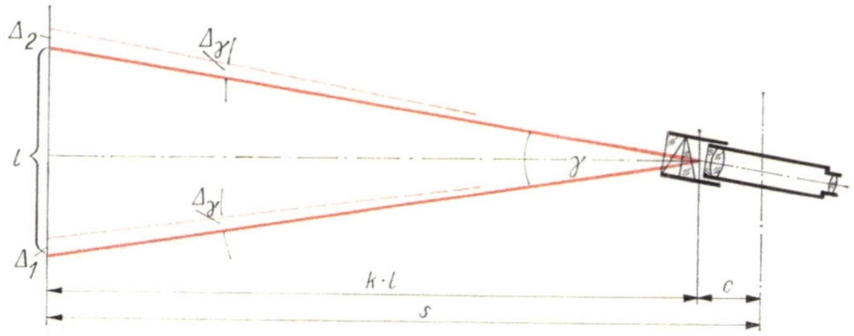

Figure 619 Operation principle of the differential distance meter

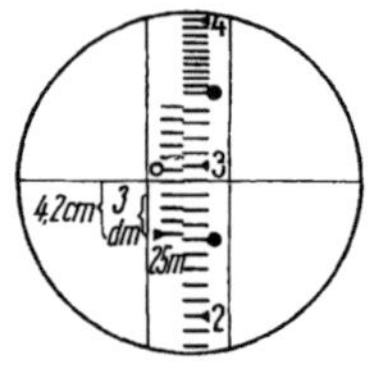

Figure 620 Reading image for the DN-04
Reading: 25.342 m

For the DN-04, 1.54 m long rods graduated on both sides with 1 resp. 2 cm graduation, similar to the Dimess rod are produced (measuring range: 10 to 125 m). The areas opposite graduation and vernier are black. The measuring accuracy is 1 : 2500. The differential distance meter DN-10 contains two half lenses instead of the glass wedges, which generate the constant parallactic angle (similar to figure 632). The measuring accuracy is 1 : 1200 for a range of 20 to 200 m.
The DNR-06 produced in U.S.S.R. is a self reducing wedge attachment. The casing contains the wedge system in the left half. One glass wedge is fixed with its refracting edge on the top, while the other is pendulum suspended with the refracting edge towards the bottom. The rays passing through the wedge system form the parallactic angle with the ones passing through the afocal lens (as optical component of the micrometer) on the right half of the casing. In the rest position ($\gamma_0 = 1 : 1000$), the telescope is tilted upwards by $\gamma_0/2$ from the horizontal plane. When tilting the telescope, the pendulum wedge retains its position in respect to the horizon. The deflection of the rays as generated by the wedge system, is reduced proportionally to the cosine of the tilt angle by proper selection of the wedge parameters (for pendulum wedge: refractive index $n = 1.6475$, $\alpha = 1°03'25.7''$; for fixed wedge: $n = 1.5187$, $\alpha = 2°25'23''$). The distance is always the rod section multiplied by 100. The tilt range is $\pm 20°$, the standard deviation of one measurement is ± 6 dm/100 m, the measuring range is 20 to 200 m. A 2.5 m long rod with 2 cm graduation and an 11-part vernier, corresponding to 10 intervals of the graduation is used. The coincidence of vernier- and graduation lines is achieved with the afocal lens of the telescopic micrometer. The reading image is similar to the one for the DN-04.

Testing and Adjusting of Double Image Tacheometers with Constant Parallactic Angles. The most rational method for the determination of the constants is the simultaneous determination of c and k as for stadia instruments (see page 240). The constant c is adjusted by the manufacturer to the value 0. The multiplication constant should be checked periodically. If it deviates from the theoretical value, the cm value is to be set at the micrometer drum, and the image of the rod vernier is to be shifted by turning the adjustment screw for the glass wedge (if available), until the corresponding lines coincide.

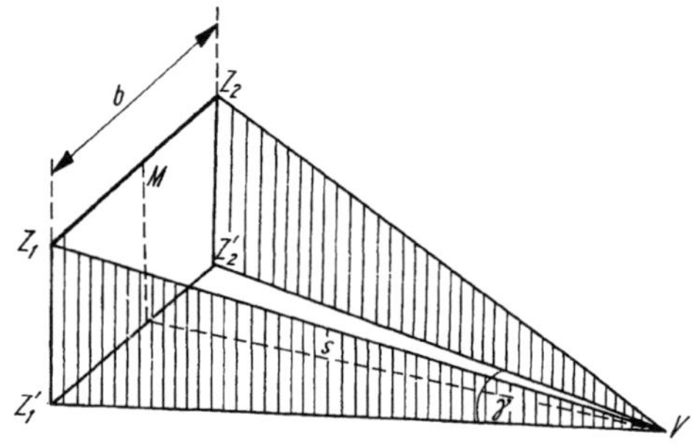

Figure 621 Parallactic triangle for subtense bar measurement

Distance Meters with Fixed Base at Target

If, at the target, a fixed base is set up horizontally and normal to the distance to be measured, and if the parallactic angle between the end points of the base is determined, the horizontal distance is directly obtained according to equation (182) (figure 621).

Subtense Bar and Precision Theodolite

The parallactic angle is generally measured with precision theodolites (see table 10). Nearly exclusively, a 2-m subtense bar is used as a base. It consists of a horizontal light metal tube, which for forced centering, is placed into a theodolite tribrack, fastened to a theodolite tripod. The ends of the tube have targets at exactly 2 m separation, which vary in shape for different manufacturers (figure 622). Usually they can be illuminated electrically. The separation between the targets is fixed by invar bars or -wires for temperature compensation. Therefore, temperature changes are practically without influence, or else their influence is eliminated with the aid of a bi-

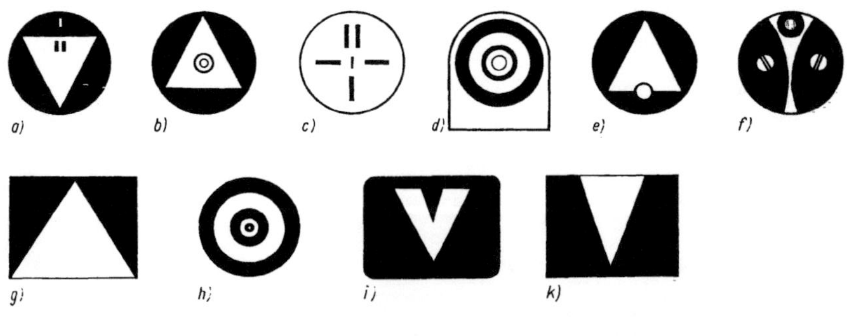

Figure 622 Shapes of targets of subtense bars
a) Zeiss, Jena
b) Keuffel & Esser, U.S.A.
c) Filotecnica Salmoiraghi, Milano; Officine Galileo, Florence
d) Zeiss, Oberkochen
e) Dietzgen, U.S.A.
f) MOM, Budapest
g) Rank Precision Industries, England
h) Kern, Aarau
i) Fennel
k) Wild, Heerbrugg
(The figures have different scales)

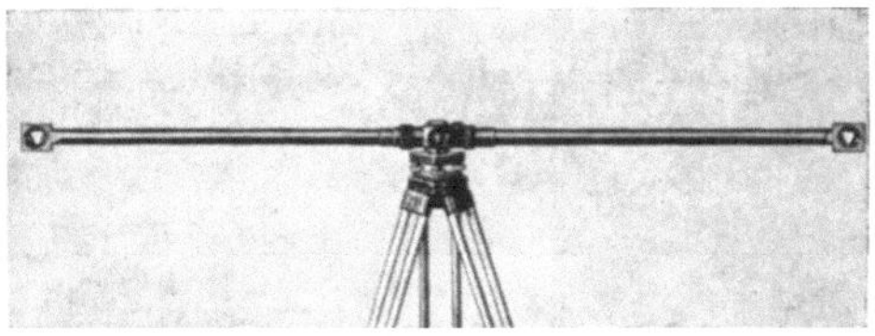

Figure 623 2-m subtense bar by Zeiss, Jena

metal compensator. In the center of the bar, there is a target for angle measurements resp. for the use of the bar as a 1 m subtense bar. The bar is levelled with the aid of a bull's eye bubble, so that it does not deviate by more than 0.3 gon from the horizontal plane. It is directed towards the theodolite with a diopter, so that it is not rotated by more than 0.3 gon from the normal to the direction of sight. The bars usually can be folded in two or taken apart for easier transportation. When reassembling the bar neither a space nor a bend can be tolerated between the two halves if an accuracy of 1 : 10,000, i.e. 0.2 mm for the bar length, shall be maintained.

The 2 m subtense bar by Zeiss, Jena (figure 623), is a 2 piece light metal tube, which can be stuck together and fastened solidly with a screw flange. The separation of the end marks is kept sufficiently constant with the aid of two invar bars which push against each other. The central mark has a rhomb-like shape. The bar can be calibrated to 25 μm.

The subtense bar by Wild, Heerbrugg, consists of a steelpipe which can be clapped together. Its halves are pressed against each other with a lever lock when ready for use. Both halves contain a free hanging invar wire whose interior end is fastened onto the stop at the joint. At the exterior end, the wire is pulled with a spring (figure 624). For a temperature increase of 10 K, the 2000 m long wire changes by 0.024 mm, while the steel pipe changes by 0.24 mm. This increases the tension in the spring by 0.2 N, which in turn pulls the wire by another 0.004 mm. The fastening device of the target mark, made of brass, counteracts this total pull of 0.028 mm by 0.018 mm, which means that the separation of the targets will increase by 0.010 mm for 10 K, resp. 0.020 m for 20 K, which is 1/100,000 of the total distance.

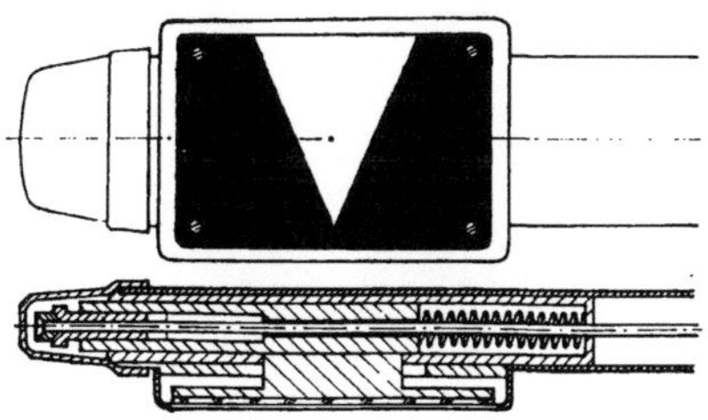

Figure 624 Target housing for Wild subtense bar

The halves of the Dietzgen subtense bar are connected with a joint-ribbon and locked with a spring lock. Inside the light metal frame an invar wire is stretched, which is fastened to the stop at the joint and to the target at the housing. One stop is plane, the other spherical so that both touch at a point. The change of the separation of the targets due to temperature is compensated by expansion in oppsoite directions for invar wire and aluminum target carrier. In the subtense bar by Kern, Aarau (figure 625), the expansion of the invar wire and the change in tension of the spring is compensated by crossing of two wire ends and by placing an aluminum pipe in between, which influences the target separation in the opposite sense. For use, the contact surfaces of the light metal tubes are pressed against each other with a spring loaded bayonet lock. The 1 m subtense bar by Kern consists of a massive invar tube of 30 mm diameter and 2 mm wall thickness, with targets rigidly fixed to it. Since the carrier of the target is 2.00 m wide, there is an addition constant of + 1.00 mm for both rod positions. Temperature changes only need to be considered if the working temperature differs by more than 10 K from the calibration temperature, because the expansion coefficient for invar is between + 1.0 and 1.6 × 10^{-5} K^{-1} at 20 °C, and the relative measuring accuracy is about ±3 × 10^{-5}.

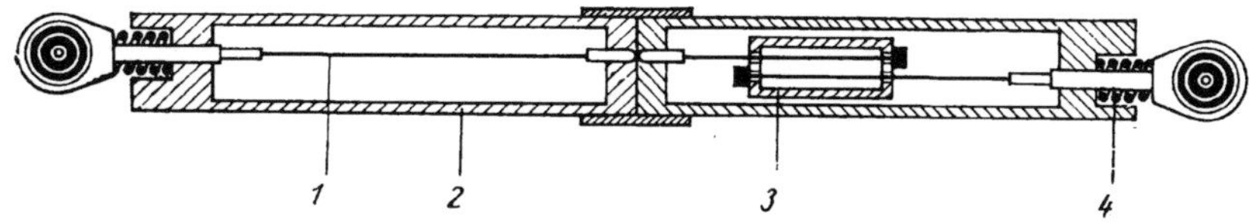

Figure 625 Compensation system for the Kern invar subtense bars
1 invar wire; *2* light metal tube; *3* compensation element (aluminum pipe); *4* tension spring for invar wire

Figure 626 2-m subtense bar by Zeiss, Oberkochen

The subtense bar by Zeiss, Oberkochen, is a snap-bar made of aluminum pipe (figure 626), where the contact surfaces touch directly. The compensation device in the righ half also neutralizes the expansion of the left half. The right target is fastened to the pipe with a lever system such that it can move (figure 627). The lever is moved inside the pipe *A* by a steel wire *S*, which is stretched by a spring and acts at a particular height on the lever. Since aluminum and steel have different expansion coefficients, the motion of the targets is monitored by matching steel wire length and transmission ratio such, that the separation from the left target remains constant for different temperature.

The subtense bar by MOM, Budapest, contains invar bars in the two sections which are screwed onto a central section. In the subtense bars by Rank

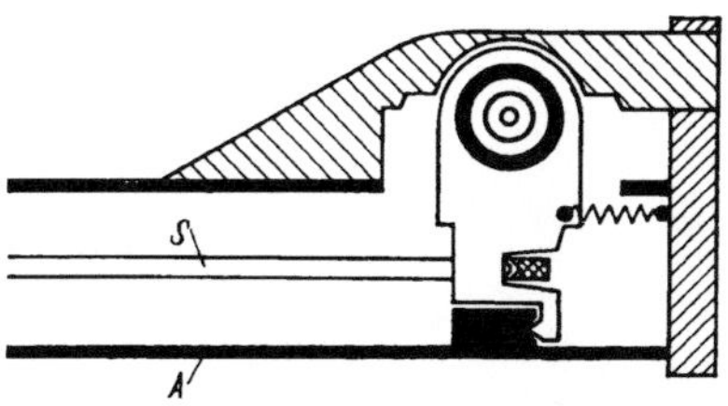

Figure 627 Compensation device for the Zeiss, Oberkochen subtense bar (schematic)
S steel; *A* aluminum

Precision Industries, Leicester, and Officine Galileo, Florence, the invar bars with target support are placed within light metal tubes. The subtense bar by Filotecnica Salmoiraghi, Milano, contains an invar wire, while the one by Fennel has quartz bars which do not require temperature compensation.

Testing of Subtense Bars. Systematic errors of the subtense bar are due to errors in the separation of the target marks and due to a deviation of the vertical plane of the targets from the vertical rotational axis. An error of 1/20,000 of the target separation, i.e. 0.1 mm, results in a distance error of, for instance 50 cm for 1000 m distance, i.e. 2,5 times the standard deviation. Therefore, the subtense bar has to be checked for its length and eccentricity, in addition to checking and adjusting bull's eye level and diopters.

The separation of the target marks of the base can be determined to ± 0.01mm with the aid of a test line or, faster, with the aid of a comparator e.g. the double image comparator by Hommelwerke, Mannheim, by Wild, Heerbrugg, or by Breithaupt & Son, Kassel, which takes about 20 minutes. The test with a comparator is similar to testing an invar levelling rod. First the comparator length is measured with test rulers, and then it is compared against the target separation of the subtense bar, placed under it.

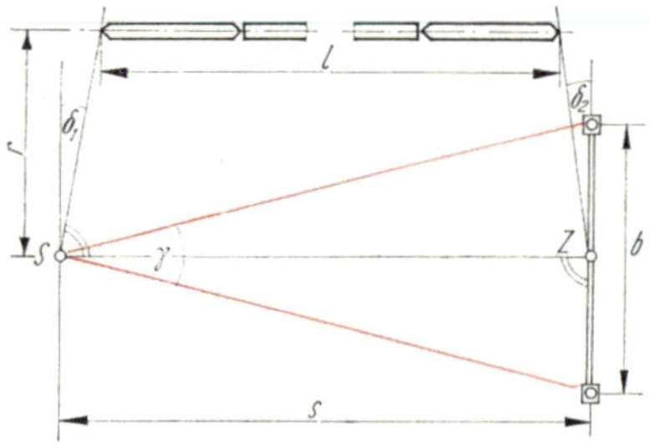

Figure 628 Determination of the separation of the targets according to *Kobold*

The method by *Kobold* utilizes a comparison distance l (figure 628) which is measured with invar test rulers and laid out parallel (i.e. at $r = 6.37$ m distance) to the test line (e.g. $s = 10$ m). With a theodolite the angles δ_1 and δ_2 to the edges of the test rulers and the angle γ to the target marks of the subtense bar are measured. From

$$s = l + \frac{r}{\varrho}(\delta_1 + \delta_2) \tag{202}$$

the separation of the targets is obtained as

$$b = 2s \tan\frac{\gamma}{2}. \tag{203}$$

For the method of *Ochsenhirt*, two theodolites are set up at a distance of 2 m (figure 629). Parallel to them the 2 m subtense bar is set up at a distance $s = 6.37$ m, and a 2 m-normal ruler is arranged independently above it (permissible deviations are ± 1 mm normal to the plane of the bar, and ± 3 mm in longitudinal direction). The separation of the target marks becomes

$$b = l + \left[\left(\frac{s}{\varrho}\right)(\delta_2 - \delta_1)\right] \tag{204}$$

where the angles δ_1 and δ_2 are measured between the target marks and the ends of the normal ruler to ± 0.3 m gon.

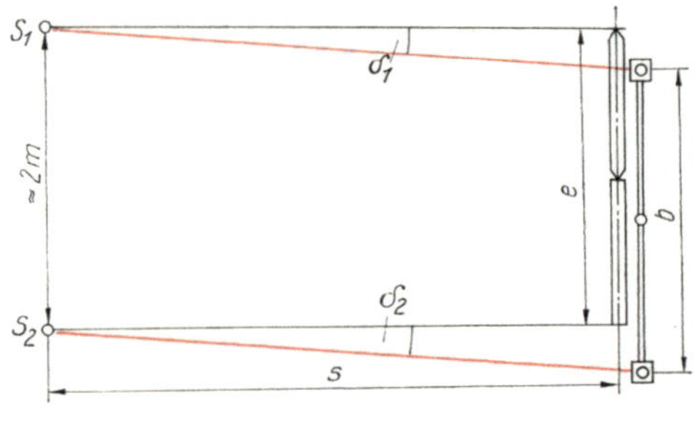

Figure 629 Determination of the separation of the targets according to *Ochsenhirt*

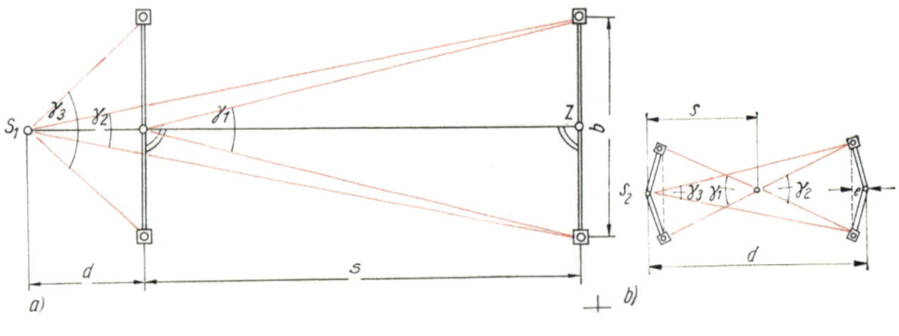

Figure 630 Determination of the separation of the targets and their eccentricity in direction of the line according a) to *Kobold* and b) with higher accuracy for to *Reiser*

The use of an invar tape rod as normal ruler is easier. The determination of the eccentricity of the subtense bar can be combined with the length determination using the Kobold method. With

$$b = \frac{2s}{\cot \gamma_2/2 - \cot \gamma_3/2} \tag{205}$$

one obtains for the eccentricity according to figure 630a

$$e_s = s - \frac{b}{2} \cot \frac{\gamma_1}{2} \tag{206}$$

where γ_i has to be determined to ± 0.2 mgon.
Standard deviations $m_{e_s} = \pm 0.3$ to 0.5 mm for the eccentricity and $m_b = \pm 0.05$ to 0.10 mm for the base length can be obtained with economically justified effort. The distance s should amount to about 4 m.
Independent of these investigations, traverses shall always be tied into known coordinated end points to eliminate the influences of systematic errors of the subtense bar. Then the errors can be adjusted mathematically.

Measuring Accuracy with Subtense Bar and Precision Theodolite. With a high precision theodolite, a parallactic angle can be determined with one measurement with a standard deviation of $m_\gamma = \pm 0.3$ to 0.6 mgon. With this, the standard deviation of the distance m_s becomes

$$m_s = \pm \frac{s^2}{b} m_\gamma . \tag{207}$$

The parallactic angle represents the difference of the horizontal circle readings for the right and left target marks. To eliminate systematic errors, one sights first to the left then to the right target, then resights that one using the fine motion screw and rotates the top part of the theodolite clockwise until the left mark is sighted again. Generally a four fold angular measurement is sufficient for an accuracy of $\pm 0{,}3$ mgon for the mean. For a 2 m-subtense bar, the following errors m_s are obtained for $m_\gamma = \pm 0.3$ mgon.

s =	40	50	60	80	100	150	200	300	[m]
m_s =	±4	±6	±9	±15	±24	±54	±95	±210	[mm]

Since the error increases proportionally to the square of the distance, one measures generally only distances up to 100 m and can then count on an accuracy of about 1 : 5000. Longer distances are determined by sub-dividing them into sections, or deriving them from measurements using auxiliary baselines.
The subtense bar has lost its importance for the determination of distances above 100 m after electronic distance measuring devices were developed. However, distances up to 30 m can be determined with it very accurately and with little effort, comparable to no other method. For distances of 3 to 8 m a relative error of $\pm 4 \times 10^{-5}$, i.e. ± 0.12 to 0.32 mm, can be reached, while for distances up to 30 m an accuracy of $\pm 10^{-4}$, i.e. ± 1 to 2 mm is possible.

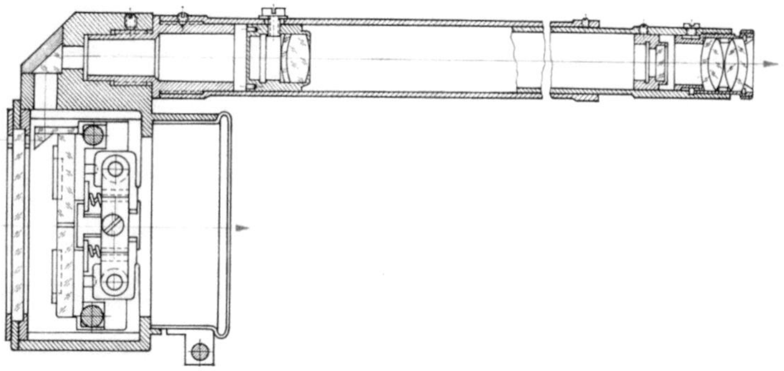

Figure 631 Distance meter DNT-2 N (longitudinal section) Further development: DN-08 (U.S.S.R.)

Double Image Distance Meters with Constant Base at the Target
The Soviet double image distance meter DN-08 (figure 631) designed by *Belizyn* is based on the heliometer principle. It is fastened onto the objective end of the telescope of a theodolite, and is used together with a constant base. The two halves, out of a positive- and a negative lens are set in their casings as shown in figure 632. If the casings are shifted by the amount Δ, which can be measured, the parallactic angle becomes $\gamma = \Delta\varrho/f_{II}$. For small

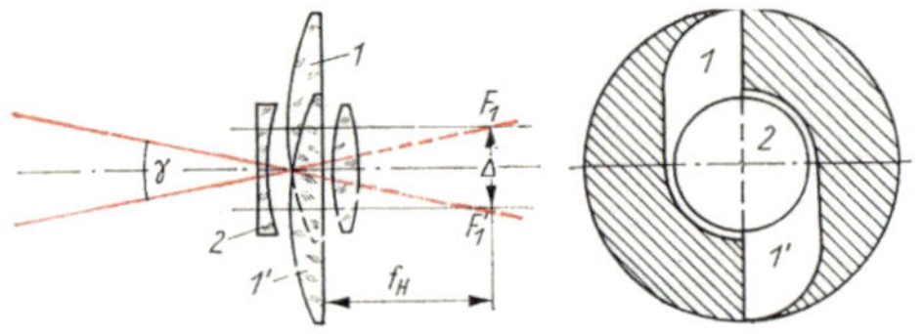

Figure 632 Principle of a distance meter with two shiftable half lenses
1, 1' half lenses;
2 negative lens

values of γ the distance becomes $s \approx b\varrho/\gamma$, or

$$s = \frac{bf_H}{\Delta} = \frac{k}{\Delta}. \tag{208}$$

The image separation is achieved with the aid of a biprism at the image plane and with a diaphragm at the eyepiece. A subtense bar (figure 633) belongs to the instrument. Its exterior target marks are separated by 1018 mm (measuring range 200 to 700 m), its interior target marks by 550 mm (measuring range 80 to 200 m). When measuring, the half lenses are shifted until the images of the target marks of the bar coincide within the field of view. When determining the distance, an addition constant of 0.10 m (distance between anallactic point and vertical axis, and distance between bar plane and target point) has to be applied. With the DN-08 distances between 50 and 700 m can be measured to an accuracy of 1 : 1500.
The optical topographic distance meter OTD (U.S.S.R.), a separate instrument, is based on the same principle. On both sides of the horizontal rod, five line pairs are placed symmetrically to its centre. With the aid of a compensator, parallactic angles of 500 to 5000″ can be measured. The telescope has 31 × magnification and can be tilted to ±22°. The sensitivity of the level bubble is 45″, the mass of the instrument is 4 kg, and of the rod 3.5 kg. The measuring range is 35 to 4000 m with an accuracy of 1 : 6000.

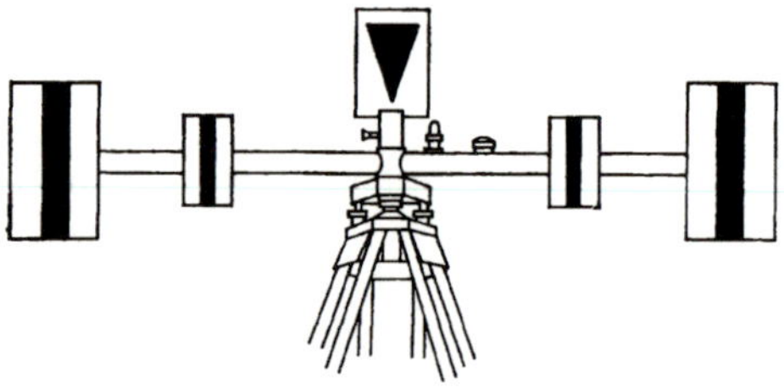

Figure 633 Subtense bar for the DNT-2 N

4.2.2. Distance Meters with Base at the Station

Distance meters with base at the station have the advantage that the target needs not to be accessible, and therefore it is not necessary to have a measuring rod in the terrain. For geodetic purposes however, it is advisable to signalize terrain points, especially if they are not clearly visible or distinguishable. The measurements are fast and easy, but it is not possible to reach the same high accuracy as with instruments with the base at the target.
Because of the mentioned advantage, instruments with base at the station serve primarily for military purposes, however a few topographic range finders were developed for civilian surveying. In these instruments (range finders) the images of the target, as simultaneously generated at the two endpoints of the base are viewed in the vertex of the parallactic angle as half- and double images with a common eyepiece or else stereoscopically. The images in monocular range finders appear as

1. sectional images (figure 634a) – the images meet at a separating edge and have to be set such, that the upper image is a continuation of the lower one.
2. reversed images (figure 634b) – the images of the target in the upright but height reversed image have to be placed on top of each other at a separating edge.
3. coincidence images (figure 634c) — two images of an object have to coincide.

All range finders for geodetic purposes have a horizontal base; a vertical base is used, for instance, for nautical purposes.

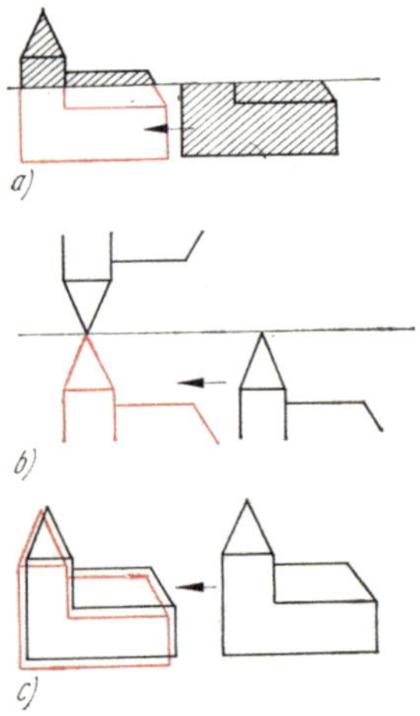

Figure 634 Half- or double images
a) section images
b) reversed images
c) coincidence images

Rangefinders with Variable Base

On a base b_i, these instruments carry two pentagon prisms P_1 and P_2, which generate two partial images. In front of one of the prisms (sometimes also of both), a glass wedge K is placed, which deviates the rays by the constant angle γ (figure 635). The rays which fall onto the prisms, always intersect with a constant angle at the target. The partial images of a target (sectional images) can be coincided by changing the base length b_i, i.e. by shifting one or both prisms. Then the distance can be computed using the parallactic triangle as:

$$s_i = b_i \cot \gamma. \tag{209}$$

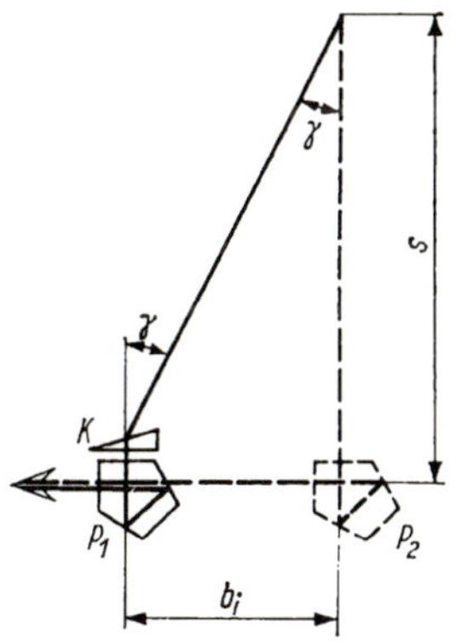

Figure 635 Distance meter with variable base at the station and constant parallactic angle at the target (schematic)

Different Types of Range Finders with Variable Base

The Teletop by Zeiss, Jena (figure 636) contains a mobile pentagon prism P_2 on a 30 cm long measuring rail, and a fixed pentagon prism P_1, situated higher as P_2, with a glass wedge K in front of it (figure 637).
A sectional image is created in the field of view of the 130 mm long telescope with 6 × magnification and 20 mm objective lens diameters. When measuring, the mobile prism P_2 has to be shifted until the two partial images generate one complete image. The image separation is accomplished with a biprism in connection to a pinhole diaphragm, which has the effect of a so-called keyhole observation. Depending on the required accuracy and the distance to be measured, measuring wedges with different deflection angles are used.

Measuring wedge	Measuring range [m]	Measuring accuracy [% of distance]
1 : 100	2...30	±0.2
1 : 250	5...75	±0.3
1 : 500	10...150	±0.5
1 : 1000	20...300	±1.0
1 : 2000	40...600	±2...3

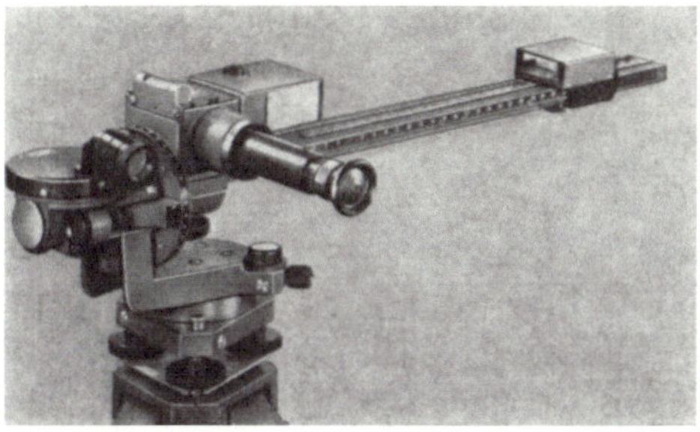
Figure 636 Teletop (Zeiss, Jena); height 200 mm, width 530 mm

The slope distance has to be reduced to horizontal with the aid of the reading on the vertical circle (85 mm diameter, 1° least count). A full circle compass (59 mm diameter) is attached to the left of the measuring rail. By shifting prism P_2 directly under prism P_1, the zero base can be set. If the vertical line of the lower image half is not directly at the edge, which was used to coincide the half images of the target, then a distance error is present, due to rotation of the biprism and of the wedge, which exceeds the unavoidable setting errors. When only measuring distance, the Teletop can be hand held. The mass of the instrument is 3 kg, with 3.5 kg for the container, and 3.2 kg for the tripod.
The rangefinder "Todis" (figure 638) by Breithaupt & Son, Kassel, is placed on the bottom part of a vernier theodolite with 6 × telescope magnification. The 90 cm long, T-shaped base rail, fastened onto two rings, carries a mobile pentagon prism each, on either side with a wedge of equal deflection angle in front of it. In the centre, there is a fixed double pentagon prism (figure 639). This symmetrical arrangement eliminates the influence of a rotation error. The measuring wedges are designed for the following range (manufacturer's data):

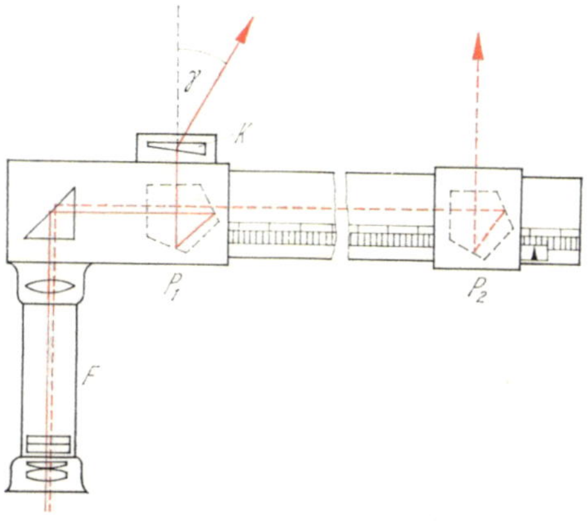

Figure 637 Sketch showing the Teletop principle

Measuring wedge	Measuring range [m]	Measuring accuracy [% of distance]
1 : 50	1...40	±0.02
1 : 100	2...80	±0.05
1 : 200	4...160	±0.1
1 : 500	10...400	±0.3

A biprism is used for image separation, just like in the Teletop. Horizontal circle (110 mm diameter, vernier reading to 20 mgon), and vertical circle (125 mm diameter, 1 gon least count) are made of metal. A full circle compass (80 mm diameter, least count 1° resp. 1 gon) is placed on the two base rings.
The mass or the instrument is 5.7 kg, with 9.0 kg for the container and 5.3 kg for the tripod.
The Soviet rangefinder DW-20 (figure 640) has a similar structure. Its telescope has a 14 × magnification. Values between 170 and 600 mm can be set on the base rail. Depending on the distance, the constants 100, 200 or 500 can be used. The level bubble has a sensitivity of 45″, the bull's eye bubble of 5′, and the least count of the compass is 1°. The maximum range is 300 m, with an accuracy of 1 : 300. The mass of the instrument is 4.2 kg, with 4.9 kg for the container.

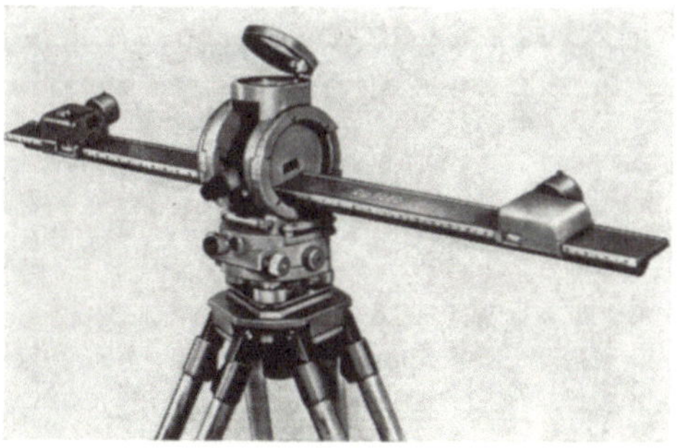
Figure 638 Rangefinder Todis (Breithaupt & Son)

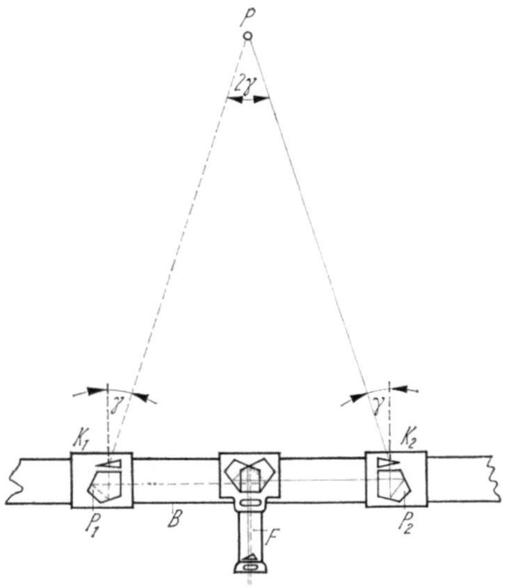

Figure 639 Sketch showing the Todis principle

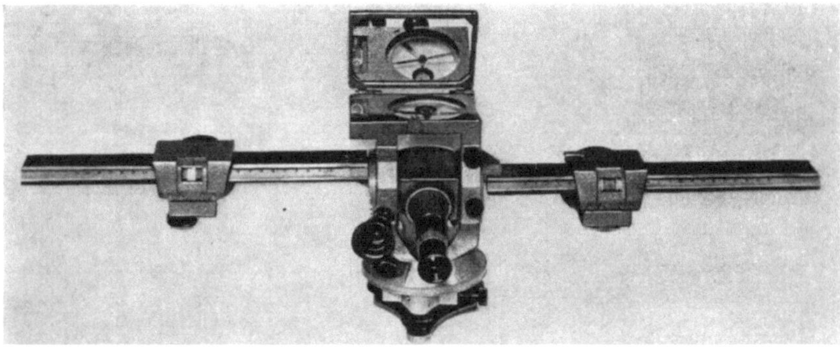

Figure 640 DW-20 (U.S.S.R.)

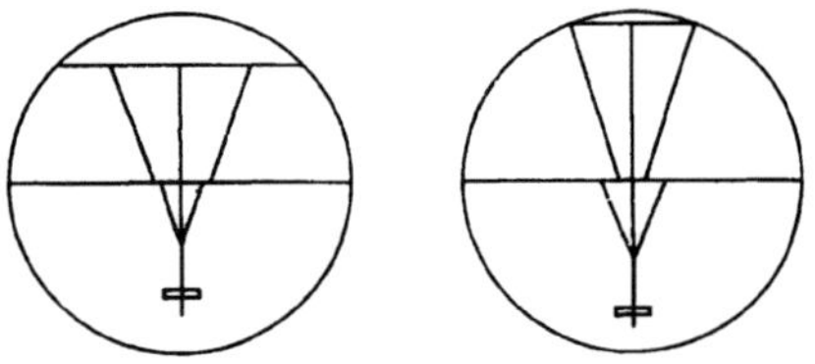

Figure 641 Height error for rangefinder with variable base

Accuracy and Applications of Range Finders with Variable Base

The measuring accuracy which can be reached indicates, that these range-finders are suitable for topographic surveys, preliminary engineering planning, distance-, direction-, and height measurements for geographic, geological, agricultural and forestry purposes. When measuring, one has to watch that a part of the target image does not appear twice or else is missing (figure 641). The half images shall meet at the separation edge such that the images of slightly inclined lines in the field of view continue exactly across the edge. Height errors in the half images have to be adjusted.

Range Finders with Constant Base

Monocular Telemeters

For monocular inverted image rangefinders (inverse-telemeters) with constant base, the prisms P_1 and P_4 at the base ends are fixed, as well as the prisms P_2 and P_3 which are used to deviate the rays into the eyepiece (figure 642). The prism P_3 reverses the image in the centre. A changeable wedge K (e.g. a scanning wedge or a pair of rotational wedges) is placed in front of one base end prism. This wedge changes the parallactic angle until the two half images of the target point are placed on top of each other. This parallactic angle is measured.
The coincidence telemeter Wild TM 10 can be used hand held for distance measurements and can be placed on a tribrack (figure 643) for topographic measurements. The measuring error for a range of 30 to 100 m is (according to manufacturer's data):

Distance	30	50	100	200	500	1000	[m]
Error	±0.05	±0.05	±0.2	±0.8	±5.0	±20.0	[m]

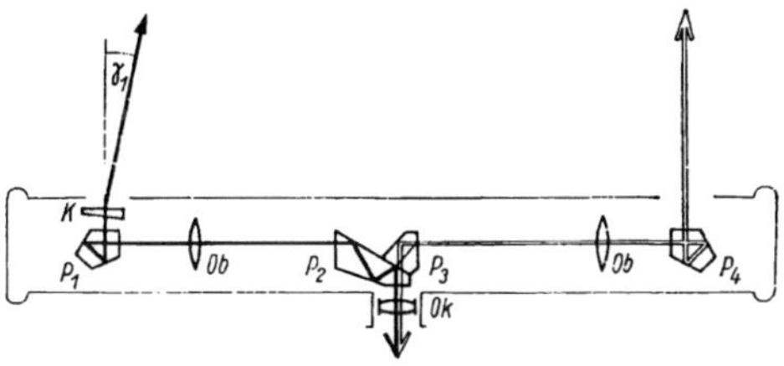

Figure 642 Sketch showing the principle of a reversed image distance meter

The telescope has a 6 × magnification and an objective lens with 21 mm diameter. The right eyepiece belongs to the telescope, the left one is used to read the distance. Horizontal and vertical circles can be read with a vernier to 1/10°. The instrument has a mass of 3.4 kg, with 1.5 kg for the metal container, 1.8 kg for the metal tribrack, and 5,7 kg for the tripod.
The rangefinder theodolite by Rank, Precision Industries, England, is similarly built. With its 1 m long base it can be used for distances up to 10,000 m with the following accuracy:

Distance	250	500	1000	2000	4000	[m]
Error	±0.3	±1.0	±3.8	±15	±61	[m]

The range finder can be placed onto the bottom part of a vernier theodolite ($\Gamma_F = 14\times$, horizontal- and vertical circle readings to 20″). The mass of instrument with tripod is 12.2 kg. SLOM, Paris, also produces a range-finder with sectional images (coincidence telemeter). Its telescope has a 10 × magnification, the base is 250 mm, and the measuring range is 10 to 500 m.

Figure 643 Coincidence telemeter Wild TM 10 on tribrack with angle measuring head (50 cm long base)

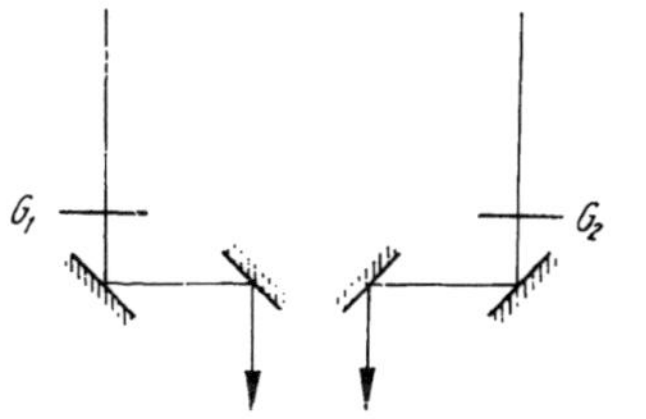

Figure 644 Sketch showing the principle of a stereo telemeter

Stereotelemeters

In the stereo telemeter, the two images are viewed separately with both eyes. Two glass plates G_1 and G_2 (figure 644), placed in the image plane, carry a number of measuring marks (figure 645a), which in matched pairs, correspond to images of spacial points at fixed distances, i.e. they represent different constant parallactic angles. When viewing with both eyes, the differentiation in depth is used for the measurement of distances. The corresponding measuring marks fuse to spatial marks giving the impression of a distance scale floating in space, which is used to estimate the distance to the target. In order to place a larger number of measuring marks within the field of view, a zig zag path leading into the depth was selected. It is also possible to utilize a spatial mark (figure 645b) ("floating mark") instead of

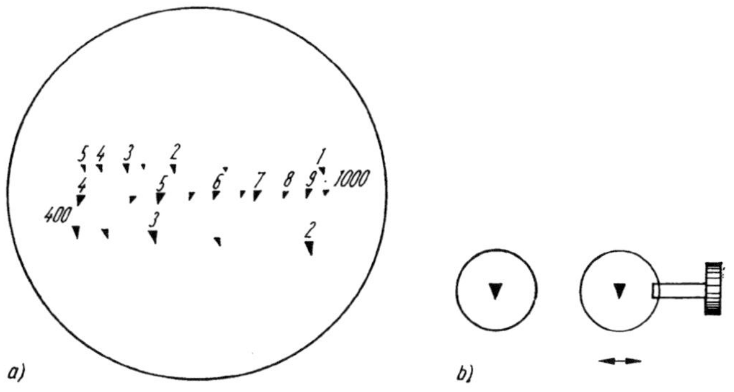

Figure 645 Floating scale (a) and floating mark (b)

the spatial scale. The distance is read on a parallax screw, after the floating mark has been set onto the object by turning it. This parallax screw is not required when a spatial scale is used. Presently, no such telemeters are produced for geodetic purposes.

4.3. Instruments for Electronic Distance Measurements

4.3.1. Principle of Electronic Distance Measurements

Electronic Distance Measurement (EDM) is based on the following principle: A transmitter placed at one end point of the distance transmits a wave resp. wave group which propagates with constant velocity, is reflected at the other end point and is received again at the original station (figure 646). This wave transmits a sharply defined signal. With the signal- and the propagation velocity c of electromagnetic waves, the length of the signal path and thus the slope distance s' becomes

$$s' = \frac{ct}{2}. \tag{210}$$

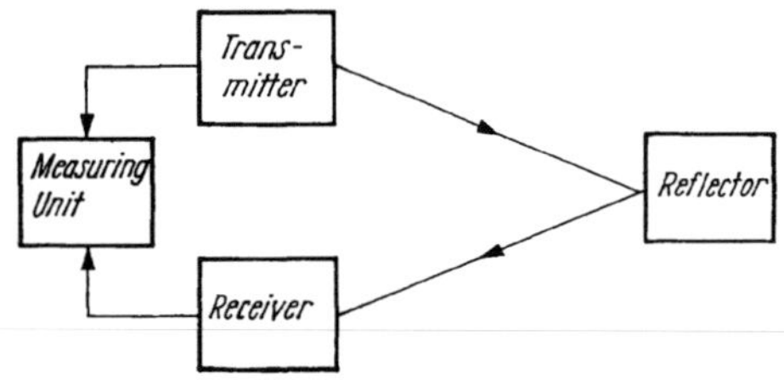

Figure 646 Principle of electronic distance measurement

The signal, for instance, can be a short impulse or the phase position of the oscillation.
For geodetic distance measurements, wave length ranges are used which have a uniquely defined path length, which can be sufficiently bundled, and for which the atmospheric absorption is not excessive. In addition to light waves ($\lambda = 400$ to 700 nm) which propagate practically along a straight line, and waves in the near infrared range beyond the visible part of the spectrum ($\lambda \approx 900$ nm). Presently microwaves with 8 mm to 10 cm wavelength are used, which have a behaviour similar- quasioptical propagation—to light waves (figure 647).
The bundle angle 2α of the transmission energy is determined approximately as

$$2\alpha = (\lambda/l)\ 55\ \text{gon} \tag{211}$$

with l being the diameter of the antenna mirror (analogue to optics). The shorter λ, the smaller it is. Since the wavelength of the microwaves is ten thousand times larger than of the light waves, the dispersion of the emitted energy is many times larger than for light waves.

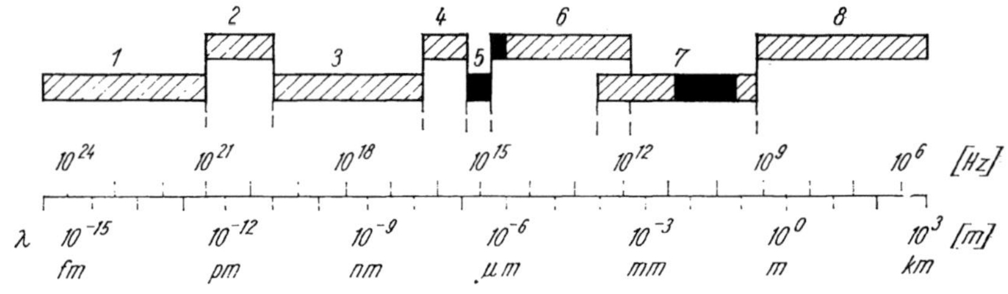

Figure 647 Ranges of wave lengths used for EDM within the electromagnetic spectrum
1 cosmic rays; *2* radiation; *3* x-rays; *4* ultra violet; *5* visible light; *6* infra red; *7* Hertz waves; *8* telegraphy

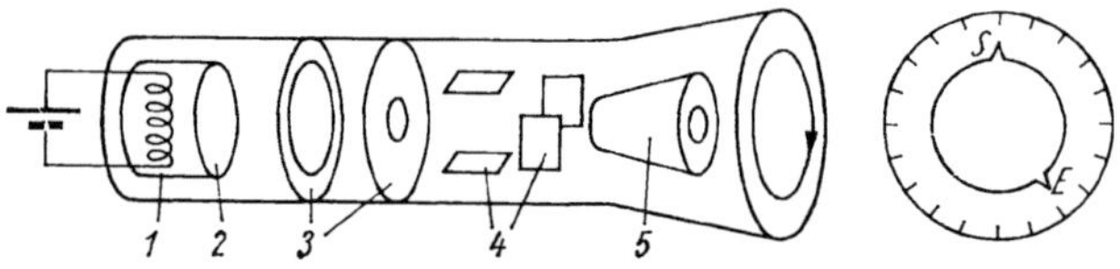

Figure 648 Cathode ray tube
1 cathode; *2* Wehnelt-cylinder; *3* anode; *4* deflection plates; *5* impulse cone

For 10 cm waves and an antenna with 60 cm diameter, the bundle angle is about 11 gon, while it is only a few arc minutes for light. When using light waves under favourable meteorological conditions, much more of emitted energy arrives at the end point than when using microwaves. Therefore, a passive reflector is sufficient, while for microwave measurements a second instrument is required, which amplifies and retransmits the received energy. The propagation velocity c of electromagnetic waves in the atmosphere is obtained from its propagation velocity c_0 in vacuum as

$$c = \frac{c_0}{n}. \tag{212}$$

The magnetic permeability μ and the dielectricity constant ε of the air is considered in the refractive index n of the medium:

$$n = \sqrt{\mu\varepsilon} \approx \sqrt{\varepsilon}. \tag{213}$$

It depends on the air temperature, air humidity, air pressure, and frequency of the waves used.
The XVIth General Assembly of the International Union of Geodesy and Geophysics (IUGG) in 1975 at Grenoble suggested for c_0 the value $c_0 = (299{,}792.458 \pm 0.0012)$ km/s for geodetic applications.

For signal transmission, a frequency band with several individual waves, i.e. a wave group, is used instead of an individual wave. Therefore, the phase- or wave velocity c has to be replaced by the group velocity c_{gr}:

$$c_{gr} = \frac{c_0}{n_{gr}}. \tag{214}$$

It has to be computed with the refractive index n_{gr} of the group and reduced to the prevailing atmospheric conditions for further use of equation (231).
Light- and microwaves propagate in straight lines and can be bundled well, but they are not suitable for path time measurements. For this, wavelengths of 10 to 30 m (f = 30 to 10 MHz) are most suitable. These waves, however, do not propagate along straight lines and are hard to bundle—compare with equation (211). The favourable characteristics of both wave groups are combined, when scale waves with, e.g. λ = 10 to 30 m are modulated onto light- and microwaves. Only these are used for the direct measurement of the distance, while the light- and microwaves only serve to transmit the signals. The frequency of the wave, which transmits the signals, is called carrier frequency, while the frequency of the modulation is called modulation- measuring- , raster- or scale frequency. The transit time t, which is needed by the emitted, reflected and received impulses for covering twice the measuring distance is directly measured with impulse methods. In phase measuring- or phase comparison methods (CW = continuous waves) it is indirectly determined from the phase shift of the received wave after covering twice the distance, in respect to the emitted one.
The accuracy of the distance measurement depends primarily on the accuracy of the determination of the transit time. A distance error of ± 3 to 5 cm corresponds to a timing error of 2 to 3×10^{-10} s. Therefore, for geodetic measurements of this accuracy, the moment of transmission and reception has to be determined to a few tenths of a nanosecond.

4.3.2. Instruments for Distance Measurements with the Impulse Method

For the impulse method, usually instruments with transmission- and reception devices (active reflectors) are used to obtain long ranges with small power output. To obtain a unique distance value, the sequence of impulses, i.e. the scale frequency, has to be selected such, that the signal transit time for twice the range is shorter or equal to the time separation of the impulses. Generally, the accuracy is not sufficient for geodetic purposes, especially because it is impossible to generate sufficiently short A/C impulses. To obtain a distance to an accuracy of 1 cm, the flanks of the impulses have to have a climbing time to 3.3×10^{-12} s, and the start of the impulse has to take place with the same accuracy. The uncertainties Δt however, have been reduced to ± 0.1 to 10 ns by newest developments. With

$$\Delta s = \frac{c}{2} \Delta t \tag{215}$$

distance errors of 0.02 to 1.5 m are obtained. The accuracy suitable for geodetic purposes, which is achieved in this manner, requires large instrumental investment.
For impulse methods, the transit time enters the measuring result. Therefore, the distance does not need to be estimated for its determination. The results are obtained uniquely, fast and easily. The method is primarily used for military purposes.
The base for geodetic application was the radar method (radio detecting and ranging). The transit time fore- and back for the pulses (sequence of short electrical impulses) is measured with a cathode ray tube (figure 648). The electron beam, emitted by the cathode is focussed with the aid of electric fields, i.e. narrowly bundled, accelerated within the electric field and thrown onto a screen, which is illuminated at the point of contact. The intensity of the illumination can be monitored with the Wehnet cylinder with the aid of variable voltage. If a high frequency A/C voltage with a phase shift of 100 gon is applied to two pairs of deflection plates (deflection condensators) normal to each other, the electron ray draws a circle unto the screen with its revolution corresponding to the voltage frequency. When transmitting the impulse, a notch S is superimposed in the impulse cone onto the illuminated circle, and a notch E upon its reception. The separation of the notches together with the voltage frequency is a measure of the transit time resp. the distance when properly graduated. Shoran (short range air

navigation) and improved Hiran (High accuracy shoran) represent geodetic applications. The distances are determined by line crossing with an aircraft, which contains the transmitter (figure 649). The ground stations receive, amplify and retransmit the impulses to the transmitter in another frequency. The range is about 750 km, with an accuracy of about $\pm 7 \times 10^{-6}$ for Hiran. The primary station has a mass of about 100 kg, with about 500 kg for each ground station. The equipment is very expensive. Numerous long distances across oceans as well as in networks in America and Australia have been measured by Shoran and Hiran.

Ranging with laser impulses is used for the determination of extremely long distances (e.g. to the moon). In spite of poor absolute accuracy a high relative accuracy is achieved. Microwave impulses are used for radio navigation. The accuracy achieved amounts to ± 1 to 3 m.

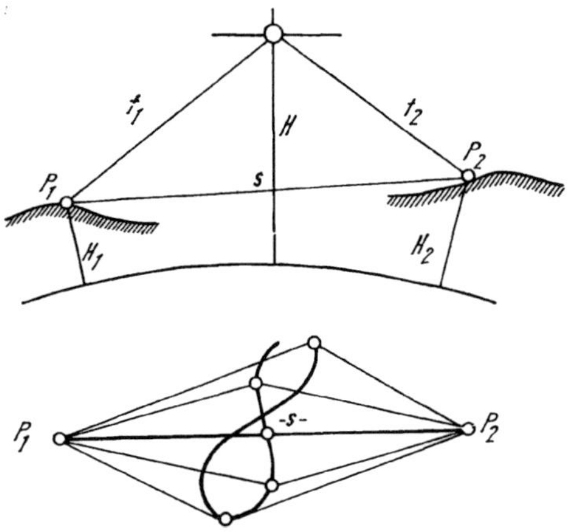

Figure 649 Determination of a distance by line crossing

4.3.3. Instruments for Distance Measurement with Phase Comparison

Principle of the Phase Comparison Approach

With the phase comparison method the distance is determined with the aid of the phase difference between the harmonic electromagnetic wave continually emitted at the station, and the wave received after reflection at the end point. A harmonic wave can be described as

$$y = a \sin(\omega t + \varphi_0) = a \sin \varphi \tag{216}$$

where a is the amplitude, $\omega = 2\pi f$ is the circular frequency, φ_0 is the initial phase at $t = 0$, and $\omega t + \varphi_0$ is the phase angle φ. The phase angle φ_2 after covering twice the distance in the time t differs from φ_1 for the emission of the wave (figure 650) by the phase difference

$$\Delta\varphi = \varphi_2 - \varphi_1 = \omega t = 2\pi f t. \tag{217}$$

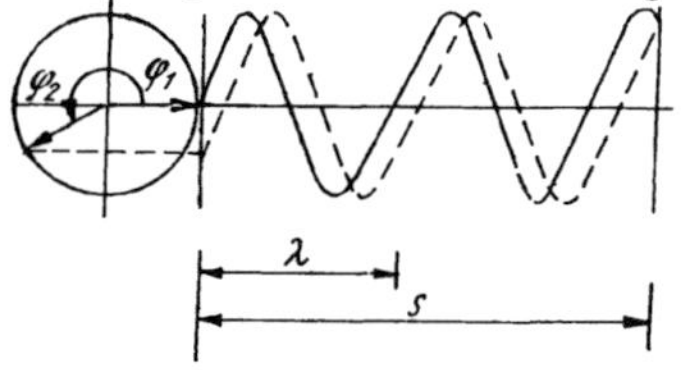

Figure 650 Distance measurement using phase comparison ($s > \lambda/2$)
1 transmitter, receiver
2 reflector

If the distance is shorter than half of the modulation wavelength, then twice the distance is immediately obtained from equation (217) with $f = c/\lambda$ (figure 651). If it is longer than $\lambda/2$, then $\Delta\varphi$ represents the distance r from the end of an integer multiple of the wavelength to the receiver (figure 650):

$$r = R\lambda = \frac{\Delta\varphi}{360°}\lambda = \frac{c\Delta\varphi}{2\pi f}. \tag{218}$$

Therefore, in order to determine the distance s', the propagation velocity of the electromagnetic waves at the meteorological conditions prevailing at the time of measurement has to be known together with the frequency and the phase difference between reference- and measuring signals. Furthermore, the number of full wavelengths has to be determined, because the measurement of the phase difference is only possible within one period. As a result, the number a of full wavelengths plus the directly read remainder R are determined:

$$s' = \frac{a\lambda + R\lambda}{2}. \tag{219}$$

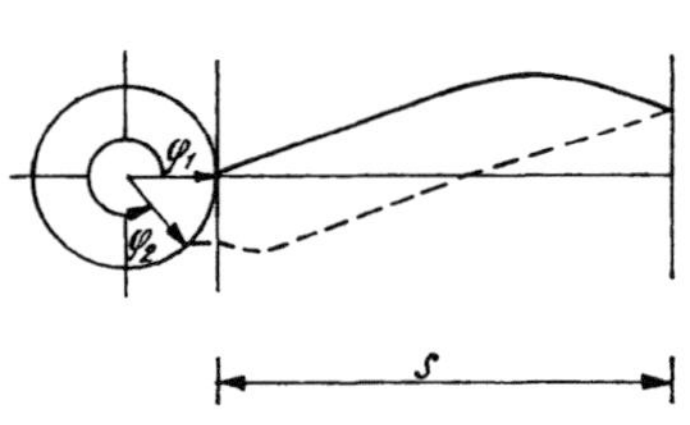

Figure 651 Distance measurement using phase comparison ($s < \lambda/2$)

Since always twice the distance is obtained because of the fore- and back path, normally the scale $U = \lambda/2$ is used in the computations:

$$s' = aU + RU. \tag{220}$$

Since a is unknown, s' becomes ambiguous. The ambiguity is most simply solved by frequency variation. The frequency of the modulating oscillator is continuously changed, e.g. with the aid of a phase displaying zero instrument, until the same phase situation φ is again obtained. These respective positions are fixed. This results in a number a of full periods, which is by 1, 2, 3 ... smaller or larger. Using both measurements and

$$s' = a\frac{c}{2f_1} = (a + 1)\frac{c}{2f_2} \tag{221}$$

the number a of full periods becomes

$$a = \frac{f_1}{f_2 - f_1}. \tag{222}$$

Inserting this into equation (221) gives

$$s' = \frac{c}{2(f_2 - f_1)}. \tag{223}$$

The exact measurement of variable frequency values to within 10^{-5} to 10^{-6} is difficult with mobile measuring units. Therefore, the measurements are preferrably performed with fixed frequencies. With two frequencies f_1 and f_2 one obtains with equations (218) and (219):

$$s' = \left(a_1 + \frac{\Delta\varphi_1}{2\pi}\right)\frac{c}{2f_1} = \left(a_2 + \frac{\Delta\varphi_2}{2\pi}\right)\frac{c}{2f_2} \tag{224}$$

where a_1 and a_2 are integer but unknown numbers of whole periods. This leads to

$$a_1 = \frac{(a_2 - a_1)\, f_1 + \frac{\Delta\varphi_2}{2\pi} f_1 - \frac{\Delta\varphi_1}{2\pi} f_2}{f_2 - f_1}. \tag{225}$$

On the right hand side of equation (225) only $(a_2 - a_1)$ is unknown. For the determination of this value, the distance has to be approximately known, i.e. for two frequencies at least to $(\lambda_{2-1}/4)\ (f_2 - f_1)$. Generally, one measures with a third frequency, which provides a check for a_1 as

$$a_1 = \frac{(a_3 - a_1)\, f_1 + \frac{\Delta\varphi_3}{2\pi} f_1 - \frac{\Delta\varphi_1}{2\pi} f_3}{f_3 - f_1}. \tag{225a}$$

It is advantageous to utilize modulation frequencies which are significantly different. If these for instance have a ratio of 1 : 10 resp. 1 : 100, then a rough measurement can first be obtained with the low frequency to give an approximation for the distance, and then a fine measurement is performed with the 10 or 100 times modulation frequency. For instance, the frequency $f_1 = 150$ kHz would provide the km-, $f_2 = 1.5$ MHz the 100 m- and $f_3 =$ 15 MHz the 10 m scale U. The exact production of such different modulation frequencies is extremely difficult. Therefore, instead of the low frequencies, one uses frequencies whose differences provide the same relationships, e.g.

Frequency	Frequency difference	Scale
$f_1 = 15.15$ MHz	$f_1 - f_3 = 150$ kHz	1000 m
$f_2 = 16.5$ MHz	$f_2 - f_3 = 1.5$ MHz	100 m
$f_3 = 15$ MHz	$f_3 = 15$ MHz	10 m

The fine measurement is therefore directly obtained from the phase difference $\Delta\varphi_3$ of the measurement with f_3.

For instruments designed for short range, generally two frequencies with 15 and 15.15 MHz are used. This results in a rough scale of 1000 m from the frequency difference, and a fine scale of 10 m. This can be compared to two tape measures of 1000 m and 10 m length respectively, which are graduated to 1/100, i.e. to 10 m and 1 dm. Then, when measuring a distance of 863.72 cm, 860 m is read on the 1000 m tape, and the remainder of 3.7 m at the 10 m tape with the 2 cm being estimated (figure 652).

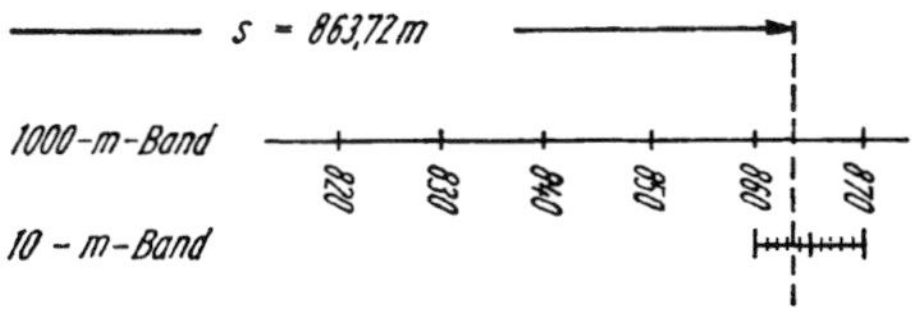

Figure 652 Solution of ambiguity by measuring with two frequencies (compare to two tapes)

The phase difference $\Delta\varphi$ can be determined to about 1/3000 of a period with analogue phase measurement and up to 1/1000 with digital phase measurement, e.g. for $f = 15$ MHz ($\lambda = 20$ m) to centimetres resp. half centimetres. The higher the modulation frequency (the smaller the scale), the better is the resolution.

However, the electronic equipment effort increases significantly when reducing the scales. For instruments built in series, having modulation frequencies between 10 and 600 MHz ($\lambda = 30$ to 0.5 m), the accuracy of phase measurements is thus between 1 cm and 0.2 mm. However, the accuracy of the distance measurement is also influenced by other factors (frequency stability, errors in the determination of the meteorological data, etc.).

For high frequencies of 10 and several tens of MHz, phase measurements are difficult. Therefore, nowadays it is always done 10 to 100 times more accurate with low frequencies of a few MHz by superimposing and using the primary frequency with a reference frequency that differs by a few kHz (see page 259).

For analogue phase measurement, the time separation between the zero points of transmitted and received signals is presented analogue with another quantity. The phase measurement is reduced to a voltage measurement. Usually, resolver-inductive goniometers of high accuracy are used, i.e.

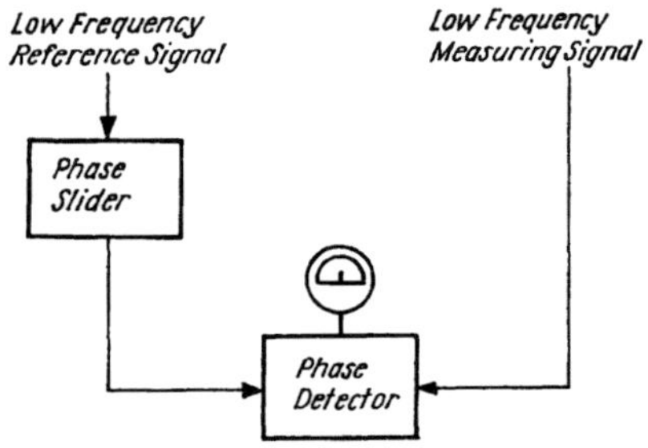

Figure 653 Principle of analogue phase measurements

rotating variable phase sliders consisting of stator and rotors, which are connected to a phase detector with null indicator (figure 653). A reference point is to be found with the phase detectors, because it indicates a zero voltage at phase differences 90 or 270° only. When turning the rotor, the voltage induced in its coil obtains a phase shift, which is exactly proportional to the mechanical angle of rotation. At the zero position of the rotor (i.e. when the signals have the same phase), the angle of the phase difference between the two signals is measured. Since the electric phase angle corresponds to a certain distance for a given modulation frequency, the ratio of transmission between resolver wave and the counter indicating the angle of rotation is to be selected such, that the values correspond to the distance to the nearest integer multiple of the modulation wavelength.

For the digital phase measurement, the low frequency signals are transformed into rectangle impulses (e.g. by triggers), because the sine waves do not produce sharply defined impulses. The phase difference is indicated by an electronic counter as impulse number. After the front flank of the triggered reference signal has opened a measuring gate, the equidistant impulses produced by the impulse given, reach the counter through the measuring gate (figure 654). Since counting- and modulation frequency are derived from the same oscillator, a calibration is not necessary. The measuring gate is closed by the triggered measuring signal in the same manner as before. The number of impulses recorded by the counter, represents the phase difference between reference- and measuring signal. Its accuracy is increased by performing several measurements and taking the mean.

Like for optical distance meters, the difference between monumented end points of the distance and the zero points of the instruments resp. reflectors as set up, has to be considered as zero error or addition constant c_k for EDM measurements.

A combination of equations (218) and (219) results in

$$s' = \frac{(a + \Delta\varphi/2\pi)\, c}{2f} + c_k\,. \tag{226}$$

If one considers, that a, $\Delta\varphi$ and c_k are independent of the distance, but that f and c are proportional to it, the standard deviation m_s' is obtained with the influences of the various terms as

$$m_s'^2 = m_a^2 + m_{\Delta\varphi}^2 + m_{ck}^2 + s^2\left[\left(\frac{mf}{f}\right)^2 + \left(\frac{m_c}{c}\right)^2\right]. \tag{227}$$

Thus the accuracy of short distances depends mainly on the accuracy of display, phase-measurement and determination of the addition constant, while for long distances it depends mainly on the accuracy of the frequency and of the determination of c (i.e. mainly on the determination of the refractive index n with the aid of the measurement of meteorological elements). With a few neglections, the terms are therefore combined to

$$m_s = a + bs\,. \tag{228}$$

Electro-Optical Distance Meters

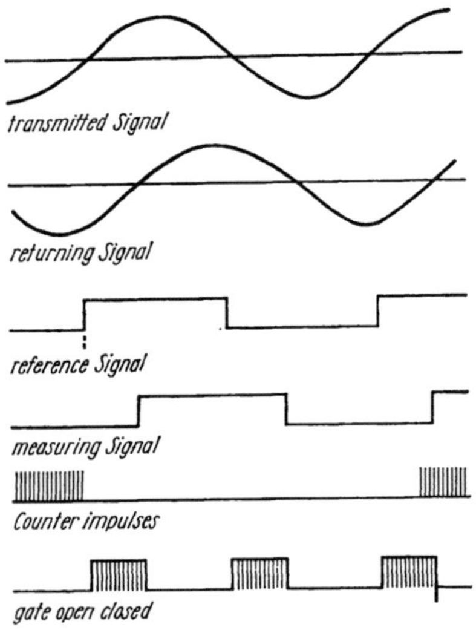

Figure 654 Principle of digital phase measurements

In principle, electro optical distance meters have the following structure (figure 655a): The continuous radiation generated by a source is led through a modulation, which is monitored by a quartz oscillator with reliably constant frequency, and which modulates the radiation. For electro optical distance meters, amplitude modulation is exclusively used. The intensity of the radiation is changed with it such, that it increases or decreases in the beat of the applied A/C voltage. With a modulation frequency of 15 MHz for instance, light becomes bright and dark 15 million times per second. The radiation is transmitted to the end point with the transmission optics. The characters modulated onto this carrier radiation propagate with its velocity, the group velocity, along the measuring path. It is reflected at the end point and, at the station, it is diverted by the receiving optics to a photo detector, which transforms the light signals into electrical current with an intensity which change sin the rhythm of the modulations. This modulation oscillation obtained after amplification is compared as measuring signal in the phase measuring element with the modulation frequency as generated by the oscillator and applied to it as a reference signal. The phase difference between measuring- and reference signals is either measured with the aid of a phase slider, with which the phase position of the reference signal are shifted

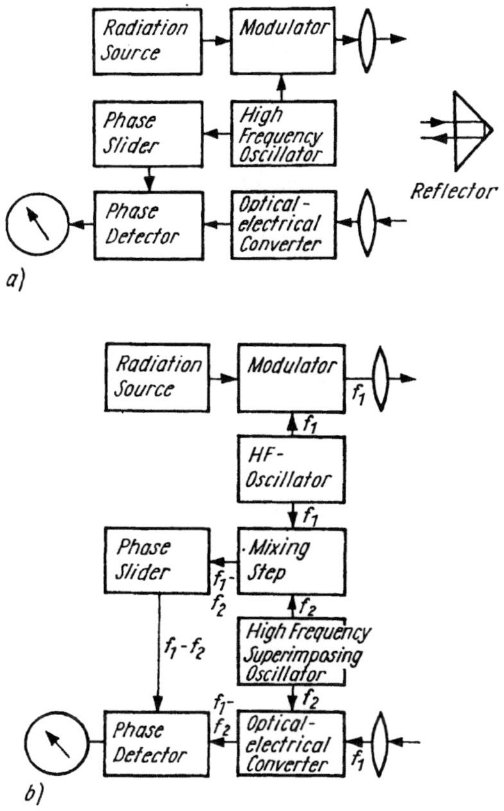

Figure 655 Block diagram of circuits for EDM instruments
a) phase measurement with modulation frequency
b) phase measurement with low frequency

until the null indicator shows no more voltage (when phase equality is achieved), and $\Delta\varphi$ can be read. It can also be obtained by digital phase measurement (see figure 654).

Since the phase positions of reference- and measuring signals drift because of the temperature depending on the electronic elements, no fixed reference point can be given with phase measurement alone. Therefore, an additional phase comparison is carried out on an internal test distance, by introducing an internal prism on a rotating diaphragm into the path of the radiation, which diverts the radiation of the transmitter directly into the receiver.

The difference between external and internal light path is practically free of drifts of the phase positions. It is automatically formed for numerous modern instruments.

Due to the reasons given on page 257, the phase measurements are carried out with low frequencies in modern instruments. In this case, the high frequency reference signal diverted from oscillator *1*, plus the returning high frequency measuring signal are converted into low frequency signals of equal phase position by superimposing and mixing them with a reference signal of different frequency (e.g. 10 kHz) generated by oscillator *2* (figure 655 b).

In older Soviet instruments the desired phase situation is obtained by variation of the modulation frequency. The frequency is then determined with an attached frequency meter, usually by direct counting of the number of oscillations within a certain time period. The modulation frequency itself is not determined, but rather the difference to a fixed frequency, which is known to a high degree of accuracy. The phase comparison between the reflected modulated light and the reference voltage can also be accomplished in these instruments with continuously changing modulation frequency by visual indication without demodulation of the light.

The distance can be computed with the measured phase difference according to equation (225) for distance meters with fixed modulation frequencies, and equation (223) for distance meters with continuously variable frequency, after c had been determined with equation (214).

The necessary index of refraction n_L is to be determined for normal conditions (dry air at 0° C, 760 Torr pressure and 0.03% CO_2 content) with the aid of the wavelength of the used radiation using the modified formula by *Edlen*

$$(n_{gr} - 1)\,10^7 = 2875.69 + 3\,\frac{16.206}{\lambda^2} + 5\,\frac{0.139}{\lambda^4} \tag{229}$$

or the formula by *Barrell* and *Sears*

$$(n_{gr} - 1)\,10^7 = 2876.04 + 3\,\frac{16.288}{\lambda^2} + 5\,\frac{0.136}{\lambda^4}. \tag{230}$$

Within the visible spectrum of light, values between 2800 and 3500 are obtained for $(n_{gr} - 1)\,10^7$. The difference between values obtained with the two formulae amounts to 10^{-7} for $\lambda = 560$ nm. In order to reduce from the normal atmosphere to the prevailing atmospheric conditions, the formula by *Barrell* and *Sears* is used in a simplified form according to *Kohlrausch*:

$$n_L - 1 = \frac{(n_{gr} - 1)}{(1 + \alpha t)}\,\frac{p}{1013{,}25} - \frac{4{,}1 \cdot 10^{-8}\,e}{(1 + \alpha t)}. \tag{231}$$

The refractive index n_L for the group under the external conditions is thus computed with the aid of the result of equations (229) or (230), the air temperature t in °C, the air pressure p in mbar, the expansion coefficient for air $\alpha = 0.003{,}661$, and the partial pressure e of water vapor. Since the values determined with equation (230) are constant for a particular instrument, n_L can be predetermined for certain meteorological conditions according to equation (231). Significant deviations from these conditions can be taken into consideration by using correction tables or settings at the instrument.

Elements of Electro-Optical Distance Meters

Source of Radiation. With minimum use of electricty, the source generating the carrier radiation shall emit a radiation of such a luminous density, that the signals returning from the reflector can be received with certainty. This is the case, if its intensity in the receiver is still sufficient to distinguish it from received interferences (dispersed light, tubal noise). Therefore, until about a decade ago, distances of several kilometres could only be measured at dusk or at night. According to a rule of thumb, the maximum range is approximately equal to the fourth root of the luminous density.

With the development of semi-conductor diodes and gas lasers, light bulbs (projection lamps) operated with low voltage (e.g. 6 or 12 V) or with accumulators and producing a luminous density of 10^3 sb for 30 W, have lost their importance just as mercury vapor lamps with more than 100 times the luminous density (for longer ranges), but also increased electric efforts and energy usage (300 W).
Of the different types of lasers, which emit largely coherent, monochromatic and sharply bundled light of high luminous density, gas lasers with continuous modulation (recently also semi-conductors) are used. The high frequency electrical voltage required to aggravate the gas mixture — usually helium and neon—amounts to 3000 to 5000 V. The frequencies are then between 30 and 40 MHz for 30 to 50 W output. The degree of efficiency for changing electrical energy into light is very low, however, this disadvantage is compensated by the very sharp bundling. The radiation emitted by He-Ne lasers is always linearly polarized. For distance meters, radiations with $\lambda = 632.8$ nm (red) and starting output of up to 5 mW are used (see section 2.6.). With them, distances of 30 km by day and 60 km at night are possible.
Semi-conductor diodes can be directly modulated. In principle, they consist of two semi-conducting layers, one of which has an electron deficiency (*p*-conducting), the other one an electron surplus (*n*-conducting) (figure 656). In the about 10 μm thick boundary zone, which acts as a barrier, electrones can be excited to a higher energy level by an outside current.
Luminence diodes with gallium-arsenide as semi-conductor material transmit this radiation spontaneously as invisible, practically monochromatic (but not coherent as for lasers) radiation of about 875 to 960 nm wavelength (infrared). Since it is only slightly bundled, optical means have to be used for bundling. The application of high intensity current leads to a laser effect, i.e. induced emission of high intensity and monochromasy.
The structure is simple and compact, and the dimensions are small. Although the efficiency of GaAs-luminescence diodes is small with about 10^{-3} at room temperature, and the output is only a few milliwatts, ranges of up to 5 km are possible in daylight.

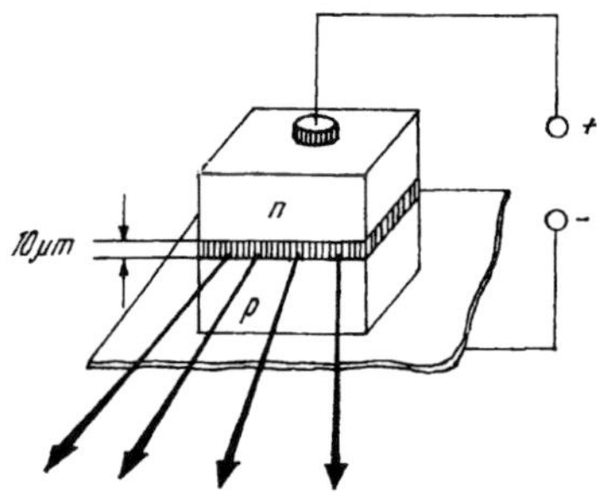

Figure 656 Systematic structure of a luminescence diode

Modulation Devices. The modulator is used to change the light intensity in the rhythm of the modulation frequency. It is desirable to have a high degree of modulation with low energy input, while amplitude and the high modulation frequency have to be constant.
Semi-conductor diodes can be directly modulated, so that a modulation is not necessary. The modulation of the radiation proportional to the injection current is achieved by modulating the latter.
For all other light sources an indirect modulation with a special modulation is necessary. A Kerr-cell is used most often. With it, the Ker-effect is utilized. The polarization plane of a linearly polarized light bundle is periodically rotated in certain liquids under the influence of an attenuating electric field. After subsequently passing through an analysor, the intensity of the light bundle, which is proportional to the square of the field strength, oscillates periodically (quadratic electrooptical effect).
The Kerr-cell consists of a glass container, filled with Nitrobenzol, with two electrodes extending into it (figure 657). Two Nicol's prisms, normal to each other, and inclined by 50 gon to the horizontal are placed on both sides of the glass container. The polarizor *P* imposes a polarization direction of 50 gon in respect to the horizontal onto the light rays coming from the light source *L*. When applying an electric voltage to the plates of the glass container, a horizontal electrical field is generated in the liquid, and the Nitrobenzol acts doubly refracting (Kerr-effect). The light, linearly polarized by the polarizor, is led normal to the electric field and separated by it into two components, the ordinary ray oscillating in direction of the electrical field vector, and the extraordinary ray normal to it. These have a phase shift depending on the voltage. The light is elliptically polarized. If a small A/C voltage is superimposed onto the applied D/C voltage, then the position of the major axis of the ellipse oscillates in the beat of the A/C voltage. The light is again linearly polarized with the analysor. Since the major axis of the ellipse is rotating, the intensity, i.e. the brightness of the emitted light, changes with the rhythm of the A/C voltage.

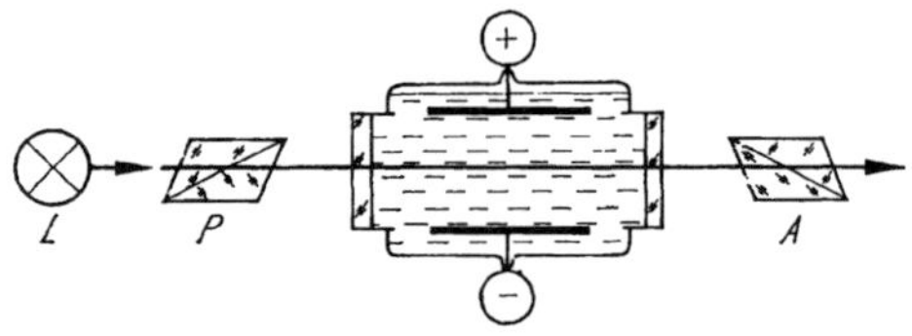

Figure 657 Kerr-cell

Because of the necessary cooling for high frequencies, due to dielectric losses, mostly modulation frequencies of 20 to 30 MHz are used. More than 50% of the supplied light is lost. Usually the modulation voltage amounts to several thousand volts. The phase position of the modulation is not

equally distributed over the light beam, so that phase shifts are possible, which lead to measuring errors. This effect is reduced if a rectangular A/C voltage is used instead of the constant D/C voltage.
For the Pockels-effect, occurring in double refracting crystals, such as KDP, ADP, LiNb, this is elliptically polarized under the influence of an electric field, like for the Kerr-cell, and subsequently obtains an amplitude modulation in the analysor. The intensity of the light bundle changes directly proportionally to the field strength (linear electro optical effect). The crystals can be modulated with frequencies up to several GHz, however, because of high dielectric losses, cooling is then necessary. Pockels modulators are only fully effective for parallel light. They are therefore especially suited for laser light. However, the costs of crystal modulators are a multiple of the costs for Kerr-cell modulators.
Ultra sound modulators are based on the diffraction effect of ultra sound waves in transparent solid or liquid media.
In a quartz Q, excited with an uneven multiple of its resonance frequency, a standing ultra sound wave is generated normal to the propagation direction of the light, and zones of different density are formed. For the penetrating light ray, they act as an optical diffraction grid, which exists and disappears again within half a period of the sound wave. By the periodic existence and disappearance, the light is modulated in connection with slit optics in the beat of twice the frequency of the ultra sound (figure 658). Since the velocity of sound is temperature dependent, the modulator M is placed inside a thermostat. The advantages are the low light loss compared to the Kerr-cell (not more than 25%), the possibility of choosing higher frequencies than for the latter, and the low energy use of the modulator. The fact, that it is only possible to work with the resonance frequency is a disadvantage, because uneven values are generated for the evaluation.

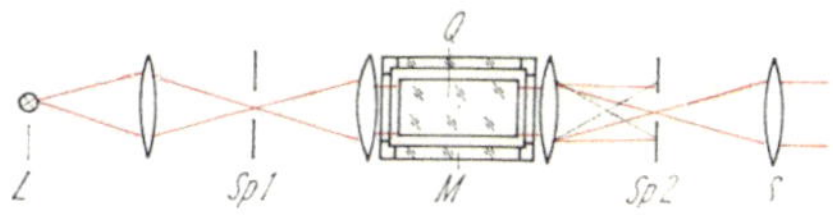
Figure 658 Ultra sound modulator

Optical Parts. Transmission- and receiving optics shall be powerful and able to narrowly bundle the light beam. Since the loss of brightness amounts to up to 30% in optical systems, they have as large an opening and focal length as possible for the given purpose. Often mirror or mirror-lens systems according to *Cassegrain* are used, which have small dimensions and favourable mass distributions in spite of large focal lengths.
The reflectors at the end of the line shall return the light without significant loss and shall be lined up fast and accurately. Besides plane mirrors, primarily reversing systems are used, whose optical axis needs only to be oriented approximately to the measuring direction, namely triple mirrors (see figure 72), reflectors with several prisms (figure 659), and combination of lens- and mirror objectives with plane and concave mirrors.

Photodetectors. The device used to transform light modulation into electrical A/C current, shall have a sufficiently small inertia (so that it can follow the light changes), shall be very light sensitive, and permit a perfect phase measurement.
For this, the photo cell is suitable, which, because of the low incident light intensity, is combined with a secondary electron multiplier (SEV, photomultiplier). The light falling onto a special photo cathode frees electrons, which migrate to the first anode (1st dynode) of the SEV when voltage is

Figure 659 Reflectors
a) for the AGA-geodimeter b) one prism reflector GDR 31 for Distomat Wild DI 3 S
c) target with TR 7 by Zeiss, Oberkochen
d) for the EOT 2000 (Zeiss, Jena)

applied, and knock out a number of secondary electrons proportional to the amount of incident light. These are accelerated and hit the 2nd dynode, etc. Since each electron frees several secondary electrodes, the photo electric current can be amplified in several steps to 10^4 to 10^7 times. While SEVs are used for visible light as carrier, Silicium-photo diodes serve as photo detectors for distance meters with infrared radiation.

Phase Measuring Instruments (see page)

HF Oscillators. The modulation frequencies are obtained from quartz oscillators, which are thermostat conrolled to keep the frequency constant to 1×10^{-6}. This causes a warm-up time. The output of the quartz oscillator is determined by the type of the modulator. The monitoring frequency shall amount to at least 10 MHz. The required accuracy of the frequency depends on the measuring method within the phase measuring element. As far as the electronic structure is concerned, the HF oscillator is similar to a common radio transmitter. For distance meters with fixed frequencies, frequency errors affect the measurements. Within the measuring range of short range distance meters, the influence of the frequency error is in the same magnitude as other instrumental errors. For these, the frequency stability of $\pm 3 \times 10^{-6}$ between -5 and $+30$ °C, which can be obtained with a temperature compensating network, is sufficient, so that no thermostat is required. Therefore the warm-up time is not needed, and measurements can commence immediately.

Types of Electro Optical Distance Meters

One distinguishes:

— electro optical distance meters for long range, with which distances up to 60 km can be measured to an accuracy of a few centimetres (application in control surveys, investigations of horizontal crustal movements and others)

— short range distance meters, which permit the measurement of distances up to 2 or 5 km to an accuracy of 1 to 2 cm (application in traversing, detail surveys, etc.)

— precision distance meters with ranges up to 3 km and accuracies of millimetres or fractions of them (applications in engineering surveys)

Among the short range distance meters, one can distinguish

— independent distance meters, which can only be used for distance measurements

— attachable instruments, which can be used independently, and, when placed onto a theodolite constitute an electronic tacheometer

— electronic tacheometers (see 5.3.).

Important functions are automized for numerous distance meters, such as

— automatic measuring procedure (tuning of the intensity of the measuring signals, measuring of external and internal light paths, switching between fine- and rough measurements, and combination of all partial results to a digitally displayed total result). The job of the observer is limited to sighting, starting of the measuring procedure, and reading of the results resp. recording onto data carriers.

— automatic consideration of the atmospheric correction by changing of the measuring frequency after setting a correction factor, or by input of a correction value to be applied by the internal computer, or with the aid of a cavity sonator (as for the Mekometer)

— automatic reduction to horizontal distance and computation of the elevation difference (especially for electronic tacheometers) via input of the visually read vertical angle into a computer or with electronic measurement of the vertical angle (see page 221) and input into the computer.

Electro Optical Distance Meters for Long Ranges (table 33)

Electrooptical distance meters for long range are independent instruments which serve for distance measurements only. Gas lasers as radiation source have filled the need for special types for medium range. The modulators are of the Pockels- or Kerr-cell type. The phase comparison is performed with low frequencies using analogue or digital phase measurements.

Table 33. Electro-optical Long Range Distance Measuring Devices (selection)

type	manufacturer	radiation source	modulation frequency: modulator: modulation fine frequency [MHz]	modulation frequency: modulator: number	modulation frequency: modulator	power requirement [W]	phase measuring method	range: day [km]	range: night [km]	mass [kg]	dimensions [cm³]	standard deviation	remarks
1	2	3	4	5	6	7	8	9	10	11	12	13	14
Geodimeter 8	AGA, Sweden	5-mW-He-Ne-Laser	30	4	KDP-crystal	75	analogue	30	60	23	56 × 30 × 26	± (5 mm + 1 · 10⁻⁶ s)	figure 660
Geodimeter 600		1-mW-He-Ne-Laser	30	4	Kerr-crystal	26	analogue	40		15	22 × 36 × 53	± (10 mm + 1 · 10⁻⁶ s)	figure 662
Geodimeter 14A		Ga-As-Diode 910 nm	15	2	–	17	digital	15		2.5	22 × 19 × 9	± (5 mm + 3 · 10⁻⁶ s)	attachment
Ranger V	Keuffel & Esser, U.S.A. (Laser Systems & Electronics Inc.)	2.2-mW-He-Ne-Laser	15	4	KDP-crystal	120	digital	60		18	32 × 23 × 43	± (5 mm + 1 · 10⁻⁶ s)	Ranger IV: 12 km range; ± (5 mm + 2 · 10⁻⁶ s)
Rangemaster II		10-mW-He-Ne-Laser	30	4	Kerr-crystal	140	analogue	30	50	25	45 × 34 × 28	± (1 cm + 2 · 10⁻⁶ s)	
Quarz	U.S.S.R.	1-mW-He-Ne-Laser	49	5		400	digital	60	80	48	86 × 51 × 40	± (1 · 10⁻⁶ s)	figure 663
Geodolite 3G	Spectra Physics, U.S.A.	10-mW-He-Ne-Laser	15	4	KDP-crystal	57	digital	25		16	32 × 23 × 43	± (1 cm + 2 · 10⁻⁶ s)	figure 666
HP 3808A	Hewlett-Packard, U.S.A.	Laser-Diode 840 nm	15	2	–	18	digital	10		9	35 × 32 × 28	± (5 mm + 1 · 10⁻⁶ s)	

Figure 660 AGA Laser-Geodimeter, model 8; dimensions: 56 cm × 31 cm × 26 cm

The Geodimeter (GEOdetic DIstance METER) model 8 (figure 660) by AGA, Stockholm, contains a 5 mW-He-Ne-Laser (the diameter of the laser beam is 0.6 mm, the beam divergency is 1.7 radians) and a Pockels modulation (KDP crystal). With the aid of a beam widening optics, the reflector is better illuminated, the divergency reduced to 1 mrad (diameter: 20″), and the light intensity is increased 200 times (figure 661).
A cylinder lens, which can be inserted into the optical train is used to search for the reflector. The influence of interferring light is reduced by an interference filter. The fine measurement is performed with 30 MHz, the phase comparison to 1.5 kHz. The range is 60 km, and the measuring time is 5 to 10 minutes. The Geodimeter 6 BL now in Geodimeter 600 has a 1 mW-He-Ne-laser as radiation source and a Kerr-cell modulator, while the models 6 A and 6 B contain light bulbs and Kerr-cells (figure 662).
Unlike the other instruments, the Geodimeter 6 A does not contain scales in decadic steps, so that the distance is obtained after a few simple computations. Otherwise, the structure of all these instruments is similar to the Geodimeter 8. Their range amounts to approximately 25 km and 40 to 60 km for Geodimeter 600.

Figure 662 AGA Geodimeter 600; dimensions: 22 cm × 36 cm × 53 cm

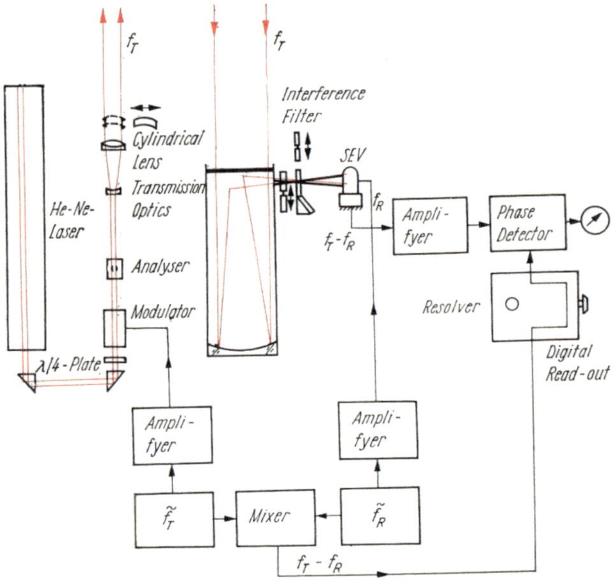

Figure 661 Block diagram for circuits of the AGA Geodimeter 8

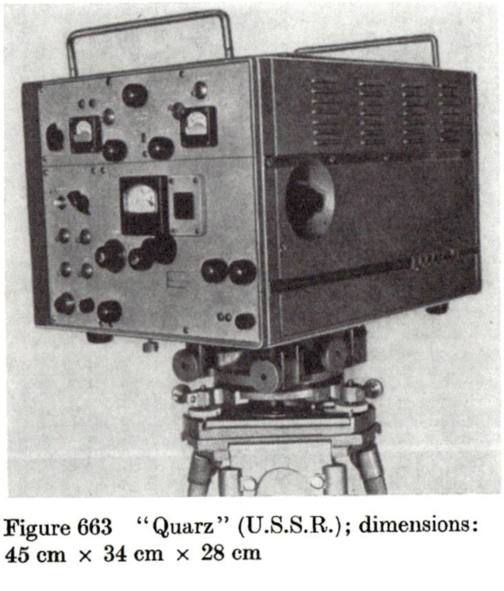

Figure 663 "Quarz" (U.S.S.R.); dimensions: 45 cm × 34 cm × 28 cm

The Soviet instrument "Quarz" (figure 663) with 50 km range operates with gas laser and Kerr-cell. Transmission- and receiving optics are arranged biaxially. Measurements are carried out with four frequencies near 30 MHz. The phase comparison is performed with an inductive phase slider to 15 kHz (figure 664).
In the electro optical distance meter EOS by Zeiss, Jena, the light of a projection lamp is modulated with the aid of an ultra sound crystal cell with twice the frequency of the ultra sound waves. The four frequencies (4.5 to 5 MHz) generate din a thermostat controlled quartz oscillator, are brought to the exciting frequencies 27 to 30 MHz via two amplifier steps, so that modulation frequencies of 54 to 60 MHz are obtained (figure 665). Since only a direct optical phase comparison is possible for ultra sound modulation, about 2% of the modulated light are deviated by the separation cube T. This radiation is directed to the SEV_2 and transformed to 10 kHz. The phase comparison between the original currents of the reference-SEV_2 and the measuring SEV_1 is carried out to 10 kHz in a phase detector D with null indicator. For the zero display at $\Delta\varphi = 90$ and 270°, the phase angle of the reference signal is changed with the aid of a goniometer G, which indicates the phase difference. One series consists of measurement of the distance and of the internal comparison path in four frequencies. The scales amount to 2, 5, 25, 200 and 3000 m. The evaluation is best be done with a computer.

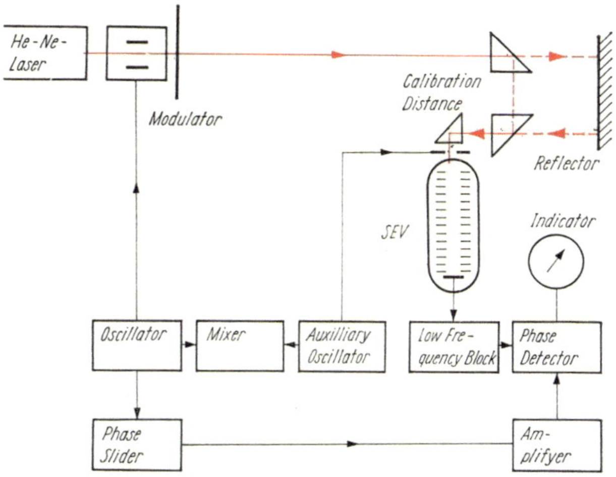

Figure 664 Block diagram for circuits of the "Quarz"

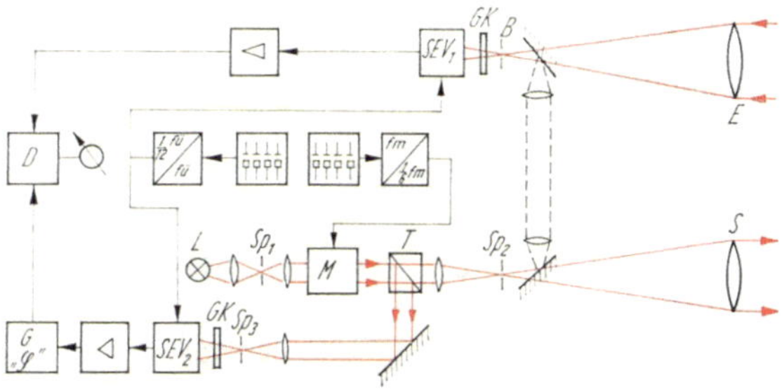

Figure 665 Block diagram for circuits of the EOS (Zeiss, Jena)

Figure 666 Geodolite 3 G (Spectra-Physics, U.S.A.)
Dimensions of telescope unit: 86 cm × 51 cm × 41 cm

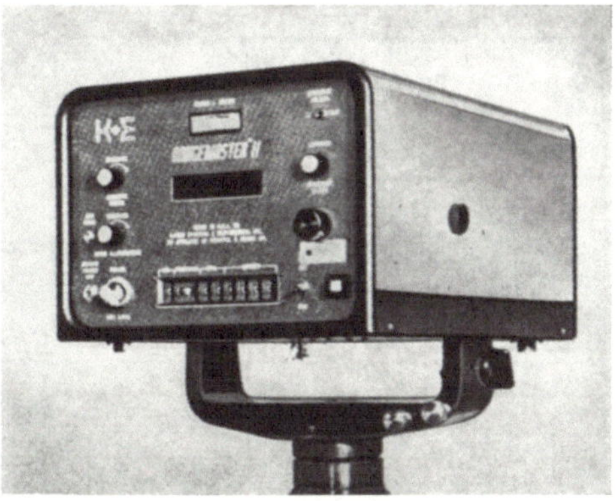

Figure 667 Range master (LSE, U.S.A.)
dimensions: 32 cm × 23 cm × 43 cm

Figure 668 EOK 2000 (Zeiss, Jena); dimensions: 30 cm × 35 cm × 29 cm

The instruments Geodolite 3 G (figure 666) by Spectra-Physics, U.S.A., as well as Ranger IV, V and Rangemaster by Laser Systems and Electronics, U.S.A. (figure 667) operate with gas laser radiation and digital phase measurement. The measuring procedure can be largely automized when digital phase measurements are utilized.

Electro Optical Distance Meters for Short Range (table 34)

Electro optical distance meters for short range are produced as individual instruments as well as attachments for wider use. Nearly exclusively, directly modulated GaAs luminescence diodes are utilized as radiation source, with *Si*-photo diodes for demodulation. Phase comparison is accomplished at low frequencies with analogue or digital phase meters. Often two frequencies are used (150 kHz, 15 MHz), with the frequency for the rough scales occasionally being derived from the frequency of the fine scale with the aid of a frequency splitter (Zeiss, Oberkochen Eldi, SM 4, Wild DI 3 S, Tellurometer CD 6).

For the EOK 2000 (figure 668) by Zeiss, Jena, the main frequency of 30 MHz provides the 5 m-fine scale, the differences to 30.3 and 33 MHz provide the rough scales 50 and 500 m. The phase comparison is performed with an analogue phase meter to 7.5 kHz.

Measuring and reference signals are led onto a discriminator (phase detector) (figure 669). The setting of the goniometer placed in between, can be read on three counters after zero tuning. The values correspond to twice the measured distance. They have to be added as follows:

$$\begin{array}{lllll} 8 & 4 & 6 & & \\ & 4 & 2 & 1 & \\ & & 2 & 3 & 3 \\ \hline 8 & 4 & 2 & 3 & 3 : 2 = 421.165 \text{ m} \end{array}$$

The calibration value for the internal light path is mechanically subtracted after pressing a button. Since the infrared radiation used for the measurements is invisible, transmission and receiving optics are placed on top of each other, coaxially combined with a search light resp. a search telescope. The measurement requires two minutes.

Table 34. Electro-optical Distance Meters for Short Range and High Precision (selection)

type	manufacturer	radiation source	modulation frequency: modulation fine frequency [MHz]	modulation frequency: number	modulator	power requirement [W]	phase measuring method	diameter of trans-mission- and receiving optics [mm]	range [km]	mass [kg]	dimensions [cm³]	standard deviation	remarks
1	2	3	4	5	6	7	8	9	10	11	12	13	14
separate Instruments													
DM 60	Cubic Corp., U.S.A.	Ga–As-Diode 900 nm	75	3	–	15	digital	biaxial 50	2	6.8	13 × 18 × 36	±(5 mm + 10 · 10⁻⁶ s)	figure 672
DM 2000	Kern & Co., Aarau	Ga–As-Diode 875 nm	15	2	–	11	digital	coaxial	2.5	10	35 × 24 × 26	±(4 mm + 4 · 10⁻⁶ s)	figure 671
HP 3805A	Hewlett-Packard, U.S.A.	Ga–As-Diode 910 nm	15	2	–	18	digital	coaxial 50/110	2	7.7	33 × 26 × 15	±(14 mm + 30 · 10⁻⁶ s)	
2 SM-2	U.S.S.R.	Ga–As-Diode 900 nm	15	3	–	15	digital	coaxial	2	4.6		±2 cm	
EOK 2000	VEB Carl Zeiss JENA	Ga–As-Diode 900 nm	30	3	–	8	analog	biaxial 32/35	2.5	12	30 × 35 × 29	±1 cm	figure 668
attachments													
Geodimeter 112A	AGA, Sweden	Ga–As-Diode 910 nm	15	2	–	17	digital	biaxial 60	6	6	22 × 19 × 9	±(5 mm + 3⁻⁶ s)	figure 677; geodimeter 110 : 5 km range
DM 501	Kern & Co., Aarau	Ga–As-Diode 900 nm	15	2	–	11	digital	biaxial	1.6	1.6	18 × 18 × 7	±(5 mm + 5 · 10⁻⁶ s)	figure 678 and 758
Autoranger	Keuffel & Esser, U.S.A.	Ga–As-Diode 910 nm	15	2	–	11	digital	coaxial 40	2	2.4	22 × 14 × 11	±(5 mm + 6 · 10⁻⁶ s)	
MND-2	Nikon, Japan	Ga–As-Diode 770 nm	15	2	–	45	digital	biaxial	1	3 +21	8 × 16 × 16	±10 mm	
SDM-1	Sokkisha, Japan	Ga–As-Diode 900 nm	1.5	1	–	8	digital	biaxial 50	0.6	1.5	15 × 12 × 6	±1 cm	
RED 1			15	2	–	12			2	3.5	16 × 9 × 18	±(5 mm + 10 · 10⁻⁶ s)	
Eldi 1	Opton Feintechnik GmbH, Ober-kochen	Ga–As-Diode 910 nm	15	2	–	5	digital	coaxial 110	5	8	14 × 12 × 33	±(5...10 mm + 2 · 10⁻⁶ s)	figure 679; 2 measuring ranges mass: battery included
Eldi 2								coaxial 60	3	4.2	14 × 12 × 16	±(5 mm + 2 · 10⁻⁶ s)	figure 680; 2 measuring ranges mass: battery included

Beetle 500	Precision Internat. Inc., U.S.A.	Ga-As-Diode 910 nm	15	2	–		digital	coaxial	0.5	2.5	24×15×13	±(6 mm + 4 · 10^{-6} s)	Beetle 1000: 1 km range; Beetle 1600: 1.6 km range S-serious with reduction computer
CI-450	Tellurometer, England	Ga-As-Diode	15	2	–		digital	biaxial	4			±(5 mm + 5 · 10^{-6} s)	with computer
CD 6		Ga-As-Diode 900 nm	15	2	–	14	digital	biaxial 60	2	3.7	24×15×8	±(5 mm + 5 · 10^{-6} s)	figure 676
Di 3S	Wild Heerbrugg AG	Ga-As-Diode 885 nm	7.5	2	–	17	digital	biaxial 35	2	2.0/4.3	19×11×7	±(5 mm + 5 · 10^{-6} s)	figure 675; with computer
Di 4			5			5			2	1.9	19×6×6	±(5 mm + 5 · 10^{-6} s)	figure 753; with micro processor with part to reduction
Precision Distance Meters													
Mekometer ME 3000	Kern & Co., Aarau	Xenon-flash lamp	500	5	KDP-crystal	18	analog	biaxial 40/34	3	18.7	46×16×22	±(0.2 mm + 1...2 · 10^{-6} s)	figure 682
MA 100	Tellurometer, England	Ga-As-Diode 930 nm	75	4	–	14	digital	coaxial 175	2	17	32×37×28	±(1.5 mm + 2 · 10^{-6} s)	figure 681
MSD-1M	U.S.S.R.	Ga-As-Diode	150	2	–	3	analog	coaxial 36	0.5	9.5		±(2 mm + 5 · 10^{-6} s)	
DVSD-1200		2-mW-He-Ne-Laser	1200	1	KDP-crystal	90	analog		0.25	27	45×22×20 (optical unit)	±(0.25 mm + 1 · 10^{-6} s)	figure 684

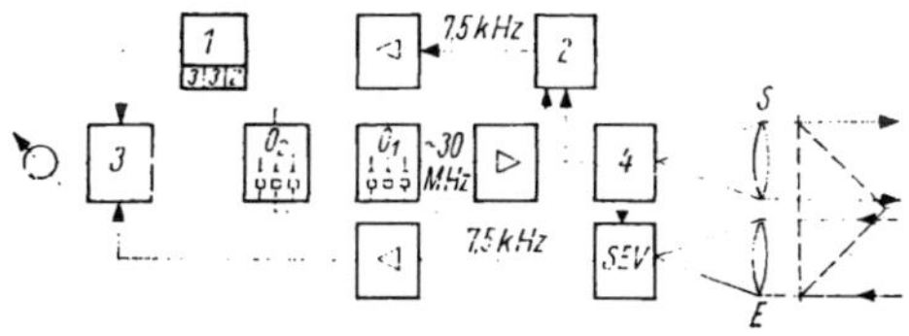

Figure 669 Block diagram of circuits for the EOK 2000

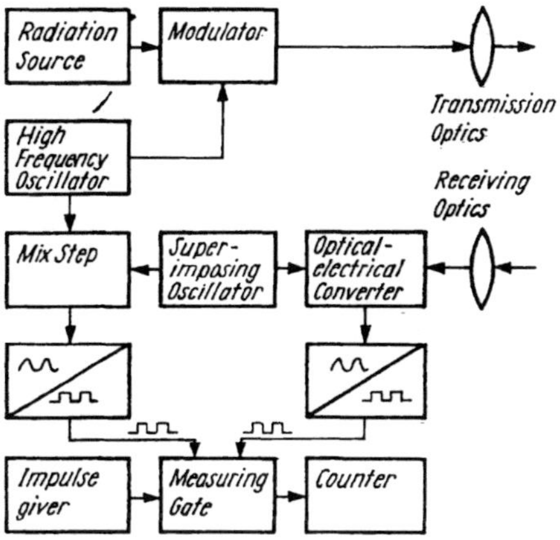

Figure 670 Block diagram of circuits for an electro-optical distance meter with digital phase measurement

Figure 671 DM 2000 (Kern, Aarau); dimensions: 35 cm × 24 cm × 26 cm

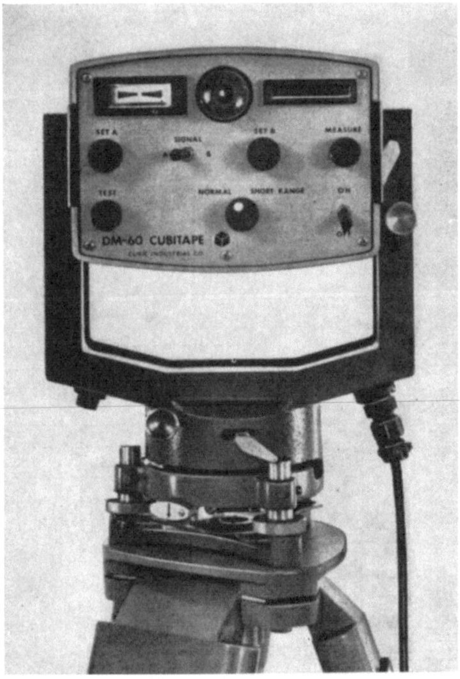

Figure 672 Cubitape DM 60 (Cubic Industries, U.S.A.)
dimensions: 13 cm × 18 cm × 28 cm

Figure 673 Wild DI 10 on seesaw

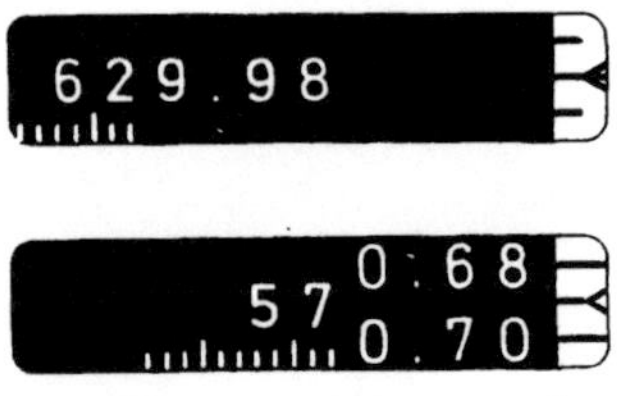

Figure 674 Reading examples for Wild DI 10: 629.98 m, 570.69 m

The Soviet instrument SM-3 also operates with analogue phase measurements.
Modern instruments have digital phase measurement and an automatic measuring procedure requiring 5 to 15 s (figure 670). At the DM 2000 (figure 671) by Kern, Aarau, transmitting and receiving systems are electronically separated. The reference signal is diverted from the transmission signal with a light splitting cube and converted into an electrical signal with the aid of a semi conductor photo diode.
The Cubitape DM-60 (figure 672) by Cubic Industrial Corp., U.S.A., is one of the lightest instruments. Unlike for these instruments and the Hewlett-Packard HP 3805, the Geodimeter 76 uses the light of a 2 mW-He-Ne laser with Kerr-cell modulation, the Geodimeter 78 the light of a 1.8-mW-He-Ne laser. The Soviet 2 SM 2 has separate sighting- and measuring parts.
In the case of attachable instruments either the whole instrument with or without battery is mounted onto the telescope or the standards of a theodolite — usually with the aid of an adapter —, or else the transmitter- and receiving system only. If it is attached to the standards, the target has to be set separately for distance- and angular measurement, or else the telescopes have to be linked with a parallel guidance. If the transmission and receiving system (measuring head) is separated from the electronic part, a cable connection between the two parts is required. If the electronic part is placed between theodolite and tripod plate, the cable does not interfere. If measuring head and electronic part, except for the battery, are combined into one unit, only a battery cable is needed. If the battery is placed inside the attachable instrument, then no cables are needed.
With the exception of the first attachable instrument manufactured in series, namely the Wild DI 10 (figure 673), all instruments utilize digital phase measurements for phase comparison. For the Wild DI-10, the ambiguity of the measurement has been resolved by an automatic gradual increase of the variable modulation frequency from a 10% lower frequency to the 15 MHz frequency of the fine scale. The resulting phase changes are automatically integrated by the phase meter, which is a trailing servomotor consisting of resolver, phase detector and D/C motor. A glass circle is placed on the resolver axis, which indicates metres, decimetres and millimetres in 2 cm steps, while a glass circle with two figures for deca- and hectometres is attached onto the axis with a 10 times reduction transmission. The reading positions of both circles are projected onto a glass screen (figure 674). The equipment consists of sighting- and measuring parts plus a reflector.
One measurement requires 60 s at the most. The Distomat Wild DI 3 S (figure 675) is a further development with digital phase measurement. After sighting and depressing of a start key, the measuring program is automatically executed in 10 s (100 fine measurements of the internal light path, 1000 fine measurements of the distance, 100 fine- and 100 rough measurements of the internal light path, 500 rough measurements of the distance, computation of the distance from single measurements, multiplication with the scale factor, display). The measuring part, which is placed on the tribrack such that it can rotate around a vertical axis, and which can be rotated with the theodolite, contains a computer. After entering the vertical angle via the key board, horizontal distance and elevation difference is displayed, while coordinate differences are displayed when entering the azimuth with the keyboard. The Wild DI 4 (figure 753) is smaller, lover and more compact.
In the Tellurometer CD 6 (figure 676) transmission and receiving systems are electronically separated. The reference signal is diverted from the signal of the transmission diode, and projected onto a semi conductor photo diode via an optical conductor.
While the AGA-Geodimeter 120 (figure 677), and his variants 110, 112 and 116 as well as the geodimeter 14A, which also can be used independently the Nikon MND-2 and Sokkisha SDM-1 C have to be mounted onto the telescope of a theodolite, the DM 502 by Kern, Aarau (figure 678) is snapped onto the telescope. A support for a pocket computer used to determine horizontal distance and elevation difference, can be fastened onto the theodolite support.
Unlike for the instruments mentioned so far, the battery is not fastened onto the tripod legs for the Eldi by Zeiss, Oberkochen, but included within the instrument together with the optical and electronic elements. The difference between the types Eldi 1 (not designed as attachment, figure 679), Eldi 2 (figure 680) — as Eldi 2 Mining it is safe for fire-damp — and Eldi 3, lies solely in the range, which requires different magnifications, telescope openings and

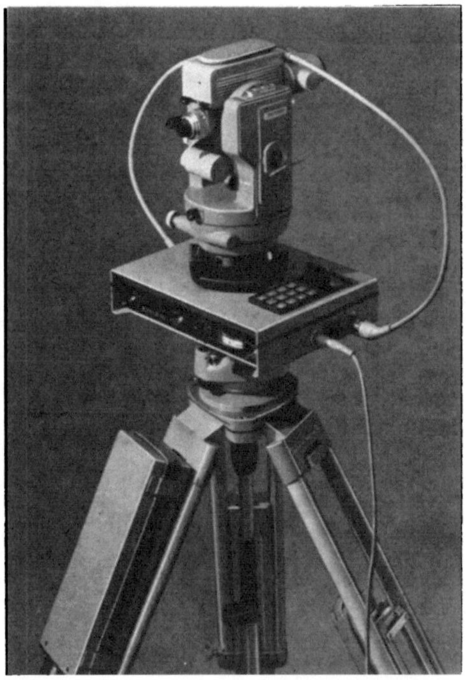

Figure 675 Distomat Wild DI 3 S with Wild T 16 theodolite

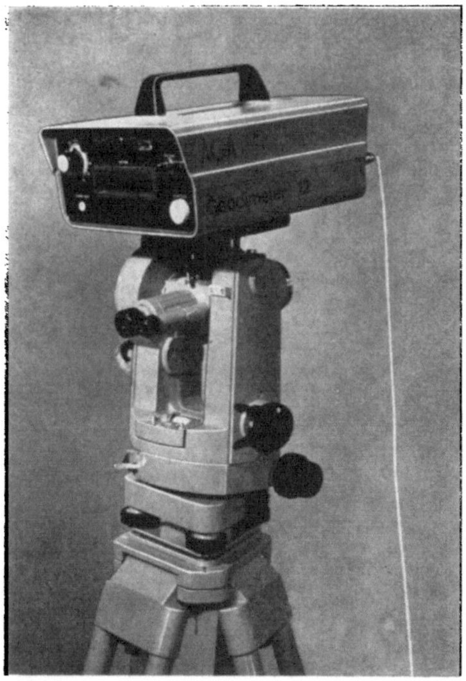

Figure 677 Geodimeter 120 (AGA, Sweden)

Figure 676 Tellurometer CD 6 on top of Breithaupt theodolite

Figure 678 DM 502 on K 1-SE (Kern, Aarau)

masses. For all types, two measuring ranges can be selected — a shorter one with high accuracy, and a longer one with somewhat lower accuracy.
The Autoranger by Keuffel & Esser, U.S.A., which ist mounted onto the telescope or the standards of a theodolite, can be combined with a digital theodolite (see figure 757). The Beetle 500 S and 1000 S by Prexision Intern. Inc., Tullahoma (U.S.A.) display the horizontal distance and elevation difference after entering the vertical angle via the keyboard the Beetle 500 is without this set-up. The AGA Geodimeter 120 contains a sensor, which gives the possibility to obtain the horizontal distance and elevation difference automatically (figure 677).

Figure 679 Eldi 1 (Zeiss, Oberkochen)

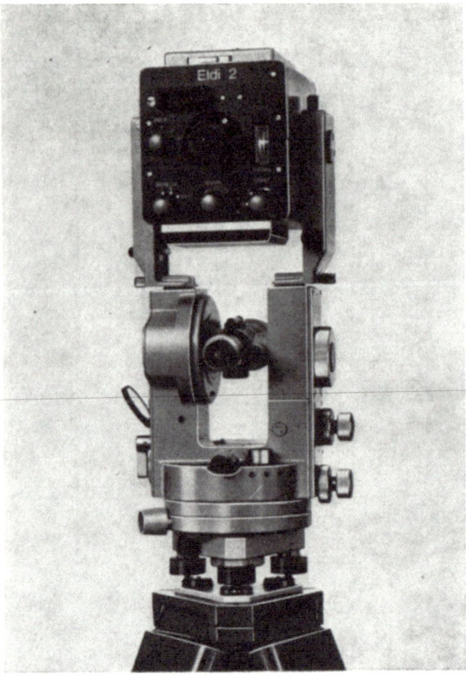

Figure 680 Eldi 2 on top of Th 2 (Zeiss, Oberkochen)

Figure 681 Tellurometer MA 100; dimensions: 32 cm × 37 cm × 28 cm
a) frontal view b) scheme of the optical system
1 receiver; *2* mirror lens; *3* silicium-photo diode; *4* light pipe; *5* mirror; *6* GaAs diode

Figure 682 Mekometer ME 3000 (Kern, Aarau); Dimensions: 46 cm × 16 cm × 22 cm

Electro Optical Precision Distance Meters (table 34)

For numerous applications in engineering surveying, accuracies in the magnitude of millimetres and fractions thereof are required. These can be reached with electro optical distance meters having the same phase measuring accuracy by reduction of the scale, i. e. increasing of the modulation frequency, which requires more effort.

For the Tellurometer MA 100 (figure 681), the frequency of the fine scale amounts to 75 MHz. To eliminate phase shifts which are independent of the frequency, and to increase the measuring accuracy, a "negatively" modulated frequency of 75 MHz is introduced. The difference of the two frequencies amounts to 150 MHz, which produces an effective frequency for the fine scale 1 m. With digital phase measurement to 2×10^{-4}, a resolution of 0.2 mm is achieved. The automatically operated measurement requires one minute of measuring time.

The Mekometer ME 3000 (figure 682) by Kern, Aarau, operates with a fine measuring frequency of 500 MHz, i.e. a 30 cm fine scale. The light flashes of a xenon gas discharge lamp (100 Hz resulting frequency, 1.5 μs pulse duration) are modulated with this frequency by linear polarization with the polarizer P_1, and elliptic polarization with a KDP- (Pockels) modulator. Then they are transmitted to the reflector (figure 683). The magnetron, which is used for the modulation and which operates with impulses, has the same impulse frequency as the xenon lamp, but has a pulse duration in such a phase position that each light flash is modulated. The returning light, which is modulated by polarization, is led through a KDP crystal which oscillates in the modulation synchronized with the transmission crystal. In the crystal, the elliptic polarization of the light is either amplified or reduced, depending on the phase difference between the electric field and the modulation of the light returning from the distance to be measured. After passing through the analysor P_2, the light generates a photo electric current in the SEV. It serves for the zero tuning with an optical mechanical phase slider in front of the receiving optics with two prisms, which can be shifted parallel and which deflect the lightrays twice (a shift by 15 cm increases the light path by 60 cm). Four further frequencies, which deviate from the main frequency by 10, 1, 0.1 and 0.01 % provide rough scales of 3, 30, 300 and 3000 m. The digital display of the measuring result is created with the aid of a calculator which transforms the numerical values, and with five clutches operated by a program switch and located on the shaft which agitates the phase slider. The phase slider is operated with the aid of a hand wheel. By monitoring the modulation frequency by the inherent frequency of a hollow sonator filled with air, the influence of refraction at the station is taken into account. The pressure-, temperature- and humidity conditions for the hollow sonator are the same as at the station. Together with a high modulation frequency, this provides the high accuracy for the measured distance. It is digitally displayed to fractions of a millimetre. The range is 3000 m, with one measurement with five frequencies requiring about 3 minutes' time.

The Soviet instrument DVSD-1200, specially manufactured in single units, uses 1200 MHz high frequency, i.e. a fine scale of only 12.5 cm. It consists of the transmission- and receiving part and the electronic part with a 2 mW-gas laser as radiation source and a KDP-modulator, which, like for the

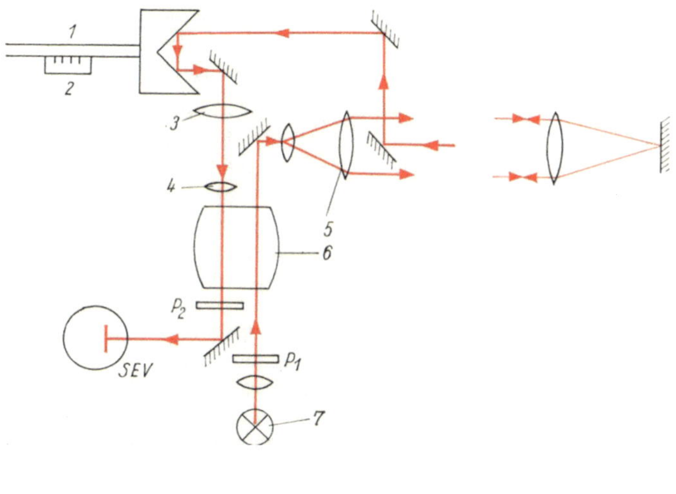

Figure 683 Block diagram of circuits of the Mekometer
1 scale; *2* vernier; *3* reception optics; *4* focussing; *5* transmission optics; *6* modulator (2 or 4 KDP-crystals); *7* light source

Figure 684 DVSD-1200 (U.S.S.R.)

Mekometer, also acts as a demodulator (figure 684). The phase difference is read at the modulator position with a micrometer dial with 0.04 mm least count, when the electric signal shows a minimum on the CRT tube. This minimum is detected with the aid of the phase slider. One measuring program with several series requires about 20 minutes.

Microwave Distance Meters

In the master- and remote stations of a microwave distance meter the carrier frequencies of 10.05 to 10.45 GHz (resp. near 3 or 35 GHz) are generated by one Klystron (microwave oscillator) each, with the carrier frequency of the supplementary station being by 33 MHz higher than the one from the main station. The carrier waves are frequency modulated with the scale frequencies ω_1 resp. ω_2 via a quartz oscillator. (It is easier to superimpose a 7.5 MHz frequency onto a 10 GHz frequency as such—rather than to impose an amplitude modulation.) Along the way between transmission station and remote station, phase delays $\varphi_M + t_1\omega_1$ resp. $\varphi_R = t_1\omega_2$ occur for the scale waves.

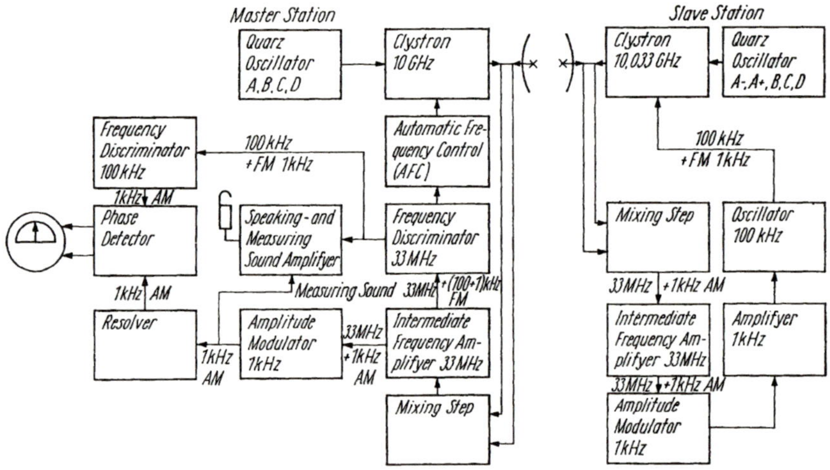

Figure 685 Block diagram of Tellurometer MRA 101

The scale frequencies of the opposite stations are received in the master and remote stations. Their phase position in respect to the emitted frequencies is determined:

$$\varphi'_M = \varphi_M - (\varphi_R + t_1\omega_2);\ \varphi'_R = (\varphi_M + t_1\omega_1) - \varphi_R.$$

The remote station continuously transmits the phase position to the master station with an additional, frequency modulated auxiliary carrier phase

$$\varphi''_M = (\varphi_M + t_1\omega_1) - \varphi_R + t_1(\omega_1 - \omega_2).$$

There, the differences of the phase positions for master- and remote stations, and thus the distance, are determined with the aid of a phase meter as:

$$\varphi''_M - \varphi'_M = (\varphi_M + t_1\omega_1) - \varphi_R + t_1(\omega_1 - \omega_2) - [\varphi_M - (\varphi_R + t_1\omega_2)] = 2\omega_1 t_1. \qquad (232)$$

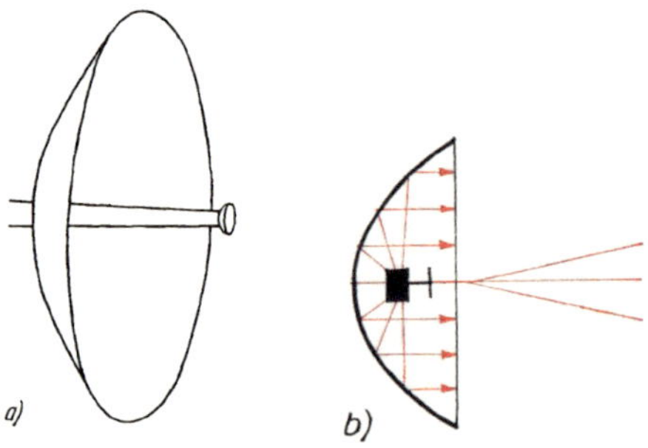

Figure 686 Parobolic antenna

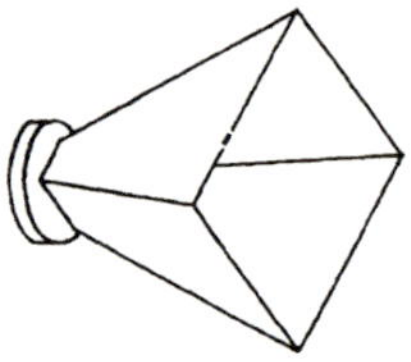

Figure 687 Horn antenna

Based on the primary circuit of the tellurometer (figure 685), the phase difference measurement of microwave distance meters is explained. The carrier of the master station is frequency modulated with the frequency A (7.492,377 MHz), while the carrier of the remote station is modulated with A + (7.491,377 MHz). Then both are transmitted via the transmission dipole of the parabolic antenna (figure 686) or horn antenna (figure 687). The signals, emitted from master- and remote stations, are superimposed at the antenna of the main station and mixed in a mixing step.
The intermediate frequency of 33 MHz, obtained at its exit, is amplitude modulated with a modulation frequency of 1 kHz (the difference of the two frequency modulations). At the remote station also a 33 MHz signal with 1 kHz amplitude modulation is obtained with the same procedure. The veil curves of the intermediate frequency signals contain the scale differences of the scale waves at the mixing steps (the mixing steps thus represent the end

points of the distance). The 1 kHz signals corresponding to the veil curves are obtained from the intermediate frequency signals via an amplitude modulator. At the master station, their 1 kHz signal is to be considered as reference signal, while the one from the remote station represents the measuring signal.
To transmit the phase position onto the master station, a frequency modulated 100 kHz auxiliary signal is generated from the 1 kHz signal at the remote station and also modulated onto the carrier. This results in an additional frequency modulation of the intermediate frequency at the master station. With the aid of a 33 MHz and of a 100 kHz frequency discriminator the 1 kHz measuring signal is derived from it. The phase comparison is performed with analogue phase measurement using resolver, phase detector and zero instrument.
To eliminate phase errors due to instrument internal path time variations, the fine measurement is repeated with the frequency $A-$, where $A-(A+) = A+(A-)$. Thus the remainder is obtained with the opposite sign, while the phase error retains its sign. Its influence is eliminated in the difference of the two measurements, while in the sum, the remainder is obtained like with double the modulation frequency.
Due to lack of narrow bundling of the microwaves, the direct ray is superimposed with rays reflected on the ground (figure 688). Since these cover a longer path, they lead to phase errors and thus to distance errors. One tries to keep these errors ("ground swing") small by repeating the fine measurement with, for instance, 10 different carrier frequencies.

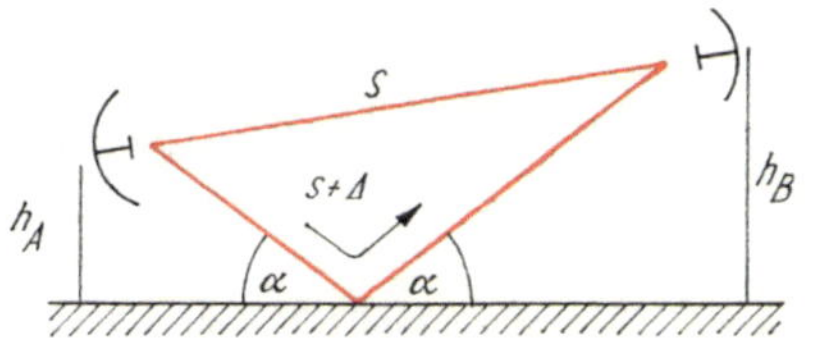

Figure 688 Influence of ground reflection

The automatic frequency control (AFC) at the master station serves to keep the difference frequency constant at 33 MHz, after the Klystron has been tuned. Furthermore, the 1 kHz-difference frequency is automatically controlled.
Like the fine scale, the rough scales are also obtained from differences:

B	7.490,879 MHz	*A–B*	1.498 kHz	100,000 m
C	7.477,392 MHz	*A–C*	14.985 kHz	10,000 m
D	7.342,529 MHz	*A–D*	149.848 kHz	1000 m
E	5.993,902 MHz	*A–E*	1.498,475 MHz	100 m
A	7.492,377 MHz	*A–(A–)*	14.984,754 MHz	10 m

Modern instruments can be switched to master- as well as to remote station. With the exception of the Klystron, most instruments are now fully transistorized, which reduced their mass. Printed circuits permit a quick exchange of the plates when defect. The antenna system consists of two dipoles, normal to each other and inclined by 50 gon to the horizontal. They are located in the axis of a parabolic mirror of 40 cm diameter. The polarization directions normal to each other, largely disconnect the dipoles, so that the antenna can be commonly used for transmitter and receiver.
Using the phase difference, the distance can be computed according to equation (210), after the value for c had been determined with equation (214). The required refractive index n_M for microwave measurements is determined with the aid of the formula by *Essen* and *Froome*:

$$(n_M - 1)\,10 = \frac{77.64}{T}(p - e) + \frac{64.68}{T}\left(1 + \frac{5748}{T}\right)e \qquad (233)$$

where $T = 273 + t$ represents the temperature in K resp. °C, p the air pressure in mbar, and e the partial vapor pressure in mbar. Since, according to experience, n is between 1.000,280 and 1.000,350, the value computed with equation (233) will amount to 280 to 350.
The vapor pressure e is determined to about 1% accuracy with Sprung's formula

$$e = E' - \frac{(t - t')}{2}\frac{p}{497} = E' - 0.000{,}1006\,p(t - t') \qquad (234)$$

using the pressure E' of saturated water vapor in mbar at the damp temperature t', the readings t and t' at the dry and damp thermometer, and the air pressure p. E' is tabulated, and t and t' are to be measured with psychrometers (Assmann's Aspiration psychrometers are especially well suited).
In modern instruments with digital readings of the distances, mean values for the meteorological conditions are taken into account. Deviations from these can be considered by switching to a corresponding refractive index or with the aid of correction tables.

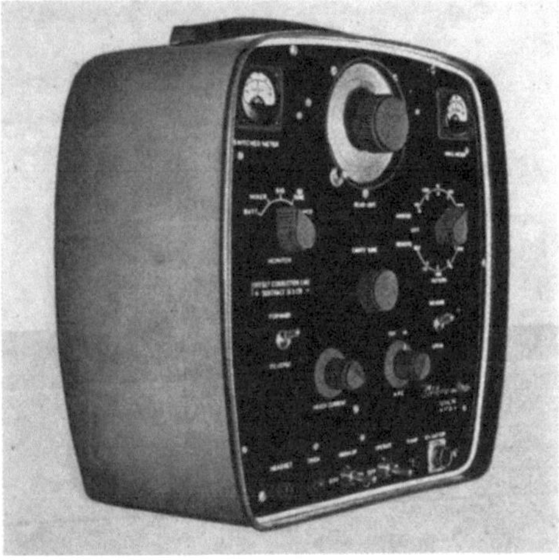

Figure 689 Tellurometer MRA 101; dimensions: 38 cm × 37 cm × 18 cm

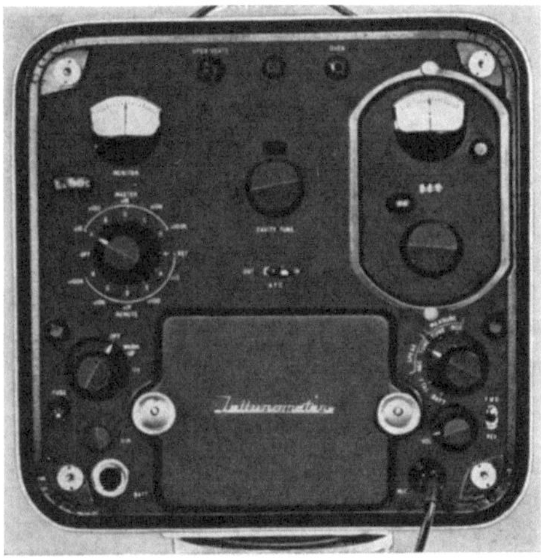

Figure 690 Tellurometer MRA 3; dimensions 36 cm × 31 cm × 31 cm

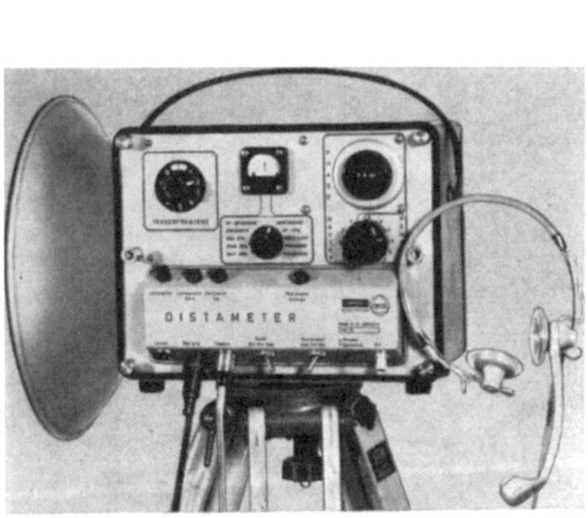

Figure 692 Distameter (Ertel, Munich); dimensions: 32 cm × 28 cm × 19 cm

Types of Microwave Distance Meters (Table 35)

The Tellurometer MRA 101 (figure 689) with only 7.3 kg mass, and its military version MRA 3 (figure 690) for extreme climates are interchangeable. The antenna is built into the broad side of the instrument. The instruments operate with 3 cm waves (carrier frequency 10 GHz). Therefore the bundle is narrower compared to 10 cm waves (3 GHz). While the bundle angle is about 20° for 3 GHz, it is about 6 to 9° for 10 GHz, and only 2° for 35 GHz, which means that smaller influences of terrain reflection and thus higher accuracies can be expected. The Tellurometer MR A 4 with 8 mm carrier waves (34.5 to 35.1 GHz) and 75 MHz modulation frequency supplies higher accuracy. In the Tellurometer MRA 5, consisting of antenna and control unit, the measuring procedure is automized. The scale frequencies do not have to be switched on manually any more. Luminous figures display the result of the distance obtained by digital phase measurement.

Figure 691 Electrotape DM-20 (Cubic, U.S.A.) dimensions: 38 cm × 36 cm × 29 cm

The Electrotape DM-20 (figure 691) by Cubic, San Diego (U.S.A.), is similar to the MRA 3. It operates with 3 cm carrier waves and five modulation frequencies between 6 and 7.5 MHz, whose difference frequencies between channel 1 and channels 3 to 6 result in exact decimal scales. The fine scale is obtained from the difference of the measured values from channels 2 and 1. The distance is obtained from the means of the single readings on a counter as follows:

Channel 6	322
Channel 5	232
Channel 4	327
Channel 3	21 4
Channel 2	1 66
	32321.66
Correction because of deviation from $n = 1.000{,}32$	+ 16
	32321.82 m

The Distameter (figure 692) by Ertel, Munich, and Grundig, Fürth, operates similar to the Electrotape. The Telemeter RG 10 (figure 693) by RAWAR, Warsaw, operates with 3 cm carrier waves and four modulation frequencies between 6 and 7.5 MHz. Three digits each are displayed digitally.

Table 35. Microwave Distance Meters (selection)

type	manufacturer	source of carrier waves: carrier frequency [MHz]	modulation frequency: modulation fine frequency [MHz]	modulation frequency: number [cm]	antenna: diameter [°]	antenna: width of half value [W]	power requirement	phase measuring method	range [km]	mass [kg]	dimensions [cm³]	standard deviations	remarks
1	2	3	4	5	6	7	8	9	10	11	12	13	14
Electro-tape DM 20	Cubic Corp., U.S.A.	clystron 10.0 up to 10.5	7.5	5	33	6		analog resolver	0.05 up to 50	11.5	38 × 36 × 29	± (10 mm + 3 · 10^{-6} s)	figure 691
Dista-meter	Ertel-Werk, Munich	clystron 10.0 up to 10.5	7.5	2	35	6	30	analog resolver	0.05 up to 50	10.8	32 × 28 × 19	± (20 mm + 5 · 10^{-6} s)	figure 692
GET-B 1	Finnomechanikai Vallalat, Budapest	clystron 2.73 up to 3.18	10	4	44	10		analog oscillo-graph	0.2 up to 50	15		± (50 mm + 3 · 10^{-6} s)	
RG 10	RAWAR, Warsaw	clystron 8.8 up to 9.5	7.5	4	29	7	32	analog	0.1 up to 40	10	33 × 18 × 38	± (30 mm + 5 · 10^{-6} s)	figure 693
SIAL MD 60	Siemens-Albis AG, Zurich	clystron 10.3	150	5	35	6	38	digital	0.02 up to 150	13.5	37 × 37 × 28	± (10 mm + 3 · 10^{-6} s)	figure 697; MD 60 S with separate antenna- and control units
MRA 3	Tellurometer, England	clystron 10.025 up to 10.450	7.5	5	25	9	32	analog resolver	0.1 up to 50	16	26 × 33 × 31	± (15 mm + 3 · 10^{-6} s)	figure 690
MRA 5		Gun-oscillator 10 up to 10.5	75	2		12	42	digital	0.1 up to 50	12/6	31 × 35 × 16 31 × 35 × 30	± (10 mm + 3 · 10^{-6} s)	separate antenna- and control units
CA 1000		Gun-oscillator 10.1 up to 10.45	25	5	horn beam emitter 20°	20	4.5	analog digital	0.05 up to 30	1.6	31 × 15 × 8	± (15 mm + 5 · 10^{-6} s)	figure 698; 4 kg with built-in batteries; instruments cannot be exchanged
Lutch	U.S.S.R.	clystron 8.6 up to 8.9	10		38	5	80	analog	0.2 up to 40	14/9.5	40 × 40 × 20 28 × 24 × 20	± (30 mm + 3 · 10^{-6} s)	figure 694; separate antenna- and control units
Distomat DI 50	Wild Heerbrugg AG	clystron 10.2 up to 10.5	15	4	36	6	50	digital	0.1 up to 50	8/14.8	44 × 25 × 20 47 × 24 × 24	± (20 mm + 5 · 10^{-6} s)	figures 695/696; separate antenna- and control units

Figure 693 RG 10 (RAWAR, Warsaw); dimensions: 33 cm × 18 cm × 38 cm

Figure 694 "Lutch" (U.S.S.R.); dimensions: 28 cm × 24 cm × 20 cm (Transmission—Receiver Unit); 40 cm × 40 cm × 20 cm (Measuring unit)

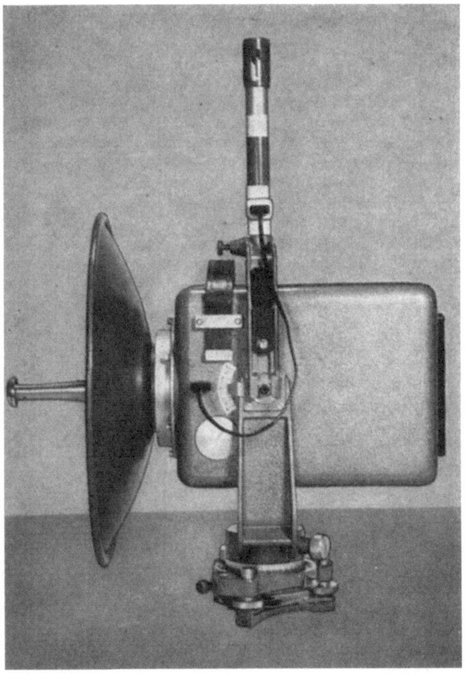

Figure 695 Wild Distomat DI 50; Transmission—receiver unit; dimensions: 44 cm × 25 cm × 20 cm

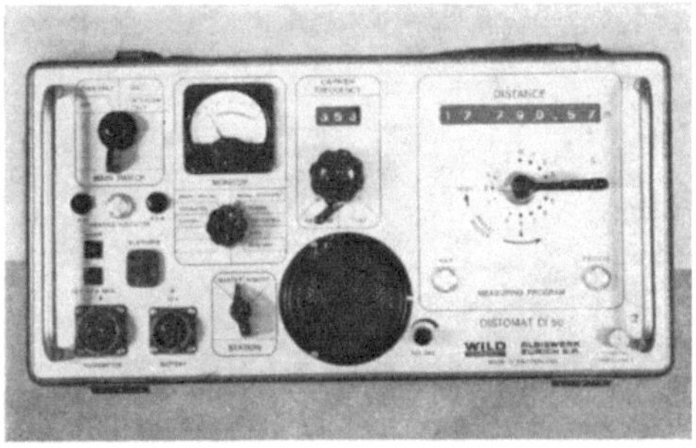

Figure 696 Wild Distomat DI 50; measuring unit; dimensions: 47 cm × 24 cm × 24 cm

For the "Lutch" (figure 694) from U.S.S.R., the control unit is connected with the antenna via cables and can be elevated to 25 m height. The carrier frequencies are between 8.6 and 8.9 GHz, the modulation frequencies around 10 MHz. The phase measurement is achieved with the aid of phase slider and -detector.

The Wild Distomat DI 50 (figures 695 and 696) by Wild, Heerbrugg, and Albiswerk Zürich, also has transmission-receiving unit and measuring unit separated and connected by cable. The units can be placed up to 30 m apart. The carrier frequencies are between 10.235 and 10.485 GHz, the modulation frequencies between 13.5 and 15 MHz. The measuring procedure is automized. After turning a program switch step by step, the distance is displayed digitally. Each display represents the mean of 20 resp. 40 electronic measurements.

In the SIAL MD-60 (figure 697) by Siemens-Albis, Zürich, a significantly higher scale frequency is utilized for fine measurements. It is obtained by multiplication of the basic frequency 15 MHz to 150 MHz. Therefore, ground reflections are largely reduced, and instead of varying the carrier frequency because of ground swing, only one carrier frequency is used (3 cm waves). Tuning and measuring process are automatic within 30 seconds of pressing a switch. In addition to the fine scale 1 m, four rough scales of 10, 100, 1000 and 20,000 m (13.5 to 15 MHz) are used, so that the display is definite up to 20 km. The instrument is supplied in two forms, as compact instrument MD 60 C and as a two part instrument MD 60 S for work on masts and

Figure 697 SIAL MD 60 (Siemens—Albis, Zurich); dimensions: 37 cm × 37 cm × 28 cm

towers; measuring- and antenna parts are connected by a 4 or 25 m long cable.

In the Tellurometer CA 1000 (figure 698), an instrument of especially small dimensions, low mass and low energy requirement, only semi conductor elements are used, mostly as integrated circuits. The carrier frequency 10.10 to 10.45 MHz is not generated by a reflex-Klystron, but by a solid state microwave source. The modulation frequencies between 19 and 25 MHz are derived from one quartz only. There is no warm-up period, because the quartz-oscillator is largely independent of the temperature, and therefore does not need to be thermostat controlled. Instead of parabolic antennas, horn beam emitters are used — for 10 km range a standard antenna horn with 20° bundling, for 30 km range a long range antenna horn. The carrier frequency provided by the master station is automatically kept by the remote station (figure 699). The fine scale is resolved to ±6 mm by a digital-analogue-phase measuring system. It is possible to have continuous verbal radio contact even when the instrument is in measuring state.

Accuracy and Application of Electronic Distance Meters Operating with Phase Comparison

The accuracy of distance measurements with the aid of phase comparison depends primarily on the accuracy of the determination of the number a of the oscillation periods of the phase difference, on the addition constant, the modulation frequency, and on the propagation velocity of the carrier frequency at given meteorological conditions.

To safely determine the number a of oscillation periods, m_a has to be smaller than 0.5. This is achieved by measuring with several modulation frequencies (resp. frequency variation). The phase difference can be determined to 1/3000 of the modulation wave length with analogue- and to 1/10,000 with digital phase measurement. Uncontrollable shifts of the instrumental zero point occur due to cyclic errors based on superimposed errors caused by electric coupling between reference- and measuring signals, and based on optical superimpositions caused by phase inhomogenities of radiation sources and photo detectors, as well as different signal strengths. Usually, they remain within the given accuracy limits, but may lead to zero errors of up to ±10 mm for fine scales of 10 to 20 m. The influence of these three factors is practically independent of the distance, while the influence of the following errors is proportional to it (see page 258).

Calibration and constant frequency determine the accuracy of the modulation frequencies taken from quartz oscillators, which when using thermostats, amount to 10^{-6} to 10^{-7}. For short distance instruments, frequency changes of 10^{-5} are permissible within a temperature range of -25 to $+50$ °C, which means that no thermostat is requierd. The accuracy of the propagation velocity of the carrier waves depends on the accuracy of its determination in vacuum ($\Delta c_0 = \pm 0.0012$ km/s, i.e. $\Delta c_0/c_0 \approx 4 \times 10^{-9}$) and of the refractive index. The accuracy, with which the empirical formulae for the computation of n represent the actual conditions, amounts to 2×10^{-7} for *Barrell* and *Sears*' formula (231) for light waves, and to 1×10^{-7} for Essen and Froome's formula (233) for microwaves. However, the accuracy of n depends primarily on the measuring accuracy of the meteorological data. Based on equations

Figure 698 Tellurometer CA 1000; dimensions: 31 cm × 17 cm × 8 cm

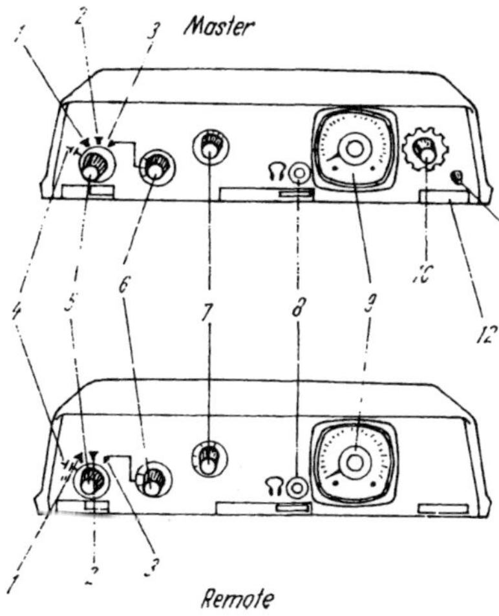

Figure 699 CA 1000 a) master station b) remote station

1 output
2 signal
3 measure
4 battery check
5 function switch
6 pattern switch
7 frequency control
8 headset socket
9 meter
10 readout switch
11 external battery input socket (below)
12 readout phase zero control

(231) and (233), errors in temperature- and pressure measurements of air under normal conditions lead, according to *Mitter*, to the following errors of the refractive index for light- (and infrared-) and microwaves as carriers:

Error	Relative error of the refractive index $\Delta n/n$	
	lightwaves	microwaves
$\Delta t = \pm 1$ K	$\pm 1.0 \times 10^{-6}$	$\pm 1.4 \times 10^{-6}$
$\Delta p = \pm 1$ mbar	$\pm 0.3 \times 10^{-6}$	$\pm 0.3 \times 10^{-6}$
$\Delta e = \pm 1$ mbar	$\pm 0.04 \times 10^{-6}$	$\pm 4.7 \times 10^{-6}$

Since an error of $\Delta p = 3.3$ mbar causes only an error of 1×10^{-6} in the distance, a calibrated aneroid barometer is sufficient for the determination of the air pressure. Since, on the other hand, an error in the dry resp. damp temperature, which causes the errors Δt and Δe, of ± 1 K for light waves, and ± 0.2 K for microwaves causes a distance error of 1×10^{-6}, the temperature has to be measured with carefully calibrated thermometers, graduated to 0.1 to 0.2 K. Temperature measurement to ± 0.2 K is hardly possible. It is possible to get a better solution using two wavelengths and the dispersion method — generally, the meteorological elements are determined at the end points only. The refractive index can therefore often be obtained to 5×10^{-6} only. It is less accurate to determine the propagation velocity of micro waves from meteorological data, than that of light waves.
Therefore, according to equation (227), under favourable conditions and knowledge of the average composition of the meteorological field, accuracies of 1 to 2×10^{-6} can be expected for electro optical distance measurements, while the values for microwaves lie at 3 to 4×10^{-6}.
Microwave distance meters have ranges up to 50 or 100 km. They are primarily used for the design of long distance traffic and supply lines, for trilateration nets with 20 to 80 km long sides, for photo control surveys and others. It is of advantage, that measurements can be taken with these instruments at any time of the day or night and under nearly any weather condition, even with poor sight or in fog, and that the instrument can be aligned relatively easily because of the cone angle of the radiation of several degrees.
With long range electro optical distance meters, distances of 30 resp. 60 km can be measured to within a few centimetres, while short range instruments permit 1 to 2 cm accuracy for distances up to 1 to 5 km. With precision distance meters an accuracy of millimetres or fractions thereof can be reached. Based on this, these instruments are used for measurements of high accuracy such as coordination of points within a geodetic network, traversing, route surveys, and different engineering applications, and others. The high accuracy and the fact that only a reflector has to be placed at the end point of the distance, are of advantage.

Testing and Adjusting of EDM Instruments

The measuring accuracy of electronic distance meters, depends primarily on the modulation frequency, the addition constant and the periodic errors. It is advisable to check the modulation frequencies about three times per year. This test is performed in a laboratory with a frequency meter.
The periodic errors have a period of half the wavelength of the modulation wave. They are comparable to micrometer errors in theodolites. They are caused by mutual influences of transmitter, receiver and photometer onto each other. A test line with well known points spread over at least one period is used for this test. The separation between instrument and reflector is not important, because the exact course of the error is determined by measuring known distance differences. For instance on a 30 m long test line consisting of a rail with a known point at every metre, instruments with $\lambda = 20$ m can be tested over three half modulation periods from any distance up to 100 m. For each period, 10 measured points are available. If one plots the errors of the measured values against the known ones as a function of the distance, a curve similar to the sine curve is obtained (figure 700). For easier adjustment, additional measuring points should be available near the maximum and minimum of the error curve (in the case presented, near 2.5 and 5.5 m).

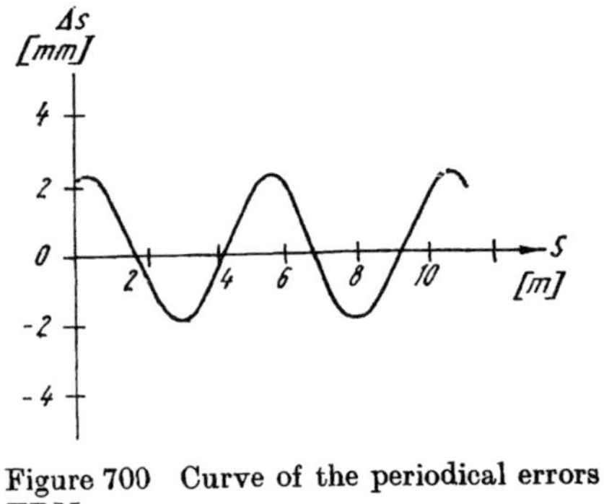

Figure 700 Curve of the periodical errors of EDM

The addition constant (difference between the separation of the two forced centering devices for instrument and reflector and the separation of the end points of the actually measured distance) should be checked regularly, e.g. three times annually. The standard deviation of the instrumental zero point can be obtained with this, which provides an overview over the condition of the instrument at various points in time for the computation of weights and for adjustments as well as for the comparison with other instruments. Periodic errors can be derived using a representative measuring arrangement. The addition constant c_k can be determined by

— measuring a distance directly and by measuring its partial distances in all combinations

— measuring an exactly known distance (test line) and comparing it with its known value.

For microwave instruments, generally the measurement on a test line is preferred.
The distances should cover the whole range, expected for practical work, and only be so long (e.g. 1 km), that frequency errors and errors in the refractive index have only insignificant influence on the result. There should be no reflections on the test lines for micro wave instruments.
When using the line segment method, the distances need not be known. If the distance $\overline{AC}$ is subdivided into $\overline{AB}$ and $\overline{BC}$ (figure 701) and the lengths s, s_1 and s_2 are measured, then

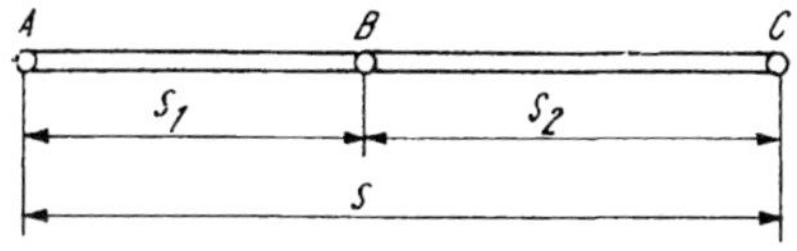

Figure 701 Determination of the addition constant using the distance segment method

$$\overline{AC} = s + c_k \quad \text{and} \quad \overline{AC} = \overline{AB} + \overline{BC} = s_1 + c_k + s_2 + c_k .$$

By combining these, c_k becomes

$$c_k = s - (s_1 + s_2) . \tag{235}$$

With several intermediate points, and distance measurements in all combinations, the accuracy can be increased. Distances with 6 points ($n = 5$ independent line segments) provide the optimum as far as accuracy and economics is concerned. In this case $\binom{6}{2} = 15$ different lengths and $N = n(n^2 - 1)/6 = 20$ equations for c_k are obtained. These are

$$c_{ki} = s_{ik} - (s_{ij} + s_{jk}) . \tag{236}$$

Using the 5 line segments and the addition constant as unknown, the addition constant c_k and its standard deviation m_{c_k} is obtained by least squares adjustment. It also follows directly from the arithmetic mean:

$$c_k = \frac{[c_{ki}]}{\frac{1}{6} n(n^2 - 1)} . \tag{237}$$

The standard deviation of its determination becomes with $v_i = c_k - c_{ki}$

$$m_{c_k} = \pm \sqrt{\frac{[vv]\,(n^3 - n - 6)}{n(n - 1)\,3(n^2 - n - 2)}} \tag{238}$$

or, with $n = 5$

$$m_{c_k} = \pm \sqrt{\frac{[vv]}{20}} \cdot 1.453 . \tag{238a}$$

When using the line segment method, it is advisable to distribute the final measurement of the segments equally over the fine scale of the instrument (figure 702).

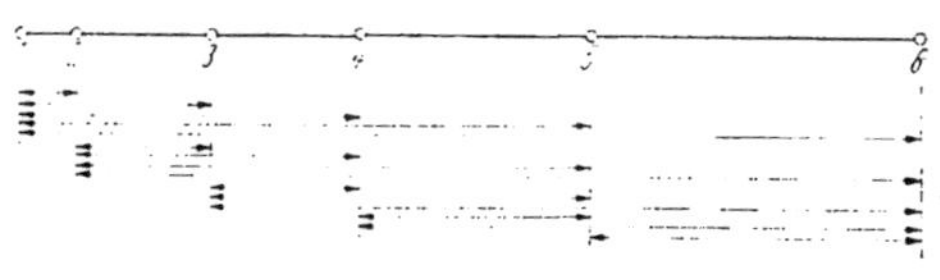
Figure 702 Determination of the addition constant by measuring distances in all combina tions

4.3.4. Interference Comparato

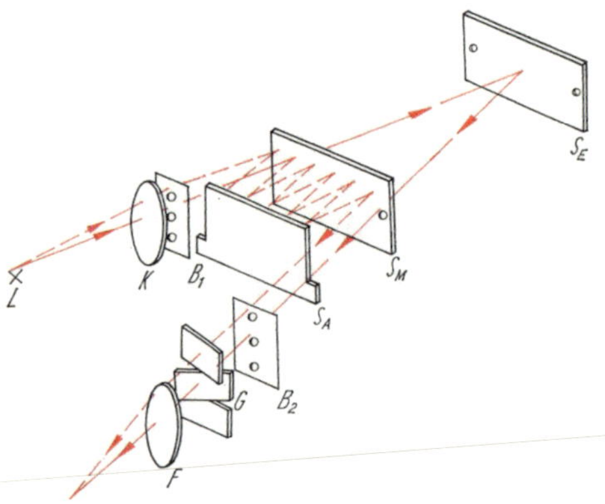

Figure 703 Interference comparator

Distance measurements with light interference do not belong to EDM, are based, however, on the characteristics of electromagnetic waves. If two wavefronts from a light source meet in one place after having covered different paths, interference occurs, if the two light paths are within a few λ of equal length. This phenomenon is used in the interference comparator designed by *Väisälä* which is used to calibrate invar wires and to measure base lines up to 864 m length with an accuracy of 10^{-7}.
The light rays, emitted from a white point source of light L (figure 703), are set parallel with the aid of the collimator objective K. The diaphragm B_1 separates two bundles of rays, one of which is reflected several times along its path between the mirrors S_A and S_M, while the other one is reflected by the mirror S_E. Both bundles of rays meet in the focal plane of the observation telescope F. If S_AS_E is an integer multiple of S_AS_M, then marked light interference occurs, and the 1 m long initial distance can be multiplied by six very precisely. Since the initial distance has to be known very accurately, it is measured with a calibrated quartz normal metre, utilizing Newton's rings. With the aid of two rotatable plan parallel glass plates G (compensators), placed directly in front of the telescope objective lens, interference can even then be detected, when the separations of the mirrors in measuring direction, deviate by amounts of up to 0.5 mm from their theoretical position. By continuous multiplication, base lines of 864 m in length can be measured to an accuracy of a few hundredth of a millimetre.
In the field comparator, developed in U.S.S.R. by *Warlich* and *Schwarz*, the 1.20 m long initial distance is multiplied to 24 m in one operation with stations, which are arranged at a distance of up to 4.80 m from the first mirror.

5. Tacheometric Instruments

All measuring methods with which position and elevation of terrain points is determined from one station with an instrument providing direction, elevation difference and distance are considered as tacheometry. Theodolites or compasses with a special optical device for distance measurement are called tacheometers, while electronic tacheometers contains an EDM. Tacheometer levels can only be used in flat and nearly plane terrain. In all other cases, stadia theodolites and -compasses, also denoted as circle tacheometers, tacheometers with base at the station, as well as equipment with plane table and alidade for graphical mapping of points, can be used. Electronic tacheometers distinguish themselves from optical ones by longer range, shorter measuring time and higher accuracy.
Each terrain point to be measured is described by polar coordinates, with the origin of the coordinate system being at the instrument station. The optically determined distances and elevation differences are reduced to horizontal, computationally for circle tacheometers, completely or partially automatically for self reducing tacheometers, and either by computation or with an internal computer for electronically determined distances.
In addition to simple tacheometers with vertical rod for fast, easy and economical terrain survey with suffificient accuracy for plotting in a common scale (for construction purposes, planning and topographic mapping), instruments with horizontal rod for precision tacheometry are in use. The optical tacheometers can be categorized with the aid of the standard deviation for a standard distance of 100 m and a measuring range of 300 m as follows:

	Standard deviations of one standard distance
Instruments of highest accuracy	$\leqq$ 1 cm
Instruments of high accuracy	$\leqq$ 3 cm
Instruments of medium accuracy	$\leqq$10 cm
Instruments of low accuracy	$\leqq$30 cm

To determine the standard deviation of distance measurements with a tacheometer, points are marked on a 200 m long test line at distances of 10, 20, 40, 60, 80, 100, 150 and 200 m from the instrument station. Each of these line segments is measured ten times with the tacheometer. For each measurement, the rod is shifted by small amounts symmetrically to the fixed marks, which provides different readings for each measurement. The 10 observations are reduced to the distance of the fixed mark, afterwards the mean value is computed. With the differences $v_i = s_m - s_i$ between the single observations s_i and the mean s_m, the standard deviation m_s of the measured distance becomes

$$m_s = \pm \sqrt{\frac{[v_i v_i]}{n-1}}. \tag{239}$$

The errors of the various line segments are transformed into the standard deviation of the standard length 100 m and expressed in cm/100 m for variable base

$$m_{100} = \frac{m_s}{s} \cdot 100 \tag{240}$$

for constant base

$$m_{100} = \frac{m_s}{s^2} \cdot 10{,}000. \tag{241}$$

The observations should be taken by three different observers. The quadratic mean value of all standard deviations m_{100} reduced to 100 m can be considered as the accuracy value of the instrument.

Electronic tacheometers can be grouped into

— instruments, where the circle reading is performed visually and the distance measurement electronically

— recording electronic tacheometers, where angle- and distance measurement is automized.

The accuracy of the distance measurement is 1 to 2 cm, the range 1 to 5 km.

5.1. Non Reducing Tacheometers

5.1.1. Stadia Theodolites

Structure and Types of Stadia Theodolites

Stadia theodolites are theodolites of low and medium accuracy (see section 2.3.3., table 8 and 9), equipped with stadia hairs to measure the distance to the target, a vertical circle to measure vertical angles, a compass (full circle compass or orientation compass), and often a levelling bubble attached to the telescope. Horizontal distance and elevation difference are computed with the aid of stadia tables, -monograms, or -slide rules, using slope distance and vertical angle. Any simple, clearly graduated levelling rod can be used as measuring rod up to about 130 m sighting distances. For longer sighting distances, rods are recommended, which are subdivided into 5 cm fields and can be read without numbers on them. For medium sighting distances, rods are most suitable, if they have the centimetre field arranged in groups of 5 cm (see figure 785).

Due to the measuring accuracy as derived from the plotting accuracy, it is usually sufficient to centre the level bubble in one position only, and to read one circle position (without using the micrometer by just estimating between the graduation lines) in only one telescope position. This leads to faster and more convenient operation.

Testing, Adjusting and Accuracy of Stadia Theodolites

Considering the measuring accuracy and the operation in only one bubble position, one circle reading and one telescope position, instrument errors are adjusted for stadia purposes only to a degree that their effect remains within the desired measuring accuracy. This is done as follows:

1. Adjustment of platen level
2. Removal of collimation- and horizontal axis errors
3. Elimination of the height index error by adjustment of the index bubble. The index error should be checked daily: For centered bubbles of index and telescope levels the vertical circle reading should amount to 100 gon.
4. Adjustment of telescope bubble, if the instrument is equipped with one.

The accuracy of distance measurement with tacheometers amounts to about 1 : 300 for distances up to 250 m. It may drop to 1 : 100 for unfavourable conditions, or reach 1 : 800 for favourable conditions (and 30 × telescope magnification).

For distances up to about 40 m a standard deviation of about 1 dm can be assumed, with 1 to 3 dm for distances of 40 to 100 m. This accuracy is sufficient for topographic surveys, but not for legal surveys.

5.1.2. Stadia Compasses

Structure and Types of Stadia Compasses

Stadia compasses are less accurate than stadia theodolites, have however a simpler structure and are easier to operate. In addition to the compass, they are equipped with a telescope with stadia hairs, a vertical circle, which is read with a vernier or line microscope, clamps and fine motion screws, as well as a separate horizontal circle. Often the telescope is supported by a one-armed standard (see figure 164). The characteristics of compass instruments have been treated in section 2.2.2. and table 4. As far as measuring rods are concerned, the remarks for stadia theodolites are also applicable for stadia compasses. Stadia compasses are primarily used for topographic surveys for scales of 1 : 5000 and smaller, and also for conservation-, forest- and agricultural surveys.

Testing and Adjusting of Stadia Compasses

Stadia compasses are tested and adjusted just like stadia theodolites and compasses. Because of the low accuracy, there is no height index bubble.

5.1.3. Stadia Levels

Stadia levels — levels with horizontal circle and stadia hairs — are easier to handle than circle tacheometers, since the horizontal distance can be directly read on the rod, and since the elevation difference is obtained from the difference between rod reading and instrument height.

5.1.4. Tacheometers with Base at the Station

Tacheometric surveys of low accuracy can be done with tacheometers with variable base at the station. These are combinations of a range finder with a telescope compass or with a simple theodolite (see figures 636, 638, 640). The coincidence telemeter Wild TM 10 can be used as tacheometer, if it is placed onto a horizontal angle measuring device. Horizontal and vertical circles are read with a vernier to 1°/10. Accuracy and applications of such instruments are mentioned in section 4.2.2.

5.2. Self Reducing Tacheometers

Stadia theodolites and compasses have the significant shortcoming, that the measuring results can only be used for further processing after time consuming and thus uneconomical computations. Therefore, for about 100 years, tacheometers have been designed, with which the horizontal distance or the elevation difference, or both simultaneously are obtained directly without intermediate readings or reduction computations (table 36).
The reduction can be achieved with mechanical, optical-mechanical and optical reduction devices. Depending on the type of device, the self reducing tacheometers are categorized into slide- or projection tacheometers, contact- and tangent tacheometers, diagram tacheometers, tacheometers with varying line separation, double image tacheometers and base reduction tacheometers.

5.2.1. Slide Tacheometers

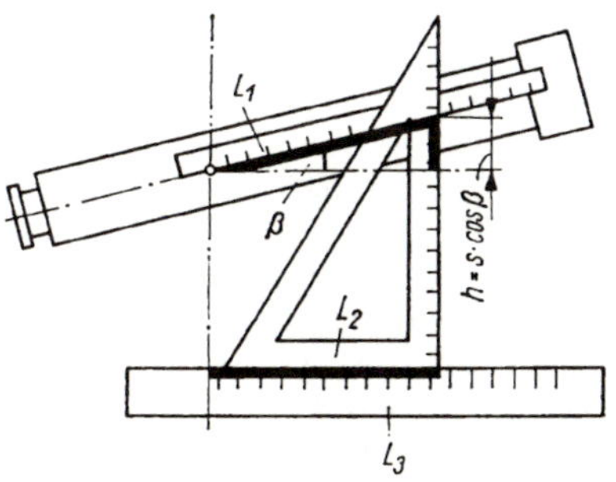

Figure 704 Slide-tacheometer (principle)

In the slide- or projection tacheometers (figure 704) the slope distance, which is directly read with the stadia hairs on a rod placed normal to the line of sight, is set in reduced scale (1 : 1000) on a ruler L_1 in direction of the line of sight. It is then automatically projected onto the scale L_3 with the sliding triangle L_2 such, that the horizontal distance and the elevation difference can be read on L_3 resp. L_2.
The standard deviation of the distance measurement amounts to about ± 3 dm for a 100 m line (1 : 300), the standard deviation of the elevation to about ± 4 cm. The instrument is very awkward in handling (about 8.5 kg mass), the operation is time consuming and the setting of the rods is cumbersome. Therefore, slide tacheometers are no longer used for practical purposes.

5.2.2. Tacheometers Based on the Tangent Principle

With tangent tacheometers the distance s to be measured is determined by comparison with a corresponding section d at the instrument (figure 705):

$$s = \frac{d}{a}\, l. \tag{242}$$

While the rod section l is variable, depending on the distance s, d and a are constant values at the instrument. They can be chosen such that for instance $d/a = k = 100$ which leads to $s = 100l$. The rod section l is obtained as difference of the rod readings in the two telescope settings, where the collimation axis passes through the end points of the section a (upper reading minus lower reading). It can also be obtained with one telescope setting if the section is reflected into the field of view of the telescope.
Only the horizontal distance is obtained by self reduction. With h' als "lower reading", the elevation difference becomes:

$$\Delta h = \frac{s}{d}\, h'. \tag{243}$$

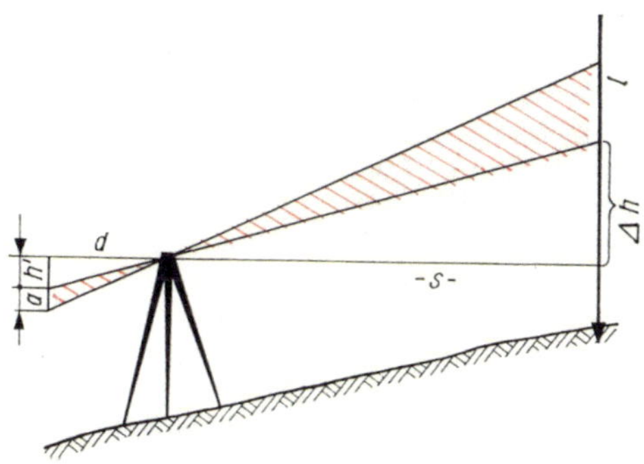

Figure 705 Measuring principle of the tangent tacheometer

For older tangent tacheometers, the measuring accuracy is 20 to 30% lower than for stadia theodolites because of the reduction, while for newer ones, it matches the accuracy of stadia theodolites.

Slope Screw Tacheometer

The slope screw tacheometer is the simplest instrument operating according to the tangent principle. It is similar to a stadia level. The tilting screw however is a precisely machined tangent screw with a pitch of $d/100$ and a graduation to count full revolutions. The tilting range amounts to only about 5% of the distance.

Contact Tacheometer

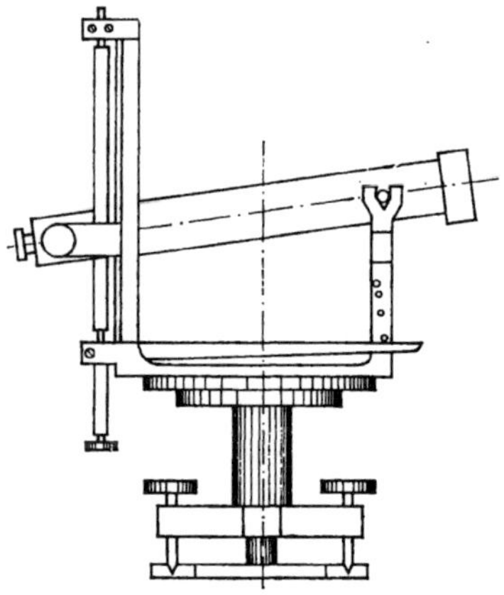

Figure 706 Contact tacheometer

In the contact tacheometer, the section d (figure 706) is limited by two contacts each. After the first reading, the telescope is clamped and its eyepiece end is tilted at the vertical tangent scale — or horizontal one, if the arm is rigidly connected to the telescope — towards the second contact with the aid of a lever. The difference between the two readings is 1/100, 1/50 or 1/20 of the distance. Contact tacheometers are produced and used in France. In the STA 32 by SLOM, Paris, the reductions system is placed in the interior of the instrument (figure 707).

Tangent Tacheometer

The tangent tacheometer by *Szepessy* contains a percentage graduation instead of the straight tangent graduation, which is engraved on a cylinder, concentric to the vertical circle, and which is reflected into the field of view of the telescope (figure 708). Two of these graduation lines serve as stadia hairs to read the rod section. In this manner, it is obtained with one telescope setting (two lines of sight), after the rod graduation has been brought to the vertical separation edge of the field of view. Depending on the interval of the lines used, various constants are applied.
In the "Tari" model 4180 and 4181 by Filotecnica Salmoiraghi, Milano, the rays of the telescope are led with fourfold reflection through the tangent graduation of the vertical circle ranging from 0 to ± 50 gon. In the field of view of the telescope the tangent lines appear like stadia lines, however with varying separation based on tangent graduation — according to the tilt of the telescope. If one sets one line onto an even value at the rod, then the horizontal distance can be directly read on the rod with the other line, when applying the proper constant (figure 709). The tangent graduation, which is visible in the right hand side of the field of view is used to determine the elevation difference. The inclination of the collimation axis is read with the horizontal cross hair as index, i.e. the tangent multiplied by 100. When multiplying this figure with the horizontal distance, the elevation difference is obtained. The horizontal circle is read with the microscope eyepiece beside the telescope. Mod. 4181 has a higher accuracy of the horizontal circle reading than Mod. 4180 (see figures 208 and 245). Under favourable circumstances the distance can be measured to an accuracy of 1 : 3000.
The Redta 002 by Zeiss, Jena, can also be used as tangent tacheometer, if its tangent graduation is used to determine the distance. The DK-RT by Kern, Aarau, can also be delivered with tangent graduation.

5.2.3. Diagram Tacheometers

Fundamentals of Diagram Tacheometers

Diagram- or Curve Tacheometers are reduction tacheometers with a diagram in the image plane. Depending on sighting distances and inclination of the telescope, a varying rod section is obtained, which provides a measure for the reduced distance or the elevation difference. They contain curves instead of fixed stadia hairs, whose separations change according to the reduction equations (190) and (192) when the telescope is tilted.
The separation p_s and p_h for the distance and height curves are obtained as follows: The slope distance s' to the zero point on the rod (at about 1.5 m height) is obtained in the parallactic triangle INM with the parallactic angle γ, and with the zenith distance ζ to the zero point of the rod, according to the sine rule as $s' = l \sin(\zeta - \gamma)/\sin\gamma$ (figure 710). The formulae for the reduced distance s and the elevation difference are in general terms:

$$s = s' \sin\zeta = l\frac{\sin(\zeta - \gamma)\sin\zeta}{\sin\gamma}, \quad \Delta h = s' \cos\zeta = l\frac{\sin(\zeta - \gamma)\cos\zeta}{\sin\gamma}$$

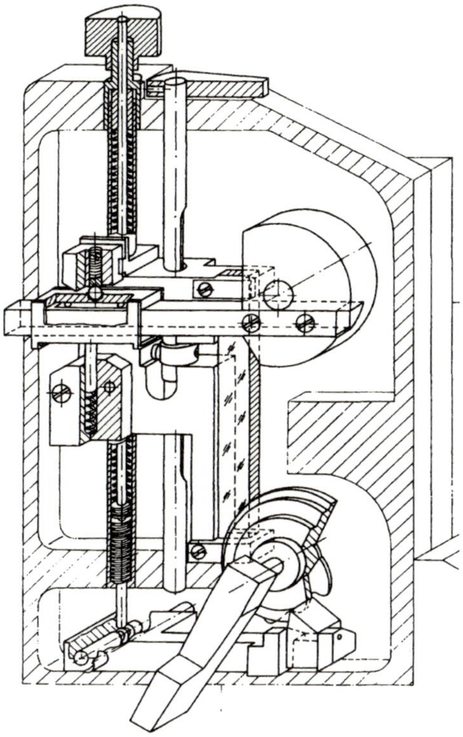

Figure 707 STA 32 (SLOM, Paris) Reduction system

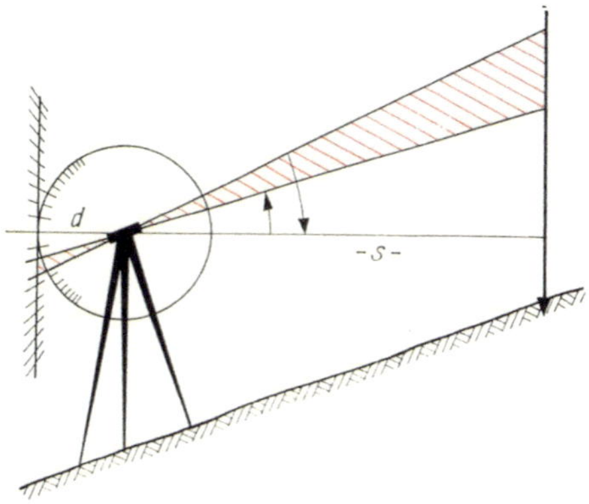

Figure 708 Principle of the tangent tacheometer according to *Szepessy*

Figure 709 Arrangement of optical elements in the reduction tacheometer Tari, model 4180 (Filotecnica Salmoiraghi, Milano); height: 310 mm

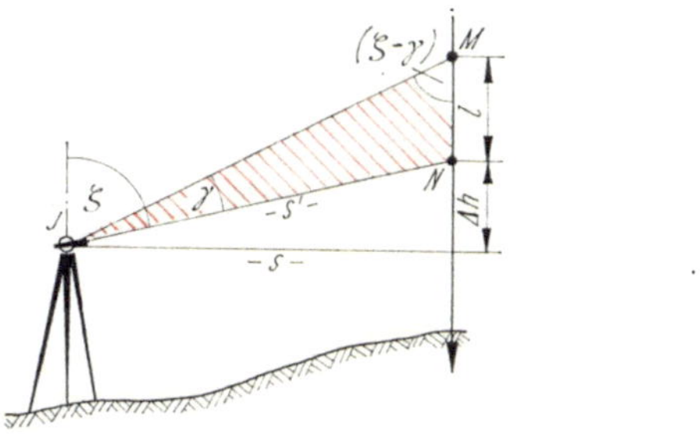

Figure 710 Parallactic traingle for a distance meter

or after rearrangement

$$s = \sin^2 \zeta (\cot \gamma_s - \sin \zeta \cos \zeta)\, l = k_s l$$

or

$$\tan \gamma_s = \sin^2 \zeta / (k_s + \sin \zeta \cos \zeta) \tag{244}$$

$$\tan \gamma_h = \sin \zeta \cos \zeta / (k_h + \cos^2 \zeta)\,. \tag{245}$$

Since $\tan \gamma_s = p_s / f_{Ob}$ with f_{Ob} an equivalent focal length of the objective lens, and p as separation of the stadia lines, the separations of p_s and similarly p_h from the base line became

$$p_s = \frac{f_{Ob} \sin^2 \zeta}{k_s \pm \sin \zeta \cos \zeta} \tag{246}$$

$$p_h = \frac{f_{Ob} \cos 2\zeta}{2(k_h \pm \cos^2 \zeta)}\,. \tag{247}$$

Table 36. Self Reducing Tacheometers

type	manufacturer	telescope					graduated circles				
							diameter		reading device		type
									least count		
		Γ_F	diameter of objective lens [mm]	field of view [']	length [mm]	shortest sighting distance [m]	*H* [mm]	*V* [mm]	*H*	*V*	
1	2	3	4	5	6	7	8	9	10	11	12
contact tacheometer											
STA 32	SLOM, Paris	32	40	82	180	1.5	80	–	10 mgon	–	optical micrometer
tangent tacheometer											
Tari 4180	Filotecnica Salmoiraghi, Milano	28	40	86	190	3.0	90	100	1′ (10 mgon)	tan	scale microscope
Tari 4181									2″ (0.5 mgon)	0.1 %	optical micrometer
diagram tacheometer											
FTRA	Fennel (FRG)	25	40	58	210	1.7	90	70	10 mgon (1′)	(1′)	scale microscope
Ta-D 41	MOM, Budapest	24	40	71	175	2.5	71	68	20 mgon (1′)	(1′)	scale microscope
RTa 4	Opton, Feintechnik GmbH, Oberkochen	25	40	83	220	2.0	98	85	10 mgon (1′)	(1′)	scale microscope
TN	U.S.S.R.	25	35	90	200	3.0	95	70	1′	1′	scale microscope
RDS	Wild Heerbrugg AG	24	40	76	190	2.4	79	70	10 mgon (1′)	(1′)	scale microscope
DAHLTA 010A	VEB Carl Zeiss JENA	25	40	76	222	3.0	86	86	10 mgon (1′)	(1′)	scale microscope
reduction tacheometer with mechanically monitored line separation											
DK-RV	Kern & Co., Aarau	27	45	90	172	1.8	80	56	0.1 gon (10″)	0.001 tan	DK with optical micrometer
K1-RA		28	45	86	172	1.8	95	75	50 mgon (1′)	0.001 tan	scale microscope
double image reduction tacheometer											
DK-RT	Kern & Co., Aarau	27	45	79	230	2.0	80	56	0.1 gon	0.1 gon	optical micrometer
TD	U.S.S.R.	27	38	90	310	2.5	95	70	1′	1′	scale microscope
Redta 002	VEB Carl Zeiss JENA	25	42	84	152	2.0	114	74	10 mgon (1′)	(1′) + tan	scale microscope
basis reduction tacheometer											
BRT 006	VEB Carl Zeiss JENA	15	25	162	–	1.8	90	90	50 mgon (5′)	(5′)	line microscope
TW	U.S.S.R.	16	25	150	–	1.5	95	85	10′	10′	line microscope

bubble sensitivity					mass				diagram				
platen level ["]	height index bubble ["]	telescope bubble ["]	bull's eye bubble [']	automatic height index, range [']	instrument [kg]	container [kg]	tripod [kg]	height [mm]	f [mm]	r' [mm]	k_s	k_h	remarks
13	14	15	16	17	18	19	20	21	22	23	24	25	26
30	–	40	–	–	5.3	3.6	5.3	210	–	–	100	–	figure 707
30	30	–	10	–	5.5	3.1	7.2	310	–	–	100, 200	tan	figure 709
													figure 245
40	–	40	8	±8′	5.4	3.8	6.7	192	250	50	100	10, 20, 50	figure 716
30	–	–	6	±5′	5.3	4.3	5.1	262	240	40	100, 200	20, 50	figure 722
30	–	–	15	±2′	5.3	4.3		350	250	46 50.5	100/50	10, 20, 50, 100	figure 717
			10		4.0	3.0	5.3	265	300	54	100	10, 20, 100	
30	30	–	8	–	7.1	2.2	6.5	289	212	52	100	10, 20, 50, 100	figure 719
20	–	(20)	8	±4′	4.7	4.4	5.8	295	300	55	100, 200	10, 20, 50, 100	figure 712
30	30	30	–	–	4.9	2.4	5.3	212	–	–	100	–	figure 726
35	–	–	–	±10′	4.5	2.2	5.3	215	–	–	100	100	figure 729
30	30	–	–	–	5.4	2.3	5.8	200	–	–	100	–	figure 747
30	–	–	10	±3′	6	3.0	5.3	255	–	–	100	100	figure 749
30	30	30	8	–	6.5	6.2	5.6	270	–	–	100	–	figure 739
60	30	–	8	–	6.7	5.8	5.6	250	–	–	200	–	figure 750
30	–	–	10	–	7.5	4.6	5.3	255	–	–	200	–	figure 752

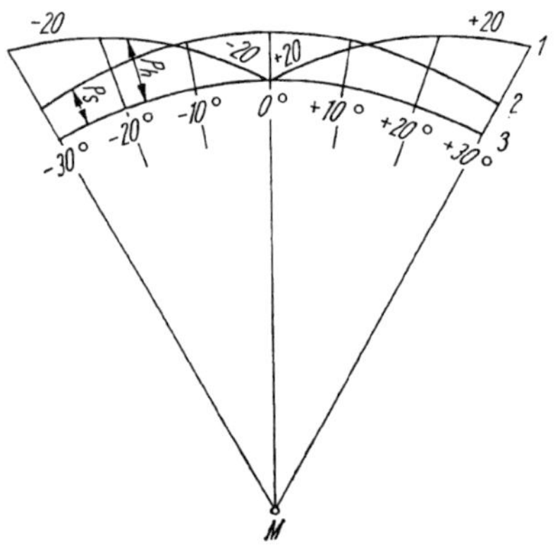

Figure 711 *Hammer-Fennel* diagram
1 height curve; *2* distance curve; *3* base curve

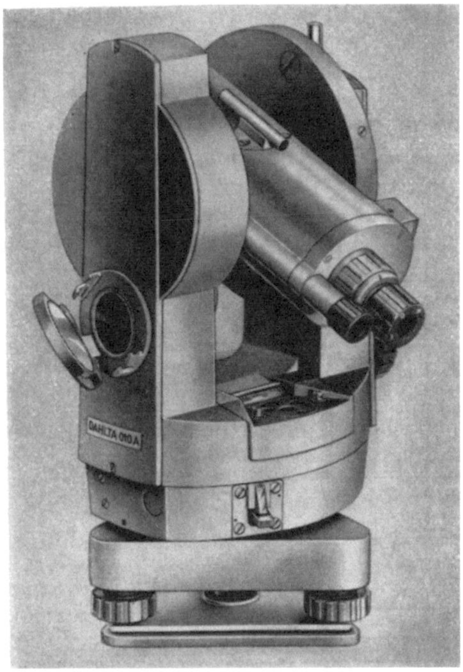

Figure 712 DAHLTA 010 A (Zeiss, Jena); height: 295 mm

The pairs of curves are engraved into glass plates with line thicknesses of about $(5 \pm <1)$ μm. They are either derived from a, e.g. 10 times enlarged curve with the aid of a pantograph resp. photographic reduction, or else (like at Kern, Aarau) continuously generated without original curve with a mechanism, which operates according to equations (246) and (247).
Most of these instruments contain a circular arc as base line for the distance- and height curves (so that only a total of three curves are needed, figure 711). For the height measurements 20, 50 and 100 is generally used as constant.

Types and Peculiarities of Diagram Tacheometers

The diagram tacheometers (table 36) are distinguished mainly in the manner in which telescope image and curves are brought into one plane and viewed together. In general structure they resemble a stadia theodolite of their manufacturer with the reduction device built in.
In the DAHLTA 010A (DAHL-tacheometer) by Zeiss, Jena, with automatic height index, the curves are engraved into a fixed diagram circle made of glass (figure 712). The telescope image which is projected onto the diagram, is imaged for a second time together with the curves into the plane of the reticule via a reversing system (figure 713). All curves can be seen in the direct telescope position. Above the base curve, the vertical line is imaged as a single line, while it is a double line below. The distance curve $k = 100$ lies above the base curve, while the one for $k = 200$ lies below. The operating parts are similar to the ones of the THEO 020 A. If needed, the image of the vertical circle can be cut off. As a complement to the instrument, a 2 m rod with telescopic foot is being produced (figure 714). By extending the leg up to 1.40 m, the instrument height can be set, and the elevation difference can be directly obtained when measuring, or else the target height on the rod can be selected such, that the computed height is an even metre. The base curve of the diagram is set onto the zero mark of the rod. Depending on the tilt of the telescope, the diagram curves have different positions. The reduced distance is read at the distance curve (nearly parallel to the base curve), while the elevation difference is read at the inclined height curves which have numbers (figure 715).

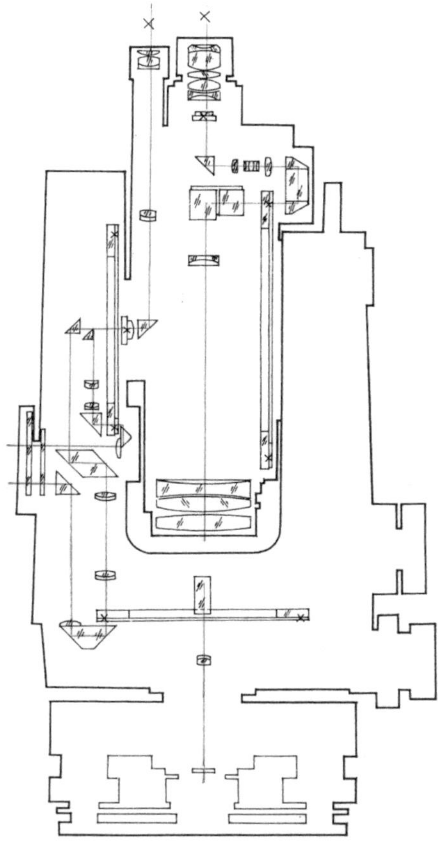

Figure 713 Optical train in the DAHLTA 010 A (The type number for Zeiss, Jena tacheometers indicates the standard deviation for distances in cm/100 m)

The reduction tacheometer FTRA (figure 716) by Fennel has an optical train similar to the Dahlta. The telescope image is upright. A rod with variable zero point belongs to the instrument.
The soviet TN is similar in structure and reading to the previously described tacheometers. The RTa 4 (figure 717) by Zeiss, Oberkochen, has an analogue optical train, and a structure similar to the Th 4, including automatic height index. Different families of curves in both telescope positions result in a short- and a long range in direct telescope position with $k = 100$, in reverse with $k = 50$ and 200. With the RTa 4 K, a rod with transversal graduation is provided for direct distance readings (figure 718). In the RDS from 1950 by Wild, Heerbrugg (figure 719), the diagram circle, attached to the horizontal axis, turns when the telescope is tilted. The inclination is four times multiplied via a planet transmission, thus the diagram circle turns in opposite direction by 3 times the angle. The planet transmission (figure 720) consists of a cogwreath (*1*) with radius $3r$ and internal cogs, which is rigidly connected to the telescope body and which rotates with it. Furthermore, it consists of the two planet wheels (*2*) with radius r, whose rotational axes are suspended in the telescope standard, and the central wheel (*3*) with radius r, with which the diagram circle is rigidly connected. If the telescope is tilted, the planet wheels (*2*) are rotated by the cogwreath (*1*) in the same direction, and move the central wheel, and the diagram circle with it. In this manner, a large path length is obtained for the curves. The curves for ± 50 gon are plotted for the total circumference of ± 200 gon, and thus spread over 300 mm length, so that they appear as being very flat. Due to favourable intersection with the rod graduation, the rod section can be read quickly and safely. With the RDS, folding rods with cm-checkerboard graduation

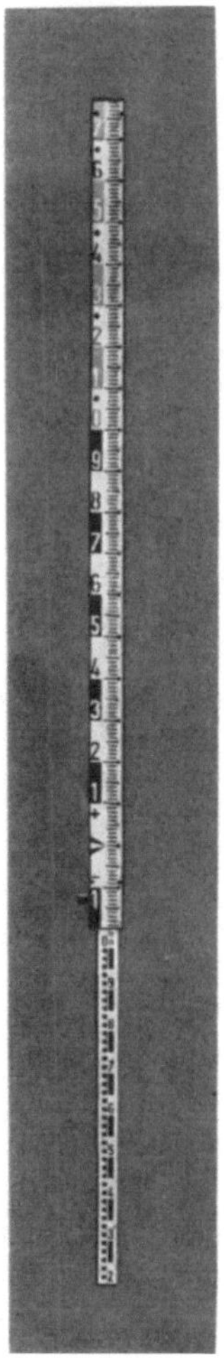

Figure 714 DAHLTA rod with telescopic leg

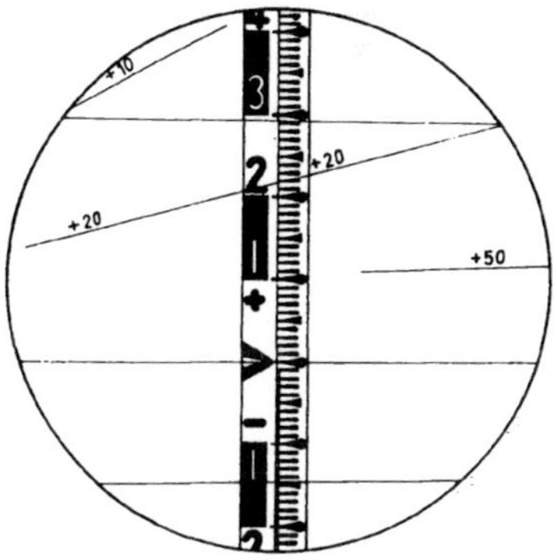

Figure 715 Field of view of the DAHLTA 010 A telescope; reading: $s = 0.292\ \text{m} \times 100 = 29.2\ \text{m}$; $\Delta h = 0.217\ \text{m}\ (+20) = +4.34\ \text{m}$

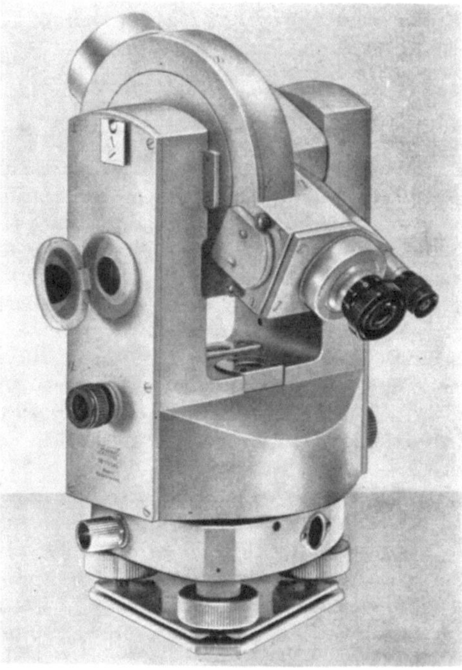

Figure 716 Fennel FTRA; height: 192 mm

Figure 717 RTa 4 (Zeiss, Oberkochen); height: 350 mm

and wedge mark are provided. They have upright numbers and a telescopic leg. 1.3 cm long short rods for a maximum sighting distance of 70 m with a transverse scale at the lower end, are produced. They have white circular marks for setting of the curves, which increase in size towards the top, so that they appear of the same size for all sighting distances and thus permit more accurate readings (figure 721).

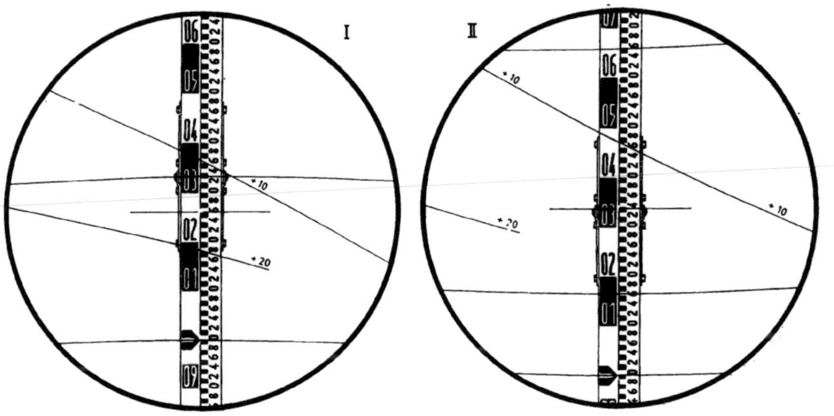

Figure 718 Field of view of the RTa 4 with rod with transversal graduation (1st reading; s = 33.2 m, Δh = 3.66 m; 2nd reading: s = 32.9 m, Δh = 4.71 m)

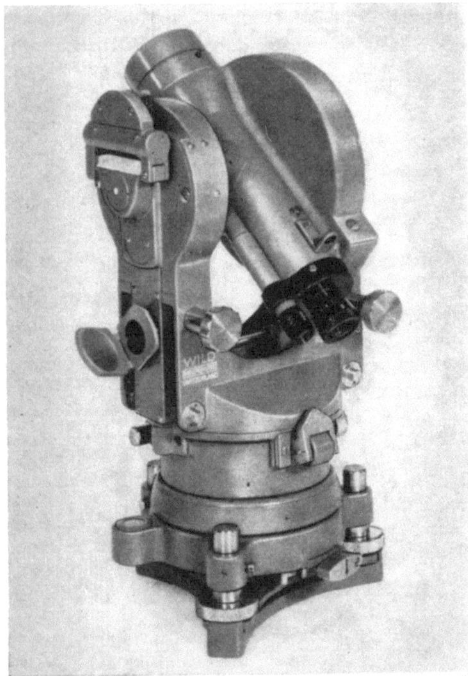

Figure 719 Wild RDS; height 289 mm

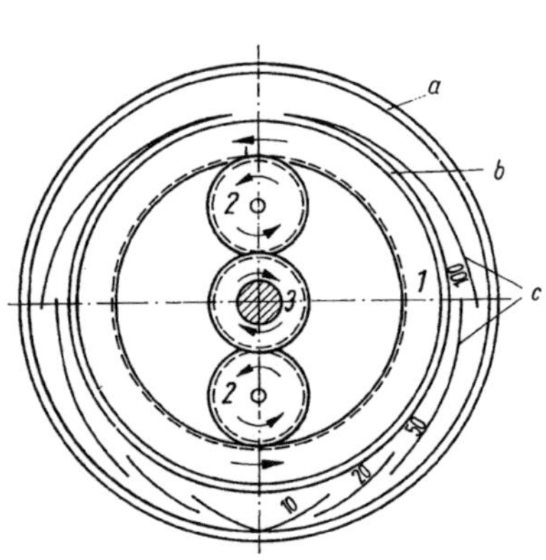

Figure 720 Reduction device for Wild RDS a) ground curve; b) distance curve; c) hight curves

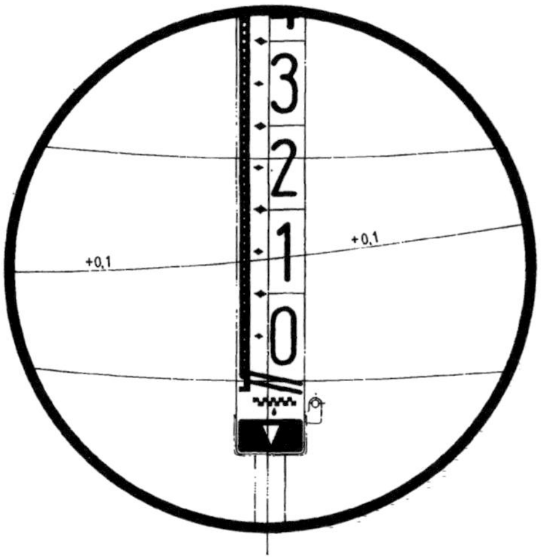

Figure 721 Field of view of the Wild RDS with short stadia rod GVL 1; reading: s = 26.32 m

In the circle diagram tacheometer Ta-D 41 (figure 722) by MOM Budapest, built according to *Bezzegh*, the diagram circle is turned in opposite direction when tilting the telescope. This is achieved with the aid of a cogwheel transmission which retains the angular velocity of the telescope. Because of the thus resulting double amount of the vertical angle, distance and height curves are circular arcs with different centers, which can be constructed relatively easily. In between the two exterior distance curves, the central distance curve is placed, which, together with the basic curve produces a pair with the constant 200 (figure 723). Therefore, distances above 200 m can be measured with a 2 m rod, however with lower accuracy. Due to the rotation of the diagram circle by 2α, the same circular arcs can be used in the reverse telescope position. The rest of the structure is similar to the MOM Te-D 41. The diagram (figure 724) of the alidade RK (see figure 777) by Kern, Aarau, is placed on a circular glass plate, eccentric within the image plane of the telescope and normal to the collimation axis (figure 725). When tilting the telescope, the glass plate rotates, four times magnified by cone wheel drive, around a center lying below the telescope axis in the reticule plane. The curve pairs needed for ± 50 gon are distributed over the whole circumference of ± 200 gon and are accordingly flatter. Instead of a common base curve for distance and height curves like in diagram tacheometers, the glass plate has two distance curves on the outside and two height curves inside. The constant for the heigth measurements is denoted by short line

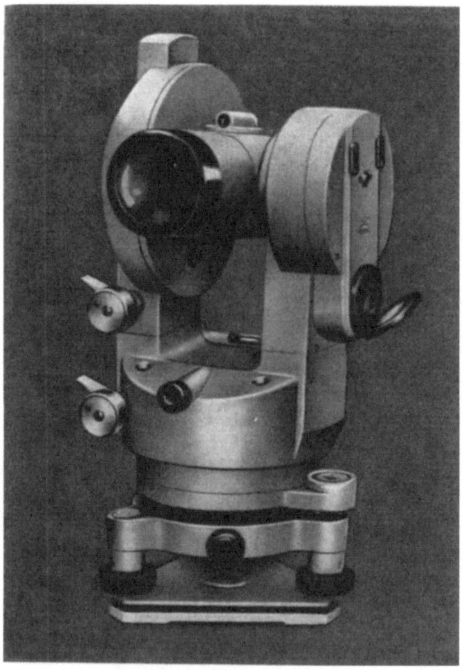

Figure 722 Ta-D 41 (MOM, Budapest)

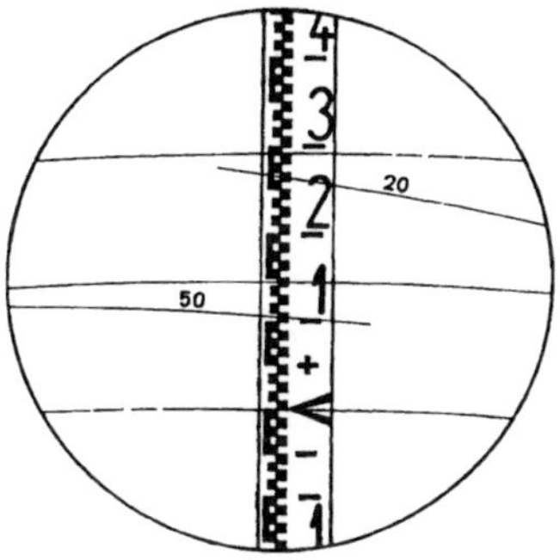

Figure 723 Field of view of the Ta-D 41; reading: $s = 29.0$ m, $\Delta h = 5.34$ m

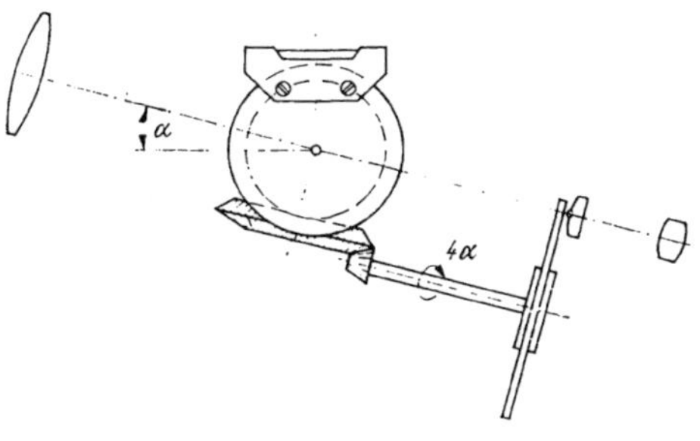

Figure 725 Reduction device of the RK

marks between the height curves. Due to the arrangement of the curves, one of the two rod sections for distance and elevation difference is to be formed from two uneven values. Otherwise, if the computation shall be simplified, two settings (with one distance- and one height curve) to an even value are necessary. The optics of the telescope is not loaded with additional prisms, however, the fixed reticule with the vertical line and the movable one with the curves are not in one plane.

Testing and Adjusting of Diagram Tacheometers

Since diagram tacheometers are similar in structure to stadia theodolites with reduction device, they are to be tested and adjusted like theodolites. In addition, the reduction device and the diagram have to be checked. Its correct position is exactly set at the factory, and it is nearly impossible that it gets out of adjustment. If an adjustment is necessary in spite of this, one has to consult the operation manual for the particular instrument.
For the Hammer-Fennel tacheometer, for example, the adjustment of the image size, i.e. the necessary separation of base- and distance curves is performed at the factory. If the image of the diagram appears to be shifted sideways in respect to the vertical line, it can be adjusted by turning a threaded bolt, after a cover plate had been unscrewed and removed. For the DAHLTA 010 A the intersection of the two height curves $+10$ and -10 should fall onto the vertical line for a zenith distance of 100 gon. If this is not the case, the coverplate of the left telescope standard has to be removed. The required position of the diagram circle can then be set for a vertical circle reading of 100 gon by using the adjustment screws. For the Wild RDS, the diagram circle is in proper adjustment, if – for horizontal sighting – the height curves + and – are symmetrical to the vertical centre line, and if this centre line bisects the short interrupted cross lines at the top and bottom of the field of view. A correction is possible by a sideways shift of the rectangular metal cover at the diagram circle housing behind the vertical clamp. This cover is fastened with two screws. This adjustment however should be left to a properly trained mechanic.

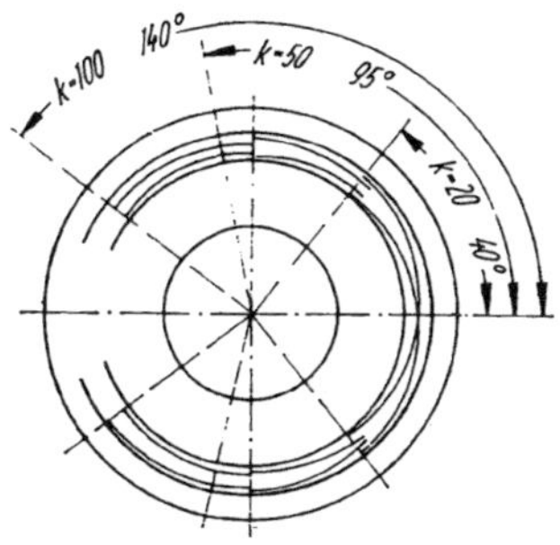

Figure 724 Diagram of the alidade RK (Kern, Aarau)

Accuracy and Application of Diagram Tacheometers

The accuracy of measurements with diagram tacheometers corresponds to the one with stadia hairs. Setting and reading accuracies are similar to stadia provided the curves are flat. For a range up to 140 m, an accuracy of 1 : 500 for distances (with $k = 100$) and a height error of ± 5 to 10 cm with curves having $k = 10$ or 20 resp. ± 10 to 20 cm for $k = 100$ can be expected. It has to be remembered, that the index bubble has to be centered prior to each reading if the instrument is equipped with one.
Diagram tacheometers are well suited for topography and engineering works which require speed and reliability. They are especially suited for difficult terrain where many terrain points have to be measured. By automatic reduction, the working speed is significantly faster than with simple stadia theodolites. Like simple stadia theodolites, these instruments can be used for trigonometric work and for traversing.

5.2.4. Tacheometers with Mechanically Monitored Variable Line Separation

In tacheometers with variable line separation (table 36), the tilt movements of the telescope are transferred by mechanical means onto the distance hairs, such that their separation changes according to the vertical angle.
In the double circle reduction tacheometer DK-RV (figure 726) by Kern, Aarau, for vertical rod, the reduction is achieved as follows: The two distance lines used for the measurements, are placed on two different reticules, facing each other. The reticule containing base curve and vertical line is

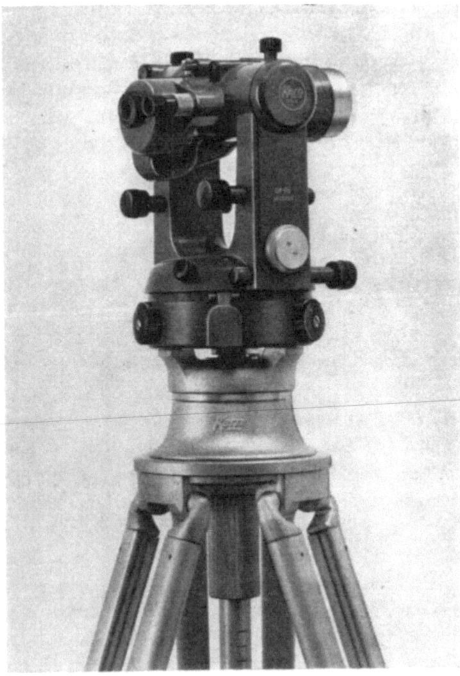

Figure 726 Reduction tacheometer DK-RV for vertical rod (Kern); height: 212 mm

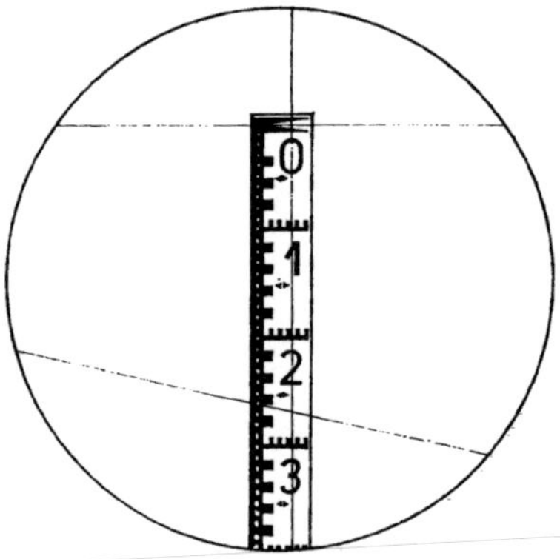

Field 728 Field of view of the DK-RV; reading: 26.64 m

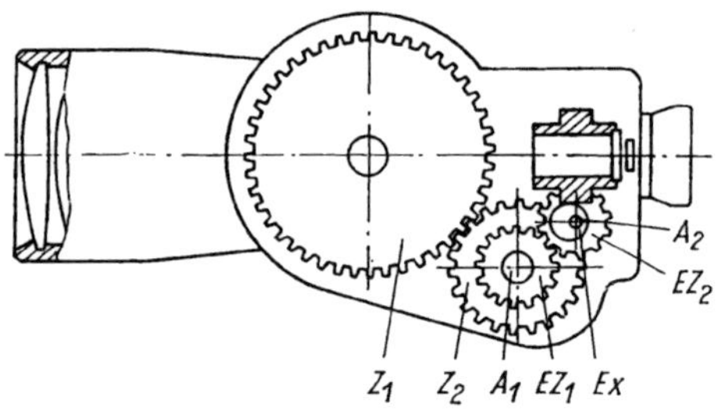

Figure 727 Reticule monitoring of the DK-RV

Figure 729 Kl-RA (Kern, Aarau); height: 215 mm

fixed, while the other one is monitored by a mechanism (figure 727). The tilting motion of the telescope is transferred twice onto the auxiliary axis A_1 via the cog wheel Z_1 on the horizontal axis and the cog wheel Z_2.

Located eccentrically on this auxiliary axis A_1, the cog wheel EZ_1 transfers it onto the axis A_2 via the cog wheel EZ_2. The eccenter disk Ex sits on the axis A_2. EZ_1 and EZ_2 are of the same size and positioned with equal eccentricity. The shortest distance between axis and cogs of one wheel is opposite to the tangent distance for the other one at horizontal positions of the telescope.

The casing of the reticule which can be rotated around a side axis, is shifted by the eccenter disk Ex, which results in lifting or lowering of the reticule. With this, the separation to the horizontal line is monitored corresponding to the tilt of the telescope.

The accuracy is increased primarily by using the transverse scale principle for the rod reading (figure 728). After setting the horizontal line onto the wedge-type zero mark (because of ground refraction at 1 m above ground), the inclined line of the movable reticule is set onto a circular mark on the rod by sideways motion of the instrument. The full metres are read on the vertical graduation, the decimetre on the rake, while the centimetres are estimated. With the instrument comes a rod, whose circular marks get larger for longer distances, so that the tilted line can be set well onto them. The range of the instrument is 200 m, with the standard deviation being ± 4 cm for a 100 m distance. The elevation difference is not obtained automatically, but has to be computed using horizontal distance and vertical angle.

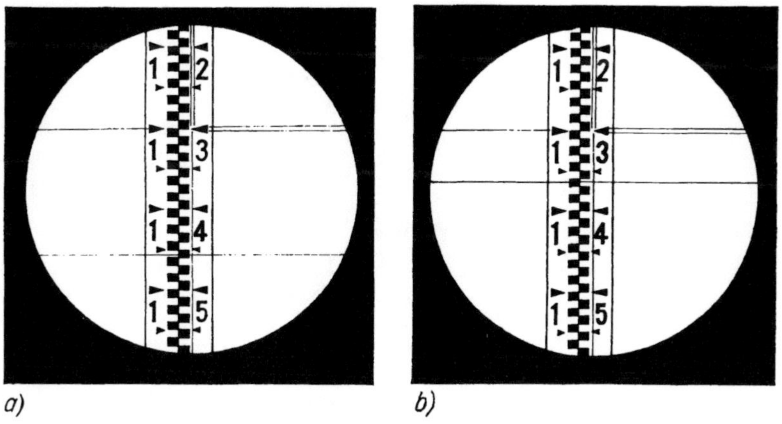

Figure 730 Field of view of the Kl-RA; readings: a) $s = 15.6$ m; b) $\Delta h = 6.4$ m

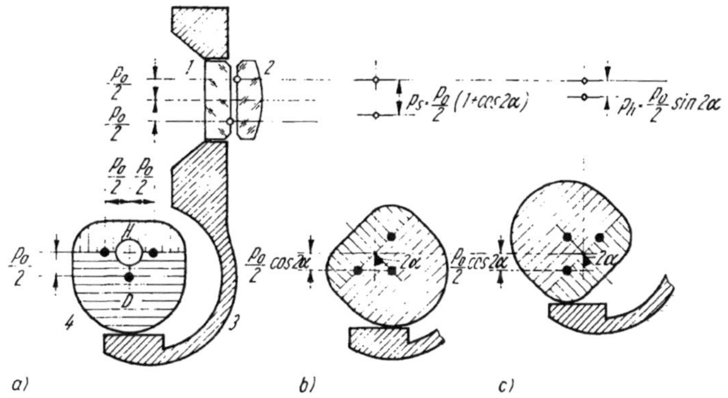

Figure 731 Reticule monitoring of the reduction device of the Kl-RA

In the self reducing engineer's tacheometer theodolite Kern Kl-RA (figures 729 and 730) the vertical line and the base line — half of it as double line — are engraved into the fixed reticule (*2*). The other horizontal line is on a movable reticule (figure 731). A cog wheel transmission monitors the mobile reticule (*1*) which is placed in the housing (*3*) depending on the tilt of the telescope via an eccenter (*4*). This is arranged such, that the rod section bounded by the horizontal lines provides the horizontal distance, and after switching with the aid of a rippled ring, it provides the elevation difference.
The reduction mechanism operates according to the following approximate formulae to determine the horizontal distance from the separation of the distance lines:

$$p_s = \frac{p_0}{2}(1 + \cos 2\alpha) \tag{248}$$

and for the elevation difference:

$$p_h = \frac{p_0}{2}\sin 2\alpha \tag{249}$$

where p_0 represents the line separation for $k = 100$, and α the vertical angle. For horizontal sight (figure 731a), $\alpha = 0$ and $p_s = p_0$. When tilting the telescope by α (figure 731b), the eccenter is rotated by 2α, which reduces the line separation p_s. If the eccenter is rotated by 180° for the same tilt of the telescope (figure 731c), the line separation p_h is obtained. As seen in the figure, the instrument has an automatic height index (see figure 543), and a circle numbering similar to figure 222. The reading positions for horizontal- and vertical circles are simultaneously visible within the field of view.

5.2.5. Reduction Tacheometers with Double Image Distance Meter

Fundamentals of Double Image Reduction Tacheometers

The reduction tacheometers with double image distance meter are based on an invention by *Boßhardt*, who arranged a pair of rotating wedges in front of half of the telescope objective lens. Like with the wedge attachement, a double image is obtained from a horizontal rod, using the double image distance meter. While the parallactic angle remains constant for the wedge attachement, the pair of rotating wedges acts like a glass wedge with varying angle of deflection.

If a single glass wedge is rotated, a ray deflected by it describes a cone, and a circle on a plane normal to it (figure 732). The intersection between the ray and the plane lies in the principal section of the wedge. If a second wedge is arranged in front of it such, that its principal section forms the angle 2α to the one of the first wedge, and that their refracting edges are located on the same side, then the ray is deflected again and hits the plane at B (figure 733). If the wedges have equal angles, then $a = b$, which means

$$c = 2a \cos \alpha . \tag{250}$$

For $\alpha = 100$ gon the principal sections of both wedges lie in one plane, and the refracting edges are opposite to each other. The two wedge effects cancel each other. The wedges act like a plan parallel plate, and c equals zero.
For $\alpha = 0$ gon, the principal sections of both wedges are also in one plane, however, the refracting edges are on the same side. In this case $c = 2a$. For wedges with a deflection angle $\delta/2$, $2a$ becomes $s \tan \delta$, with s being the separation of vertex and plane. Thus

$$c = s \tan \delta \cos \alpha . \tag{251}$$

For the varying parallactic angle γ, we get according to figure 733
$c = s \tan \gamma$, so that

$$\tan \gamma = \tan \delta \cos \alpha \tag{252}$$

and since γ and δ are small angles

$$\gamma = \delta \cos \alpha . \tag{253}$$

Starting from the position of maximum deflection (figure 734), the deflection decreases proportionally to the cosine of half the rotation angle of the two wedges.
If a pair of rotatable wedges is placed in front of one half of the telescope objective lens, the telescope tilt can be used to rotate the wedges by coupling the telescope movement with the wedges. Thus the deflection is reduced mechanically corresponding to the cosine of the tilt angle (figure 735).
Since for this approach, two images are generated in a common image plane, mixed images would be the result. For the double image tacheometers the images are separately led to the eye with some auxiliary devices, and can be brought together along a sharp edge.
The image separation can be obtained with a separation wall S between objective lens Ob and image plane B (figure 736). To eliminate the effect of eye deficiencies because of irregular curvature of the eye lens, the two half images are superimposed for the DK-RT by Kern, with the aid of a double wedge Dk.
With a slightly roof shaped double wedge, whose separation edge is parallel to the separating line of the two objective halves, image separation is possible without separating wall. The double wedge diverts the rays from the upper image half downwards, and the ones from the lower half upwards (figure 737).

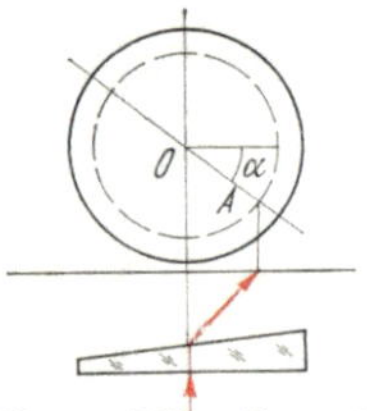

Figure 732 Operational principle of a rotating wedge

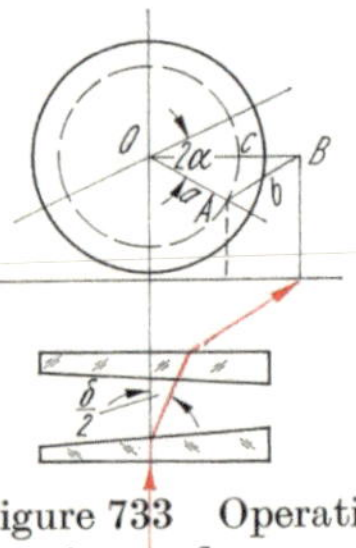

Figure 733 Operational principle of a pair of rotating wedges

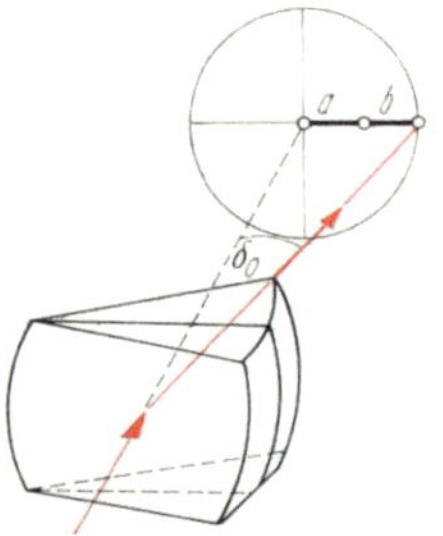

Figure 734 Operational principle of a pair of rotating wedges in zero position

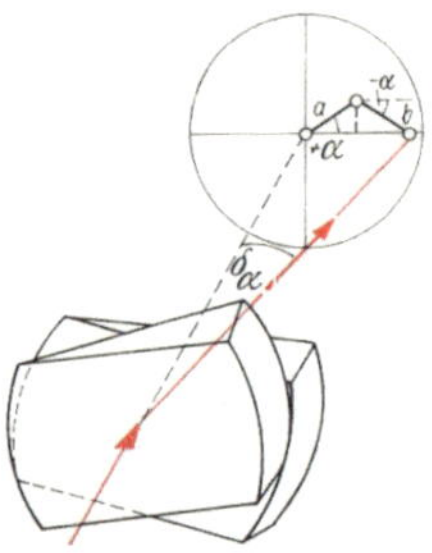

Figures 735 Operational principle of a pair of rotating wedges for inclined sight

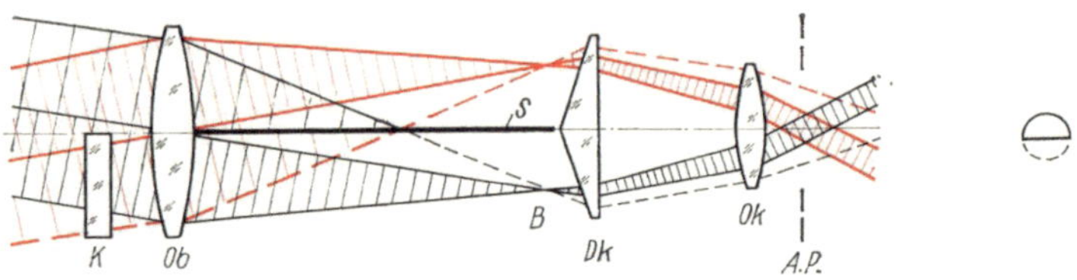

Figure 736 Image separation for double image reduction tacheometers with the aid of a mechanical separating wall

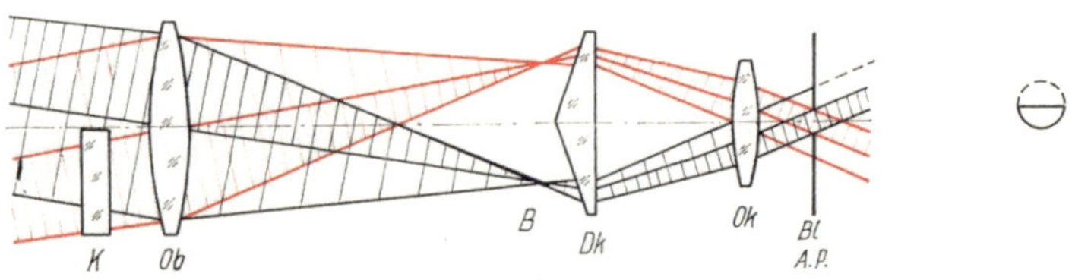

Figure 737 Image separation for double image reduction tacheometers with the aid of a double wedge (bi-prism)

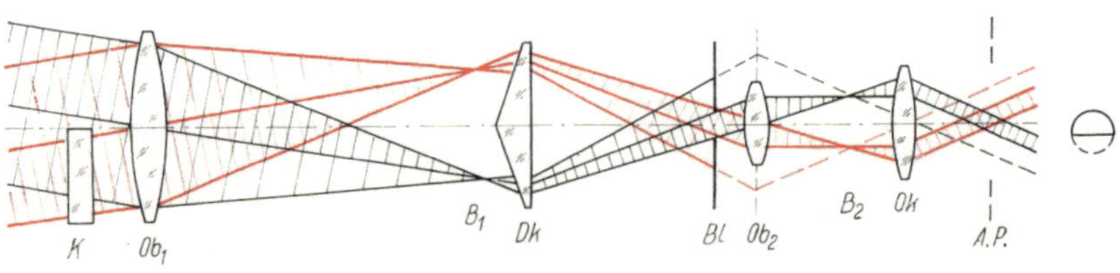

Figure 738 Image separation for double image reduction tacheometers with the aid of a double wedge and intermediate imaging

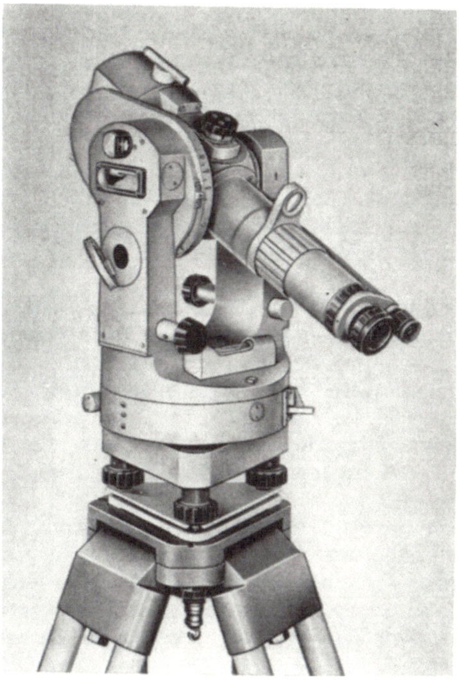

Therefore, the collecting lens, placed directly behind the double wedge in front of the objective opening does not represent one uniform exit pupil, but two different ones. By properly dimensioning the wedge, the exit pupils can be superimposed. The parts which are not coincident are caught by a pinhole or slot diaphragm *Bl* (about 2 mm behind the focal plane of the eyepiece). Due to the superimposition of the two exit pupils, eye deficiences practically do not have an influence in the measurements. When viewing into the telescope, the observer sees two different image halves, separated by the horizontal centre edge of the double wedge. When measuring, the separation edge is set onto the centre of the rod. Since the diaphragm lies in the exit pupil, the observer cannot bring his eye into this position — especially if he wears glasses — and a so-called keyhole observation is obtained. For image separation in the Teletop (with pinhole diaphragm) and in the Wild RDH (with slot diaphragm) a double wedge is used.
In order to avoid interference of the diaphragm with the eye position, an intermediate image is created at the Zeiss, Jena, Redta. Behind the pinhole diaphragm *Bl*, there is a reversing system Ob_2 with generates a second upright separated image B_2. Due to this arrangement, the eye can be brought into the plane of the exit pupil (figure 738). However, this requires a longer telescope.

Types of Double Image Reduction Tacheometers (table 36)

The Redta 002 (figure 739) by Zeiss, Jena, is as far as its structure is concerned, an optical theodolite with repetition capability, stadia hairs and double image distance meter.
In front of the lower half of the telescope objective lens there is a rotating pair of wedges, which provides the reduction. For horizontal telescope, the principal sections of the wedges are horizontal, and the deflection angle is $\gamma_0 = 1/200 + 1/200$. If the telescope is tilted by α, the wedges are rotated against each other by the same angle $+\alpha$ with the aid of cog wheels with fixed axes, around an axis parallel to the collimation axis (figure 740). While the cogwheel Z_1 is connected to the height index bubble, the cogwheels Z_5 and Z_6, turning on ball bearings (with balls of 2 mm diameter), are connected to the casings of the wedges. The cog wheel Z_7 is only used to reduce backlash. Since $+(\sin\alpha/200) - (\sin\alpha/200) = 0$, no height deflection is obtained, but only a horizontal deflection of

$$\gamma = (\cos\alpha/200) + (\cos\alpha/200) = \cos\alpha/100. \tag{254}$$

The upper half of the bundle of rays is shifted 22 mm parallel in height by a rhomb prism and generates a reversed intermediate image, just like the bundle which passes through the pair of wedges. This intermediate image which was generated after the rays had passed through the objective lens, the focusing lens and the biprism (for image separation), is combined with a collecting lens. By passing through a reversing lens, it becomes upright and is then imaged at the image plane of the eyepiece. Thus an upright image is observed through the eyepiece (figure 741).
Two sideways shifted images of a horizontal rod are generated within the field of view. The zero point of one image serves as index for reading the rod section b. With $\gamma = b/s' = \cos\alpha/100$ the slope distance becomes $s' = 100b/\cos\alpha$, and the horizontal distance becomes $s = s'\cos\alpha$ or

$$s = \frac{100\,b}{\cos\alpha}\cos\alpha = 100\,b. \tag{255}$$

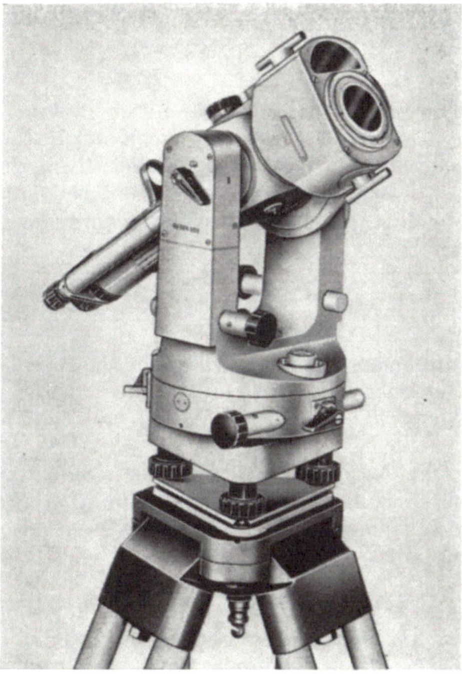

Figure 739 Redta 002 (Zeiss, Jena); height 270 mm

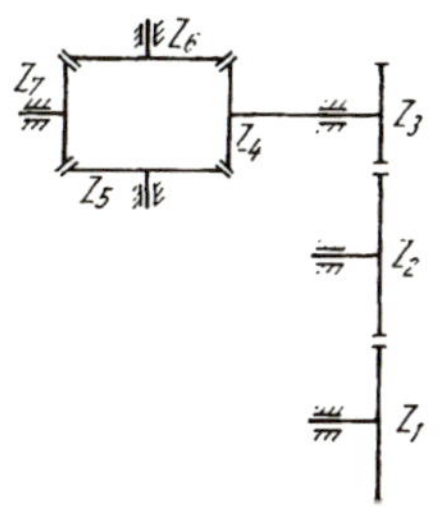

Figure 740 Reduction device for the Redta (schematic)

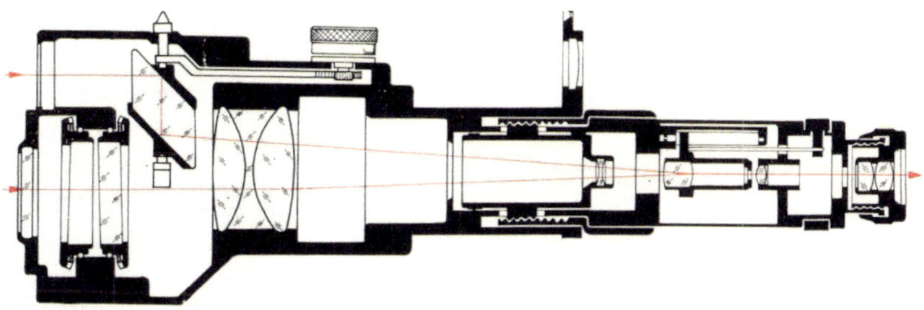

Figure 741 Section through the Redta telescope (length: 310 mm)

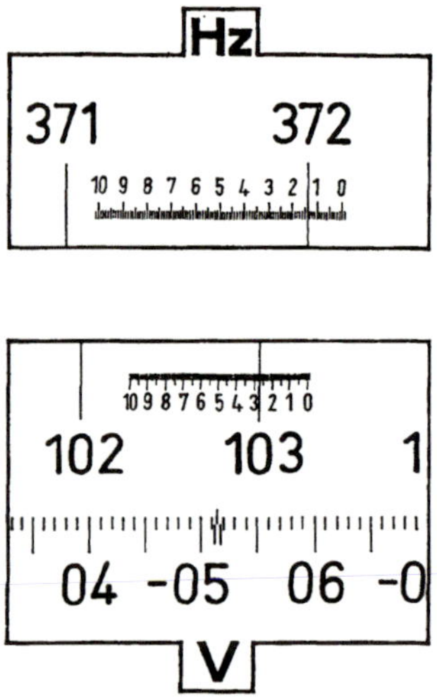

Figure 742 Field of view of the Redta 002 reading microscope. Reading: H: 372.136 gon; V: 103.271 gon; tangent: −0.0515

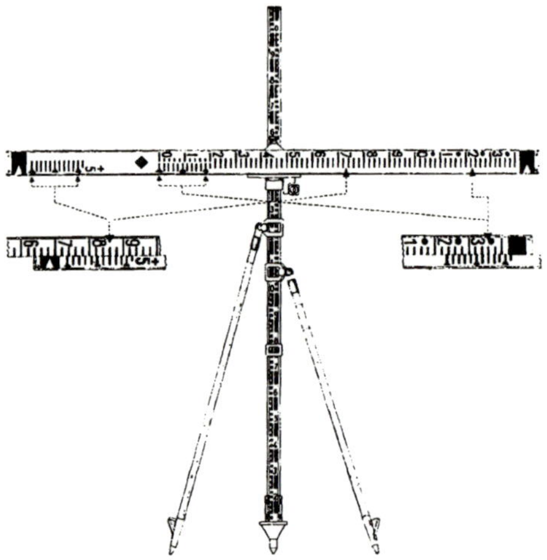

Figure 743 Redta rod with large stand

The rod section, multiplied by 100 results immediately in the horizontal distance s. The addition constant, comprised of the separation between the vertex of the angle and the vertical axis of the Redta, and between the graduated surface of the rod and the centre of the target, is eliminated by shifting the zero position of the parallel plate micrometer and by shifting the verniers of the rod graduation. Only for extremely steep sights, small reductions have to be applied, which can be read in centimetres from a correction graduation at the housing of the vertical circle, and which have to be subtracted from the determined horizontal distance. The reductions amount to 0.5, 1.0 and 2.0 cm for zenith distances of 77, 68 and 54 gon respectively.

The rhomb prism of the upper objective half can be rotated around an axis, which is normal to the plane of the measuring angle. It simultaneously serves as plan-parallel plate to obtain coincidence between vernier line and graduation line of the two rod images. With this parallel plate effect, the sighting ray can be shifted at the rod by the amount of the vernier value, i.e. by 0.2 cm. The micrometer drum is graduated into 20 parts, which means, that its graduation unit of 0.01 cm corresponds to one centimetre of the distance, because of $k = 100$.

By turning a rippled ring in front of the eyepiece, one can switch from the double image to a complete single image, which makes direction measurements easier. With this, the biprism, combined with the collecting lens, and the diaphragm are bypassed, and a lens without biprism and a screen covering the lower half of the objective lens with the diverted image are introduced instead.

The vertical circle of the Redta 002 contains a tangent graduation with a least count of 0.001 (figure 742) in addition to the regular graduation, so that the elevation difference can be easily obtained as

$$\Delta h = s \tan \alpha \,. \tag{256}$$

The tangent graduation can also serve for distance measurement with a vertical rod, with the Redta being used like a tangent tacheometer in this case. With the readings at the tangent graduation $\tan_o$ and (after tilting the telescope) $\tan_u$—for simpler computations, $\tan_o$ and $\tan_u$ should be round values and with the corresponding rod readings a_o and a_u at the horizontal cross hair, the distance s becomes

$$s = \frac{a_o - a_u}{\tan_o - \tan_u} \,. \tag{257}$$

For lower accuracy distance measurements, the stadia hairs of the reticule (one pair each for horizontal and vertical rod) with $k = 100$ can be used after switching to complete single rays. The addition constant $c = -0.3$ m has to be taken into account.

The Redta rod with 2.1 m long tripod legs with bull's eye level and fine braces (figure 743) or else with measurable telescopic 1.45 m long tripod legs, is used for the measurements. For measurements with forced centering it is used with short 0.3 m standing pipes with plugs, which can be set and clamped into tribracks.

The slit between decimeter bar and centimetre graduation of the standing pipe serves as target for direction measurements. The horizontal rod is aligned normal to the line of sight with the aid of a diopter with collimator.

With the interior vernier of the 2.09 m long wooden rod with a metal tape carrying the graduation, horizontal distances of 2.50 to 122 m can be measured, while the exterior vernier is used for distances between 122 m (72 m + 50 m) and 172 m (122 m + 50 m). For the distance measurement, the telescope has to be tilted until equally long graduation lines from rod graduation

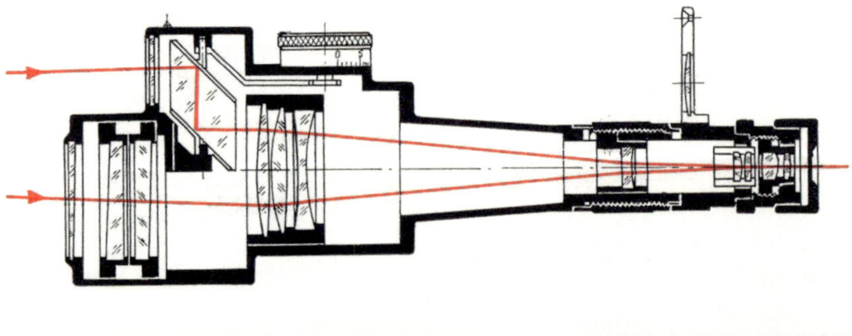

Figure 744 Section through the Wild RDH telescope; length: 272 mm

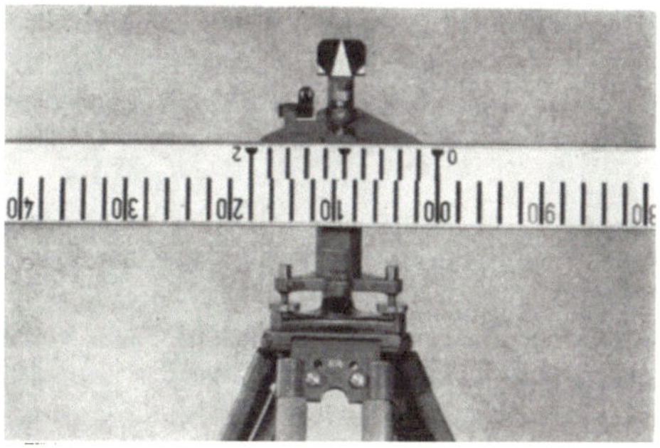

Figure 745 Rod for the Wild RDH

and vernier appear above and below the graduation line. The standing pipe should lie within the used rod section. The zero line of the vernier with a triangle at its lower end is used as reading index.

The Redta rod with triangular marks at its ends can be used as 2 m subtense bar. Since the graduated surface lies 4.4 cm in front of the plug axis, the measured distances have to be lengthened by that amount. The separation of the end marks of the base is 2 m $\pm$ 0.1 mm at $t = +20\,°C$. Double image measurements do not require temperature corrections, because the influence of temperature changes on the rod, is compensated by a corresponding change of the refractive index of the rotating wedges. For detail surveys (buildings) and isolated topographic surveys, shorter rods of 1.52 m length (measuring range 2 to 128 m) and manual rods ar available.

The reduction tacheometer Wild RDH distinguishes itself from the Zeiss, Jena-Redta primarily by the optical system and the possibility of directly obtaining the elevation difference from the readings on the horizontal rod. The image separation is accomplished with a biprism without intermediate imaging (figure 744).

When turning the switch on top of the right telescope standard, the cog-wheel Z_1 (figure 740) is changed by 100 gon, which causes a rotation of the wedges by 100 gon such, that their principal sections lie vertical for a horizontal telescope and that their refracting edges lie opposite each other. The parallactic angle, which has its maximum value if the principal sections are horizontal, now equals zero. If the telescope is tilted, the wedges are rotated by the same amount in opposite directions. Instead of $b = (\cos\alpha/100)\,s'$ according to equation (255) the rod section becomes now $b = [(\cos 100\,\text{gon} \pm \alpha)/100]\,s'$ which equals $(\sin\alpha/100)\,s'$, which means that the elevation difference is directly obtained as

$$\Delta h = s' \sin\alpha = \frac{100b}{\sin\alpha} = 100b. \tag{258}$$

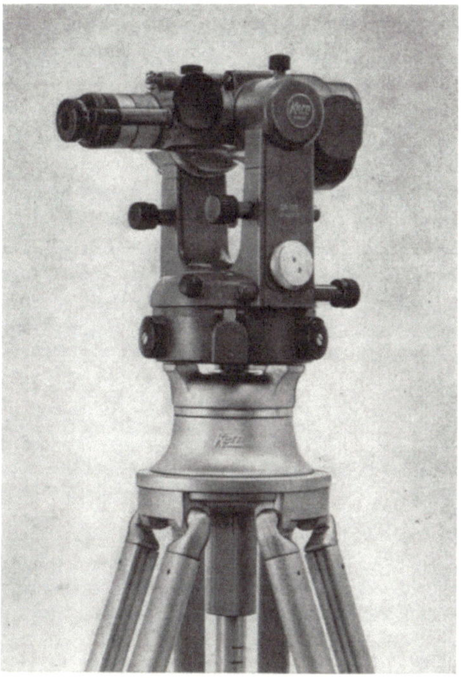

Figure 746 Double circle reduction tacheometer DK-RT (Kern, Aarau); height: 200 mm

For negative elevation differences, the area of red numbers is reached (figure 745). In this case, 100.00 m has to be subtracted from the reading. The rod contains three verniers, of which the two outside ones are shifted by 70 cm to the central one. The left vernier, numbered in black from 70 to 72 is to be used for distances beyond 70 m, while the right one, numbered in red with 30, is to be used for elevation differences of more than -70 m. The vernier readings have to be added to the rod readings. The range is 164 m.

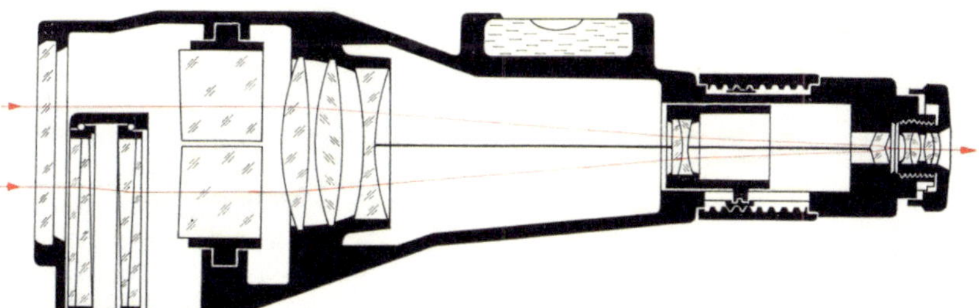

Figure 747 Section through Kern DK-RT telescope; length 230 mm

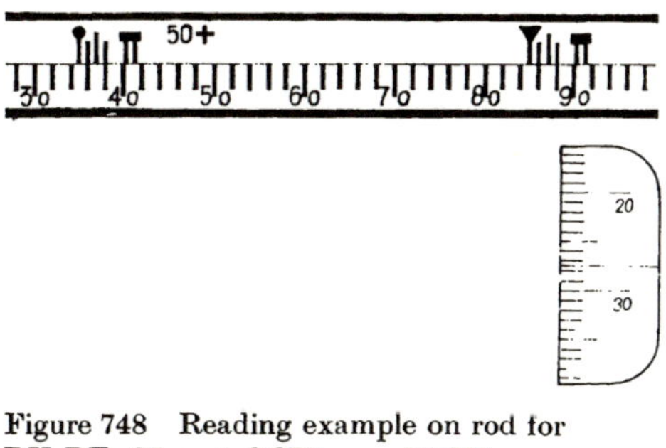

Figure 748 Reading example on rod for DK-RT: 85 m + 0.275 m = 85.275 m

The double circle reduction tacheometer DK-RT (figure 746) by Kern, Aarau, differs from the Bosshardt-Zeiss, Jena-Redta primarily by the structure of the optical system. None of the bundles of rays which generate the two half images is shifted in height (figure 747). The image separation is mechanically arranged with a separation wall inside the telescope. Near the image plane, a roof prism eliminates a horizontal strip, thus getting rid of the disturbing influence of the separation wall and superimposing the two half pupils. A switch-ring between eyepiece and focussing drive serves to switch from complete single image to double image, with a reading magnifiying glass being attached to the ring for reading of the micrometer drums. In front of the objective lens, there is a double parallel plate micrometer, which shifts the two bundles of rays by up to 5 mm against each other. Therefore two index graduations with four graduation lines each are placed on the rod instead of the verniers. The origin of the graduation is used to read the full metres, while the other lines are aids for the coincidence setting. If the second and fourth line are brought to coincidence (figure 748),

Figure 749 Optical reduction tacheometer TD (U.S.S.R.)

an odd number of metres has to be read. The reading at the check-index with red crossbar provides a distance which is constantly longer by 5.27 m. The measuring range of the 1.7 m long standard rod of light metal with graduation on a steel band is 2.5 to 166 m, and 206 m with the 2.1 m long special rod.
Braced tripods with a 2 m long standing pipe or tripods with telescopic legs and telescopic standing pipe, or centering tripods for forced centering are used.
The addition constant of the telescope of 73 mm (separation of parallactic angle and vertical axis) is accounted for with the position of the index graduation on the rod. Since the error influence of the multiplication constant counteracts the one for the addition constant, the remaining errors are below 1 cm for slopes up to 30°. Therefore no correction graduation is used.
The elevation difference has to be computed by multiplying the read horizontal distance with the tangent of the vertical angle. The instrument can also be obtained with a tangent graduation instead of the degree graduation for the vertical circle.
The Soviet self reducing tacheometer TD (figure 749) contains a double image distance meter with two fixed glass wedges. A wedge compensator, which covers half the object, is used for reduction. A Porro-system creates a double path of the rays through the compensator. When tilting the telescope with the prism system, the principal section of the compensator remains horizontal, and the parallactic angle changes proportionally to the tilt angle. When turning the compensator by 100 gon, elevation differences can be read on the rod. A telescopic half lens serves as micrometer, and a pendulum prism with air dampening as automatic height index. The central of the three rod verniers is designed to measure distances from 20 to 180 m and elevation differences up to ± 90 m, with the right one for distances from 90 to 180 m and elevation differences between $+90$ and $+180$ m, and the left one for elevation differences between -90 and -180 m.

Testing and Adjusting Double Image Reduction Tacheometers

The testing of the reduction tacheometers consists of checking the instrument as angle measuring device and as distance meter, as well as checking the reduction device and the rod with its setting up equipment. Testing and adjusting of the angle measuring device follows the methods given in section 2.3.4.

Testing and Adjusting of the Rods and their Setting Equiment

Rods with irregular graduation errors <0.05 mm are checked like levelling rods. The regular graduation errors as well as changes in the rod length are covered when checking the multiplication constant. In order to check the bull's eye level on the standing pipe, it is set plumb with the aid of a plumb bob. The bubble can then be adjusted with its adjustment screws. The sighting plane of the diopter is normal to the rod, if in an angle prism, held directly above the diopter, a target appears in line with the ends of the rods as seen in the prism. A deviation is eliminated with the aid of the adjustment screws of the diopter.

Testing of the Distance Meter and of the Reduction Device

Errors in the reduction device, i.e. deviations of the rotating wedges from their zero positon, appear only for steep sights. Therefore an inclined distance is measured several times by exchanging instrument and rod. If the results of fore- and back measurements are the same, the reduction device is in proper working condition. If not, the mean of the two measurements provides the correct reading free of the error of the reduction device. After setting this value, the reduction device is adjusted.
At the Redta 002, the adjustment screws are accessible after removal of the cover of the left telescope standard. For the DK-RT they are visible to the

left of the illimination mirror after removal of the cover. Generally changes in the adjustment of the reduction devices are nearly impossible.
For horizontal sighting, the addition constants shall be zero. Therefore, the results of the optical measurement of a 5 to 10 m long line are compared with the results of taping. Since an error of the multiplication constant is without influence for this distance, the difference between the two values represents the addition constant. After loosening its support screws, the micrometer drum is adjusted to the desired reading.
The multiplication constant shall equal 100. It is tested by comparison of a 60 to 100 m long distance, measured optically and by calibrated tape. Differences of up to 5 cm for 100 m are always mathematically corrected, because they may be caused by personal errors. For larger differences, proportional to the distance, the multiplication constant can be changed. To do this, the wedge shaped cover glass in front of half of the objective lens is rotated until coincidence with the respective rod lines is obtained. This is applicable for the Redta 002 and the Wild RDH after the correct distance had been set and the index bubble had been centered. At the ring of the casing of the Redta 002, there is a graduation for resetting, because it has to be set newly when changing observers because of personal angular errors (which may change). At the DK-RT, the multiplication constant is changed by adjusting with two adjustment screws with cross-holes, which become accessible after removing the switch ring.
The establishment of an accurately determined, for instance 100 m long test line with intermediate points is recommended (see page 240). Then the constants can be checked weekly or every two weeks during the field season. A correction table can be made based on the results of the check measurements, with which the measuring results can be corrected. The instrument should be adjusted only if a significant error is apparent. For larger misadjustments, e.g. because of a fall, it is advisable to send the instrument to its manufacturer.

Figure 750 Base reduction tacheometer BRT 006 (Zeiss, Jena)

Accuracy and Application of Double Image Reduction Tacheometers

Among the self reducing tacheometers, the double image reduction tacheometers occupy first place as far as accuracy is concerned. 100 m distances can be determined to ± 1 to ± 2 cm in fore- and back sight. The errors in the automatic determination of the elevation difference amount to ± 4 to ± 5 cm for 100 m with the Wild RDH and the Russian TD. The quality of the measurement depends on the exact reading and the correct operation of the reduction devices. Air movements can make reading difficult or even impossible. The influence of ground refraction is avoided by setting the rod horizontally at about 1.5 m height above ground, and by watching that the sighting rays between instrument and rod are never closer than 1 m to the ground. Because of side refraction they also have to have clearance from heated walls etc.
Because of their high accuracy of 1 : 5000, double image reduction tacheometers are well suited for traversing, precision tacheometry and detail surveys using the polar method. For traversing, a saving of 50 to 70% for the distance measurement can be expected, while for detail surveys it amounts to 20 to 50% as compared to measurements with tape and angle prism. The awkward rod is however, a disadvantage. The measuring range is about 150 m.

5.2.6. Self Reducing Tacheometers with Base at the Station

The awkward horizontal rod of double image reduction tacheometers can often only be set up with difficulties or not at all in urban areas. To utilize the advantages of the polar method of detail surveying together with optical distance measurements in spite of this, Zeiss, Jena developed a reduction tacheometer with the base at the station (table 36) according to the teletop principle, namely the base-reduction tacheometer BRT 006 (figure 750). On the base rail (*13*) (figure 751a) with a 300 mm long 0.5 mm graduation, a mobile pentagon prism (*11*) with index (*12*) is arranged in front of the lower part of the objective lens (*7*), and a fixed pentagon prism (*10*) in front of the upper part. They deflect the rays coming from the target by 100 gon resp. 100 gon $\pm \gamma_0$. Like for the teletop, a parallactic triangle with constant parallactic angle at the target is obtained. With the variable base b' at the station, the slope distance can be determined after shifting prism (*11*) until

Figure 751 Optical train in the BRT 006

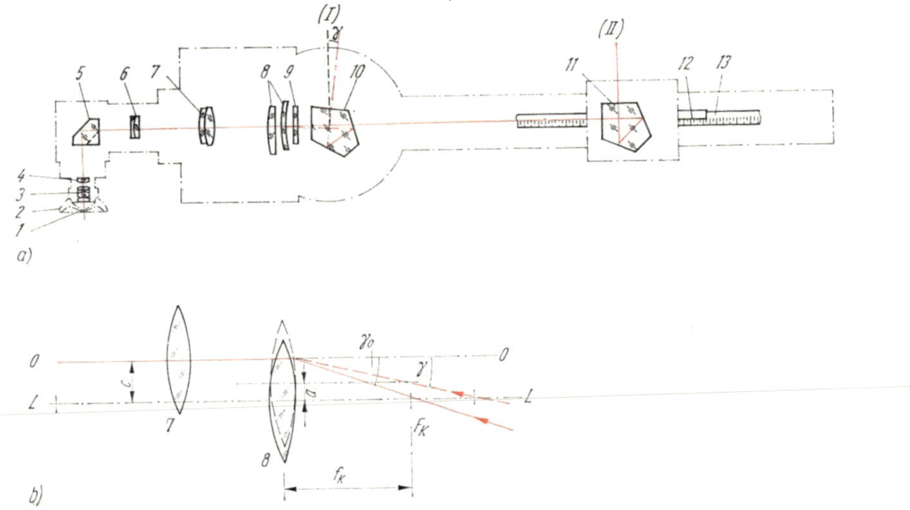

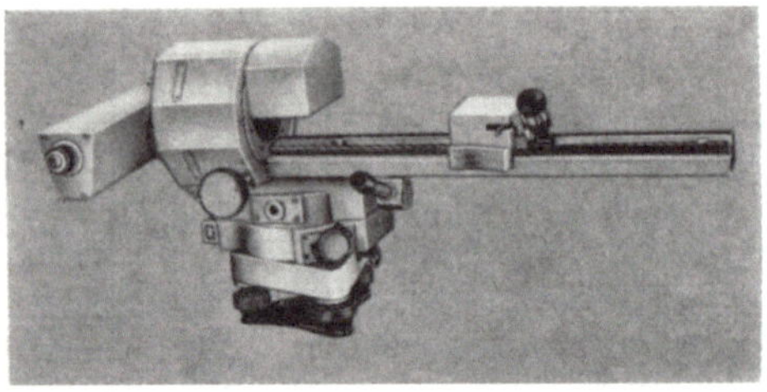

Figure 752 Base reduction tacheometer TW (U.S.S.R.)

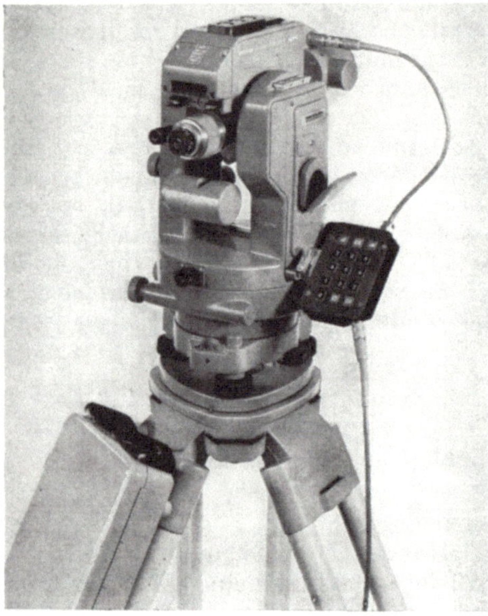

Figure 753 Wild Distomat DI 4

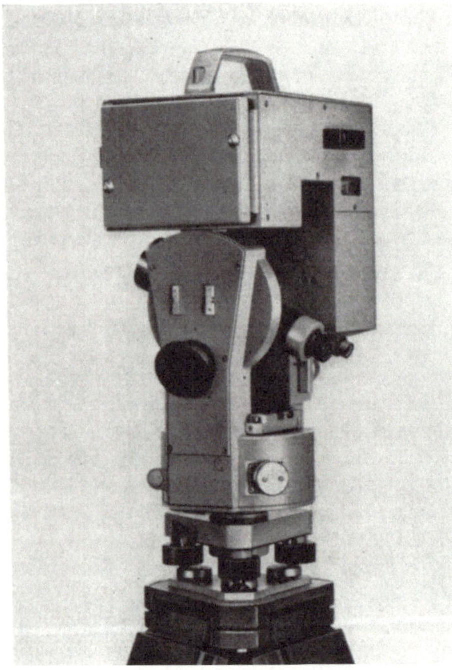

Figure 754 Electronic tacheometer theodolite SM 4 (Zeiss, Oberkochen); height: 336 mm

the two images coincide. The distance is obtained according to equation (209) as $s' = b' \cot \gamma_0 = b'k$, with $k = 200$. The necessary variation of γ_0 for the reduction is accomplished with a shiftable half lens system. The half lens (*8*) acts like one half of a heliometer.
The optical axis LL of this lens (*8*) (figure 751 b) is located at the separation c from the optical axis 00 of the objective lens (*7*). Since this lens unites in the focal point all rays entering parallel to the optical axis, the ray passing along the optical axis of the telescope is deflected in (*8*) though its front focal point F_K by the amount γ_0:

$$\gamma_0 = \frac{c}{f_K} \varrho . \tag{259}$$

After the rays pass through the focussing lens (*6*) the roof prism (*5*) deflects them into the eyepiece (*3*). Like for the teletop, image separation is carried out with a biprism (*4*) and a pinhole diaphragm (*2*) with opening (*1*) (key hole observation) such, that the upper and lower half images meet at a fine separation edge. If the half lens (*8*) is shifted from the position given in figure 751 b normal to the optical axis by the amount a, the parallactic angle γ_0 changes by

$$\Delta\gamma = \gamma_0 - \gamma = a/f_K . \tag{260}$$

This possibility is used for the reduction from the slope distance to the horizontal plane. The parallactic angle is changed depending on the vertical angle α such, that a section b, proportional to the horizontal distance is read on the base rail. Since $s = s' \cos \alpha = b \cos \alpha k$, b has to equal $b' \cos \alpha$. With $b \approx s\gamma$ and $b' \approx s\gamma_0$, γ becomes $\gamma_0 \cos \alpha$. For automatic reduction, γ_0 has to be reduced by

$$\Delta\gamma = \gamma_0(1 - \cos \alpha) \tag{261}$$

and the half lens has to be shifted according to equations (260) (261) by

$$a = f_K \gamma_0 (1 - \cos \alpha) . \tag{262}$$

The reduction system can be switched off, so that one can chose to measure horizontal or slope distances.
Since the position of the anallactic point is determined by the object principal plane of the lens system *7, 8* an addition constant $c = -75$ mm is present. It does not have to be taken into account when measuring slope distances, because of corresponding adjustment of the reading index (*12*). When measuring with automatic reduction, its influence is determined with the aid of a correction graduation near the eyepiece at the top part of the instrument. It can be neglected most of the time, since its influence is only 1 cm for vertical angles of $\alpha = 33$ gon. The swing error, which is possible at the Teletop because of setting of the target outside the centre of the field of view is largely eliminated in the BRT 006 by variable different magnification of the half images. Two short vertical lines denote the range of the swing error elimination within the field of view (1/8 of the width of the field of view).

The measuring range is up to 90 m for $b = 15$ m, and up to 120 resp. 180 m with $b = 30$ resp. 60 cm (auxiliary rod). Horizontal and vertical circles are read in the microscope eyepiece, located besides the telescope eyepiece. For a good coincidence target (e.g. range pole), the relative distance error is at least 1 : 1700, which means a standard deviation of ± 3 cm and less for 50 m distance. It is of advantage, that no rod is required at the target. The BRT is especially suited for detail surveys in densely built up areas. The TW (figure 752), which from the outside resembles the BRT, is manufactured in U.S.S.R. Rigidly built-in wedges, which do not follow the tilt motion of the telescope, are used for reduction instead of the half lens.

5.2.7. Auxiliary Instruments for Automatic Reduction

The Ewing distance- and height meter by Rank Precision Industries, Leicester is an optical mechanical device. It consists of the curve drum, attached to the telescope standard and the reading microscope, directly coupled with the telescope (figure 15). The curve drum contains two curves of different colour, one for distance reduction the other for the elevation difference. After setting the rod section as determined, distance reduction and elevation difference resp. the elevation can be read on the curve drum.

5.3. Electronic Tacheometers

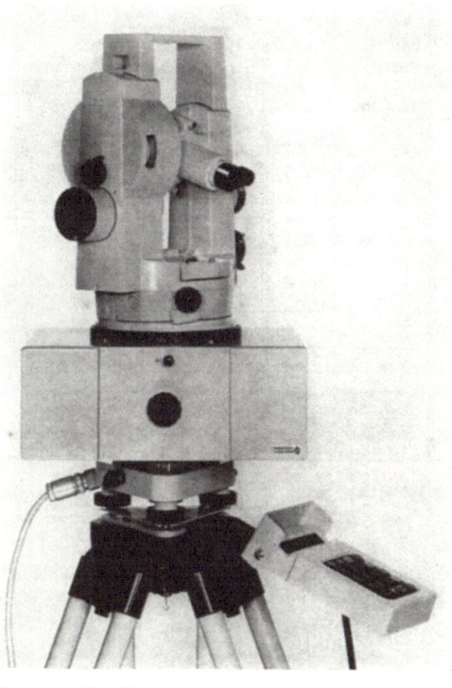

Figure 755 Electro optical tacheometer EOT 2000 (Zeiss, Jena)

Electronic tacheometers (table37) are instruments which combine theodolite and electro-optical distance measuring device. One distinguishes between

— electronic tacheometers with visual circle reading and EDM, and
— recording electronic tacheometers (with automated angular and distance measurements).

The operation of an electronic tacheometer with visual circle reading is also obtained by mounting an attachment for EDM onto a theodolite, e.g. AGA Geodimeter 10, 12 resp. 110, 112, 116 or 120 (figure 677) Tellurometer CD 6 (see figure 676); Nikon MND-2; Sokkisha SDM-1 and RED-1; Zeiss, Oberkochen Eldi 2 and 3 (see figures 679 and 680); Cubitape DM 60 (see figure 672); Beetle 500, 1000 or 1600, Autoranger by Keuffel & Esser (see figure 757). If the attachment is fastened to the standards, however, generally two separate sightings for direction- and distance measurements are necessary (see page 268). The reduction of observed inclined distances to horizontal distances, together with the determination of the elevation difference is carried out mathematically, e.g. with an electronic hand calculator. Such a calculator can be directly attached to the telescope of the Kern DM 502 (see figure 678). A process computer, which is part of the distance measuring part of the Wild DI 3 S, and wich sits on the tribrack and carries the theodolite, is also used to compute horizontal distances and elevation differences (see figure 675). All that is required, is the vertical angle, which is

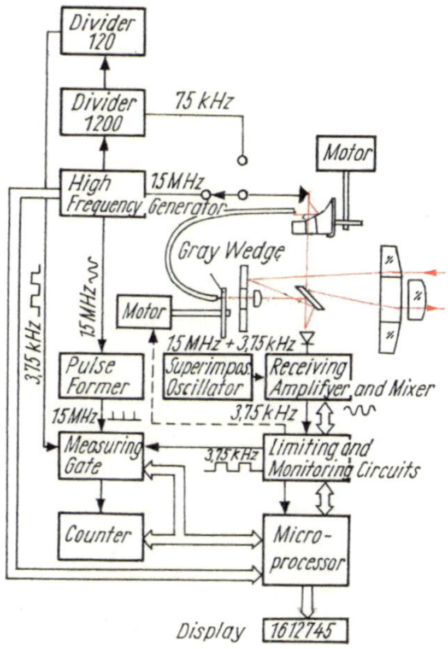

Figure 756 Block diagram for circuits of the HP 3810 A (Hewlett-Packard)

Table 37. Electronic Tacheometers

type	manufacturer	distance measuring part						
		radiation source	modulation fine frequency [MHz]	modulator	power use [W]	phase measuring method	range [km]	standard deviation
1	2	3	4	5	6	7	8	9
with visual circle reading								
HP 3810A	Hewlett Packard, U.S.A.	Ga-As-Diode	15	–	18	digital	3	$\pm(5\text{ mm} + 10 \cdot 10^{-6}\text{ s})$
SM 4	Opton Feintechnik GmbH, Oberkochen	Ga-As-Diode 910 nm	15	–	5	digital	2	$\pm(5 \ldots 10\text{ mm} + 2 \cdot 10^{-6}\text{ s})$
EOT 2000	VEB Carl Zeiss JENA	Ga-As-Diode 910 nm	15	–	13	digital	2	±10 mm
SDM 3	Sokkisha, Japan	Ga-As-Diode 900 mm	15	–	10	digital	1	±1 cm
mith automatic recording								
Geodimeter 710	AGA, Sweden	1-mW-He-Ne-Laser 632.8 nm	30	Kerr-cell	30	analog (resolver)	5	$\pm(5\text{ mm} + 10^{-6}\text{ s})$
HP 3820A	Hewlett Packard, U.S.A.	Ga-As-Diode 830 nm	15	–	15	digital	5	$\pm(5\text{ mm} + 1 \cdot 10^{-6}\text{ s})$
Reg Elta 14	Opton Feintechnik GmbH, Oberkochen	Ga-As-Diode 910 nm	15	–	10	digital	1.2	$\pm(5 \ldots 10)$ mm
Elta 2		Ga-As-Diode 910 nm	15	–		digital	4	$\pm(5\text{ mm} + 2 \cdot 10^{-6}\text{ s})$
Elta 4								
SDT-1	Sokkisha Ltd., Japan	Ga-As-Diode 900 nm	15	–	10	digital	1	±1 cm
Wild TC1	Wild Heerbrugg AG	Ga-As-Diode 885 nm	15	–	16	digital	2	$\pm(5\text{ mm} + 5 \cdot 10^{-6}\text{ s})$

Figure 757 Autoranger on top of digital theodolite Vectron (Keuffel & Esser, U.S.A.)

entered with the keyboard (or else the coordinate differences after entering of the azimuth). In the Wild DI 4 (figure 753) with microcomputer the vertical angle is handed in by a separate keyboard. The version „S" of the Beetle instruments displays horizontal distance and elevation difference after setting the vertical angle. The calculation is carried out automatically with sensors and microcomputer in AGA 120 (figure 677).

Electronic tacheometers require only one sighting for both angular and distance measurement, because the transmission- and receiving optics has a common optical axis with the theodolite telescope. Zeiss Oberkochen combined a scale microscope theodolite (see figure 302) with the elements of the Eldi (see figure 679) to get the SM 4 (figure 754). Theodolite, distance meter and power supply are combined into one compact unit.

The electro-optical tacheometer EOT 2000 (figure 755) of Zeiss, Jena, is a theodolite THEO 010 A (see figure 316) combined with an electro-optical distance measuring device. A spot light is used to find the reflector, just like for the EOK 2000. 10 s after starting the measurement, the slope distance is displayed at the freely turnable keyboard (with 20 input keys). After entering the vertical angle and possible atmospheric corrections, the microcomputer calculates and displays horizontal distance, elevation difference, resp. the mean of the vertical angle from direct and reverse observations. The battery is fastened to the tripod, which can also carry a recording unit.

The "Total Station" HP 3810 A from Hewlett-Packard displays immediately horizontal distance and elevation difference. The whole measuring process is automatic with the exception of direction measurements, with visual reading of the directions in one telescope position only. Everything else is monitored by a built-in microcomputer which checks the proper functioning

angle measuring part													
telescope				graduated circles			bubble sensitiv.		automatic height index/range	standard deviation of a direction measured direct and reverse	height	mass	remarks
Γ_F	diameter of objective lens [mm]	field of view [']	shortest sighting distance [m]	least count H	least count V	material resp. angle measuring method	platen level [″]	bull's eye level [']			[cm]	[kg]	
10	11	12	13	14	15	16	17	18	19	20	21	22	23
18			3	20″	–	glass	30	–	±33 gon	±3 mgon	33	11.9	figure 756
25	40	80	3	10 mgon (1′)		glass circles	30	15	±40 mgon	±1 mgon (±3″)	34	9	figure 754
30	40	78	2.5	0.2 mgon (1″)		glass circles	20	8	±80 mgon	±0.3 mgon	43	11	figure 755
				10″		glass circles							
30	60	96	1.7	incremental method			20	–	±40 mgon	±0.5 mgon	44	25	figure 760
30	66	90	5	code scanning					±45 mgon	±0.2 mgon	34	9.5	figure 765
25	35		3	code scanning			20	10	±30 mgon	±1 mgon	44	21.5	figure 759
30	60		2.5	code scanning			15	10	±26 mgon	±0.2 mgon	38	13.5	figure 764
25	40		23	incremental method			15	10	–	±1 mgon	37	6.5	figure 763
28				(magnetic scale drum)						±3″			
26	34	108	2	incremental method			30	8	±100 mgon	±0.6 mgon	37	9.8	figure 762

Figure 758 DM 501 on E 1 (Kern & Co.)

of the instrument (such that even with interruptions of the ray, always the necessary 100 observations are made, and that the given internal accuracy is reached), determines the quality of measurements, applies earth curvature and — after manual input of the refractive index — refraction corrections, and calculates horizontal distance and elevation difference. Depending on the setting of a switch, slope- or horizontal distance, elevation difference or zenith distance can be selected as output. For staking purposes, consecutive horizontal distances can be displayed every 3 seconds. There are no cables, as the battery fits into the unit.

The vertical angles are determined electronically by measuring resistance (see page 221). The distance measurement is based on the fine scale 10 m (15 MHz) and the rough scale 2000 m (75 kHz) with the low frequency of 3.75 kHz. The frequencies 75 kHz and 3.75 kHz are derived from the modulation frequency of the fine scale (15 kHz) with a frequency divider (figure 756). A diaphragm rotating with 10 Hz, alternatively opens the external and internal light path for the radiation generated by a GaAs diode. The signals of both paths are mixed in the receiver with a superimposed frequency of an oscillator which is 3.75 kHz larger. An optical cable transfers the 15 MHz and 75 kHz reference signals of the transmission diode directly to the receiver diode. The amplitudes of the signals of the internal and external light paths are checked via a microcomputer, which causes a servo motor to shift a grey wedge, located in the internal light path until both amplitudes coincide. The phase comparison is done digitally.

The operation of a recording electronic tacheometer can be approximated by mounting an attachment for EDM onto a digital theodolite, e.g. the Autoranger onto the Vectron (figure 281) from Keuffel & Esser (figure 757) — in

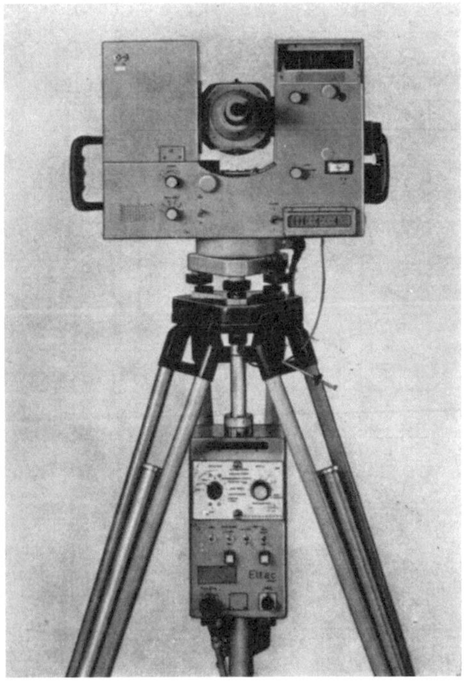

Figure 759 Recording electronic tacheometer Reg Elta 14 (Zeiss, Oberkochen); height: 32 cm

this case, however, two sightings for distance- and direction measurements are necessary, or the DM 501 resp. 502 on to the electronic theodolite E 1 by Kern (figure 758).

Electronic tacheometers require one sighting only, because theodolite, transmission- and receiving optics are arranged coaxially. The recording electronic tacheometer Reg Elta 14 (figure 759) from Zeiss, Oberkochen carries the complete electronic equipment for direction — and distance measurement in the telescope standards. By pressing a button, an electronic monitoring program for the whole measuring and recording process is automatically released, which performs the following: zero setting of counter and mean calculator; zero display; light modulation with f_1; 1000 phase measurements on f_1, computation of mean and storage of result; switch to frequency f_2, computation of mean; checking of the means; display of the result. Directions and slope distances are measured in about 30 seconds and displayed at the 6 digit tube display, or recorded on punched tape, like the zenith distance, together with code numbers for station, target, and type of measurement, which are entered with a 12 digit input keyboard. The direction measurement is achieved by code-scanning.

Horizontal- and vertical axes are equipped with a cogwheel coupling (cog circle theodolite, see page 107).

In order to scan the fine reading of the vertical circle, the eyepiece is built as an automatic height index for the projection of the code strip. Distances are measured with digital phase measurements. For staking purposes, the special computer Eltac (figure 759), monitored by a program, can be attached to the RegElta 14. The computer automatically receives the measuring results of the tacheometer. Additional data can be entered via the keyboard of the RegElta. The results of the Eltac are displayed digitally and recorded if desired. To accelerate staking with the Eltac, the reading instrument ASG was developed. The corrections as computed with the Eltac are measured resp. set with the ASG.

The geodimeter 710 (figure 760) from AGA, Stockholm, consists of the theodolite shaped measuring unit (with theodolite telescope, transmission and receiving systems, direction measuring system, and most of the internal computer), the reading and control unit, and the tape punch Geodat 700. Horizontal and vertical angles are measured with an incremental scanning system (see pages 107 and figure 280). For distance measurement, a 1-m-He-Ne-laser as readiation source, and a Kerr cell modulator are used (figure 761). The divergence of the ray can be varied between 0.2 and 10 m/km. An automatic intensity control system is used to always obtain the signal as received by SEV with equal amplitude. The frequencies (rough scale: 500 m, fine scale: 5 m) are manually set. The end result is displayed automatically. The phase positions are determined automatically with separate digital resolvers, whose stepwise rotation is monitored by impulses, given with pulse generators. The phase detector causes the pulse generator to provide faster impulses for larger phase differences. The impulses go simultaneously to an electronic counter, and the result of the analogue phase measurement is converted into a digital value. The reduction of the distances and the determination of elevation differences takes place in a mini computer, which, together with other electronic devices, is placed in the reading unit. On Numitran tubes, there is a continuous display of either horizontal- and vertical angles, or horizontal angle and- distance up to 4.7 km, or vertical angle and slope distance. The measuring time is 15 seconds. With the geodimeter 700 a distance reduction was only possible up to 500 m with no determination of elevation differences.

Figure 760 Geodimeter 710 (AGA, Stockholm)

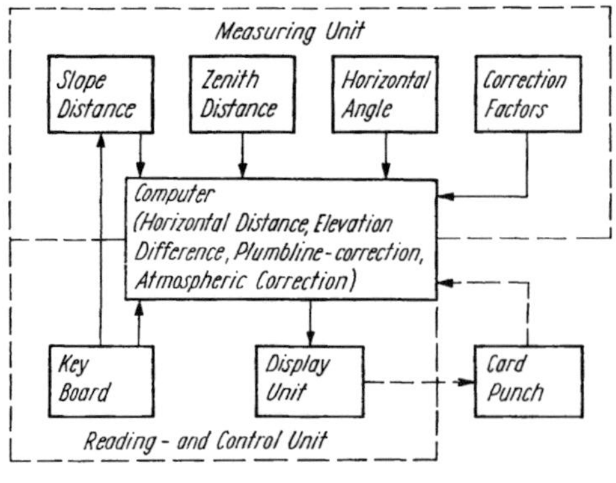

Figure 761 Block diagram of circuits of the AGA-Geodimeter 710

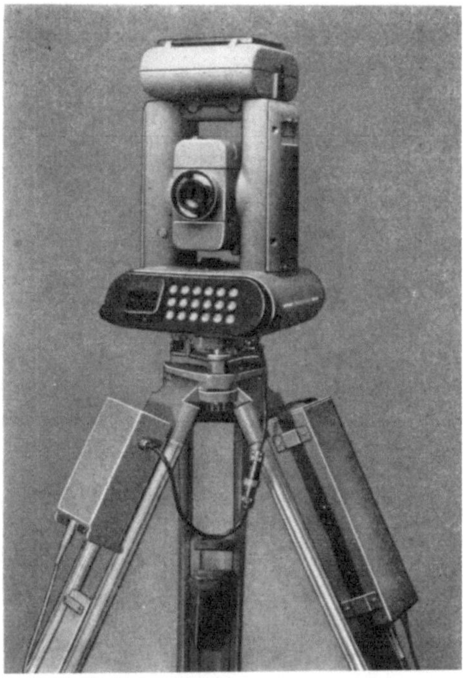

Figure 762 Electronic reduction tacheometer TC 1 (Wild, Heerbrugg)

Figure 764 Elta 2 (Zeiss, Oberkochen)

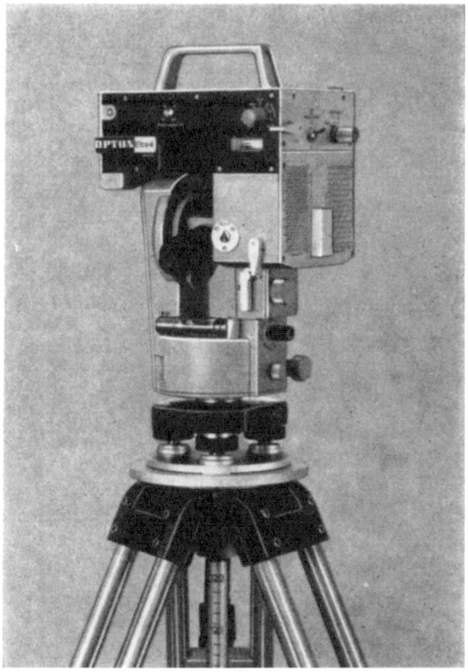

Figure 763 Elta 4 (Zeiss, Oberkochen)

Figure 765 Total-Station HP 3820 A (Hewlett-Packard)

The electronic reduction tacheometer Tachymat TC 1 (figure 762), developed by Wild Heerbrugg in cooperation with Sercel, France, contains telescope and EDM part as compact unit. Transmission- and receiving optics, as well as telescope are coaxial. By depressing a key, three measuring programs can be selected: measurement under normal conditions with 10 s duration, inclusive reduction; long distances with 15 s; staking with automatic repetition of the measurement until the stop command for 5 s. The angular measurement is carried out with incremental scanning (see page 108). A pendulum, supported with crossed threads ($\pm 5'$ operational range) acts as automatic height index for the vertical circle. The program sequence is

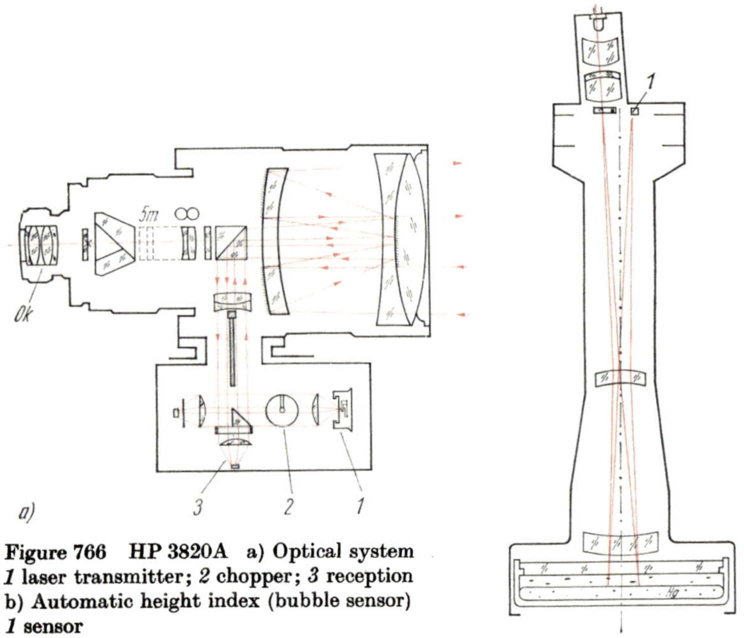

Figure 766 HP 3820A a) Optical system
1 laser transmitter; *2* chopper; *3* reception
b) Automatic height index (bubble sensor)
1 sensor

stored in the microcomputer such, that zenith distances are displayed. The micro-computer monitors centrally the measuring systems and prepares the measured values for the display, namely: computation of the slope distance, including scale correction and addition constant; reduction to the horizon of the station; determination of the elevation difference with earth curvature- and refraction corrections; computation of coordinates and elevation of the targets after input of starting values; selective display of the angles in gon or degrees; selective display of distances, coordinates, and elevations on two 7 digit LED displays. The field data can furthermore be recorded on magnetic tape cassettes. The recording unit is also monitored and controlled by the microcomputer.
The self reducing electronic tacheometers Elta 4 (figure 763) and Elta 2 (figure 764) by Zeiss, Oberkochen contain the EDM part of the Eldi 2. The measurement is checked by means of microcomputers. While the angles in the Elta 4 are measured with an incremental scanning system, the angular measuring system of the Elta 2 operates with code-scannig and interpolation. In similar way operates the Recota by Zeiss, Jena.
The Hewlett-Packard "Total Station" 3820 A (figure 765) is the combination of a 0.1 mgon digital theodolite, and EDM instrument, and a microcomputer into one unit. The optical system (figure 766a) serves as transmission- and receiving optic, as well as the aiming telescope. The angular measuring system operates with code scanning and interpolation (see page 107). The plate- and index levels (resp. automatic height index), usually found on theodolites, are replaced by a bubble sensor with two axes. The plumbline is given by a mercury horizon, dampened with silicone oil (figure 766b). The system is a biaxial collimator. A GaAs-Laser-Diode is used for distance measurement. This provides a range of 5 km, and is used with three modulation frequencies (15 MHz, 375 kHz, 3.75 kHz). As before, the microprocessor monitors the measuring systems, and prepares the measured values for display, resp. recording on a peripheral unit.

5.4. Plane Table and Alidade

While only numbers are obtained for fieldwork with theodolite, level and tacheometer, from which a map can be produced independent of the survey, in any scale by numerical and graphical evaluation, the map is generated during measurements with the plane table, because directions and distances are represented graphically. It is of advantage, that the terrain points are mapped and the contour lines plotted in direct view of the terrain. The relatively large mass of the equipment consisting of plane table and alidade, is a disadvantage.

Table 38. Plane Table Equipment (selection)

type	manufacturer	telescope					vertical circle			bubble sensitivity						mass				remarks
								reading device												
		Γ_F	diameter of objective lens [mm]	field of view [']	length [mm]	shortest sighting distance [m]	diameter [mm]	least count	type	levelling bubble ['']	telescope bubble ['']	height index bubble ['']	bull's eye bubble [']	range; automatic height index [']	height [mm]	alidade [kg]	container [kg]	plane table [kg]	tripod [kg]	
1	2	3	4	5	6	7	8	9	10	11	12	13	14	15	16	17	18	19	20	21
non-reducing alidades																				
MEHAN	F. W. Breithaupt & Son, Kassel	50	40	83	190	1.2	78	10′ (0.1 mgon)	scale microscope	60	30	50	20	–	180	4.3	5.5	7.9	5.0	figure 771
Expl. Alid.	W. & L. E. Gurley, U.S.A.	13	38		200	1.8	51	1′	vernier	50	–	60	15	–	120	2.2	1.7		5.0	telescope on ruler; compass
Stand. Alid.		17	38		266	1.2	51	1′	vernier	50	–	60	15	–	210	2.5	2.9		5.0	compass
Geol. Kr.	Keuffel & Esser Co., U.S.A.	16	34		270	3.0	102	1′	vernier	45	–	60	15	–	180	3.2	3.0		5.0	compass
Micr. Alid.	Rank Prec. Ind., England	15	31		239	1.5	51	10′ (0.2 mgon)	line microscope	60	45	40	15	–	216	2.0	3.7	5.5	5.0	
self-reducing alidades																				
MERED	F. W. Breithaupt & Son, Kassel	25	40	83	190	1.7	78	10′ (0.1 gon)	scale microscope	60	30	60	20	–	180	5.2	6.2	7.9	5.0	
RK	Kern & Co., Aarau	27	40	86	130	2.8	50	10′ (0.1 gon)	line microscope	60	–	–	–	–	205	3.1	2.2		5.7	figure 777
Self-ind. Alid.	Keuffel & Esser Co., U.S.A.	16	35		230	2.5	64	2′	scale microscope	–	–	–	15		160	3.2			5.0	figure 772; compass
Ma-5	MOM, Budapest	24	45	86	195	3.0	59	10′	scale microscope	60	30	50	6	–	215	3.2	8.1	7.1	5.5	figure 774; looking into the telescope from the top
SAOR	SLOM, Paris	18	40	103	160	1.5	–	–	–	40	–	40	3	–	150	2.4	2.5	4.4	4.9	figure 776; broken eyepiece
RK 1	Wild Heerbrugg AG	25	40	80	195	2.4	76	10′ (0.1 gon)	line microscope in the field of view of the telescope	120	60	60	8	–	180	2.5	9.2	3.1	4.2	figure 780
KN	U.S.S.R.	25	46	80	265	5.0	110	5′	line microscope	–	–		6		230	2.8	2	7.9	4.3	

5.4.1. Plane Table Equipment

Figure 767 Plane table

Plane Table

The plane table (figure 767) is the plotting table for the alidade. Basically, it consists of table top, frame and tripod (figure 768). The table top *M*, which serves as measuring- or drawing board, about 60 cm × 60 cm in area and 2 to 4 cm thick, is made either of dry softwood or else of light metal. Its plane surface is covered under tension with drafting material.
The plane table head is the frame *F* with which the table top *M* is fastened onto the tripod *St*, levelled and oriented. It consists of a tribrack with socket (resp. plug) which rests with its three footscrews on the plate of the tripod. Into (or onto) it, fits the plug (resp. socket) of the top part, which can be fastened with a clamp and moved sideways with a fine motion screw. The table top is connected with the top part by three fastening screws at the triangular plate *D* of the top part and can easily be removed. For the tripods with ball joint head, as used in the USA and Great Britain, just a disk is screwed onto the table top which contains a recepticle with thread to screw onto the tripod head.
Generally, plate tripods are used for plane tables. Kern, Aarau manufactures a plane table tripod with plane table head specifically suited for the alidade RK.
The connection between plane table head and table top is handled with a bayonet coupling. In the USA and Great Britain, tripods with ball joint heads according to *Johnson* (figure 769) are common, because they permit ast levelling and orienting by manually moving the ball joint head.

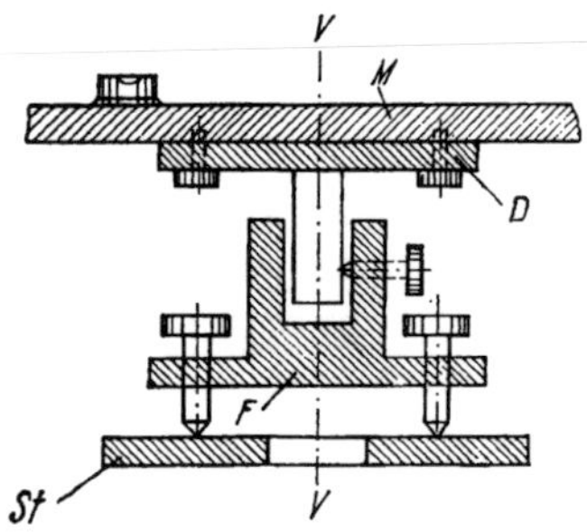

Figure 768 Section through plane table

Alidade

The alidade is an instrument for the graphical survey of terrain points. Its main parts are a stand with plotting device, a telescope standard rigidly connected to it, and a telescope which can be rotated around a horizontal axis, and which may be equipped with a vertical circle resp. -segment, but always with a distance measuring device.
The alidade stand together with the plotting device serves for the graphical representation of the sighting direction, and for the plotting of the measured distance to scale on the plane table. The plotting device is a 40 to 60 cm long ruler with a sharpened edge (figure 770), which usually can be swivelled and shifted parallel to the collimation axis of the telescope. Usually it is equipped with a scale and a plotting pin.
Connected to this alidade *U* stand is the telescope standard, which carries the vertical axis *KK* with the telescope, and the vertical circle *V*. The horizontal axis is accurately levelled on the plane table *M*, which is only roughly levelled. This is achieved with the aid of a screw and the cross bubble fixed to the stand. The bubble of the height index level *H* has to be accurately centered for all readings, like for the theodolite. Occasionally, an automatic height index is used instead of the index bubble. Since it is sufficient to read the vertical angle to 1′, measurement in one telescope position and reading at one circle position with vernier, line-, or scale microscope is sufficient. The telescope is clamped in its position with the height clamp *Hk* and sensitively tilted with the fine motion screw *Hf*. It is equipped with a telescope bubble for surveys in flat terrain.
The telescope, which can be plunged has a 20–30 times magnification. In der alidades, stad ia hairs were used for distance measurements. Modern olidades are self reducing, utilizing reduction devices as mentioned with the aliagram tacheometers.

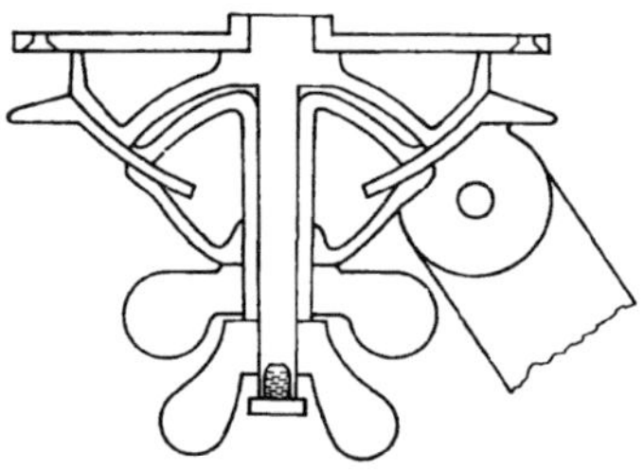

Figure 769 Section through tripod head of the Johnson type for plane tables

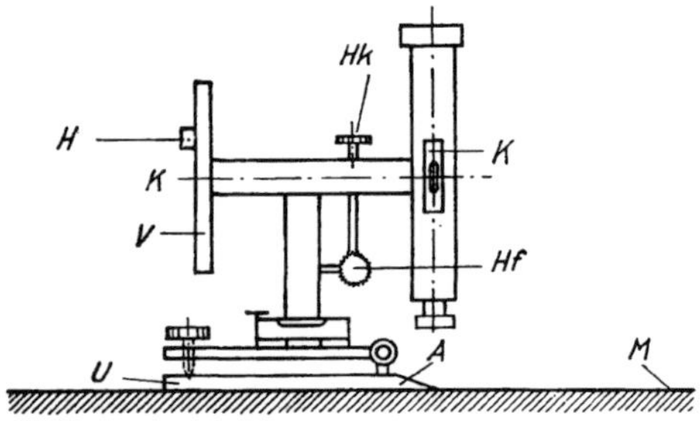

Figure 770 Alidade

Auxiliary Equipment for Plane Table Work

An orientation compass, which is often fastened to the alidade ruler like the cross bubble, is used to orient the plane table to magnetic north. A plummet fork is used to center the plane table. It consists of a metal- or wooden angle, whose horizontal leg lies on the table top with the point at the projection of the instrument station. The other leg supports a plumb bob.

5.4.2. Overview of Plane Table Equipment

Plane table equipment

Compared to alidades with common stadia hairs (figure 771), self reducing alidades are easier to operate and faster. For the Paragon alidades (figure 772) by Keuffel & Esser, Hoboken, the reduction is not automatic, but is obtained by simple computation — with uneven numbers.

Figure 771 Normal alidade No. 5050 MEHAN (Breithaupt & Son, Kassel); height 180 mm

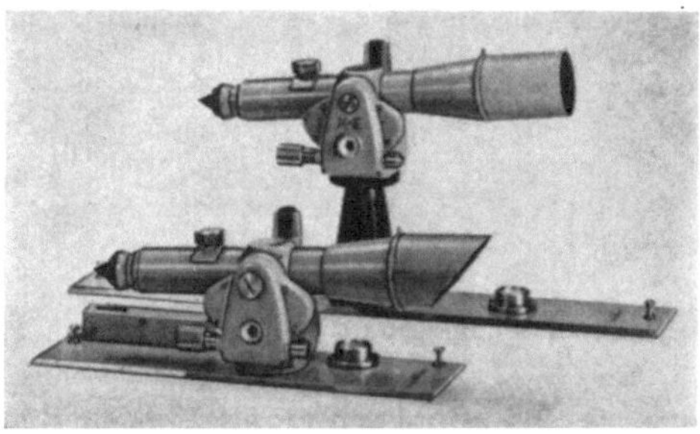

Figure 772 Paragon alidades with automatic indices (Keuffel & Esser, U.S. A)

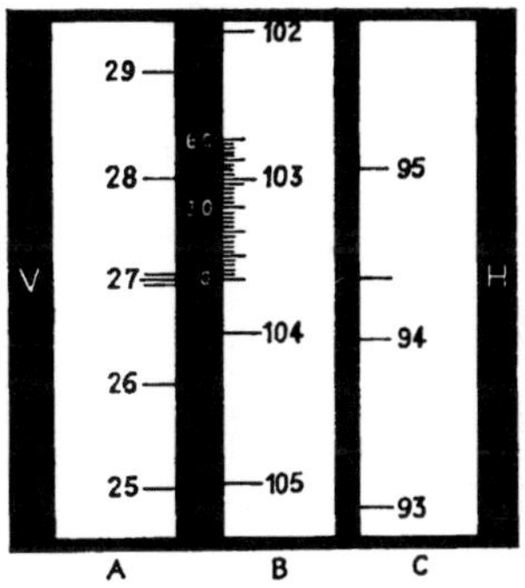

Figure 773 Field of view of the Paragon alidades A. Vertical scale: 27.0
B. Zenith distance: 103°42′
C. Multiplication constant for distance: 94.4

Figure 774 Self reducing alidade Ma-5 (MOM, Budapest); height of alidade 215 mm

With an automatic height index and analogue automatic indices, one can read the multiplication constants for distance and height, as well as error free vertical angles (figure 773). The view into the telescope is inclined from the top.
The alidade Ma-5 (figure 774) by MOM Budapest contains a diagram. When setting the ground curve onto the zero mark, distance and elevation difference can be directly read (figure 775).
The self reducing alidades by Fennel and by Breithaupt & Son (the latter looks like the normal alidade, figure 771), the Russian alidade KN and the "Alidade optoreductrice SOAR" (figure 776), by SLOM, Paris have viewing into the telescope from the top and contain diagrams.
Kern, Aarau manufactures the alidade RK (figure 777) with the diagrams shown in figure 724), reduction devices, fixed viewing into the eyepiece at an angle of 33 gon, and upright telescope image.
Instead of the telescope, a prism in front of the objective lens is tilted. This prism is equipped with a view finding collimator, and its horizontal axis can be clamped in any position with a clamping mechanism (figure 778). The alidade has a ruler with a guiding parallelogram connected to it by ball bearing points. Longitudinal and cross bubbles are the only elements which have to be adjusted. The plane table head is built into the tripod. For the RK, a topographic rod is produced (figure 779), whose even decimetres are marked by circles. The marks of the even decimeters have a 10 mm diameter and serve for distances above 100 m, the ones for the odd decimetres have a 5 mm diameter and are designed for short sighting distances.
The reduction alidade Wild RK 1 (figure 780) operates according to the principle of the Wild RDS. The rotatable telescope eyepiece at 50 gon to the optical axis permits confortable viewing with the telescope. The telescope does not contain a height clamp, and tilts with the aid of a breaking spring. The vertical circle reading appears in the field of view of the telescope (figure 781), and in reverse positions there is a distance curve with $k = 200$.

Plane Tables as Tacheometers

Several companies manufacture a plane table in a form that stadia theodolites or self reducing tacheometers take over the role of the alidade.
The plotting table KARTI 250 mm (figure 782) by Zeiss, Jena is an attachment for theodolites and tacheometers, which can be utilized as a plane table. The instrument plug fits into the socket of the plotting table and is clamped there with the screw which is located at the left side of the tribrack. The top part of the instrument can then be coupled with the rotatable part of the plotting table.
The plotting table rotates together with the instrument. A transmission rotates the plotting surface of the table (250 mm diameter) in opposite direction, so that it remains oriented to the terrain. The bridge, containing ruler, pricking device and reading index, which is suspended above the circular plotting surface, and which is parallel to the collimation axis, rotates with it around the vertical axis. Thus the direction to a rod set up on the terrain is transferred onto the plotting surface. After the optically measured distance is set at the ruler (in 1 : 1000, 1 : 2000, 1 : 2500 or 1 : 5000), the position of the target is plotted by depressing the pricking device. Thus for each instrument station a partial map is obtained, which together with the ones from other stations, can be combined to one map. The plotting table has a mass of 4.8 kg. The tacheometer plane table by Breithaupt & Son, Kassel (figure 783) consists of two centrically arranged separate systems of vertical axes, which are connected with the instrument arm. The plane table is levelled with the lower axis system and oriented with the aid of clamp and fine motion screw. The table top can be transposed by 200 gon. The telescope column, equipped with clamp and fine motion screw, which carries the tacheometer telescope, rotates around the upper axis system. Connected to it is the parallel ruler guide, which can be adjusted and which accepts four exchangeable rulers for the scales 1 : 1000, 1 : 2000, 1 : 5000 and 1 : 25,000. The instrument has a horizontal- and a vertical circle and also a telescope bubble. The direction to the target is automatically transferred from telescope to map. The tacheometer plane table has a mass of 2.5 kg.
The Topometer Geo 06 (figure 784) by Topomatic Denmark, is used to directly transfer terrain data within a 50 m surrounding area onto a map. It is however, a different type and consists of a recording and a plotting unit. The target is sighted with a diopter, while the distance is measured with

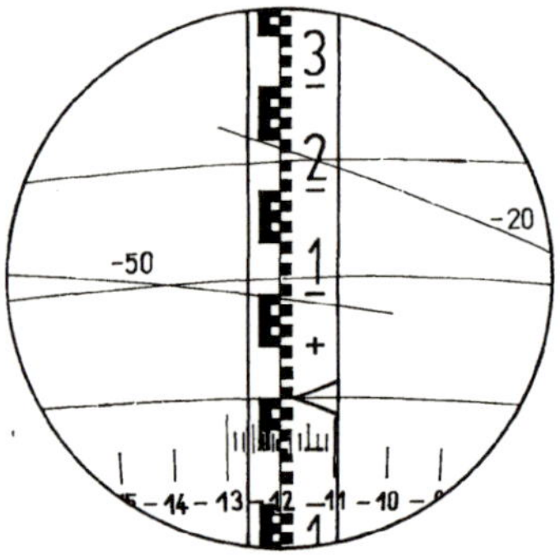

Figure 775 Field of view of alidade Ma-5
Reading: $s = 23.0$ m; $\Delta h = -4.85$ m

Figure 776 Self reducing alidade SAOR (SLOM, Paris); height: 150 mm

Figure 777 Self reducing alidade RK (Kern & Co.); height: 205 mm

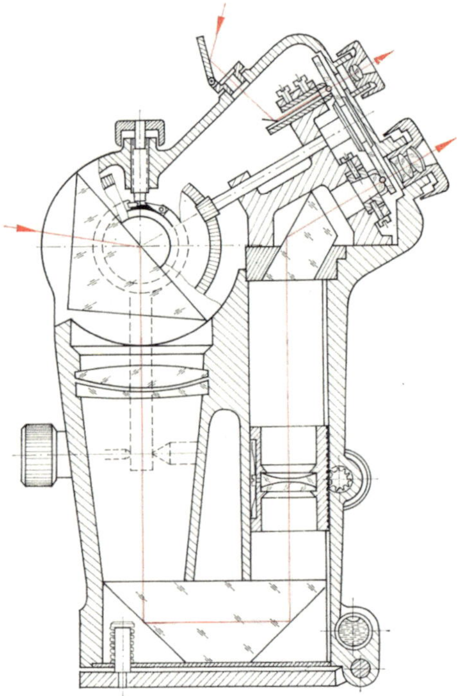

Figure 778 Section through the alidade RK

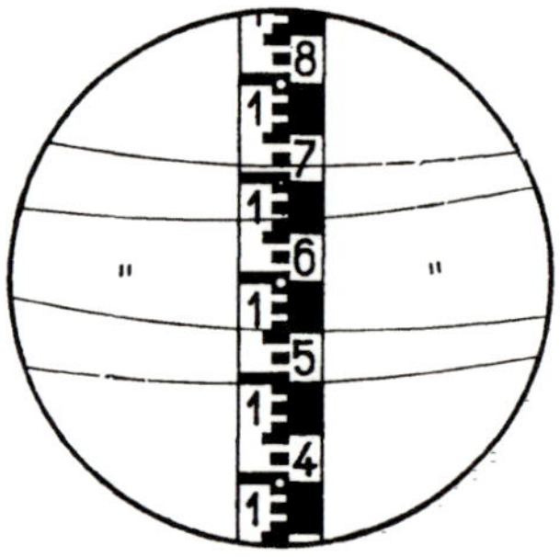

Figure 779 Field of view of the RK
($s = 21.7$ m)

Figure 780 Self reducing alidade Wild RK 1; height: 180 mm

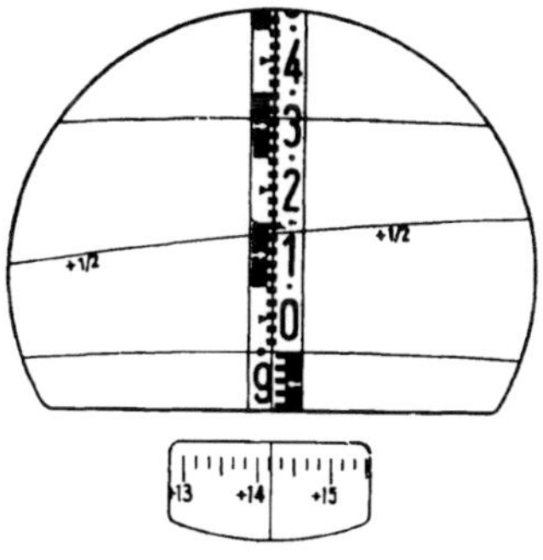

Figure 781 Field of view of Wild RK 1
Reading = s = 36.2 m, Δh = 18.2 m · (+1/2) = +9.1 m, α = 14°11′

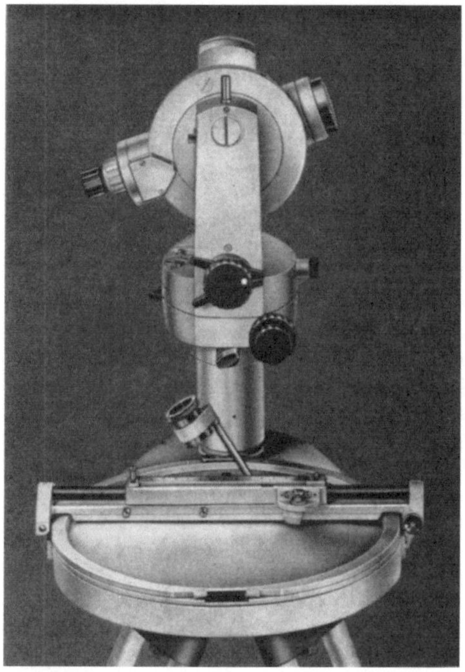

Figure 782 Plotting table KARTI 250 (250 mm draughting surface) with DAHLTA 010 A (Zeiss, Jena)

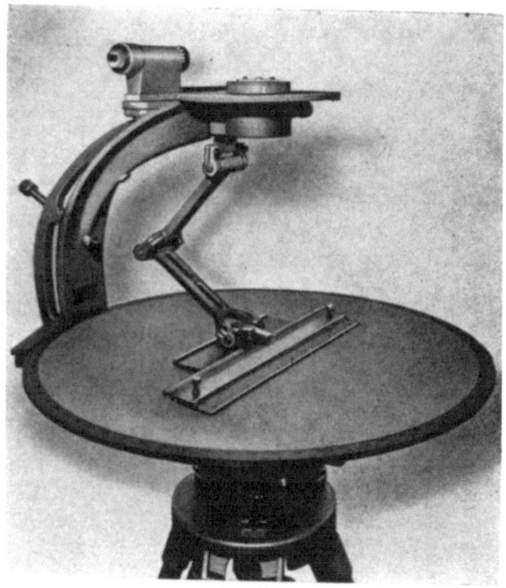

Figure 783 Tacheometer plane table (MEWES), Breithaupt & Son

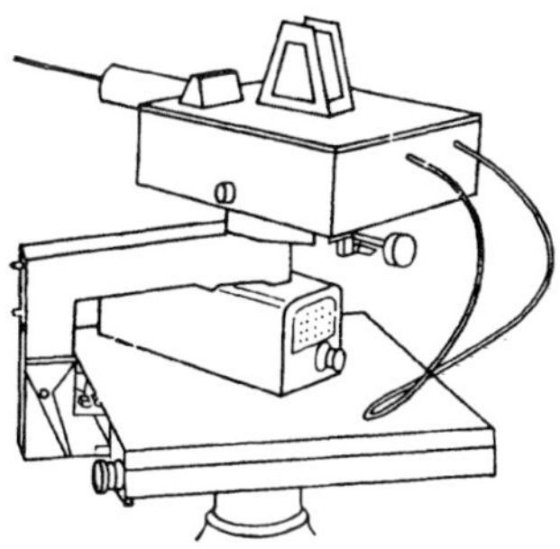

Figure 784 Topometer Geo 06 (Topomatic, Denmark)

a perforated measuring tape, built into the instrument. The tape activates the recording mechanism of the instrument. After initiating the plotting, the point is transferred onto the plotting table, which is covered with draughting foil.

5.4.3. Testing and Adjusting of Plane Table Equipment

Testing and Adjusting of the Plane Table

The requirements for proper plane table operation are checked as follows

1. The surface of the plane table has to be plane.

A calibrated ruler is moved with its edge across the table top in various directions. There should be no spaces between ruler edge and table top. Irregularities have to be corrected in the carpenter's shop by planing.

2. The plane table has to be rigid.

After setting the target with the alidade, the edge of the plane table top is depressed lightly in two orthogonal directions. When released, the target has to appear again in the cross hair centre of the alidade. An adjustment can only be performed in a work shop.

Testing and Adjusting of the Alidade

The alidade is similar in structure to the top part of a theodolite and is checked and adjusted similarly. The following requirements have to be met and checked after the plane table is levelled:

1. The edge of the ruler has to be a straight line.

The ruler is laid onto two points, and its position is drawn with a sharp line. This is repeated after transposing the ruler. If the edge is straight, both lines coincide.

2. The axis of the bull's eye level has to be normal to the ruler surface, while the axis of the cross bubble has to be parallel to it.

This is checked according to the method used for reversible striding levels.

3. The vertical line of the cross hair has to be vertical.

A target set onto one end of the vertical line shall remain on it when tilting the telescope. If this is not the case, the reticule has to be rotated by half the amount of the deviation.

4. The collimation axis has to be normal to the horizontal axis: $ZZ \perp KK$ (checking for collimation error).

A distant object is sighted in the height of the horizontal axis, and the direction is plotted with the ruler. Then, after plunging the telescope, the alidade is transposed, and the ruler is set onto the direction to the target as plotted. A horizontal deviation of the target from the vertical cross hair line indicates twice the collimation error. It is eliminated by shifting the reticule by half the amount and by rotating the plane table for the other half.

5. The horizontal axis has to be parallel to the bottom surface of the ruler (checking for horizontal axis error).

After the collimation error has been eliminated, the collimation axis is normal to the horizontal axis, and remains in one plane when tilting the telescope. When placing the alidade onto a horizontal surface, this plane has to be vertical. One sights to a high target and plots the direction with the edge of the ruler. After plunging the telescope and transposing the alidade, the ruler is set onto the direction to the target as plotted. The deviation of the target from the vertical cross hair line represents twice the influence of the horizontal axis error. It is eliminated by raising or lowering the telescope support for half the amount, and by rotating the plane table for the other half.
The elimination of the horizontal axis error can also be achieved according to a method used for theodolites, where a high target point is projected onto the horizontal plane in both telescope positions, and the cross hair is shifted to the mean point between the two. If the alidade is equipped with a cross bubble, it has to be carefully centred with the tilting screw when sighting to the targets (figure 770). Half of the lateral deviation between target and vertical cross hair line is eliminated with the tilting screw. Then the bubble is centered again with the aid of the vertically acting adjustment screws. Both collimation- and horizontal axis errors have to be carefully tested and adjusted, because they are not eliminated in plane table work by the measuring procedure.

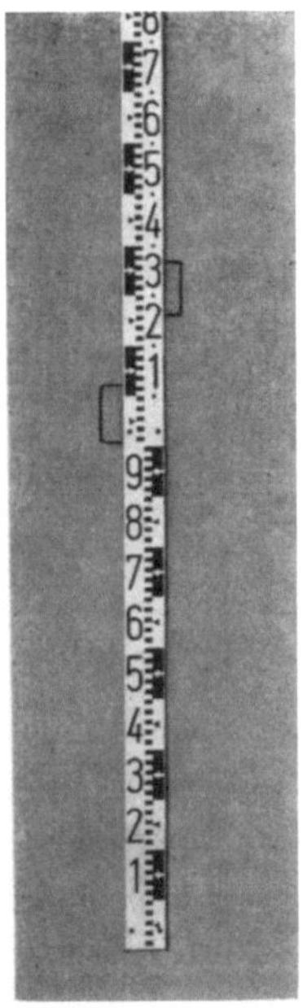

igure 785 Topographic rod by Wild, Heer-rugg

6. The vertical plane through the collimation axis has to be parallel to the edge of the ruler (checking for alidade crossing error).
A crossing error of an alidade is easily recognized if one looks from the objective lens with a small openings onto the ruler. Another method would be, to set the telescope to a range pole, some 50 m distant. When looking along the edge of the ruler, this point should be in line. For several alidades the parallelity can be obtained by loosening the screws on the telescope standard and rotating it accordingly. Nonparallelity between target plane and ruler edge is without influence in plane table work, provided the same alidade was used for the whole map.

7. The height index error should be zero.

The index error of the height index bubble is tested and adjusted like for the theodolite.

8. Testing of the constants of the distance meter and of the reduction device of self reducing alidades follows the same procedure as for theodolites and tacheometers. The telescope bubble is treated like at a level with fixed telescope.

Testing and Adjusting of Incidentials Belonging to the Plane Table Equipment

Compass and bull's eye bubble are checked according to the methods described in sections 2.2.2 and 1.2.2. The testing of the plumb fork can be done by holding a point on the plane table and transposing the plumb fork. The plumb bob should point to the same ground point for both positions.

5.4.4. Topographic Rods

For large scale topographic surveys, usually levelling rods with cm-graduation are sufficient. For small scale surveys however, topographic rods with larger intervals (e.g. 5 cm or 1 dm graduation) are used, which permit safe reading of decimetres even at larger ranges (e.g. up to 800 m). The rods are usually 3 or 4 m long, 8 to 12 cm wide and 2 to 3 cm thick folding rods. The 3 or 4 m long topographic rod by Wild, Heerbrugg, (figure 785), for instance, has cm- or dm graduation. The graduation of the rods by Kern, Aarau can be seen in figure 779. The topographic rods of other manufacturors are of similar design.

References

(selection of text books)

Genike, A. A., Larin, B. A., Nasaroff, W. M.: Geodetic EDM (in Russian). Nedra, Moscow 1974

Großmann, W.: Vermessungskunde. Slg. Göschen, W. de Gruyter, Berlin/ New York – I. Stückvermessung und Nivellieren. Nr. 2160, 15. Aufl., 1976 – II. Winkel- und Streckenmeßgeräte, Polygonierung, Triangulation und Trilateration. Nr. 7469, 12. Aufl., 1975 – III. Trigonometrische und barometrische Höhenmessung, Tachymetrie und Ingenieurgeodäsie. Nr. 2162, 11. Aufl., 1979

Gruber, O. von, a. o.: Optische Streckenmessung und Polygonierung. H. Wichmann, Berlin 1955

Gusjeff, N. A.: Mining-Surveying Instruments (in Russian). 2nd ed. Nedra, Moscow 1968

Jordan/Eggert/Kneißl: Handbuch der Vermessungskunde. 10. Aufl., J. B. Metzler, Stuttgart – Band III. *Kneißl, M.:* Höhenmessung und Tachymetrie. 1956 – Band IV/1. *Kneißl, M.:* Mathematische Geodäsie (Landesvermessung). 1958 – Band VI. *Rinner, K., Benz, F.:* Die Entfernungsmessung mit elektromagnetischen Wellen und ihre geodätische Anwendung. 1966

Kahmen, H.: Elektronische Meßverfahren in der Geodäsie. 2. Aufl., H. Wichmann, Karlsruhe 1978

König, A., Köhler, H.: Die Fernrohre und Entfernungsmesser. 3. Aufl., J. Springer, Berlin/Göttingen/Heidelberg 1959

Kondrashkoff, A. W.: Electronic Distance Measurement. VEB Verlag für Bauwesen, Berlin 1961

Laurila, S. H.: Electronic Surveying and Navigation. J. Wiley and Sons, New York 1976

Litwinoff, B. A., Lobatcheff, W. M., Woronkoff, N. N.: Surveying Instruments (in Russian). Nedra, Moscow 1971

Michejetcheff, W. S.: Geodetic EDM (in Russian). Nedra, Moscow 1979

Riekher, R.: Fernrohre und ihre Meister. VEB Verlag Technik, Berlin 1957

Stipa, V.: Optik der geodätischen Instrumente. Techn. Hochschule Aachen, 1957

Szymonski, J.: Surveying Instruments (in Polish). Warsaw. Part I: 1968, part. II: 1971, part III: 1972

Volquardts, H., Matthews, K.: Vermessungskunde. Teubner, Stuttgart. Teil I, 23. Aufl., 1972; Teil II, 12. Aufl., 1973

Wittke, H.: Einführung in die Vermessungstechnik. 4. Aufl. F. Dümmler, Bonn 1971

Yelissejeff, S. W.: Surveying Instruments (in Russian). 3rd ed. Nedra, Moscow 1973

Zetsche, H.: Elektronische Entfernungsmessung (EDM). K. Wittwer, Stuttgart 1979

Index

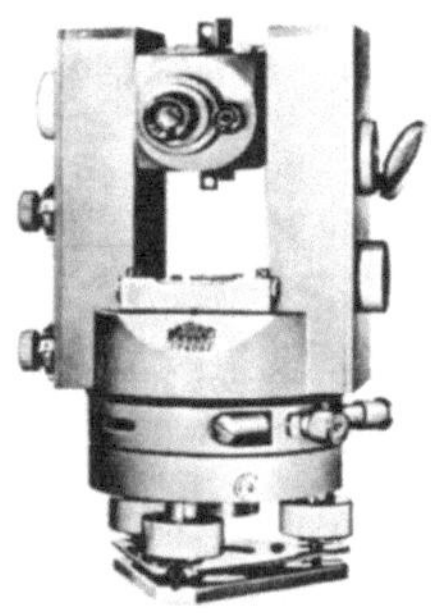

Bei Fragen zur Produktsicherheit wenden Sie sich bitte an:
If you have any questions regarding product safety,
please contact:

Walter de Gruyter GmbH
Genthiner Straße 13
10785 Berlin
productsafety@degruyterbrill.com